Springer Series on
Wave Phenomena

4

Editor: Leopold B. Felsen

Springer Series on
Wave Phenomena

Editors: L. M. Brekhovskikh   L. B. Felsen   H. A. Haus
Managing Editor: H. K. V. Lotsch

V. M. Babič   V. S. Buldyrev

Short-Wavelength Diffraction Theory

Asymptotic Methods

Translated by E. F. Kuester

With 76 Figures

Springer-Verlag

Berlin Heidelberg New York London
Paris Tokyo Hong Kong Barcelona

Professor Dr. Vasili M. Babič

Mathematical Institute,
Academy of Sciences of the USSR,
191011 Leningrad, USSR

Professor Dr. Vladimir S. Buldyrev

Faculty of Physics,
Leningrad State University,
198904 Leningrad, USSR

Translator:

Professor Edward F. Kuester, Ph. D.

Department of Electrical and Computer Engineering, University of Colorado, Campus Box 425,
Boulder, CO 80309-0425, USA

Series Editors:

Professor Leonid M. Brekhovskikh, Academician

P. P. Shirsov Institute of Oceanology, Academy of Sciences of the USSR, Krasikowa Street 23,
SU-117218 Moscow, USSR

Professor Leopold B. Felsen, Ph. D.

Department of Electrical Engineering, Polytechnic University, Route 110,
Farmingdale, NY 11735, USA

Professor Hermann A. Haus

Department of Electrical Engineering & Computer Science, MIT,
Cambridge, MA 02139, USA

Managing Editor: Helmut K. V. Lotsch

Springer-Verlag, Tiergartenstrasse 17,
W-6900 Heidelberg, Fed. Rep. of Germany

Title of the original Russian edition: *Asimptoticheskie metody v zadachakh difraktsii korotkikh voln*
© Nauka, Moscow 1972

ISBN-13: 978-3-642-83461-5   e-ISBN-13: 978-3-642-83459-2
DOI: 10.1007/978-3-642-83459-2

Library of Congress Cataloging-in-Publication Data. Babich, V. M. [Asimptoticheskie metody v zadachakh difraktsii korotkikh voln. English] Short-Wavelength Diffraction Theory: Asymptotic Methods / V. M. Babič, V. S. Buldyrev ; translation by E. F. Kuester. p. cm. – (Springer series on wave phenomena ; v. 4) Translation of: Asimptoticheskie metody v zadachakh difraktsii korotkikh voln. Bibliography: p. Includes index. ISBN 0-387-19189-5 (U.S.) 1. Waves-Diffraction. 2. Eigenfunctions. 3. Asymptotic expansions. I. Buldyrev, V. S. (Vladimir Sergeevich) II. Title. III. Title: Shortwave diffraction theory. IV. Series: Springer series on wave phenomena ; 4. QC482.D5B3313 1991 539.7'222–dc20 89-11374

Preface

In the study of short-wave diffraction problems, asymptotic methods – the ray method, the parabolic equation method, and its further development as the "etalon" (model) problem method – play an important role. These are the methods to be treated in this book. The applications of asymptotic methods in the theory of wave phenomena are still far from being exhausted, and we hope that the techniques set forth here will help in solving a number of problems of interest in acoustics, geophysics, the physics of electromagnetic waves, and perhaps in quantum mechanics. In addition, the book may be of use to the mathematician interested in contemporary problems of mathematical physics.

Each chapter has been annotated. These notes give a brief history of the problem and cite references dealing with the content of that particular chapter. The main text mentions only those publications that explain a given argument or a specific calculation.

In an effort to save work for the reader who is interested in only some of the problems considered in this book, we have included a flow chart indicating the interdependence of chapters and sections.

The authors consider it their pleasant duty to thank M.M. Popov, who, at the authors' request, wrote Sects. 10.1–7, and also I.A. Molotkov, who was involved in all stages of preparing this book. I.A. Molotkov wrote Sect. 7.5 and most of Chap. 11, and was our coauthor in writing Chap. 6. We are indebted to participants at the seminar of the Leningrad Section of the V.A. Steklov Institute of Mathematics and Leningrad State University on the mathematical theory of diffraction, but especially to V.F. Lazutkin, for their constructive criticism. We owe a great deal to the editors of the book. Their careful work, at times going far beyond the limits of direct editorial duties, enabled us to eliminate a number of shortcomings in the manuscript.

Leningrad, April 1990

V.M. Babič
V.S. Buldyrev

Contents

Interdependence of Chapters and Sections
(Flow Chart)

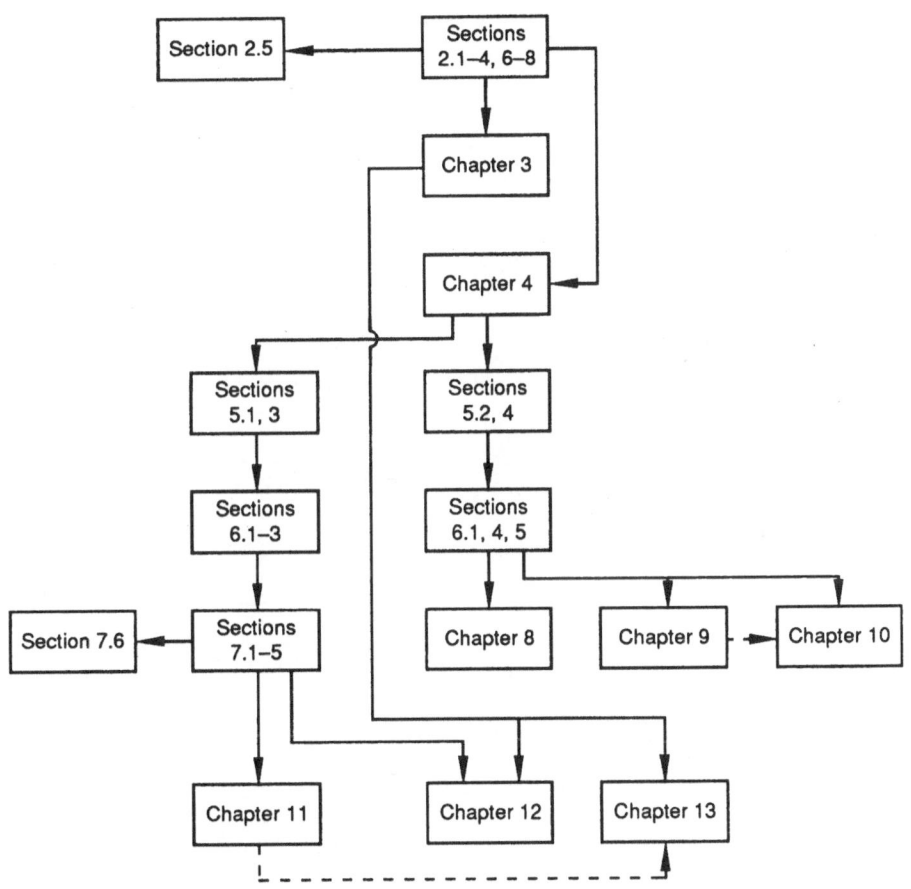

1. Introduction

Wave motion of one kind or another is studied in various branches of physics (acoustics, elasticity theory, electrodynamics, quantum mechanics, and so on). Such motion is described mathematically by certain hyperbolic differential equations. The basic properties of these equations are evident from a study of their simplest representative – the wave equation

$$\frac{\partial^2 W}{\partial x^2} + \frac{\partial^2 W}{\partial y^2} + \frac{\partial^2 W}{\partial z^2} - [c(x, y, z)]^{-2}\frac{\partial^2 W}{\partial t^2} = 0 \quad , \tag{1.1}$$

where $c(x, y, z)$ is the velocity of wave propagation.

In many problems it is natural to assign a harmonic time dependence using the factor $\exp(-i\omega t)$ in which ω is the angular frequency of oscillations:

$$W(x, y, z) = e^{-i\omega t}u(x, y, z) \quad .$$

The function $u(x, y, z)$ satisfies Helmholtz's equation

$$\frac{\partial^2 u}{\partial x^2} + \frac{\partial^2 u}{\partial y^2} + \frac{\partial^2 u}{\partial z^2} + \left[\frac{\omega}{c(x, y, z)}\right]^2 u = 0 \quad . \tag{1.2}$$

Equation (1.2) must be augmented by boundary conditions which result from the physical content of the problem.

Eigenfunction problems require nontrivial solutions of (1.2) that satisfy one of the following conditions on the boundary S of a three-dimensional bounded region Ω:

$$u|_S = 0 \quad \text{or} \tag{1.3}$$

$$\left.\frac{\partial u}{\partial n}\right|_S = 0 \quad , \tag{1.4}$$

or finally

$$\frac{\partial u}{\partial n} + i\omega g(M)u|_S = 0 \quad . \tag{1.5}$$

Here n is the outward normal to S and $g(M)$ is a function that is defined at points M of the surface S.

The field of a point source satisfies the inhomogeneous equation

$$\frac{\partial^2 u}{\partial x^2} + \frac{\partial^2 u}{\partial y^2} + \frac{\partial^2 u}{\partial z^2} + \left[\frac{\omega}{c(x, y, z)}\right]^2 u = -\delta(M - M_0) \quad , \tag{1.6}$$

where $\delta(M - M_0)$ is a delta function concentrated at the point M_0. The field of a point source is ordinarily studied in an unbounded region Ω. In addition to the boundary conditions (1.3, 4 or 5), the function $u(x, y, z)$ must satisfy certain conditions at infinity that ensure uniqueness of the solution. The principle of limiting absorption, for instance, might be used for this purpose [1.1][1] The main purpose of the book is the investigation of the asymptotic behavior of solutions of boundary value problems for (1.2 or 6) as $\omega \to \infty$. Let us proceed to give a more detailed description of the contents of the book and of its logical outline.

Physicists have found in numerous and varied experiments that wave fields, i.e. solutions of (1.2 or 6), are governed by definite geometric laws at high frequencies. It is natural to expect that these geometric principles characterize the asymptotic behavior of the fields as $\omega \to \infty$, and consequently that they can be derived directly from (1.2 or 6).

In fact, there is a formal technique for constructing asymptotic expansions with respect to inverse powers of ω (or a large parameter proportional to ω) of solutions to boundary value problems for (1.2 and 6). This technique, which was first outlined for the simplest case by A. Sommerfeld and C. Runge, was later given the name of the *ray method*.

The principal term of a ray expansion contains not only the ray description of the wave motion (which is peculiar to geometrical optics) but also gives the amplitude characteristics of the wave.

Construction of a high-frequency asymptotic relation by the ray method is possible only when certain conditions are satisfied. These conditions are formulated as follows: Let α, β, τ be "ray" coordinates, of which α and β characterize the ray, i.e. the extremal of the Fermat functional $\int d\sigma/c(x, y, z)$, where $d\sigma$ is a differential arc length, while τ (the so-called *eikonal*) is the time of propagation along the ray (Sect. 2.3). The ray method can be applied only if the cartesian coordinates (x, y, z) are smooth functions of the ray coordinates (α, β, τ), and the functional determinant $D(x, y, z)/D(\alpha, \beta, \tau)$ differs from zero. When this condition is satisfied the field of rays is called regular.

However, a rigorous mathematical proof that the formulas for the ray method are asymptotic expansions of solutions of the given boundary-value problems in the case of regular ray fields is difficult, and in most cases is still an unsolved problem. Despite the lack of rigorous proofs, neither physicists nor mathematicians doubt the validity of the asymptotic formulas of the ray method. This confidence stems first of all from the fact that such proofs have been demonstrated in a number of cases (Sect. 2.6 and the introductory remarks to Chap. 11, Sect. 11.1 and notes to Chap. 2, where the appropriate references are cited), secondly, that the derivation of the formulas of the ray method agrees with the physics of short-wave propagation, and thirdly, that these formulas have been satisfactorily confirmed by experiment.

The simplicity and clarity of the ray method make it an irreplaceable tool for calculating fields at high frequencies. Chapter 2 deals with the principles of the ray method.

---

[1] It should be mentioned that the principle of limiting absorption has been rigorously proved only with special assumptions on the behavior of the function $c(x, y, z)$ and the boundary S at infinity [1.2].

The study of fields corresponding to irregular ray fields is an important problem. Assume that the field of rays has an envelope. At points of the envelope we have $D(x, y, z)/D(\alpha, \beta, \tau) = 0$, and therefore the ray method in its original form is inapplicable in the vicinity of the envelope. In physics, the envelope of the field of rays is usually called a *caustic*. The ray field near a caustic is a very simple, but in practice very important case of an irregular ray field. An asymptotic expansion containing Airy functions can be derived for a wave field in the vicinity of a caustic surface that has no singularities. Outside of a certain layer surrounding the surface, this expansion reduces to an expansion of the ray method. The thickness of the layer decreases with increasing frequency. The third chapter describes the wave field in the vicinity of a caustic surface. Our presentation is based on papers by *Kravtsov* [1.3] and *Gazaryan* [1.4]. In this chapter the reader will find not only formulas for the first-order approximations, but also an investigation of the analytical properties of subsequent approximations.

The ray method supplemented by certain results of Chap. 3 enables asymptotic formulas for eigenfunctions and the corresponding eigenvalues of boundary value problems associated with Helmholtz's equation (1.2) to be found.

Ray representations for finding the asymptotics of eigenfunctions of the Laplacian operator have been used by *Keller* and *Rubinow* [1.5]. The Keller-Rubinow method assumes the existence of ray congruences that are invariant relative to reflections at the boundary of the region and continuously dependent on a number of parameters. Such congruences have been constructed in the two-dimensional case for a circle and an ellipse, and in the three-dimensional case for a sphere and a triaxial ellipsoid, i.e. in problems with separable variables. Similar techniques have also been used by *Vainshtein* [1.6–8] and *Bykov* [1.9, 10] to determine the asymptotic characteristics of eigenfunctions and eigenvalues. The fourth chapter outlines the ray method for constructing asymptotic formulas for eigenfunctions and eigenvalues (the Keller-Rubinow method).

In pondering how best to present the Keller-Rubinow method, the authors came to the conclusion that the exposition would be made clearer by using some elementary concepts of the topology of manifolds. We hope that this has not detracted from the physical clarity of the Keller-Rubinow construction.

Unfortunately, the Keller-Rubinow method in its original form has a very restricted range of application that reduces, in practice, to problems with separable variables. Attempts to use this method in more general cases have run up against fundamental difficulties involving the existence of zones of instability of solutions of dynamic systems. Meanwhile, developments in laser technology have brought about an urgent need for asymptotic expansions of eigenfunctions concentrated around certain closed curves — one-dimensional cycles or closed chains.[2]

---

[2] A function is said to be concentrated in the vicinity of a curve *l* if this function decreases exponentially with increasing distance from curve *l* beyond the limits of a fairly narrow layer that contains curve *l* (for a more precise and detailed discussion, see Chaps. 5 and 6).

The examples of the ellipse and ellipsoid show that eigenfunctions do not concentrate near any arbitrary one-dimensional closed chain. In order for eigenfunctions to exist in the vicinity of a closed chain, this chain must satisfy certain conditions. We will call a closed chain stable in the first approximation if rays that are initially close to the chain are still located in a small neighborhood of the chain after propagating far enough in the inhomogeneous medium and after a sufficiently large number of reflections from the boundary of the region have occurred (for exact definitions, see Chap. 5). It is to be expected that the requirement that a closed chain be stable in the first approximation must be satisfied in order for eigenfunctions concentrated in the vicinity of that chain to exist. The naturalness of this requirement is implied by an intuitive physical consideration: if the waves corresponding to rays that have traversed a closed chain many times, are to form an eigenfunction, they must constructively interfere with each other a sufficient number of times before leaving the neighborhood of the chain.

It turns out that the asymptotic properties of eigenfunctions of the whispering-gallery and bouncing-ball types can be found by a technique that is a modification of the Keller-Rubinow method. Since this modification of the Keller-Rubinow method deals with rays that belong to a fairly small neighborhood of the closed chain, we will call it the *ray method in the small.*

It should be pointed out that the ray method in the small is applicable when and only when the corresponding closed chain is stable in the first approximation. This situation shows that the requirement for stability of a chain is not only sufficient, but apparently also necessary for the existence of eigenfunctions of the whispering gallery and bouncing-ball types. In Chap. 5 the ray method in the small will be used to construct asymptotic relations for eigenvalues corresponding to eigenfunctions of the whispering gallery and bouncing ball types in various two-dimensional problems with non-separable variables and a variable wave propagation velocity $[c(x, y) \neq \text{const.}]$.

By analogy with fluid mechanics it is natural to use the term diffraction boundary layer for the neighborhood of curves in which the solutions of (1.2 and 6) are appreciably different from zero. The fields in boundary layers, in the first approximation, are not described by simple eikonal and transport equations (the basic equations of the ray method), but rather by a more complicated equation of the Schrödinger type. This equation, which is usually termed parabolic in diffraction theory, is an analog of the conventional boundary layer equations of fluid mechanics. A parabolic equation for describing wave fields has been proposed by *Leontovič* and *Fock* [1.11] (see also the notes to Chaps. 6 and 11).

A detailed exposition of boundary-layer methods in diffraction theory has been given in the book by *Babič* and *Kirpičnikova* [1.12].

The parabolic-equation method is examined in Chap. 6. In this chapter the parabolic equation is derived and solved for the case of smooth boundaries.[3]

---

[3] The parabolic equation is applicable as well to problems with boundaries that are not smooth. For instance, the high-frequency asymptotic properties of a field diffracted from a corner can be found not only from the exact solution of the problem, but also by the parabolic equation method [1.13]. However, problems with boundaries that are not smooth will not be taken up in the present book.

Replacing the original (elliptic) Helmholtz equation with a parabolic equation, as we have already pointed out, gives only the principal terms of the asymptotic expansion as $\omega \to \infty$. However, in many cases we are interested not only in the principal terms, but also in the subsequent terms of the expansions.

The construction of higher-order terms of asymptotic expansions has led, in recent years, to the development of the so-called *"etalon" (model) problem method*, which can be treated as a further natural development of the parabolic-equation method.

The etalon-problem method is an extension of the method of comparison equations, currently extensively used for deriving asymptotic expansions for solutions of ordinary differential equations, to boundary value problems of wave diffraction and propagation. The etalon-problem method is based on the principle that similar ray geometry leads to similar asymptotic formulas (as $\omega \to \infty$) for wave fields.

Just as the comparison equation in the asymptotic theory of differential equations is the simplest equation with the same singularities in its coefficients as the original equation, so an etalon problem in the theory of wave diffraction and propagation is the simplest problem in which the field of rays has the same singularities as in the original problem. The essence of the etalon problem method is as follows. The initial problem under consideration is replaced by the simplest comparison (etalon) problem that can be solved exactly, usually by the method of separation of variables. The exact solution of the etalon problem is investigated as $\omega \to \infty$, and from this an asymptotic expression is obtained which describes the field in the region of interest where the ray field has the singularities peculiar to the original problem. This expression is usually a product of special functions or a contour integral of special functions with arguments that are asymptotic series with respect to fractional negative powers of the large parameter ω. The field in the original problem is sought in the same analytical form, but with different coefficients in the asymptotic series appearing in the arguments of the special functions. In other words, the analytical expression for the field found in the investigation of the etalon problem is taken over to the original problem. The coefficients of the asymptotic series are determined recursively by substituting this analytical expression into the Helmholtz equation and boundary conditions that correspond to the original problem. The more terms that are found and retained in the asymptotic series, the faster the discrepancy in the equation and boundary conditions (i.e., the degree to which they are not satisfied) should approach zero with increasing ω. When the point of observation is moved into a region where the field of rays is regular, the formulas found by the etalon-problem method reduce to those of the ray method. This can be considered a justification of the asymptotic formulas thus constructed (not yet at the level of theorems, of course). The etalon-problem method is demonstrated for two-dimensional problems of eigenfunctions of the whispering gallery-type (Chap. 7) and the bouncing-ball type (Chap. 8).

The reader should readily notice a hierarchy in the presentation of eigenfunctions concentrated in the neighborhood of one-dimensional closed chains. The problem is considered first by the ray method in the small, then by the parabolic equation method, and finally by the etalon problem method, which even gives some rigorous results (Sect. 7.6). By means of this kind of a presentation we hope not only to shed more light on the asymptotic properties of eigenfunctions, but also to illustrate on simple examples some important methods of the mathematical theory of diffraction. Chapters 9 and 10 examine eigenfunction problems for three-dimensional and multidimensional regions.

In their investigation of the natural oscillations of the ellipsoid

$$\frac{x^2}{a^2} + \frac{y^2}{b^2} + \frac{z^2}{c^2} = 1 \; , \quad a>b>c>0 \; . \tag{1.7}$$

Vainshtein [1.7] and *Bykov* [1.9] came to the conclusion that certain subsequences of the eigenfunctions of the ellipsoid may be concentrated in the vicinity of the largest and smallest ellipses that are formed from the intersection of coordinate planes with the surface of the ellipsoid. These ellipses are closed geodesics on the ellipsoid surface that are stable in the first approximation. It turns out [1.14] that the Laplacian operator defined on the ellipsoid surface also has subsequences of eigenfunctions concentrated in the neighborhood of these geodesics.

It is natural to consider how to generalize these results. This is the topic considered in Chap. 9, where the surface of the ellipsoid is replaced by an arbitrary $(m+1)$-dimensional compact Riemannian manifold. Here the basic mathematical apparatus is the theory of Hamiltonian systems of linear ordinary differential equations with periodic coefficients, and certain differential operators with properties that are quite similar to the well-known creation and annihilation operators in quantum field theory.

The application of creation operators to a function that has properties analogous to those of the ground state in quantum-field theory produces the desired eigenfunctions in the first approximation. The complete asymptotic expansions for these eigenfunctions are constructed here in the form of series in inverse powers of the frequency: the coefficients of the series are expressed fairly simply in terms of the eigenfunctions of the first approximation.

Another problem considered in Chap. 9 is the construction of asymptotic relations for natural oscillations of a three-dimensional region Ω that are concentrated in the vicinity of a closed geodesic of the surface S bounding the region Ω. This problem is a distinctive combination of that of eigenfunctions concentrated in the neighborhood of a closed geodesic on a two-dimensional surface, and of whispering gallery eigenfunctions. The algorithm for constructing the coefficients of their asymptotic expansions is also a combination of the algorithms used for constructing the asymptotic solutions of these two problems.

In Sect. 9.6, the solution of a second-order elliptic system of differential equations which is concentrated in the neighborhood of the characteristics of the Hamiltonian system corresponding to this elliptic one is constructed. The apparatus used in Sect. 9.6, just as in the rest of Chap. 9, is based on the parabolic equation method. The important point here is the fact that the solution of the parabolic equation which describes the "concentrated solution in the first approximation" is expressed in term of the solution of the linearized Hamiltonian system. Thus, what has been accomplished is the description of purely a wave phenomenon in the classical terms of rays in the first approximation.

The mathematical techniques developed in Chap. 9 are also applicable to the problems considered in Chaps. 7 and 8, but in Chap. 10 we concern ourselves with a more interesting application – the natural oscillations of a multiple-mirror resonator. This problem is treated both by the parabolic-equation method and by the ray method in the small. The latter requires recourse to some rather subtle techniques from analytical mechanics.

Problems involving the construction of Green's functions (or in the language of physics, point-source problems) are of considerable physical and mathematical interest. The classical Fourier method leads to an expression for the Green's function in the form of a superposition of eigenfunctions. The methods presented in this book enable one to construct asymptotic forms of the eigenfunctions.

It is natural to expect that by superposition of the asymptotic expressions for the appropriate analogs of eigenfunctions, one might be able to find the asymptotic expressions for solutions of corresponding point source problems as well. The asymptotic versions of the whispering gallery waves found in Chap. 7 are used in Chaps. 11 and 12 to construct asymptotic formulas for the Green's function.

Chapter 11 examines the two-dimensional problem of diffraction of a wave from a point source by a smooth reflecting obstacle. Many papers have dealt with this classical problem of diffraction theory; of fundamental importance among these is the work of V.A. Fock. Important results have also been obtained by W. Franz and J.B. Keller.

The etalon problem method takes the next step, giving not only the principal term of the asymptotic expansion, but all subsequent ones. Chapter 11 concentrates mainly on the construction of asymptotic expansions for the Green's function in a boundary layer adjacent to a reflecting surface S. Any of the boundary conditions (1.3–5) can be assigned on surface S (without any special assumptions relative to $g(M)$ in the case of the mixed boundary condition). The case of the Dirichlet condition is considered in the most detail. The expansions constructed in Chap. 11 are fairly simple formal series in fractional powers of the wave number k. However, these expansions in their original form are not applicable beyond the limits of the boundary layer. To obtain formulas that are suitable beyond the limits of the boundary layer a transition must be made from the boundary layer coordinates to so-called evolutional coordinates. In this way asymptotic formulas are derived and written out which are valid to an error of $O(k^{-2/3})$ at any distance from the boundary of the obstacle.

In Chap. 12 the asymptotic properties of eigenfunctions of the whispering-gallery type are applied to the problem of the field of a source located on the surface of a concave body. In this problem we encounter the whispering gallery effect and the existence of a surface wave of interference type. In the case of a source located on the surface, there are in any arbitrarily small neighborhood of the boundary of the solid, an infinite number of caustic curves. These are

envelopes of rays multiply reflected from the boundary. Until recently, ali
no consideration has been given to the asymptotic behavior of fields in the
of non-isolated singularities of the ray field. The method of normal waves
pansion of the field in a series of certain special solutions of the wave equat
which is ordinarily used in problems of this kind has, in addition to its
doubted advantages, the following disadvantage: representation of the wave
by a sum of normal waves precludes tracing the formation of waves that (
the usual geometric-optics principles when the observation point moves in
region where the field of rays is regular.

In Chap. 12 the interference field of a surface wave is described by a
tour integral of special functions. This kind of approach to the descriptio
wave fields with non-isolated singularities in the ray field derives from pa
by V.A. Fock dealing with investigation of the field in a penumbra. Cor
integrals that describe the surface-wave field are constructed in this book b;
etalon-problem method. The integrands of these integrals have quite pronou:
maxima along the integration contour, which appreciably facilitates the comp
tabulation of the integrals and the compilation of tables. On the other hand, t
contour integrals can be calculated by the method of steepest descents whei
observation point goes from a region where a large number of rays overlap
a region where the field of rays is regular. The formulas derived in this
coincide with those of the ray method. The matching of the contour inte;
which describe interference wave fields with formulas of the ray method m
the method developed in the book superior to the normal-wave method.

Chapter 13 examines the diffraction of a wave given by its ray expan:
by a smooth reflecting surface S. It is assumed that the surface is convex
non-zero Gaussian curvature. The constructions in this chapter are based o
Ludwig's idea of finding the field in the form of a superposition of express
analogous to the caustic solutions of Chap. 3. In this way a function $I(M,$
constructed (k is the wave number $k = \omega/c$, c = const.) that satisfies the Diri(
boundary condition and Helmholtz's equation in a penumbral region wit:
arbitrarily small discrepancy (in orders of magnitude in k^{-1}, as $k \to \infty$). Ir
illuminated region $I(M, k)$ is transformed into a ray formula. In the shadow :
the behavior of $I(M, k)$ agrees with formulas proposed at one time by J.B. K
on the basis of heuristic considerations.

By using the procedure of Chap. 13 one can construct the asymptotic ex
sion of the Green's function (in both the two-dimensional and three-dimensi
cases) if the source is removed from the reflecting surface S by a distance
is independent of the wave number. On the other hand, if the source is f
close to S, then the formulas of Chap. 13 are not applicable. The construci
of Chap. 11 can then be used in the two-dimensional case. Thus the resul:
Chaps. 11 and 13 complement each other.

A number of supplementary topics (such as the theory of Airy functions
examined in the appendices. A schematic diagram showing the interdepend
of the chapters and sections of this book can be found after the table of conti

Most of the formulas in the book have been derived on the basis of heuristic considerations; in other words, they stem from assumptions that are supplementary to the mathematical formulation of the problem. These assumptions are usually simple and graphic. For instance, they include Fock's localization principle in the theory of high-frequency diffraction, the requirement of existence of outgoing-wave phase ("radiation" condition) in the solution, and a number of others. All mathematical constructions leading from these assumptions to the final results have, to the best of our ability, been carried out so that they meet the requirements of mathematical rigor. Some of the results found on the basis of heuristic consideration can be rigorously substantiated, but the book does not include these proofs because of the cumbersome estimations. An exception is the theorem that establishes the asymptotic nature of the expansions for eigenvalues in the problem of eigenfunctions concentrated near the boundary of a region. The proof of this short and beautiful theorem is given in Chap. 7.

The justification for using other (not yet rigorously proven) results of the asymptotic theory of diffraction is similar in many ways to the justification for treating as fact theoretical hypotheses in the experimental sciences. In analyzing the explicit solution of an etalon diffraction problem, one can frequently arrive via certain constructions at an asymptotic formula that pertains to a general case. This more general formula is then tested on other problems that permit an explicit solution. In the experimental sciences the experiment is the analog of our explicit solution. The fresh mathematics graduate rarely has mastered the art of reasoning on a physical level of rigor; but nevertheless, heuristic considerations are absolutely necessary for solving the mathematical problems encountered in physics. We hope that our book will be equally useful in showing the young mathematician the utility and importance of less rigorous guiding considerations.

2. The Ray Method

In this chapter we present the geometrical-optics approximation (or *ray method*) for the wave equation and for Maxwell's equations. The ray method lies at the heart of all of the constructions in this monograph.

2.1 The Basic Principles

The ray method, as we have indicated in Chap. 1, is an asymptotic method for solving diffraction problems. In mathematical physics, an asymptotic method is often merely one version or another of a perturbation technique, and the ray method is no exception. At its basis is the idea of seeking a solution of the wave equation (1.1) in the form of some formal expansion. Usually, we employ an expansion of the form

$$W = e^{-i\omega t + i\omega\tau(M)} \sum_{j=0}^{\infty} \frac{u_j(M)}{(-i\omega)^{j+\gamma}} \quad ; \quad u_0 \neq 0 \tag{2.1.1}$$

where M is a point in three-space[1], while $\tau(M)$ and $u_j(M)$ are functions to be determined. Infrequently, we have occasion to use an expansion of the form

$$W = \exp\left[-i\omega t + i\omega\tau(M) + \sum_{j=0}^{\infty} \frac{v_j(M)}{(i\omega)^j}\right] \quad ; \quad v_0 \neq 0 \tag{2.1.2}$$

where $\tau(M)$ and $v_j(M)$ are to be determined.

Both these expansions, usually called ray solutions, are modelled after the plane wave – classically the most elementary solution of the wave equation. To convince ourselves that these expansions are indeed perturbations of plane waves, let x, y, z be the Cartesian coordinates of a point M. Replacing the function $\tau(M)$ in (2.1.1, 2) by its Taylor expansions in the neighborhood of some point $M_0(x_0, y_0, z_0)$, we obtain to first order

$$W(M, t) \simeq W(M_0, 0)e^{-i\omega t}\exp[\tau_x(M_0)(x - x_0) + \tau_y(M_0)(y - y_0)$$
$$+ \tau_z(M_0)(z - z_0)] \tag{2.1.3}$$

where

---

[1] Our arguments do not depend significantly on the number of dimensions. Certain modifications which have to be made for the planar case will be indicated below.

$$W(M_0, 0) = u_0(M_0)e^{i\omega\tau(M_0)} \quad \text{for the expansion (2.1.1) or}$$

$$W(M_0, 0) = \exp[v_0(M_0) + i\omega\tau(M_0)] \quad \text{for the expansion (2.1.2)} .$$

For a fixed point M_0, (2.1.3) is indeed a plane wave with the wave vector

$$\vec{k} = \omega\nabla\tau(M_0) .$$

The plane wave (2.1.3) satisfies the wave equation

$$\frac{1}{c^2(M)}\frac{\partial^2 W}{\partial t^2} - \Delta W = 0$$

at $M = M_0$, provided that $|\nabla\tau(M_0)| = c^{-1}(M_0)$. As the point M_0 is varied, (2.1.3) can be considered a plane wave whose wave vector $k(M_0)$, amplitude $|W(M_0, 0)|$, and initial phase $\omega\tau(M_0)$ change from point to point. Thus the ray solutions (2.1.1, 2) can be thought of as *locally-plane* waves with continuously changing characteristic, to first-order accuracy. Clearly, the more slowly the medium varies with distances on the order of a wavelength $\lambda = 2\pi c(M_0)/\omega$ (i.e., the smaller the quantity $\lambda|\nabla c(M)|/c(M) = 2\pi|\nabla c(M)|/\omega$), the more accurately the plane wave (2.1.3) and hence also the ray expansions (2.1.1, 2) will describe a solution of (1.1). This simple physical consideration leads to the following condition for the applicability of the ray expansions (2.1.1, 2)

$$\frac{2\pi}{\omega}|\nabla c(M)| \ll 1 .$$

Note that (2.1.1, 2) are expansions with respect to a "dimensioned" parameter ω. This might evoke a certain dissatisfaction in the reader. In fact, a calculation will show that each term in (2.1.1) has the same dimensions, while each term in (2.1.2) is dimensionless. Thus, (2.1.1, 2) both have a dimension-free character, and can always be written in some other form not explicitly containing a power of the dimensioned quantity ω. Despite this, the use of an expansion in inverse powers of the dimensioned quantity ω has a number of advantages, formal and otherwise. A formal advantage is that these expansions can be treated as one would treat any power series; a less formal one is that this type of expansion gives a clear indication of the frequency dependence of the wave process, and this dependence is not obscured by incidental details of a specific problem. In what follows, we shall often use expansions with respect to a large or small dimensioned parameter.

Let us now obtain the equations which the functions $\tau(M)$, $u_j(M)$ and $v_j(M)$ of (2.1.1, 2) must satisfy. To do this, we substitute (2.1.1 or 2) into (1.1) or (what amounts to the same thing in this case) into (1.2). Setting the coefficient of each succeeding power of ω equal to zero in the resulting expression, we arrive at a system of recurrence relations

$$(\nabla\tau)^2 - 1/c^2(M) = 0 , \tag{2.1.4}$$

$$2\nabla u_j \cdot \nabla\tau + u_j\Delta\tau - \Delta u_{j-1} = 0 \quad \text{with}$$
$$j = 0, 1, 2, \ldots ; \quad u_{-1} \equiv 0 \tag{2.1.5}$$

in the case of (2.1.1), and

$$(\nabla \tau)^2 - 1/c^2(M) = 0 \quad , \tag{2.1.6}$$

$$2\nabla \tau \cdot \nabla v_0 + \Delta \tau = 0 \quad ,$$

$$2\nabla \tau \cdot \nabla v_j + \sum_{i=0}^{j-1} \nabla v_i \cdot \nabla v_{j-i-1} + \Delta v_{j-1} = 0 \quad \text{with} \quad j = 1, 2, \ldots \tag{2.1.7}$$

in the case of (2.1.2).

Equation (2.1.4 or 6) is one of the most important equations in mathematical physics, and it bears the special name, the *eikonal equation*. Equations (2.1.5, 7) are known as *transport equations*. If the eikonal equation is multiplied by $\omega^2 c^2(M)$, and if $\omega > 0$, it can be written in the form

$$\omega = |\vec{k}| c(M) \quad . \tag{2.1.8}$$

A relation of the form $\omega = \omega(\vec{k})$ which gives the dependence of the frequency of a wave process on the wave vector $\vec{k}$ is known in physics as dispersion equation. Equation (2.1.8), which applies to the wave equation (1.1), is the simplest of all dispersion equations. For media in which the wave propagation is described by more complicated equations than the wave equation (for instance, elastic media, magnetohydrodynamic media, etc.), the dispersion equation will take a more complicated form, and consequently the equation for the function will be more complicated than the eikonal equation (2.1.4).

The factor $\exp[-i\omega t + i\omega \tau(M)]$ entering into (2.1.1, 2) is called the phase factor: the expression $\omega[t - \tau(M)]$ is termed the phase of the wave. This phase is constant along the moving surfaces $t - \tau(M) = \text{const}$, which are usually referred to as wave fronts.

We next introduce the concept of the displacement velocity of a moving surface, and show that the displacement velocity of a wave front (the phase velocity) is equal to $c(M)$. Let a surface S which is moving in space be described, at a moment of time t, by the equation $\psi(x, y, z, t) = \text{const}$. At the time $t + \Delta t$, $\Delta t > 0$, it will be described by the equation $\psi(x, y, z, t + \Delta t) = \text{const}$. Choose a point $M_0(x_0, y_0, z_0)$ on the surface $\psi(x, y, z, t) = \text{const}$, construct the normal to that surface at M_0, and extend it up to its intersection with the surface $\psi(x, y, z, t + \Delta t) = \text{const}$. Denote the point of intersection by $M_1(x_1, y_1, z_1)$ (Fig. 2.1). The displacement velocity v of the surface S at the point (x_0, y_0, z_0, t) will be defined

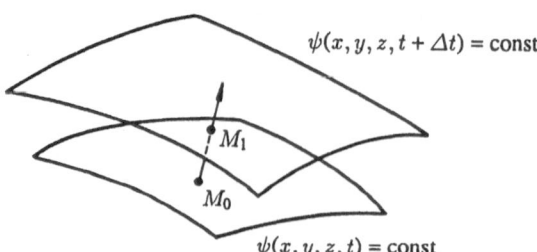

$\psi(x, y, z, t + \Delta t) = \text{const}$

M_1

M_0

$\psi(x, y, z, t) = \text{const}$

Fig. 2.1. Adjacent wavefront

as the limit of the length of the segment $M_0 M_1$ divided by Δt as $\Delta t \to 0$, i.e.,

$$v = \lim_{\Delta t \to 0} \frac{|M_0 M_1|}{\Delta t} \quad .$$

To evaluate this limit, we use Taylor's formula

$$\psi(x_1, y_1, z_1, t + \Delta t) = \psi(x_1, y_0, z_0, t) + \overrightarrow{M_0 M_1} \cdot \nabla \psi(x_0, y_0, z_0, t)$$
$$+ \Delta t \frac{\partial \psi(x_0, y_0, z_0, t)}{\partial t}$$
$$+ \text{second-order terms} \quad . \tag{2.1.9}$$

(Here, $\nabla \psi$ is the gradient with respect to the spatial variables x, y, z only). Since $\psi(x_0, y_0, z_0, t)$ and $\psi(x_1, y_1, z_1, t + \Delta t)$ are equal to the same constant, we have to first order from (2.1.9)

$$\overrightarrow{M_0 M_1} \cdot \nabla \psi(x_0, y_0, z_0, t) + \Delta t \frac{\partial \psi(x_0, y_0, z_0, t)}{\partial t} = 0 \quad .$$

Hence,

$$v(M_0) = \frac{-\partial \psi / \partial t}{\vec{n} \cdot \nabla \psi} \bigg|_{M_0, t} \tag{2.1.10}$$

where $\vec{n}$ is the unit vector normal to the surface S in the direction of its motion, i.e., along $\overrightarrow{M_0 M_1}$. Since by definition $v > 0$, the numerator and denominator in (2.1.10) must have the same sign. Thus, if $\partial \psi / \partial t > 0$, then $\vec{n} \cdot \nabla \psi < 0$ and the movement of the surface S is in the direction opposite to that of $\nabla \psi$. If $\partial \psi / \partial t < 0$, then the motion of S is along the direction of $\nabla \psi$.

To apply (2.1.10) to the calculation of the wave front velocity, we set $\psi = t - \tau(M)$, and thus the wave front moves in the direction of $-\nabla \psi = \nabla \tau$ (that is, along the gradient of the eikonal function τ), with the velocity

$$v = \frac{1}{|\nabla \tau|} = c(M) \quad .$$

It can likewise be stated that the wave front moves in the direction of the wave vector $\vec{k} = \omega \nabla \tau$.

2.2 Variational Theory of the Fermat Functional

The reader undoubtedly recalls Fermat's principle from physics: light (or more accurately, a wave disturbance described by (1.1)) propagates in an inhomogeneous medium between two given points M_0 and M_1 along a path which requires the least amount of time compared to all other possible paths.

We can supply the Fermat principle with a mathematical formulation. Let $c(M)$ be the velocity of propagation of a disturbance at a point M on a curve l, ds be a differential arc length on l, and dt be the time it takes for the disturbance

13

to travel the length ds of arc. Clearly, $dt = ds/c(M)$. We can calculate the time required for a wave disturbance to travel between two points M_0 and M_1 lying on l by integrating along l

$$\int_{M_0}^{M_1} \frac{ds}{c(M)} \quad . \tag{2.2.1}$$

For various curves l which contain the points M_0 and M_1, the value of integral (2.2.1) will be different. A function defined on such a set of curves and taking different values is known in mathematics as a functional. The integral in (2.2.1) is usually called the *Fermat functional* and denoted by $\Phi(l)$.

Let $\{l\}$ be the set of all smooth curves containing the points M_0 and M_1. Fermat's principle as stated above requires that the wave disturbance be propagated between points M_0 and M_1 along the path for which the Fermat functional

$$\Phi(l) = \int_{M_0}^{M_1} \frac{ds}{c(M)} \tag{2.2.2}$$

takes on the smallest value of any curve in the set $\{l\}$. We assume that the curves of $\{l\}$ are given in some curvilinear coordinate system

$$\vec{r} = \vec{r}(q^1, q^2, q^3)$$

in the parametric form

$$q^1 = q^1(\sigma) \ , \quad q^2 = q^2(\sigma) \ , \quad q^3 = q^3(\sigma)$$

where σ is a parameter specifying a point on the curve. Then

$$ds = \sqrt{\frac{\partial \vec{r}}{\partial q^i} \cdot \frac{\partial \vec{r}}{\partial q^j} dq^i dq^j} = \sqrt{\frac{\partial \vec{r}}{\partial q^i} \cdot \frac{\partial \vec{r}}{\partial q^j} \dot{q}^i \dot{q}^j} \, d\sigma \quad .$$

Here and in what follows we will adopt the summation convention: repeated indices i or j are to be summed from 1 to 3. In addition, we have put $\dot{q}^i = dq^i/d\sigma$.

Introducing the metric tensor

$$g_{ij} = \frac{1}{c^2(M)} \frac{\partial \vec{r}}{\partial q^i} \cdot \frac{\partial \vec{r}}{\partial q^j}$$

we can write the Fermat functional in the form

$$\Phi(l) = \int_{\sigma_0}^{\sigma_1} \mathcal{F}(q^i, \dot{q}^j) d\sigma \tag{2.2.3}$$

where

$$\mathcal{F}(q^i, \dot{q}^j) = \sqrt{g_{ij}(q^1, q^2, q^3) \dot{q}^i \dot{q}^j} \tag{2.2.4}$$

is a positively-homogeneous function of first degree with respect to the variables $\dot{q}^j$; σ_0 and σ_1 are the values of the parameter σ corresponding to the points M_0

14

and M_1. If the coordinates q^1, q^2, q^3 are cartesian, then

$$\frac{\partial \vec{r}}{\partial q^i} \cdot \frac{\partial \vec{r}}{\partial q^j} = \delta_{ij} = \begin{cases} 1, & i = j \\ 0, & i \neq j \end{cases}$$

and

$$\mathcal{F}(q^i, \dot{q}^j) = \frac{1}{c(q^1, q^2, q^3)} \sqrt{\sum_{i=1}^{3} (\dot{q}^i)^2} \quad .$$

In the more traditional notation for cartesian coordinates, $q^1 = x$, $q^2 = y$, $q^3 = z$, and

$$\mathcal{F} = \frac{1}{c(x, y, z)} \sqrt{(\dot{x})^2 + (\dot{y})^2 + (\dot{z})^2} \quad .$$

We now introduce the idea of the first variation of the functional (2.2.3) at a curve l [the specific form (2.2.4) of the function $\mathcal{F}$ is not essential for this purpose; only the fact that $\mathcal{F}$ is a homogeneous function of the $\dot{q}^j$ of first degree is important]. Consider a curve near l

$$l + \delta l : \quad q^i = q^i(\sigma) + \delta q^i(\sigma) \quad . \tag{2.2.5}$$

Here it is assumed that the parameter σ varies on the interval $(\sigma_0 + \delta\sigma_0, \sigma_1 + \delta\sigma_1)$, over which the functions $q^i(\sigma)$ and $\delta q^i(\sigma)$ are assumed to be given.

The first variation $\delta\Phi$ of the functional Φ is defined to be the linear (first-order) part of the increment $\delta\Phi = \Phi(l + \delta l) - \Phi(l)$ with respect to δq^i, $\delta\sigma_0$ and $\delta\sigma_1$, if these latter are regarded as first-order infinitesimals. In variational calculus, one proves the formula [2.1]

$$\delta\Phi = \int_{\sigma_0}^{\sigma_1} \left(\frac{\partial \mathcal{F}}{\partial q^i} - \frac{d}{d\sigma}\frac{\partial \mathcal{F}}{\partial \dot{q}^i} \right) \delta q^i \, d\sigma + \left[\left(\mathcal{F} - \frac{\partial \mathcal{F}}{\partial \dot{q}^i}\dot{q}^i \right) \delta\sigma \right]_{\sigma=\sigma_0}^{\sigma_1} + \frac{\partial \mathcal{F}}{\partial \dot{q}^i}\delta q^i \Big|_{\sigma=\sigma_0}^{\sigma=\sigma_1} .$$

Here $\delta q^i|_{\sigma_0, \sigma_1}$ are the so-called variations of the end points of the curve l, i.e., the differences between the coordinates of the end points of the curve $l + \delta l$ and the corresponding end points of the curve l.

Since $\mathcal{F}(q^i, \dot{q}^j)$ is homogeneous of first degree in the variables $\dot{q}^j$, we have the relationship

$$\frac{\partial \mathcal{F}}{\partial \dot{q}^i}\dot{q}^i = \mathcal{F}$$

and consequently

$$\delta\Phi = \int_{\sigma_0}^{\sigma_1} \left(\frac{\partial \mathcal{F}}{\partial q^i} - \frac{d}{d\sigma}\frac{\partial \mathcal{F}}{\partial \dot{q}^i} \right) \delta q^i \, d\sigma + \frac{\partial \mathcal{F}}{\partial \dot{q}^i}\delta q^i \Big|_{\sigma=\sigma_0}^{\sigma=\sigma_1} . \tag{2.2.6}$$

If the end points M_0, M_1 of the curve l are fixed, then the curve (2.2.5) near l must pass through M_0 and M_1 as well. In this event $\delta q^i|_{\sigma=\sigma_0, \sigma_1} = 0$, and (2.2.6) for the first variation becomes

$$\delta\Phi = \int\limits_{\sigma_0}^{\sigma_1} \left(\frac{\partial\mathcal{F}}{\partial q^i} - \frac{d}{d\sigma}\frac{\partial\mathcal{F}}{\partial\dot{q}^i} \right) \delta q^i\, d\sigma \quad . \tag{2.2.7}$$

On the curve l which extremizes the functional $\Phi(l)$, the first variation $\delta\Phi$ will vanish [2.1]. However, $\delta\Phi = 0$ does not in turn guarantee that l will extremize Φ.

Curves for which the first variation of a functional Φ vanish, i.e., for which $\delta\Phi = 0$, are called extremal curves (or extrema) of Φ. The extremal curves of the Fermat functional (2.2.3, 4) are called rays.

To obtain the equation for a ray, we note that

$$\delta\Phi = 0 \tag{2.2.8}$$

on a ray l, and consequently the right side of (2.2.6) must vanish for arbitrary δq^i. If this occurs for all possible neighboring curves, it must certainly occur for those neighboring curves whose end points coincide with those of l, i.e., when $\delta q^i|_{\sigma=\sigma_0} = \delta q^i|_{\sigma=\sigma_1} = 0$. Thus on the ray we must have

$$\int\limits_{\sigma_0}^{\sigma_1} \left(\frac{\partial\mathcal{F}}{\partial q^i} - \frac{d}{d\sigma}\frac{\partial\mathcal{F}}{\partial\dot{q}^i} \right) \delta q^i\, d\sigma = 0 \quad . \tag{2.2.9}$$

In view of the arbitrary nature of δq^i, it follows that

$$\frac{\partial\mathcal{F}}{\partial q^i} - \frac{d}{d\sigma}\frac{\partial\mathcal{F}}{\partial\dot{q}^i} = 0 \;, \quad i = 1, 2, 3 \tag{2.2.10}$$

must hold on the ray l.[2]

Equations (2.2.10) for a ray l are called Euler's equations. For each i the left-hand side of the Euler equation is called a variational derivative and is denoted by

$$\frac{\delta\mathcal{F}}{\delta q^i} \equiv \frac{\partial\mathcal{F}}{\partial q^i} - \frac{d}{d\sigma}\frac{\partial\mathcal{F}}{\partial\dot{q}^i} \;, \quad i = 1, 2, 3 \quad .$$

In cartesian coordinates Euler's equations (2.2.10) take the form

$$\frac{\partial}{\partial q^j} \left(\sqrt{\sum_{i=1}^{3}(\dot{q}^i)^2} \Big/ c \right) - \frac{d}{d\sigma} \left(\dot{q}^j \Big/ c\sqrt{\sum_{i=1}^{3}(\dot{q}^i)^2} \right) = 0, \; j = 1, 2, 3. \tag{2.2.11}$$

Let us choose the parameter σ to coincide with the arc length s along the ray, so that the square roots in (2.2.11) are equal to 1. The three equations which thus result from (2.2.11) can be written in the vector form:

$$\frac{d}{ds}\left(\frac{\vec{s}_0}{c} \right) = \nabla\left(\frac{1}{c} \right)$$

---

[2] The passage from (2.2.9 to 10) is the essence of the so-called fundamental lemma of the variational calculus, whose proof involves certain definite assumptions about the function $\mathcal{F}$ and the curve l [2.1].

where $\vec{s}_0$ is the unit vector tangent to the ray.

Let us denote by τ the arc length in this Riemannian metric [2.2][3]

$$\tau = \tau_0 + \int_{\sigma_0}^{\sigma_1} \sqrt{\sum_{i,j=1}^{3} g_{ij} \dot{q}^i \dot{q}^j} \, d\sigma \quad .$$

If σ is chosen to be identical to τ, then

$$\sum_{i,j=1}^{3} g_{ij} \dot{q}^i \dot{q}^j = 1$$

and the Euler's equations can be written as

$$\frac{d}{d\tau}(g_{ij}\dot{q}^j) - \frac{1}{2}\frac{\partial}{\partial q^i}(g_{kj}\dot{q}^k\dot{q}^j) = 0 \ , \quad i = 1, 2, 3 \quad . \tag{2.2.12}$$

Clearly, (2.2.12) are Euler's equations for the functional

$$\frac{1}{2}\int g_{ij}\dot{q}^i\dot{q}^j \, d\tau = \frac{1}{2}\int \mathcal{F}^2 \, d\tau \quad .$$

Equations (2.2.12) are three equations, of second order in the quantities q^j, $j = 1, 2, 3$. By introducing three new unknown quantities p_j, $j = 1, 2, 3$ which are functions of τ, the system (2.2.12) can be reduced to one of 6 equations of the first order.

Next, we introduce the canonical variables. The canonical variables for an Euler equation of the form

$$\frac{\partial G}{\partial q^i} - \frac{d}{d\tau}\frac{\partial G}{\partial \dot{q}^i} = 0$$

are the unknown functions $q^j(\tau)$ and $p_j(\tau)$, $j = 1, 2, 3$, and the known function $H(q^j, p_j)$ of the variables q^j and p_j, of which $p_j(\tau)$ and $H(q^j, p_j)$ are introduced by the expressions

$$p_j(\tau) = \frac{\partial G}{\partial \dot{q}^j} \quad \text{and} \quad H(q^j, p_j) = \dot{q}^i p_i - G \quad .$$

It is not possible to introduce canonical variables into the Euler equations (2.2.10) directly, inasmuch as $H \equiv 0$ in this case because $\mathcal{F}$ is a homogeneous function of first degree. We must introduce the canonical variables by proceeding from the form (2.2.12) of the Euler equations. We put

$$p_i(\tau) = g_{ij}\dot{q}^j \ , \quad i = 1, 2, 3 \ , \quad H(q^i, p_i) = \dot{q}^j p_j - \frac{1}{2}g_{kl}\dot{q}^k\dot{q}^l \quad .$$

Differentiating the function H with respect to q^i and p_i, and remembering to account for the dependence of $\dot{q}^j$ on q^i and p_i, we have

---

[3] We will show in the next sectiion that this τ is indeed the same as that introduce in (2.1.1); i.e., it is a solution of the eikonal equation.

$$\frac{\partial H}{\partial q^i} = \frac{\partial \dot{q}^j}{\partial q^i} p_j - \frac{1}{2}\frac{\partial}{\partial q^i}(g_{kl})\dot{q}^k \dot{q}^l - g_{kl}\dot{q}^k \frac{\partial \dot{q}^l}{dq^i} \quad,$$

$$\frac{\partial H}{\partial p_i} = \frac{\partial \dot{q}^j}{\partial p_i} p_j + \dot{q}^i - g_{kl}\dot{q}^k \frac{\partial \dot{q}^l}{\partial p_i} \quad.$$

The first and last terms on the right-hand sides of these equations cancel each other, since by definition $p_j = g_{jk}\dot{q}^k$ and $g_{jk} = g_{kj}$. The second term of the first equation is equal to $-dp_i/d\tau$ by virtue of the Euler equations (2.2.12) and the definition of p_i. Thus, the functions $q^i(\tau)$ and $p_i(\tau)$ must satisfy the system of equations

$$\frac{dq^i}{d\tau} = \frac{\partial H}{\partial p_i} \quad ; \quad \frac{dp_i}{d\tau} = -\frac{\partial H}{\partial q^i} \quad . \tag{2.2.13}$$

The first-order system of (2.2.13) is known as the canonical Hamiltonian system, and the function H is called the Hamiltonian function, or simply the Hamiltonian. It is not difficult to find the explicit dependence of the Hamiltonian on q^i and p_i. Denoting the inverse to the matrix g_{ij} by g^{ij}, we have $\dot{q}^j = g^{ji}p_i$. Substituting this expression into the function $H(q^i, p_i)$, and using the identity $g_{kl}g^{li} = \delta^i_k$, where δ^i_k is the Kronecker delta, we obtain

$$H(q^i, p_i) = \tfrac{1}{2}g^{jl}p_j p_l \quad . \tag{2.2.14}$$

Now let us consider (2.2.8) on curves whose end points are movable. Since Euler's equation must be satisfied on l, the integral in (2.2.6) vanishes and (2.2.8) takes the form

$$\frac{\partial \mathcal{F}}{\partial \dot{q}^i} \delta q^i \bigg|_{\sigma=\sigma_0}^{\sigma=\sigma_1} = 0 \quad .$$

Considering in turn neighboring curves fixed first at one of the end points and then at the other, we arrive at the conditions

$$\frac{\partial \mathcal{F}}{\partial \dot{q}^i} \delta q^i \bigg|_{\sigma=\sigma_0} = 0 \quad \text{and} \quad \frac{\partial \mathcal{F}}{\partial \dot{q}^i} \delta q^i \bigg|_{\sigma=\sigma_1} = 0. \tag{2.2.15}$$

Conditions (2.2.15) must be satisfied at the ends of the ray l, if they are not assumed to be fixed points.

Suppose that one of the end points of l is moved through a surface described by the equation $\phi(q^1, q^2, q^3) = 0$. In this case the variations δq^i corresponding to the motion of the end point must satisfy the equation

$$\phi_{q^i} \delta q^i = 0 \tag{2.2.16}$$

and can be considered as components of an arbitrary vector $\vec{\chi} = (\chi^1, \chi^2, \chi^3)$ tangent to the surface $\phi(q^1, q^2, q^3) = 0$.

Equation (2.2.16) and conditions (2.2.15) pertaining to variable end points will be satisfied simultaneously for arbitrary δq^i if the derivatives ϕ_{q^i} and $\partial \mathcal{F}/\partial \dot{q}^i$ are proportional, i.e., if

$$\frac{\partial \mathcal{F}/\partial \dot{q}^1}{\phi_{q^1}} = \frac{\partial \mathcal{F}/\partial \dot{q}^2}{\phi_{q^2}} = \frac{\partial \mathcal{F}/\partial \dot{q}^3}{\phi_{q^3}} \quad . \tag{2.2.17}$$

Condition (2.2.17) is known as the transversality condition. For the function $\mathcal{F}$ defined by (2.2.4), the transversality condition can be rewritten in the form

$$g_{ij}\dot{q}^j\chi^i = 0 \quad . \tag{2.2.18}$$

In the form (2.2.18), the transversality condition can be treated as an orthogonality condition (in the sense of the Riemannian geometry) of the ray l to the surface $\phi = 0$. In cartesian coordinates, $g_{ij} = \delta_{ij}/c$, and condition (2.2.18) takes the form of an ordinary orthogonality condition

$$\vec{q} \cdot \vec{\chi} = 0 \quad . \tag{2.2.19}$$

The vector $\dot{\vec{q}} = (\dot{q}^1, \dot{q}^2, \dot{q}^3)$ is tangent to the ray l at its end points, and the vector $\vec{\chi}$ is an arbitrary tangent to the surface $\phi = 0$.

In concluding this section, let us note that all of our constructions have been carried out for a three-dimensional space, but carry over without the slightest variation to a space of n-dimensions.

2.3 The Solution of the Eikonal Equation; Ray Coordinates and the Geometrical Divergence

Let us now put together a family of rays (i.e., curves satisfying the Euler equations) leaving some given fixed point M_0 in all directions. It can be shown [2.1] that with certain restrictions on the function $\mathcal{F}(q^i, \dot{q}^i)$, the family of rays thus constructed forms a field in some neighborhood of the point M_0; in other words, one and only one ray from this family passes through any point M in this neighborhood. Since all rays in this family pass through the same point M_0, such a ray field is called a central ray field.

Consider the Fermat functional (2.2.2) or (2.2.3, 4) on rays of a central field leaving the point M_0 and terminating at a point M

$$\int_{M_0}^{M} \frac{ds}{c(M)} = \int_{\sigma_0}^{\sigma} \mathcal{F}(q^i, \dot{q}^i)d\sigma \quad . \tag{2.3.1}$$

For a fixed point M_0 the integral (2.3.1) is a function of the point M. Denote this function by $\tau(M)$, so that by definition

$$\tau(M) = \int_{M_0}^{M} \frac{ds}{c(M)} \tag{2.3.2}$$

where the integral is taken along the ray between M_0 and M. We will now show that $\tau(M)$ satisfies the eikonal equation (2.1.4). Evaluating the differential of τ, we see that it is precisely the first variation of the functional (2.3.1), and thus

$$dr = \frac{\partial \tau}{\partial q^i} \delta q^i = \int_{M_0}^{M} \frac{\delta \mathcal{F}}{\delta q^i} \delta q^i \, d\sigma + \frac{\partial \mathcal{F}}{\partial \dot{q}^i} \delta q^i \Big|_{M_0} - \frac{\partial \mathcal{F}}{\partial \dot{q}^i} \delta q^i \Big|_{M} \ .$$

Since the integral is evaluated between points M_0 and M along a ray, and the point M_0 was fixed, then

$$\frac{\delta \mathcal{F}}{\delta q^i} = 0 \quad \text{and} \quad \delta q^i \Big|_{M_0} = 0 \ , \quad \frac{\delta \mathcal{F}}{\delta q^i} \equiv \frac{\partial \mathcal{F}}{\partial q^i} - \frac{d}{d\sigma} \frac{\partial \mathcal{F}}{\partial \dot{q}^i} \ .$$

Hence $(\partial \tau / \partial q^i) \delta q^i = (\partial \mathcal{F} / \partial \dot{q}^i) \delta q^i$. Since the δq^i are linearly independent, we have $\partial \tau / \partial q^i = \partial \mathcal{F} / \partial \dot{q}^i$. Taking (2.2.4) into account, we arrive at

$$\partial \tau / \partial q^i = \frac{\partial \mathcal{F}}{\partial \dot{q}^i} = g_{ij} \dot{q}^j \mathcal{F}^{-1} \tag{2.3.3}$$

whence

$$\dot{q}^j = g^{ij} (\partial \tau / \partial q^i) \mathcal{F}$$

where (g^{ij}) denotes the inverse of the matrix (g_{ij}), see (2.2.14). Multiplying the last relation by $\partial \tau / \partial q^j$ and summing over j, we obtain

$$\frac{\partial \tau}{\partial q^j} \dot{q}^j = g_{ij} \dot{q}^i \dot{q}^j \mathcal{F}^{-1} = g^{ij} \frac{\partial \tau}{\partial q^i} \frac{\partial \tau}{\partial q^j} \mathcal{F} \ , \quad \left(\mathcal{F} = \sqrt{g_{ij} \dot{q}^i \dot{q}^j} \right) \ .$$

If we utilize (2.2.4) and cancel $\mathcal{F}$ from both sides of the last equation, there then follows

$$g^{ij} \frac{\partial \tau}{\partial q^i} \frac{\partial \tau}{\partial q^j} = 1 \ .$$

In Cartesian coordinates we have

$$\frac{\partial \tau}{\partial q^i} = \dot{q}^i \left(c \sqrt{\sum_{j=1}^{3} (\dot{q}^j)^2} \right)^{-1} ; \quad i = 1, 2, 3 \quad \text{or}$$

$$\nabla \tau = \frac{s_0}{c(M)} \tag{2.3.4}$$

where, as in Sect.2.2, s_0 is a unit vector tangent to the ray passing through M. Consequently, the function τ satisfies the eikonal equation

$$(\nabla \tau)^2 = \frac{1}{c^2(M)} \ . \tag{2.3.5}$$

Equation (2.3.3) is equivalent to (2.3.5), and is the eikonal equation expressed in curvilinear coordinates. We have constructed τ — the solution of the eikonal equation (2.3.4) — using a central field of rays. Another solution of the eikonal equation can be constructed using a so-called general ray field. The most general ray field is constructed as follows. Let S_0 be a sufficiently smooth surface in 3-space. From each point M_0 of S_0 a ray is emitted such that the orthogonality condition (2.2.19) is satisfied at M_0. The family of rays thereby formed constitutes a field in some neighborhood of S_0. Now, consider the function

$$\tau(M) = \int\limits_{M_0}^{M} \frac{ds}{c(M)} \tag{2.3.6}$$

where the integral is evaluated along the ray between M_0 and M. Given a point M, we can, by the same token, determine the ray along which (2.3.6) is to be carried out, and thus also the point M_0. Therefore, the function τ is in fact a function only of the point M — the upper limit of the integral (2.3.6). The derivatives $\partial\tau/\partial q^i$ of (2.3.6) are calculated just as in the case of (2.3.2), by using (2.3.3). The term arising from the endpoint M_0 now vanishes by virtue of the orthogonality between the family of rays we have constructed and the surface S_0, and so $\partial\tau/\partial q^i = \partial\mathcal{F}/\partial\dot{q}^i$ as before. Thus (2.3.6) satisfies the eikonal equation (2.3.5 or 3), if matters are being examined in the curvilinear coordinates (q^1, q^2, q^3) just as (2.3.2) did.

The wave front $t = \tau(M) = \mathrm{const}$ (Sect. 2.1) is a surface S which is motionless in space at any fixed instant of time t. We will now show that the rays intersect T orthogonally. Since t is fixed,

$$d\tau(M)\Big|_{M \in S} = 0$$

and from (2.3.3) there follows

$$\frac{\partial\mathcal{F}}{\partial\dot{q}^i}\delta q^i\Big|_{M \in S} = 0 \quad .$$

This equation can clearly be rewritten in the form $\vec{s}_0 \cdot \vec{\chi} = 0$, where $\vec{\chi}$ is an arbitrary vector tangent to the surface S, i.e., at each fixed moment in time, the rays are in fact orthogonal to the wave fronts. This is one of the fundamental laws of the optics of isotropic media.

Equation (2.3.6) permits the wave front Σ at a given instant t to be readily constructed if its position Σ_0

$$t_0 - \tau(M_0) = \mathrm{const} \quad , \quad M_0 \in \Sigma_0$$

is known at some instant t_0. From a point M_0 of the wave front Σ_0, we launch a ray orthogonal to Σ_0. On each ray we choose a point M so that the integral

$$\int\limits_{M_0}^{M} \frac{ds}{c(M)}$$

taken along the ray joining M_0 and M is equal to $t - t_0$. The equation

$$\int\limits_{M_0}^{M} \frac{ds}{c(M)} = t - t_0 \quad , \tag{2.3.7}$$

thereby obtained is also the equation for the wave front Σ at the instant t. In fact, (2.3.7) can be rewritten in the form $\tau(M) - \tau(M_0) = t - t_0$ or, since $t_0 - \tau(M_0) = \mathrm{const}$, in the form $t - \tau(M) = \mathrm{const}$.

So it is that the solution of the eikonal equation (2.1.4) is constructed. As we will show in the next subsection, the transport equations (2.1.5) are most conveniently solved in a special coordinate system, the so-called ray coordinates.

Let us introduce the ray coordinates for the case of a central ray field emanating from a point M_0. A ray leaving M_0 will be completely determined by specifying a unit vector $\vec{s}_0$ tangent to the ray at M_0. This vector can be characterized by two parameters α and β, which, for example, we could take to be the coordinates of the endpoint of $\vec{s}_0$ on the surface of the unit sphere $|\vec{s}_0| = 1$ centered at M_0. A point M in the vicinity of M_0 can now be specified by the parameters α, β of the ray passing through M, and the value of the integral $\tau = \int_{M_0}^{M} ds/c(M)$ taken along the ray $\overline{M_0 M}$. These very coordinates α, β, τ that we have introduced in the neighborhood of M_0 are the ray coordinates.

Now, suppose a ray field is formed by rays orthogonal to an orientable surface Σ, and that α, β are coordinates on that surface. We can introduce the ray coordinates α, β, τ in a neighborhood of Σ in the following manner. To each point M in a neighborhood of Σ there corresponds a point $M_0 \in \Sigma$ such that the ray passing through M_0 perpendicular to Σ passes through M. The first two ray coordinates are chosen to be the coordinates α, β of M_0 on Σ, and the third ray coordinate is taken to be

$$\tau = \pm \int_{M_0}^{M} \frac{ds}{c(M)}$$

where the integral is taken over the ray $\overline{M_0 M}$. The $\pm$ signs correspond to points M located on one side or the other side of the surface Σ.

In the case of a central ray field or of a more general ray field, the ray coordinate

$$\tau = \int_{M_0}^{M} \frac{ds}{c(M)} \qquad (2.3.8)$$

could obviously be replaced by the arc length s of the ray joining M_0 and M. Differentiating (2.3.8) with respect to s for fixed α and β, we obtain the relation

$$\frac{d\tau}{ds} = \frac{1}{c(M)} \qquad (2.3.9)$$

which we will often have occasion to use in what follows.

Let $\vec{r}$ be the position vector of a point M and x, y, z its cartesian coordinates. The transition from ray coordinates (α, β, τ) to cartesian coordinates can be written as a vector equation

$$\vec{r} = \vec{r}(\alpha, \beta, \tau) \quad . \qquad (2.3.10)$$

The properties of the vector-function $\vec{r}(\alpha, \beta, \tau)$ either in the case of a central ray field or of a general one are similar: fixing τ in (2.3.10), we obtain an equation for a wave front at some moment of time; fixing α and β we obtain the equation for a ray orthogonal to the wave fronts. Clearly, the vector $\vec{r}_\tau$ is tangent to a ray, and $\vec{r}_\alpha$ and $\vec{r}_\beta$ are tangent to a wavefront, and so

$$\vec{r}_\alpha \cdot \vec{r}_\tau = 0 \ , \vec{r}_\beta \cdot \vec{r}_\tau = 0 \quad . \tag{2.3.11}$$

Since the rays are orthogonal to the wave fronts, the vectors $\vec{r}_\tau$ and $\nabla \tau$ are collinear. It is not difficult to establish a more precise relation between these vectors. With the aid of (2.3.9 and 4), we obtain

$$\vec{r}_\tau = \vec{r}_s \left(\frac{d\tau}{ds} \right)^{-1} = c \vec{s}_0 = c^2 \nabla \tau \quad . \tag{2.3.12}$$

A ray field is called regular in a given region if the Jacobian of the transformation from cartesian to ray coordinates is nonzero in this region, i.e., if

$$\frac{D(x, y, z)}{D(\alpha, \beta, \tau)} \neq 0 \quad .$$

The quantity

$$J = \frac{1}{c(M)} \left| \frac{D(x, y, z)}{D(\alpha, \beta, \tau)} \right| = \frac{\pm 1}{c(M)} \begin{vmatrix} \frac{\partial x}{\partial \alpha} & \frac{\partial x}{\partial \beta} & \frac{\partial x}{\partial \tau} \\ \frac{\partial y}{\partial \alpha} & \frac{\partial y}{\partial \beta} & \frac{\partial y}{\partial \tau} \\ \frac{\partial z}{\partial \alpha} & \frac{\partial z}{\partial \beta} & \frac{\partial z}{\partial \tau} \end{vmatrix} \neq 0 \tag{2.3.13}$$

plays an important role, and bears the name *geometrical spreading*. In order to understand the motivation for this terminology, we square (2.3.13), multiplying the determinants column-by-column and using (2.3.11, 12) to obtain

$$J = \frac{1}{c(M)} \sqrt{ \begin{vmatrix} \vec{r}_\alpha \cdot \vec{r}_\alpha & \vec{r}_\alpha \cdot \vec{r}_\beta & \vec{r}_\alpha \cdot \vec{r}_\tau \\ \vec{r}_\beta \cdot \vec{r}_\alpha & \vec{r}_\beta \cdot \vec{r}_\beta & \vec{r}_\beta \cdot \vec{r}_\tau \\ \vec{r}_\tau \cdot \vec{r}_\alpha & \vec{r}_\tau \cdot \vec{r}_\beta & \vec{r}_\tau \cdot \vec{r}_\tau \end{vmatrix} } = \sqrt{ \begin{vmatrix} \vec{r}_\alpha \cdot \vec{r}_\alpha & \vec{r}_\alpha \cdot \vec{r}_\beta \\ \vec{r}_\beta \cdot \vec{r}_\alpha & \vec{r}_\beta \cdot \vec{r}_\beta \end{vmatrix} }$$

$$= \sqrt{EG - F^2}$$

where

$$E = |\vec{r}_\alpha|^2 \ ; \quad G = |\vec{r}_\beta|^2 \ ; \quad F = \vec{r}_\alpha \cdot \vec{r}_\beta$$

are the coefficients of the first fundamental form of the surface $\vec{r} = \vec{r}(\alpha, \beta, \tau)$ for τ fixed, i.e., of the wave front at τ. In differential geometry [2.2–4] is derived the formula

$$dS = \sqrt{EG - F^2} \, d\alpha \, d\beta$$

for a differential element dS of surface area. Consequently, an element of surface area on the wave front $\vec{r} = \vec{r}(\alpha, \beta, \tau)$ with τ fixed can be cast in the form

$$dS = J \, d\alpha \, d\beta \quad . \tag{2.3.14}$$

A collection of rays whose parameters α and β satisfy the inequalities

$$\alpha' \leq \alpha \leq \alpha'' \quad ,$$
$$\beta' \leq \beta \leq \beta''$$

is called a *ray tube*. Hence, the expression $J\,d\alpha\,d\beta$ gives the cross-sectional area of an infinitely thin $(\alpha'' - \alpha' = d\alpha,\ \beta'' - \beta' = d\beta)$ ray tube at a wave front, and consequently the larger J is, the more will rays in the neighborhood of some fixed ray with parameters α, β diverge from each other.

Note that the parameters α, β might have been chosen differently. If α and β are replaced by the parameters α_1 and β_1, then by basic properties of the Jacobian

$$J(\alpha_1, \beta_1, \tau) = \frac{1}{c(M)}\left|\frac{D(x,y,z)}{D(\alpha_1,\beta_1,\tau)}\right| = \frac{1}{c(M)}\left|\frac{D(x,y,z)}{D(\alpha,\beta,\tau)}\frac{D(\alpha,\beta,\tau)}{D(\alpha_1,\beta_1,\tau)}\right|$$

$$= \frac{1}{c(M)}\left|\frac{D(x,y,z)}{D(\alpha,\beta,\tau)}\right|\left|\frac{D(\alpha,\beta)}{D(\alpha_1,\beta_1)}\right| = J(\alpha,\beta,\tau)\left|\frac{D(\alpha,\beta)}{D(\alpha_1,\beta_1)}\right| \quad .$$

The factor $|D(\alpha,\beta)/D(\alpha_1,\beta_1)|$ is constant along a ray. Thus, the geometrical divergence is determined for any ray field up to a factor which is constant along a fixed ray.

When the velocity $c(M) = c$ is a constant, then the rays are straight lines. Let us obtain a formula for the geometrical divergence in this case. Let the surface Σ_0 be the wave front $t - \tau(M) = \text{const}$ at the instant $t = t_0$. The parametric equation for Σ_0 has the form

$$\vec{r} = \vec{r}(\alpha, \beta, t_0 - \text{const}) = \vec{r}(\alpha, \beta) \quad .$$

At the time $t = t_1 > t_0$, the wave front $t - \tau(M) = \text{const}$ will occupy some surface Σ_1 whose parametric equation is written as

$$\vec{r} = \vec{r}(\alpha, \beta, t_1 - \text{const}) = \vec{R}(\alpha, \beta) \quad .$$

Points on Σ_1 can be obtained from those on Σ_0 by moving the latter along the normal by a distance $l = c(t_1 - t_0)$. Denote the unit normal to Σ_0 directed along the motion of the wave front by $\vec{n} = \vec{n}(\alpha, \beta)$. Then

$$\vec{R}(\alpha, \beta) = \vec{r}(\alpha, \beta) + l\vec{n}(\alpha, \beta)$$

and we arrive at

$$J_1^2 = (\vec{r}_\alpha + l\vec{n}_\alpha)^2(\vec{r}_\beta + l\vec{n}_\beta)^2 - [(\vec{r}_\alpha + l\vec{n}_\alpha)\cdot(\vec{r}_\beta + l\vec{n}_\beta)]^2 \qquad (2.3.15)$$

for the geometrical divergence J_1 at points on the wave front Σ_1. Since $\vec{n}\cdot\vec{n} = 1$, we have $\vec{n}_\alpha\cdot\vec{n} = \vec{n}_\beta\cdot\vec{n} = 0$. Consequently, the vectors $\vec{n}_\alpha$ and $\vec{n}_\beta$ lie in the plane tangent to Σ_0 and can be expressed in terms of $\vec{r}_\alpha$ and $\vec{r}_\beta$. From differential geometry we have the formulas

$$\vec{n}_\alpha = \frac{1}{\sqrt{EG - F^2}}[(MF - LG)\vec{r}_\alpha + (LF - ME)\vec{r}_\beta] \ ,$$

$$\vec{n}_\beta = \frac{1}{\sqrt{EG - F^2}}[(NF - MG)\vec{r}_\alpha + (MF - NE)\vec{r}_\beta] \ , \qquad (2.3.16)$$

where E, F, G are the coefficients in the first fundamental form and

$$L = \vec{r}_{\alpha\alpha}\cdot\vec{n} \ ; \quad M = \vec{r}_{\alpha\beta}\cdot\vec{n} \ ; \quad N = \vec{r}_{\beta\beta}\cdot\vec{n}$$

are the coefficients of the second fundamental form of the surface Σ_0. We also require formulas expressing the principal radii of curvature R_1 and R_2 of the surface in terms of the coefficients of the first and second fundamental form [2.2–4]

$$\frac{1}{R_1 R_2} = \frac{LN - M^2}{EG - F^2} \; ; \; \frac{1}{R_1} + \frac{1}{R_2} = \frac{2MF - EN - GL}{EG - F^2} \qquad (2.3.17)$$

where R_1 and R_2 are reckoned positive if the centers of curvature of the intersections of Σ_0 with the corresponding normal planes are located on the side of the surface associated with $-\vec{n}$.

Substituting (2.3.16) for $\vec{n}_\alpha$ and $\vec{n}_\beta$ into (2.3.15) and using some simple transformations of (2.3.17) in the last step, we obtain

$$J_1 = \frac{(R_1 + l)(R_2 + l)}{R_1 R_2} J_0 \qquad (2.3.18)$$

where $J_0 = \sqrt{EG - F^2}$ is the geometrical divergence at points on the wave front Σ_0. Equation (2.3.18) has a simple geometrical interpretation (Fig. 2.2): the ratio of the areas of the curvilinear rectangles $A_1 B_1 C_1 D_1$ and $A_0 B_0 C_0 D_0$ is, in the first approximation, the product of the ratios of $O_1 A_1 = R_1 + l$ to $O_1 A_0 = R_1$ and $O_2 B_1 = R_2 + l$ to $O_2 B_0 = R_2$, the sides of the similar triangles $A_1 B_1 O_1 \sim A_0 B_0 O_1$ and $B_1 C_1 O_2 \sim B_0 C_0 O_2$.

We conclude this section with a few words about the planar case. All of the basic laws formulated above for the three-dimensional case carry over to the planar case. Here, a ray is specified by a single parameter α only, and the transition from ray to cartesian coordinates now has the form $\vec{r} = \vec{r}(\alpha, \tau)$, with

$$\vec{r}_\alpha \cdot \vec{r}_\tau = 0 \; , \; |\vec{r}_\tau| = c(M) \quad .$$

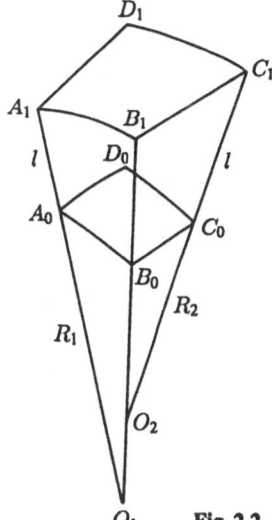

O_1 **Fig. 2.2.** Divergence of a ray tube

25

For each fixed α, the equation $\vec{r} = \vec{r}(\alpha, \tau)$ describes a ray, and for each fixed τ it describes a wave front. The geometrical divergence is given by

$$J = \frac{1}{c}\left|\frac{D(x, y)}{D(\alpha, \tau)}\right| = |\vec{r}_\alpha| \quad .$$

Setting $R_1 = R$ and $R_2 = \infty$ in (2.3.18), we obtain the expression for the geometrical divergence for constant velocity

$$J = \frac{R+l}{R}J_0 \quad . \tag{2.3.19}$$

Here R is the radius of curvature of a wave front at the instant t_0, while $l = c(t_1 - t_0)$ is the distance between wave fronts.

2.4 Integration of the Transport Equations

Let us begin with (2.1.5) for $j = 0$

$$2\nabla u_0 \cdot \nabla \tau + u_0 \Delta \tau = 0 \quad . \tag{2.4.1}$$

Multiplying (2.4.1) by u_0 ($u_0 \neq 0$), we obtain an equivalent equation which can be written in the form

$$\nabla \cdot (u_0^2 \nabla \tau) = 0 \quad . \tag{2.4.2}$$

Next, we integrate (2.4.2) over the volume Ω bounded by the sides of a ray tube $\alpha' \leq \alpha \leq \alpha''$, $\beta' \leq \beta \leq \beta''$ and the two wave fronts $\tau = \tau_0$ and $\tau = \tau_1$. By the divergence theorem,

$$\int\int_{\partial\Omega} u_0^2 \nabla \tau \cdot \vec{n} \, dS = 0 \tag{2.4.3}$$

where the boundary $\partial\Omega$ of Ω consists of the sides S_δ of the ray tube, and its two endcaps S_0 and S_1; $\vec{n}$ is the outward unit normal to $\partial\Omega$. Since the vector $\nabla\tau$ is tangent to the rays, on the surfaces S_δ, S_0 and S_1 we have

$$\nabla\tau \cdot \vec{n}\Big|_{S_\delta} = 0 \; , \quad \nabla\tau \cdot \vec{n}\Big|_{S_0} = -\frac{1}{c(M_0)}\Big|_{M_0 \in S_0} \; , \quad \nabla\tau \cdot \vec{n}\Big|_{S_i} = \frac{1}{c(M)}\Big|_{M \in S_1} \; .$$

Thus, (2.4.3) reduces to

$$\int\int_{S_1} \frac{u_0^2(M_1)}{c(M_1)}dS_1 - \int\int_{S_0} \frac{u_0^2(M_0)}{c(M_0)}dS_0 = 0 \quad .$$

Changing to ray coordinates in the latter integrals and using (2.3.14), we obtain

$$\int_{\alpha'}^{\alpha''}\int_{\beta'}^{\beta''} \left[\frac{u_0^2(\alpha, \beta, \tau_1)}{c(\alpha, \beta, \tau_1)}J_1(\alpha, \beta, \tau_1)\right.$$

$$-\frac{u_0^2(\alpha, \beta, \tau_0)}{c(\alpha, \beta, \tau_0)} J_0(\alpha, \beta, \tau_0)\Bigg] d\alpha\, d\beta = 0. \tag{2.4.4}$$

Equation (2.4.4) must hold for arbitrary α', α'', β' and β'', and so the integrand in (2.4.4) must vanish. Equating the integrand to zero, we have

$$u_0(\alpha, \beta, \tau_1) = u_0(\alpha, \beta, \tau_0)\sqrt{\frac{J(\alpha, \beta, \tau_0)}{c(\alpha, \beta, \tau_0)}} \sqrt{\frac{c(\alpha, \beta, \tau_1)}{J(\alpha, \beta, \tau_1)}} \quad . \tag{2.4.5}$$

We shall regard the value of τ_1 as variable and drop the subscript "1". On the other hand, τ_0 will be taken to be fixed, and we can introduce the function

$$\psi_0(\alpha, \beta) = u_0(\alpha, \beta, \tau_0)\sqrt{\frac{J(\alpha, \beta, \tau_0)}{c(\alpha, \beta, \tau_0)}}$$

which does not depend on the choice of τ_0. Then, (2.4.5) will take the form

$$u_0(\alpha, \beta, \tau) = \psi_0(\alpha, \beta)\sqrt{\frac{c(\alpha, \beta, \tau)}{J(\alpha, \beta, \tau)}} \quad . \tag{2.4.6}$$

Equation (2.4.6) gives the desired solution to (2.1.5) for $j = 0$. An important formula for $\Delta\tau$ follows from (2.4.1,6). Using (2.3.4,9), we can rewrite (2.4.1) as

$$\frac{2}{c^2}\frac{\partial u_0}{\partial \tau} + u_0\Delta\tau = 0 \quad .$$

Hence,

$$\Delta\tau = -\frac{2}{c^2}\frac{1}{u_0}\frac{\partial u_0}{\partial \tau}$$

and from (2.4.6)

$$\Delta\tau = \frac{1}{Jc}\frac{\partial}{\partial \tau}\left(\frac{J}{c}\right) \quad . \tag{2.4.7}$$

For $j \geq 1$, (2.1.5) are easily integrated by means of the substitution $u_j = w_j u_0$, where w_j is a new unknown function which is commonly used to find solutions of an inhomogeneous equation when a nontrivial solution of the corresponding homogeneous equation is known. For w_j, we have

$$\frac{2}{c^2}\frac{\partial w_j}{\partial \tau} = \frac{1}{u_0}\Delta u_{j-1}$$

and as a result

$$u_j = u_0 w_j = u_0\left[\int^\tau \frac{1}{2}\frac{c^2}{u_0}\Delta u_{j-1}d\tau + \tilde{\psi}_j(\alpha, \beta)\right] \quad .$$

Taking (2.4.6) into account, the expression for u_j can be rewritten in the form

$$u_j = \sqrt{\frac{c}{J}}\left[\psi_j(\alpha, \beta) + \int^\tau \sqrt{\frac{J}{c}}\frac{c^2}{2}\Delta u_{j-1}d\tau\right] \tag{2.4.8}$$

27

where $\psi_j(\alpha, \beta)$ is a function constant along each ray. Following *Keller* [2.5], the functions $\psi_j(\alpha, \beta)$ are sometimes referred to as diffraction coefficients.

The transport equations (2.1.7) are likewise easily integrated. Setting $j = 0$ and taking account of (2.4.7), we obtain

$$2\nabla\tau \cdot \nabla v_0 + \Delta\tau = \frac{2}{c^2}\frac{\partial v_0}{\partial \tau} + \frac{1}{Jc}\frac{\partial}{\partial \tau}\left(\frac{J}{c}\right) \quad .$$

Hence

$$v_0 = \frac{1}{2}\ln\frac{c}{J} + \chi_0(\alpha, \beta)$$

where $\chi_0(\alpha, \beta)$ is some function constant along each ray. The integration of the succeeding equations does not present any serious difficulty, since (2.1.7) for $j > 1$ take the form

$$\frac{2}{c^2}\frac{\partial v_j}{\partial \tau} + \text{ functions already determined } = 0 \quad .$$

As a rule, we will be dealing with (2.1.1) in the sections which follow.

We note in conclusion that (2.4.2) has a clear interpretation in terms of energy. In any case where a wave process is described by the wave equation

$$\frac{1}{c^2(M)}W_{tt} - \Delta W = 0$$

the energy density w associated with W is given by the expression

$$w = \text{const}\left[\frac{1}{2}\frac{1}{c^2(M)}(W_t)^2 + \frac{1}{2}(\nabla W)^2\right] \quad , \tag{2.4.9}$$

where the constant multiplier depends on the physical nature of the process.

We substitute the leading term of the ray expansion into (2.4.9): $W \approx \text{Re}\{u_0\exp[-i\omega t + i\omega\tau(M)]\}$. Assuming u_0 to be real, taking the eikonal equation for τ into account, and retaining only the dominant terms for $\omega \to \infty$, we obtain

$$w \sim \frac{\omega^2}{c^2}u_0^2 \sin^2\psi \quad ; \quad \psi = \omega[t - \tau(M)]$$

for the energy density. Averaging this expression over a period $T = 2\pi/\omega$,

$$\overline{w} \equiv \frac{1}{T}\int_0^T w\, dt \sim \frac{1}{T}\int_0^T \frac{\omega^2}{c^2}u_0^2\left(\frac{1}{2} - \frac{1}{2}\cos 2\psi\right)dt = \frac{\omega^2}{2c^2}u_0^2$$

and introducing a vector propagation velocity

$$\vec{v} = c\vec{s}_0 = c^2\nabla\tau$$

we find that (2.4.2), or equivalently the transport equation (2.4.1) can be written as

$$\nabla \cdot (\overline{w}\vec{v}) = 0 \quad . \tag{2.4.10}$$

Relation (2.4.10) is analogous to the classical continuity equation of hydrome-chanics

$$\frac{\partial \varrho}{\partial t} + \nabla \cdot (\varrho \vec{v}) = 0$$

for the case of a stationary flow ($\partial \varrho / \partial t = 0$). As is well known, the continuity equation is equivalent to the law of conservation of fluid mass in such a context. Thus (2.4.10) and consequently the transport equation (2.4.1) guarantee that the law of conservation of energy in the wave process will be satisfied.

2.5 Maxwell's Equations

The derivation of the basic formulas of the ray method can be extended to Maxwell's equations without any fundamental changes. Let us consider Maxwell's equations for the case of a nonabsorbing inhomogeneous isotropic medium

$$\nabla \times \vec{\mathcal{H}} = \frac{\varepsilon}{c} \frac{\partial \vec{\mathcal{E}}}{\partial t} \ ,$$

$$\nabla \times \vec{\mathcal{E}} = -\frac{\mu}{c} \frac{\partial \vec{\mathcal{H}}}{\partial t} \ . \tag{2.5.1}$$

Here and elsewhere in this section c is the speed of light in vacuum, ε is the relative permittivity, μ is the relative permeability, and Gaussian units are em-ployed.

We will seek the solution of this system in the form of the series

$$\vec{\mathcal{H}} = \sum_{s=0}^{\infty} \vec{\mathcal{H}}_s(x, y, z) \frac{\exp\{-\mathrm{i}\omega[t - \tau(x, y, z)]\}}{(-\mathrm{i}\omega)^{s+\gamma}} \ ,$$

$$\vec{\mathcal{E}} = \sum_{s=0}^{\infty} \vec{\mathcal{E}}_s(x, y, z) \frac{\exp\{-\mathrm{i}\omega[t - \tau(x, y, z)]\}}{(-\mathrm{i}\omega)^{s+\gamma}} \ , \quad \text{with}$$

$$(-\mathrm{i})^{s+\gamma} = \exp\left[-\frac{\pi}{2}\mathrm{i}(s+\gamma)\right] \ . \tag{2.5.2}$$

Substituting (2.5.2) into Maxwell's equations and setting the coefficients of iden-tical powers of $1/\omega$ equal to zero, we get

$$\frac{\varepsilon}{c} \vec{\mathcal{E}}_{s+1} + (\nabla \tau \times \vec{\mathcal{H}}_{s+1}) = \nabla \times \vec{\mathcal{H}}_s \ ,$$

$$-\frac{\mu}{c} \vec{\mathcal{H}}_{s+1} + (\nabla \tau \times \vec{\mathcal{E}}_{s+1}) = \nabla \times \vec{\mathcal{E}}_s \tag{2.5.3}$$

$$(s = -1, 0, 1, 2, \ldots; \quad \vec{\mathcal{H}}_{-1} = \vec{\mathcal{E}}_{-1} \equiv 0) \ .$$

Consider the case $s = -1$:

$$\frac{\varepsilon}{c} \vec{\mathcal{E}}_0 + (\nabla \tau \times \vec{\mathcal{H}}_0) = 0 \ , \quad -\frac{\mu}{c} \vec{\mathcal{H}}_0 + (\nabla \tau \times \vec{\mathcal{E}}_0) = 0 \ . \tag{2.5.4}$$

The homogeneous system (2.5.4) gives us the eikonal equation for the problem under consideration. In fact, (2.5.4) implies that $\nabla\tau$, $\vec{\mathcal{H}}_0$ and $\vec{\mathcal{E}}_0$ are mutually orthogonal vectors. Then moving the second terms in (2.5.4) to the right-hand sides and setting the lengths of the corresponding vectors equal to each other, we get

$$(\nabla\tau)^2 = \frac{\varepsilon\mu}{c^2} = \frac{1}{c_1^2(M)} \;, \quad c_1(M) = \frac{c}{\sqrt{\varepsilon\mu}} \;, \tag{2.5.5}$$

i.e., the eikonal equation. The function $c_1(M)$ defines the propagation velocity of wave fronts $t = \tau$ (Sect. 2.1).

We now write out the general solution of system (2.5.4). As before, we introduce the ray-extremals of the Fermat functional

$$\int \frac{\sqrt{\varepsilon\mu}}{c}\,ds = \int \frac{ds}{c_1(M)} \tag{2.5.6}$$

and the ray coordinates $\vec{r} = \vec{r}(\alpha, \beta, \tau)$.

Let $\vec{n}$ and $\vec{\zeta}$ be the principal normal and binormal, respectively, of a ray. In virtue of the fact that $\vec{\mathcal{E}}_0 \perp \nabla\tau$ and $\vec{\mathcal{H}}_0 \perp \nabla\tau$ there will be scalars A', A'', B', B'' such that

$$\vec{\mathcal{E}}_0 = A'\vec{n} + A''\vec{\zeta} \;, \quad \vec{\mathcal{H}}_0 = B'\vec{n} + B''\vec{\zeta} \;.$$

Substituting the expressions for $\vec{\mathcal{E}}_0$ and $\vec{\mathcal{H}}_0$ into (2.5.4) we get two relations for A', A'', B', B'' from which it is readily apparent that the general solution of (2.5.4) takes the form

$$\vec{\mathcal{E}}_0 = \vec{n}\Phi_0(\tau, \alpha, \beta)\sqrt{\mu} + \vec{\zeta}\Psi_0(\tau, \alpha, \beta)\sqrt{\mu} \;,$$
$$\vec{\mathcal{H}}_0 = -\vec{n}\Psi_0(\tau, \alpha, \beta)\sqrt{\varepsilon} + \vec{\zeta}\Phi_0(\tau, \alpha, \beta)\sqrt{\varepsilon} \;, \tag{2.5.7}$$

where Φ_0, Ψ_0 are scalar functions to be determined.

Let us now set $s = 0$ in (2.5.3). Then

$$\frac{\varepsilon}{c}\vec{\mathcal{E}}_1 + \nabla\tau \times \vec{\mathcal{H}}_1 = \nabla \times \vec{\mathcal{H}}_0 \;,$$
$$-\frac{\mu}{c}\vec{\mathcal{H}}_1 + \nabla\tau \times \vec{\mathcal{E}}_1 = \nabla \times \vec{\mathcal{E}}_0 \;. \tag{2.5.8}$$

Expressions (2.5.8) comprise a system of linear algebraic equations for finding $\vec{\mathcal{E}}_1$ and $\vec{\mathcal{H}}_1$. The determinant of the system is equal to zero since the corresponding homogeneous system coincides with (2.5.4), and (2.5.4) has two linearly independent solutions that can be found by first setting $\Phi_0 = 1$, $\Psi_0 = 0$, then $\Phi_0 = 0$, $\Psi_0 = 1$ in (2.5.7).

A system of linear algebraic equations with zero determinant is not always solvable. The following theorem holds:

Theorem. *In order for the system of linear algebraic equations below to be solvable, it is necessary and sufficient that the vector f be orthogonal to all solutions of the conjugate homogeneous system $A^*y = 0$.*

$$Ax = f$$

$$A = \begin{bmatrix} a_{11} & \cdots & a_{1n} \\ \cdots & \cdots & \cdots \\ \cdots & \cdots & \cdots \\ a_{n1} & \cdots & a_{nn} \end{bmatrix} \ , \ x = \begin{pmatrix} x_1 \\ \cdot \\ \cdot \\ \cdot \\ x_n \end{pmatrix} \ , \ f = \begin{pmatrix} f_1 \\ \cdot \\ \cdot \\ \cdot \\ f_n \end{pmatrix} .$$

Here A^* is the Hermitian-conjugate matrix of A. (To get A^* we must replace all elements a_{ik} of the matrix A by their complex conjugates, and interchange the rows and columns of the matrix).

It can be readily shown that a necessary and sufficient condition for solvability of system (2.5.8) is that for any pair of vectors $\vec{l}$ and $\vec{m}$ such that

$$\frac{\varepsilon}{c}\vec{l} + \vec{m} \times \nabla\tau = 0 \ ,$$

$$-\frac{\mu}{c}\vec{m} + \vec{l} \times \nabla\tau = 0 \ , \tag{2.5.9}$$

the equality

$$\vec{l} \cdot \nabla \times \vec{\mathcal{H}}_0 + \vec{m} \cdot \nabla \times \vec{\mathcal{E}}_0 = 0 \tag{2.5.10}$$

must be satisfied.

The fundamental system of solutions of (2.5.9) takes the following form

$$\vec{l}_1 = \sqrt{\mu}\vec{\zeta} \ , \ \vec{l}_2 = \sqrt{\mu}\vec{n} \ ,$$

$$\vec{m}_1 = \sqrt{\varepsilon}\vec{n} \ , \ \vec{m}_2 = -\sqrt{\varepsilon}\vec{\zeta} \ .$$

Therefore (2.5.10) is equivalent to the two equalities

$$\sqrt{\mu}\vec{\zeta} \cdot \nabla \times \vec{\mathcal{H}}_0 + \sqrt{\varepsilon}\vec{n} \cdot \nabla \times \vec{\mathcal{E}}_0 = 0 \ ,$$

$$\sqrt{\mu}\vec{n} \cdot \nabla \times \vec{\mathcal{H}}_0 - \sqrt{\varepsilon}\vec{\zeta} \cdot \nabla \times \vec{\mathcal{E}}_0 = 0 \ . \tag{2.5.11}$$

Further transformations of (2.5.11) are based on the two identities

$$\vec{\zeta} \cdot \nabla \times \vec{n} - \vec{n} \cdot \nabla \times \vec{\zeta} = \nabla \cdot \vec{s}_0 \ ,$$

$$\vec{n} \cdot \nabla \times \vec{n} + \vec{\zeta} \cdot \nabla \times \vec{\zeta} = 2/T \ ,$$

where $\vec{s}_0 = c_1\nabla\tau$ is the unit vector tangent to the ray, and T is the radius of torsion of the ray. The first identity is a simple corollary of the classial equality $\vec{s}_0 \times \vec{n} = \vec{\zeta}$ [2.2, 4]. The proof of the second identity is cumbersome, and we will omit it.

Let us divide the left-hand sides of (2.5.11) by c — the speed of light in vacuum — and substitute the expressions for $\vec{\mathcal{E}}_0$ and $\vec{\mathcal{H}}_0$ from (2.5.7). After rather complicated calculations we get

$$-2\nabla\tau \cdot \nabla\Psi_0 - \Psi_0\Delta\tau + \frac{2}{c_1(M)T}\Phi_0 = 0 \ ,$$

$$-2\nabla\tau \cdot \nabla\Phi_0 - \Phi_0\Delta\tau - \frac{2}{c_1(M)T}\Psi_0 = 0 \ , \tag{2.5.12}$$

where again T is the radius of torsion of a ray.

We multiply the first of (2.5.12) by Ψ_0, the second by Φ_0, and combine the results. After division by $-\sqrt{\Phi_0^2 + \Psi_0^2}$ we arrive at

$$2\nabla\tau \cdot \nabla\sqrt{\Phi_0^2 + \Psi_0^2} + \sqrt{\Phi_0^2 + \Psi_0^2}\,\Delta\tau = 0 \quad,$$

which, when solved, gives (Sect. 2.4)

$$\sqrt{\Phi_0^2 + \Psi_0^2} = \psi_0(\alpha,\beta)\sqrt{\frac{c_1(M)}{J}} \quad, \tag{2.5.13}$$

where $J = |\vec{r}_\alpha \times \vec{r}_\beta|$ is the divergence of the ray field. Equation (2.5.13) has the same energy meaning as its scalar analog, namely (2.4.6). Of course, instead of the energy density (2.4.9), here we must take the energy density of the electromagnetic field

$$\frac{1}{8\pi}(\varepsilon|\vec{E}|^2 + \mu|\vec{\mathcal{H}}|^2) \quad.$$

The quantity $\sqrt{\Phi_0^2 + \Psi_0^2}$ determines the lengths of vectors $\vec{\mathcal{E}}_0$ and $\vec{\mathcal{H}}_0$, see (2.5.7). In order to ascertain how the directions of these vectors change, we introduce an angle θ via the relations

$$\Phi_0 = \sqrt{\Phi_0^2 + \Psi_0^2}\cos\theta \quad,$$

$$\Psi_0 = \sqrt{\Phi_0^2 + \Psi_0^2}\sin\theta \quad. \tag{2.5.14}$$

Substituting (2.5.14) into (2.5.12), and after a few transformations, we get the important equality

$$\frac{d\theta}{ds} = \frac{1}{T} \quad, \tag{2.5.15}$$

where T is the radius of torsion, and ds is a differential arc length. Equations (2.5.4, 7, 13–15) determine the change in the vectors $\vec{\mathcal{E}}_0$ and $\vec{\mathcal{H}}_0$ as the wave front moves along a ray [except for the yet undefined function $\psi_0(\alpha,\beta)$].

In fact, (2.5.4) imply that $\vec{\mathcal{E}}_0$, $\vec{\mathcal{H}}_0$, $\nabla\tau$ from a right-handed set of mutually orthogonal basis vectors (if the coordinate system x, y, z is right-handed). This triad moves along the ray with the wave front velocity $c_1(M)$.

The lengths of the vectors $\vec{\mathcal{E}}_0$ and $\vec{\mathcal{H}}_0$ vary in accordance with the formulas

$$|\vec{\mathcal{E}}_0| = \sqrt{\mu}\sqrt{\Phi_0^2 + \Psi_0^2} = \psi_0(\alpha,\beta)\sqrt{\mu}\sqrt{\frac{c_1(M)}{J}} \quad,$$

$$|\vec{\mathcal{H}}_0| = \sqrt{\varepsilon}\sqrt{\Phi_0^2 + \Psi_0^2} = \psi_0(\alpha,\beta)\sqrt{\varepsilon}\sqrt{\frac{c_1(M)}{J}} \tag{2.5.16}$$

which are easily derived from (2.5.7, 13).

In order to find the law of rotation of the vector $\vec{\mathcal{E}}_0$ around a vector tangent to $\nabla\tau$, we project $\vec{\mathcal{E}}_0$ onto the normal $\vec{n}$ and the binormal $\vec{\zeta}$ to the ray. Equations (2.5.7, 15) will give

$$\mathcal{E}_{0n} = |\vec{\mathcal{E}}_0| \cos\theta \ , \quad \mathcal{E}_{0\zeta} = |\vec{\mathcal{E}}_0| \sin\theta \ , \tag{2.5.17}$$

$$\frac{d\theta}{ds} = \frac{1}{T} \ . \tag{2.5.18}$$

Relations (2.5.17, 18) show the rate at which angle θ changes between the principal normal to the ray and the vector $\vec{\mathcal{E}}_0$.

We briefly mention one interesting geometric interpretation of (2.5.18). If we use the term *distance between two arbitrarily close points* $M_0(x_0, y_0, z_0)$ and $M(x, y, z)$ for the quantity $\tau(M, M_0) = \int_{M_0}^{M} c_1^{-1}\, ds$ (the integral is taken over a ray connecting the points M_0 and M), then the region where the wave field is being investigated, is treated as a Riemannian manifold with metric tensor

$$(g_{ij}) = \begin{bmatrix} \frac{1}{c^2} & 0 & 0 \\ 0 & \frac{1}{c^2} & 0 \\ 0 & 0 & \frac{1}{c^2} \end{bmatrix}$$

where the rays will obviously coincide with geodesic lines. Any (covariant or contravariant) vector in euclidean space can be treated as a vector (covariant or contravariant) on this manifold. If now an arbitrary vector perpendiular to the ray is translated [2.2, 3] parallel to itself along this ray (parallel in the sense of Riemannian geometry), it will remain everywhere perpendicular to the ray, and the angle between this vector and the Euclidean principal normal to the ray will vary in accordance with (2.5.18). Therefore the equalities (2.1.17, 18) mean that in the propagation of an electromagnetic field the direction of the field vectors, in the zero approximation, correspond to parallel translation (in the sense of Riemannian geometry) along a ray.

To find $\vec{\mathcal{E}}_1$ and $\vec{\mathcal{H}}_1$ we note that in virtue of the fact that conditions (2.5.11, 12) are satisfied, system of algebraic equations (2.5.8) is solvable, although nonuniquely. The general solution of system (2.5.8) may be written in the form

$$\vec{\mathcal{E}}_1 = \vec{\mathcal{E}}_1^0 + \vec{\mathcal{E}}_1^1 \ , \quad \vec{\mathcal{H}}_1 = \vec{\mathcal{H}}_1^0 + \vec{\mathcal{H}}_1^1 \ , \tag{2.5.19}$$

where

$$\vec{\mathcal{E}}_1^0 = \frac{c}{2\varepsilon}[(\nabla \times \vec{\mathcal{H}}_0 \cdot \vec{s}_0)\vec{s}_0 + \nabla \times \vec{\mathcal{H}}_0] \ ,$$

$$\vec{\mathcal{H}}_1^0 = -\frac{c}{2\mu}[(\nabla \times \vec{\mathcal{E}}_0 \cdot \vec{s}_0)\vec{s}_0 + \nabla \times \vec{\mathcal{E}}_0] \ , \tag{2.5.20}$$

$$\vec{s}_0 = \frac{c}{\sqrt{\varepsilon\mu}}\nabla\tau = \nabla\tau/|\nabla\tau| \ ; \quad |\vec{s}_0| = 1 \ ,$$

$$\vec{\mathcal{E}}_1^1 = \vec{n}\Phi_1(\tau, \alpha, \beta)\sqrt{\mu} + \vec{\zeta}\Psi_1(\tau, \alpha, \beta)\sqrt{\mu} \ ,$$

$$\vec{\mathcal{H}}_1^1 = -\vec{n}\Psi_1(\tau, \alpha, \beta)\sqrt{\varepsilon} + \vec{\zeta}\Phi_1(\tau, \alpha, \beta)\sqrt{\varepsilon} \ . \tag{2.5.21}$$

The vectors $\vec{\mathcal{E}}_1^0$, $\vec{\mathcal{H}}_1^0$ are a particular solution of the nonhomogeneous system of equations, while $\vec{\mathcal{E}}_1^1$, $\vec{\mathcal{H}}_1^1$ are a general solution for the corresponding homogeneous system which is obviously given by (2.5.7) where the subscript "0" is to be replaced by "1".

Let us now set $s = 1$ in (2.5.3). We get a system of linear algebraic equations for determining the components of the vectors $\vec{\mathcal{E}}_2$ and $\vec{\mathcal{H}}_2$. The determinant of this system is equal to zero since the corresponding homogeneous equation has non-zero solutions.

Setting up the conditions of solvability we get, see (2.5.11),

$$\sqrt{\mu}\,\vec{\zeta}\cdot\nabla\times\vec{\mathcal{H}}_1 + \sqrt{\varepsilon}\,\vec{n}\cdot\nabla\times\vec{\mathcal{E}}_1 = 0 \quad,$$

$$\sqrt{\mu}\,\vec{n}\cdot\nabla\times\vec{\mathcal{H}}_1 - \sqrt{\varepsilon}\,\vec{\zeta}\cdot\nabla\times\vec{\mathcal{E}}_1 = 0 \quad.$$

Substituting (2.5.19) here we arrive at the equations

$$\sqrt{\mu}\,\vec{\zeta}\cdot\nabla\times\vec{\mathcal{H}}_1^1 + \sqrt{\varepsilon}\,\vec{n}\cdot\nabla\times\vec{\mathcal{E}}_1^1$$
$$= -\sqrt{\mu}\,\vec{\zeta}\cdot\nabla\times\vec{\mathcal{H}}_1^0 - \sqrt{\varepsilon}\,\vec{n}\cdot\nabla\times\vec{\mathcal{E}}_1^0 \quad,$$

$$\sqrt{\mu}\,\vec{n}\cdot\nabla\times\vec{\mathcal{H}}_1^1 - \sqrt{\varepsilon}\,\vec{\zeta}\cdot\nabla\times\vec{\mathcal{E}}_1^1$$
$$= -\sqrt{\mu}\,\vec{n}\cdot\nabla\times\vec{\mathcal{H}}_1^0 + \sqrt{\varepsilon}\,\vec{\zeta}\cdot\nabla\times\vec{\mathcal{E}}_1^0 \quad. \tag{2.5.22}$$

In place of $\vec{\mathcal{E}}_1^1$ and $\vec{\mathcal{H}}_1^1$ we substitute their expressions from (2.5.21). Making use of the fact that the left-hand sides of (2.5.22) coincide with the left-hand sides of (2.5.11) with the subscript "1" replaced by the subscript "0", and that the left-hand sides of (2.5.12) differ from the left-hand sides of (2.5.11) only by the factor $1/c$, we get

$$-2\nabla\tau\cdot\nabla\Psi_1 - \Psi_1\Delta\tau + \frac{2}{c_1(M)T}\Phi_1$$
$$= [-\sqrt{\mu}\,\vec{\zeta}\cdot\nabla\times\vec{\mathcal{H}}_1^0 - \sqrt{\varepsilon}\,\vec{n}\cdot\times\vec{\mathcal{E}}_1^0]\frac{1}{c} \quad,$$

$$-2\nabla\tau\cdot\nabla\Phi_1 - \Phi_1\Delta\tau - \frac{2}{c_1(M)T}\Psi_1$$
$$= [-\sqrt{\mu}\,\vec{n}\cdot\nabla\times\vec{\mathcal{H}}_1^0 + \sqrt{\varepsilon}\,\vec{\zeta}\cdot\nabla\times\vec{\mathcal{E}}_1^0]\frac{1}{c} \quad. \tag{2.5.23}$$

To solve this system we multiply the first of equations (2.5.23) by $i = \sqrt{-1}$ and add the result to the second equation. Then

$$-\frac{2}{c_1^2}\frac{d}{d\tau}\Lambda_1 - \Lambda_1\left[\frac{1}{c_1 J}\frac{d}{d\tau}\left(\frac{J}{c_1}\right) - i\frac{2}{c_1 T}\right] = C_1 i + D_1 \tag{2.5.24}$$

where $\Lambda_1 = \Phi_1 + i\Psi_1$, while C_1 and D_1 denote the right-hand members of (2.5.23).

Equation (2.5.24) is obviously equivalent to system (2.5.23). In virtue of the fact that (2.5.24) is an ordinary first-order linear equation, it can be solved in quadratures to within an arbitrary constant. When we have found $\vec{\mathcal{E}}_1$ and $\vec{\mathcal{H}}_1$ from (2.5.19–24), we can find $\vec{\mathcal{E}}_2$, $\vec{\mathcal{H}}_2$, $\vec{\mathcal{E}}_3$, $\vec{\mathcal{H}}_3$ and so forth in a completely analogous fashion.

Let us note that (2.5.2, 7) give

$$\vec{\mathcal{E}} \approx \vec{\mathcal{E}}_0 e^{-i\omega(t-\tau)}(-i\omega)^{-\gamma}\perp\nabla\tau \ ,$$
$$\vec{\mathcal{H}} \approx \vec{\mathcal{H}}_0 e^{-i\omega(t-\tau)}(-i\omega)^{-\gamma}\perp\nabla\tau$$

in the zero approximation, i.e., electromagnetic waves are transverse in this approximation. However, the coefficients of subsequent approximations of the ray method $\vec{\mathcal{E}}_s$, $\vec{\mathcal{H}}_s$, $s = 1, 2, \ldots$ are, in general, no longer perpendiular to $\nabla\tau$.

2.6 Determining the Short-Wavelength Asymptotic Solution of a Diffraction Problem Using the Ray Method – An Example

Let us assume that the function u that characterizes a wave field depends only on x and y, i.e., we will be dealing with a two-dimensional problem. Let us assume further that in the medium under consideration $c(x, y) = 1$, and therefore $\omega/c = \omega = k$.

Consider the problem where a wave emanating from a point source located at M_0 (Fig. 2.3) is incident onto a smooth convex curve S that bounds a finite convex region Ω. This problem is rigorously formulated as follows: find the function $u(M, M_0)$ [$M = M(x, y)$ and $M_0 = M_0(x_0, y_0)$ lie outside of the region Ω] that satisfies the conditions

$$u = \frac{i}{4}H_0^{(1)}(kR) + u_r(M, M_0; k)$$
$$\left(\frac{\partial^2}{\partial x^2} + \frac{\partial^2}{\partial y^2} + k^2\right)u_r = 0 \ , \tag{2.6.1}$$

$$u|_s = 0 \ , \quad \sqrt{R}\left(\frac{\partial u}{\partial R} - iku\right) \xrightarrow[R\to\infty]{} 0 \ ,$$
$$R = |\overline{M\,M_0}| = \sqrt{(x - x_0)^2 + (y - y_0)^2} \ , \tag{2.6.2}$$

where $(i/4)H_0^{(1)}(kR)$ ($H_0^{(1)}$ being the Hankel function of the first kind) describes

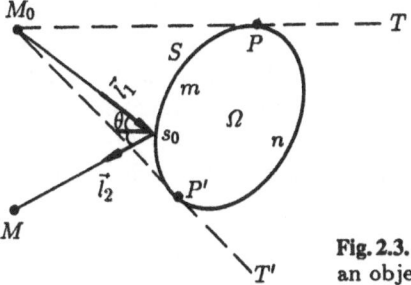

Fig. 2.3. Incident and reflected rays at the boundary of an object

35

a wave propagating from a point source located at M_0, and $u_r(M, M_0)$ is the reflected wave, having no singularities outside of region Ω. The function $u(M, M_0)$ that is the solution of (2.6.1,2) is called the *Green's function* for Helmholtz' equation.

First, let us draw the tangents to S from M_0. If a light source were placed at M_0 the region $TPnP'T'$ would be in shadow, and the rest of the outside of region Ω would be illuminated. It is in this illuminated region that we will find the asymptotics of $u(M, M_0; k)$ as $k \to \infty$.

The incident wave $(i/4)H_0^{(1)}(kR)$ that corresponds to a point source of oscillations in an infinite plane as $k \to \infty$ has an asymptotic expansion of the form [2.6]

$$\frac{i}{4} H_0^{(1)}(kR) \sim e^{ikR} \sum_{s=0}^{\infty} \frac{u_s^{\mathrm{inc}}}{(-ik)^{s+1/2}} \quad , \tag{2.6.3}$$

where (with Euler's gamma function Γ)

$$u_s^{\mathrm{inc}} = \sqrt{\frac{2}{\pi R}} \cdot \frac{1}{4} \frac{1}{(2R)^s} \frac{\Gamma(\tfrac{1}{2} + s)}{s! \Gamma(\tfrac{1}{2} - s)}$$

$$(-i)^{s+1/2} = \exp[-i\tfrac{\pi}{2}(s + \tfrac{1}{2})] \quad .$$

We will look for the reflected wave in the form

$$u_r = e^{ik\tau(M, M_0)} \sum_{s=0}^{\infty} \frac{u_s^r(M, M_0)}{(-ik)^{s+1/2}}$$

where $\tau(M, M_0)$ and u_s^r must satisfy (2.1.4,5) and (2.4.6,8). We now require that the boundary condition $u|_S = 0$ be satisfied. Then

$$R = |\overline{M\,M_0}| \Big|_{M \in S} = \tau(M, M_0) \Big|_{M \in S} \quad , \tag{2.6.4}$$

$$u_s^r(M, M_0)|_S + u_r^{\mathrm{inc}}(M, M_0)|_S = 0 \quad . \tag{2.6.5}$$

Equation (2.1.4) together with condition (2.6.4) is a Cauchy problem for the eikonal equation (Sect. 2.3). This problem has two solutions, which is not surprising as the eikonal equation is of second degree. One solution, $\tau(M, M_0) \equiv R \equiv |\overline{M\,M_0}|$, corresponds to the incident wave, while the other solution corresponds to the reflected wave.

In order to construct the function $\tau(M, M_0)$ that corresponds to the reflected wave, let us first find the value of $\nabla \tau$ on S. From (2.6.4) we know that the projections of $\nabla \tau$ and $\nabla |\overline{M\,M_0}|$ on S are identical. By virtue of the eikonal equation, (2.6.4) and the fact that the inequality $\nabla \tau \ne \nabla |\overline{M\,M_0}|$ must be satisfied on S (otherwise, as we can easily see, we would arrive at $\tau = |\overline{M\,M_0}|$, and we are interested in the eikonal that corresponds to the reflected rather than the incident wave), we get

$$\frac{\partial \tau}{\partial n} = -\frac{\partial}{\partial n} |\overline{M\,M_0}|$$

($\partial/\partial n$ is the derivative with respect to the normal to S). In this way, the projections of $\nabla\tau$ onto the tangent and normal to S are defined, and thus the vector $\nabla\tau$ as S is uniquely determined. Rays corresponding to the eikonal τ must emanate from S in the direction of $\nabla\tau$. The unit vector $\nabla\tau$ is directed in such a way that its projection and the projection of the vector $\nabla|\overline{M\,M_0}| = \overline{M_0 s_0}/|\overline{M s_0}|$ on the tangent to S coincide. We have thus arrived at the well known law of geometrical optics: the incident and reflected rays make equal angles with the reflecting curve, or more succinctly, the angle of incidence is equal to the angle of reflection. The derivative of the eikonal τ along the reflected ray is equal to unity (since the velocity in this case is equal to unity), and at the reflection point s_0 the eikonal $\tau(M, M_0)$ is equal to the value of $|\overline{M_0 s_0}|$; therefore the function $\tau(M, M_0)$ is equal to the length of ray $M_0 s_0 M$.

By means of simple variational arguments, we can show that the function $\tau(M, M_0)$ thus constructed actually satisfies the eikonal equation. To this end we evaluate

$$d\tau(M, M_0) = d\left(\int_{M_0 s_0 M} ds\right) = d\left(\int_{M_0 s_0} ds\right) + d\left(\int_{s_0 M} ds\right) \ .$$

From the formula for variation of an integral with movable end points (Sect. 2.2) [2.7]

$$d\tau(M, M_0) = (\vec{l}_1 - \vec{l}_2) \cdot \delta\vec{r}\,\big|_{s_0} + l_{2x}dx + l_{2y}dy\big|_M \ .$$

Here $\vec{l}_1$ and $\vec{l}_2 = (l_{2x}, l_{2y})$ are unit vectors directed along the rays $M_0 s_0$ and $s_0 M$, respectively; $\delta\vec{r} = (\delta x, \delta y)$ is an infinitesimally small displacement along S. The fact that the incident and reflected rays form identical angles with S implies that

$$(\vec{l}_1 - \vec{l}_2) \cdot \delta\vec{r}\,\bigg|_{S_0} = 0 \ ,$$

whence

$$d\tau(M, M_0) = l_{2x}dx + l_{2y}dy = \tau_x dx + \tau_y dy \ .$$

Consequently

$$\tau_x = l_{2x} \ , \quad \tau_y = l_{2y} \ ,$$

in virtue of which

$$(\nabla\tau)^2 = \tau_x^2 + \tau_y^2 = l_{2x}^2 + l_{2y}^2 = |\vec{l}|^2 = 1 \ ,$$

i.e., the function τ does in fact satisfy the eikonal equation[4]. All the $\psi_s(\alpha)$ ($s = 0, 1, \ldots$) − the "arbitrary" functions in (2.4.8) − are found from boundary condition (2.6.5) for the reflected wave, and are uniquely determined. In particular, for the zeroth-order approximation of u_r we get

---

[4] The fact that $\tau(M, M_0)$ satisfies the eikonal equation also follows readily from the theory of characteristics for first-order nonlinear differential equations [2.7].

$$u_r \approx \frac{e^{ik\tau} u_0^r}{\sqrt{-ik}} = \frac{e^{ik\tau(M,M_0)}}{\sqrt{-ik}} \sqrt{\frac{2}{\pi}} \cdot \frac{1}{4} \frac{1}{\sqrt{l_0 + l + 2l_0 l/(\varrho \cos \theta)}}$$ (2.6.6)

with

$$\sqrt{-i} = e^{-\pi i/4} \quad .$$

Here $l_0 = |\overline{M_0 s_0}|$, $l = |\overline{M s_0}|$, $\varrho = \varrho(s_0)$ is the radius of curvature of the boundary at the point $s_0 \varepsilon S$, and θ is the angle of incidence of the ray.

The formulas obtained in this section have been rigorously proven to be the asymptotic expansion of the exact solution.

2.7 Determination of the Function ψ_0 by Using the Localization Principle

The derivation of asymptotic formulas for solutions of diffraction problems requires expressions for the functions $\psi_0(\alpha, \beta)$, $\psi_1(\alpha, \beta)$, ... (Sect. 2.4), which are not always easy to find.

For the zeroth approximation we need only to know the function $\psi_0(\alpha, \beta)$. This function can often be found from what we might call *the localization principle*. For example, in the case of a point source in a nonhomogeneous medium

$$\left(\Delta + \frac{\omega^2}{c^2(M)} \right) u = -\delta(M - M_0)$$

we have

$$u \approx \sqrt{\frac{c}{J}} \psi_0(\alpha, \beta) \frac{e^{i\omega\tau(M,M_0)}}{(-i\omega)^\gamma} \quad , \quad \tau(M, M_0) = \int\limits_{M_0}^{M} \frac{ds}{c}$$

where α, β are parameters that characterize rays emanating from the point M_0. For α and β we may take, for instance, the spherical coordinates of the end of the unit vector tangent to the ray at M_0. In the two-dimensional case $\psi_0 = \psi_0(\alpha)$, where α is the angle between the tangent to the ray at M_0 and the x-axis.

The localization principle in this problem amounts to the statement that as $M \rightarrow M_0$ and $\omega \rightarrow \infty$, the leading term of $u(M, M_0)$ (the zeroth approximation) is the same as it would be for the problem of a point source in a homogeneous medium, with a propagation velocity identically equal to $c(M_0)$. In the two-dimensional case for such a homogeneous medium

$$u = \frac{i}{4} H_0^{(1)}(\omega\tau) \quad , \quad \tau = \frac{R}{c(M_0)} \quad , \quad R = |\overline{M M_0}| \quad ,$$

$$u \approx \frac{1}{4} \sqrt{\frac{2}{\pi\tau\omega}} e^{\pi i/4} e^{i\omega\tau} \quad .$$

In the three-dimensional case

$$u = \frac{1}{4\pi R}e^{i\omega\tau} \quad , \quad \tau = \frac{R}{c(M_0)} \quad , \quad R = |\overline{M\,M_0}| \quad .$$

Hence in the two-dimensional case $\gamma = 1/2$, and in the three-dimensional case $\gamma = 0$; ψ_0 is easily determined from the relations

$$\sqrt{\frac{c(M)}{J}}\psi_0(\alpha)\frac{e^{\pi i/4}}{\sqrt{\omega}} \underset{M\to M_0}{\sim} \frac{e^{\pi i/4}}{4}\sqrt{\frac{2}{\pi}\frac{c(M_0)}{R\omega}}$$

in the two-dimensional case, and

$$\sqrt{\frac{c}{J}}\psi_0(\alpha,\beta) \underset{M\to M_0}{\sim} \frac{1}{4\pi R}$$

in the three-dimensional case. Accordingly we get

$$\psi_0(\alpha) = \lim_{M\to M_0} \frac{1}{4}\sqrt{\frac{2c(M_0)}{\pi c(M)}}\sqrt{\frac{J}{R}} = \frac{1}{4}\sqrt{\frac{2}{\pi}} \quad ,$$

$$\psi_0(\alpha,\beta) = \lim_{M\to M_0} \frac{1}{4\pi}\frac{1}{R}\sqrt{\frac{J}{c}} = \frac{\sqrt{\sin\beta}}{4\pi\sqrt{c(M_0)}} \quad , \quad 0\le\beta\le\pi \quad .$$

Here $J = |\vec{r}_\alpha \times \vec{r}_\beta|$, and the coordinates x', y', z' of the end of the unit vector tangent to a ray at point M_0 are expressed in terms of α and β by the formulas

$$x' = \cos\alpha\sin\beta \quad , \quad y' = \sin\alpha\sin\beta \quad , \quad z' = \cos\beta \quad .$$

2.8 Caustics

In applications of the ray method it frequently happens that a family of rays has an enveloping curve in the two-dimensional case, and an enveloping surface in the three-dimensional case. Such envelopes of a family of rays are called *caustics*.

Let the parameters α,β characterize a ray, and let $\tau = \tau(\alpha,\beta)$ be the value of the parameter τ on the ray at a point M of tangency with a caustic. Thus the equation of the caustic in vector form is

$$\vec{r} = \vec{r}[\alpha,\beta,\tau(\alpha,\beta)] \quad .$$

The vectors $\vec{r}_\tau$, $\vec{r}_\alpha + \vec{r}_\tau\tau'_\alpha$, $\vec{r}_\beta + \vec{r}_\tau\tau'_\beta$ are coplanar since they lie in the plane tangent to the caustic; consequently their scalar triple product is

$$(\vec{r}_\alpha + \vec{r}_\tau\tau'_\alpha) \times (\vec{r}_\beta + \vec{r}_\tau\tau'_\beta)\cdot\vec{r}_\tau = 0$$

or, since $\vec{r}_\tau \times \vec{r}_\tau = 0$,

$$\vec{r}_\alpha \times \vec{r}_\beta \cdot \vec{r}_\tau = 0 \quad .$$

In virtue of the fact that $\vec{r}_\alpha\perp\vec{r}_\tau$, $\vec{r}_\beta\perp\vec{r}_\tau$, $|\vec{r}_\tau| = c>0$

$$\vec{r}_\alpha \times \vec{r}_\beta \cdot \vec{r}_\tau = |\vec{r}_\tau| |\vec{r}_\alpha \times \vec{r}_\beta| = c |\vec{r}_\alpha \times \vec{r}_\beta| = 0 \quad,$$

i.e., $J = 0$ on the caustic. Let us recall that (Sect. 2.3)

$$J = \frac{1}{c} \left| \frac{D(x, y, z)}{D(\alpha, \beta, \tau)} \right| \quad,$$

and therefore the Jacobian $D(x, y, z)/D(\alpha, \beta, \tau)$ vanishes on a caustic. Hence we find that in the vicinity of a caustic, ray coordinates are no longer single-valued functions of the cartesian coordinates. This fact is also clear from geometrical considerations. The fact that $J = 0$ implies that even in the zeroth approximation (Sect. 2.4), the ray method is not valid in the neighborhood of a caustic.

The behavior of a field in the vicinity of a caustic will be studied in Chap. 3. We present here some of the results[5] that derive from the considerations of Chap. 3.

The ray approximation is not valid in the layer $n = O(\omega^{-2/3})$ (n being the distance along the normal to the caustic). Passing through each point N of region II (Fig. 2.4) are two rays: one going toward the caustic, and the other going away from it. The zeroth approximation of the ray method at N takes the form

$$u \approx e^{i\omega \tau_1(N)} \psi_0(\alpha_1, \beta_1) \sqrt{\frac{c(N)}{J(\alpha_1, \beta_1, \tau_1)}} \frac{1}{(-i\omega)^\gamma}$$
$$+ e^{i\omega \tau_2(N) - (\pi i/2)} \psi_0(\alpha_2, \beta_2) \sqrt{\frac{c(N)}{J(\alpha_2, \beta_2, \tau_2)}} \frac{1}{(-i\omega)^\gamma} \quad.$$

Here $\psi_0(\alpha, \beta)$ has the same meaning as in Sect. 2.4: α_1, β_1 are parameters that characterize a ray going toward the caustic and passing through N; $\tau_1 = \tau_1(N)$ is the value of the eikonal on this ray at point N; and $\alpha_2, \beta_2, \tau_2$ are the corresponding parameters for a ray going away from the caustic. The point to remember is that on a ray going away from the caustic the phase factor is $\exp[i\omega \tau_2(N) - \pi i/2]$

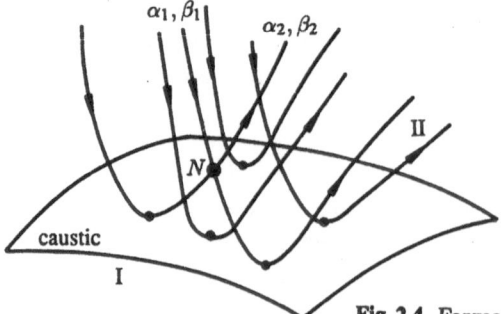

Fig. 2.4. Formation of a caustic as an envelope of rays

---

[5] After becoming familiar with these results, the reader may omit Chap. 3, and go on to study Chaps. 4, 5 and so on (see the flowchart showing the interdependence of the chapters and sections).

rather than $\exp[i\omega\tau_2(N)]$. In other words, the phase of a wave that has passed through a caustic undergoes a phase jump of $-\pi/2$.

In region I outside of the immediate vicinity of the caustic $[n = O(\omega^{-2/3})]$ the wave field damps out exponentially as $\omega \to +\infty$. We point out that the tendency of the wave field to vanish as $\omega \to +\infty$ could have been predicted from physical considerations: region I is the geometrical shadow region.

2.9 Notes on the Literature

As it would be impossible to present anything close to a complete review of work in the area of the ray method here, we will be content to list the following monographs which deal to one extent or another with ray techniques: [2.8–15]. Let us also mention the paper of *Keller* et al. [2.16], and the reviews by *Kravtsov* [2.17] and *Kouyoumjian* [2.18], and Chap. 9 of the book edited by *Mikhlin* [2.19], which contains a substantial bibliography of work involving ray techniques up to 1964.

The variant (2.1.2) of the ray expansion seems to have been first suggested by *Rytov* [2.20], and is sometimes called a *Rytov* expansion as a result. It has also been discussed by *Bochenek* and *Plebański* [2.21, 22], *Pogorzelski* [2.23, 24] and *Car* et al. [2.25].

The ray method for Maxwell's equations (in the zeroth approximation) was first suggested by *Rytov* [2.26]. It is discussed at some length by *Kline* and *Kay* [2.27] and by *Born* and *Wolf* [2.28], to whom the reader is referred for further references, notably the paper of *Kline* [2.29], who gave the complete formal expansion (2.5.2), due originally to *Luneberg* (cf. [2.30]).

An extension of ray techniques to more general partial differential equations has been given by *Trjitzinsky* [2.31].

The examples given in Sects. 2.6, 7 have been studied by many authors. In particular, the asymptotic expansion of the Green's function for the exterior of a bounded convex region was considered by *Keller* [2.32].

A number of papers exist devoted to the rigorous substantiation of the ray method. *Buslaev* [2.33] proved that the leading terms of the ray expansions do indeed give the asymptotic expansion of the function u in the case of the *Dirichlet* boundary condition $u|_S = 0$. Using a method employed by *Ursell* [2.34] for a different problem, *Babič* [2.35, 36] showed that the first two terms of the ray method for the *Neumann* problem $(\partial u/\partial n)_S = 0$ give the asymptotic expansion for the Green's function in this case as well. Later *Grimshaw* [2.37] showed that the formally constructed series is actually an asymptotic series for u. All of these works deal exclusively with the substantiation of the planar problem. The corresponding proofs for the three-dimensional case are to be found in the papers of *Babič* [2.38] and *Buslaev* [2.39, 40]. Other work devoted to the rigorous justification of the ray method is referred to in Sect. 13.9.

3. The Field Near a Caustic

An important example of an irregular ray field (one for which the formulas of Chap. 2 are not valid at every point) is the ray field near a caustic (Sect. 2.8). Chapter 3 will be devoted to the construction of an analog of the ray method for this case.

3.1 Preliminary Remarks

We remind the reader that a *caustic* is an envelope of a family of rays. As an example, consider the plane problem for the case where the velocity $c(M) \equiv 1$. Let AB (Fig. 3.1) be the position of some wave front. In this case, the family of rays is the set of normals to AB, and the caustic curve coincides with the envelope CD of this family of normals. We recall that the envelope of normals to AB is called the *evolute* of curve AB, and the curve AB itself is called the *involute* of curve CD. Note that the curve CD has an infinite set of involutes, and that they all intersect the tangents to CD at right angles, i.e., they are wave fronts. The classical theory of evolutes and involutes is thus identical to the geometry of the rays and wave fronts [in the case $c(M) \equiv 1$].

Let the parametric equations of CD be $x = f(\alpha)$, $y = g(\alpha)$. The equation for each involute of the curves CD can be put into the following well-known parametric form:

$$x = f(\alpha) + (\tau - s(\alpha))\gamma_1 \quad ,$$
$$y = g(\alpha) + (\tau - s(\alpha))\gamma_2 \quad . \tag{3.1.1}$$

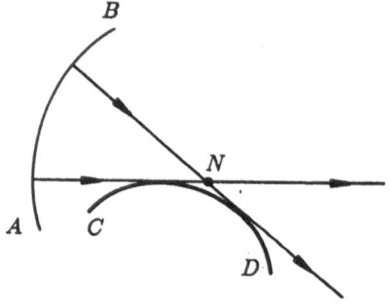

Fig. 3.1. Relationship of a family of rays to a caustic

Here

$$\gamma_1 = \gamma_1(\alpha) = \frac{f'}{\sqrt{f'^2 + g'^2}} \quad , \quad \gamma_2 = \gamma_2(\alpha) = \frac{g'}{\sqrt{f'^2 + g'^2}}$$

are the components of a unit vector tangent to CD, $s(\alpha)$ is the arc length along the curve CD measured (with appropriate sign) from a fixed point on CD to the point $x = f(\alpha)$, $y = g(\alpha)$. The quantity τ is constant on each involute, i.e. on each wave front. By fixing α and varying τ we get a ray, and by fixing τ we get a wave front (involute of the curve CD). In virtue of the fact that $\gamma_1^2 + \gamma_2^2 = 1$, the distance $|\overline{M_1 M_2}|$ between two arbitrary points M_1 and M_2 on the same ray is equal to $|\tau_1 - \tau_2|$ (τ_1 and τ_2 correspond to the points M_1, M_2). Therefore, τ plays the part of the eikonal (recall that when $c \equiv 1$ the rays are straight lines and $\tau(M_1, M_2) = \int_{M_1}^{M_2} (ds/c) = \int_{M_1}^{M_2} ds = |\overline{M_1 M_2}|$). We can find τ as a function of x, y from (3.1.1). This function will be double-valued since two rays pass through each point N (Fig. 3.1) in the vicinity of the caustic, and the values of τ on each such ray will not be the same at N.

To illustrate this ambiguity in τ, let us consider the example of a circle. Let CD be an arc of a circle of radius a. In polar coordinates r, ϕ ($x = r \cos \phi$, $y = r \sin \phi$) the circle has the equation $r = a$. Formulas (3.1.1) then give the well known equation for the involute of a circle. Solving (3.1.1) for τ, we get

$$\tau = \pm \left(\sqrt{r^2 - a^2} - a \arccos\frac{a}{r} \right) + a(\phi - \phi_0)$$

where r, ϕ are the coordinates of the point where the eikonal is calculated, and $\phi_0 = \text{const}$.

3.2 The Etalon Problem for Caustics

There is a simple solution of Helmholtz's equation $(\Delta + k^2)u = 0$ such that the rays corresponding to this solution are tangent to some circle. By analyzing its shortwave asymptotic properties, we can "guess" the asymptotic properties of the shortwave field near a caustic of more general configuration as well.

In polar coordinates this solution has the form

$$u = J_{ka}(kr)e^{ika\phi} \quad . \tag{3.2.1}$$

Here $J_{ka}(kr)$ is a Bessel function, while $a = \text{const} > 0$.

In order to avoid multivalued expressions, we will consider (3.2.1) in the "cut" plane $\pi \geq \phi > -\pi$. The Debye asymptotic form of the Bessel function [3.1] for $r > a$ gives

$$e^{ika\phi} J_{ka}(kr) = \frac{1}{2}\sqrt{\frac{2}{\pi ka}} \frac{1}{\sqrt[4]{(r/a)^2 - 1}}$$
$$\times \exp\left[ik\left(\sqrt{r^2 - a^2} - a \arccos\frac{a}{r} + a\phi \right) - \frac{\pi i}{4}\right]$$

$$\times \left(1 + O\left(\frac{1}{k}\right)\right) + \frac{1}{2}\sqrt{\frac{2}{\pi ka}} \frac{1}{\sqrt[4]{(r/a)^2 - 1}}$$

$$\times \exp\left[ik\left(-\sqrt{r^2 - a^2} + a\arccos\frac{a}{r} + a\phi\right) + \frac{\pi i}{4}\right]$$

$$\times \left(1 + O\left(\frac{1}{k}\right)\right) , \qquad\qquad (3.2.2)$$

where $O(1/k)$ are asymptotic series in integral powers of $1/k$. Expression (3.2.2) has the form of a ray expansion. The two terms in this expansion correspond to the two rays passing through a point N with coordinates (r, ϕ) (Fig. 3.2). The eikonals of these two terms are equal to

$$\tau_\pm(r, \phi) = a\phi \pm \left(\sqrt{r^2 - a^2} - a\arccos\frac{a}{r}\right) .$$

The wave fronts described by the equations $\tau_\pm = $ const are involutes of the circle $r = a$. The rays corresponding to the eikonal $\tau_-(N) = \tau_-(r, \phi) = a\phi + a\arccos\frac{a}{r} - \sqrt{r^2 - a^2}$, are half-lines tangent to the circle $r = a$, the directions being taken as shown in the figure (from infinity to the circle $r = a$). The first two terms in the formula for $\tau_-(N)$ represent respectively the arc lengths KQ and QM, and the third term $-\sqrt{r^2 - a^2}$ — is the length of segment MN. This quantity is subtracted because the direction on the ray is taken from N towards M. In an analogous way the rays corresponding to the eikonal

$$\tau_+(r, \phi) = a\phi - a\arccos\frac{a}{r} + \sqrt{r^2 - a^2} ,$$

are shown as broken half-lines directed as in the figure from the circle $r = a$ to infinity.

Let us call attention to the following point: the wave going toward the circle $r = a$ has phase $k\tau_-(r, \phi) + \pi/4$, whereas the phase of the wave going away from the circle is $k\tau_+(r, \phi) - \pi/4$. Thus the phase abruptly changes by $-\pi/2$ upon passage through the caustic. This will also be so in the general case (Sect. 3.9).

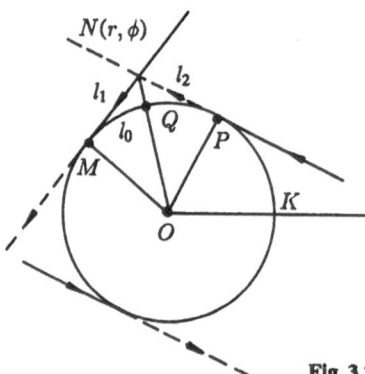

Fig. 3.2. Rays tangent to a circular caustic

The amplitude factor $1/\sqrt{a}[(r/a)^2 - 1]^{1/4}$ in (3.2.2), as was to be expected, is precisely $J^{-1/2}$, where J is the geometric divergence of the ray field (see Sect. 2.3). Naturally on the caustic $r = a$ this factor becomes infinite when $J = 0$, so that (3.2.2) is not valid as $r \to a$.

The shadow zone, i.e. the region into which rays do not penetrate, is the region $r < a$. For $r < a$ the Debye asymptotic form of the Bessel function gives

$$u = J_{ka}(kr)e^{ika\phi} = \frac{1}{2}\sqrt{\frac{2}{\pi k}}\frac{1}{\sqrt[4]{a^2 - r^2}}$$

$$\times \exp\left\{-k\left[a\ln\left(\frac{a}{r} + \sqrt{\left(\frac{a}{r}\right)^2 - 1}\right) + \sqrt{a^2 - r^2}\right] + ika\phi\right\}$$

$$\times \left(1 + O\left(\frac{1}{k}\right)\right) . \tag{3.2.3}$$

Thus in the shadow zone the wave field decays exponentially with increasing k, as we might have expected.

Our greatest interest at the moment is in asymptotic ($k \to \infty$) formulas for u that are uniformly valid in the neighborhood of the circle $r = a$, since, as $r \to a$, (3.2.2) is no longer applicable, as we have said. Using uniform asymptotic formulas for the Bessel function [3.2–4], we get

$$u = J_{ka}(kr)e^{ika\phi}$$

$$\approx (kr)^{-1/3}\left[\sqrt{\frac{2}{\pi}}\frac{\chi^{1/4}(a/r)}{\sqrt[4]{(a/r)^2 - 1}} + \sum_{l=2}^{\infty}\frac{b_l(a/r)}{(kr)^{2l}}\right]v\left((kr)^{2/3}\chi\left(\frac{a}{r}\right)\right)e^{ika\phi}$$

$$+ (kr)^{-5/3}\sum_{l=0}^{\infty}\frac{d_l(a/r)}{(kr)^{2l}} \cdot v'\left((kr)^{2/3}\chi\left(\frac{a}{r}\right)\right)e^{ika\phi} . \tag{3.2.4}$$

Here $v(t)$ is one of the Airy functions (Appendix A.1), and the function $\chi = \chi(s)$ is defined by the equation

$$\frac{2}{3}\chi^{3/2}(s) = \begin{cases} s\ln(s + \sqrt{s^2 - 1}) - \sqrt{s^2 - 1}, & s \geq 1, \\ i(s\arccos s - \sqrt{1 - s^2}), & 0 \leq s \leq 1 . \end{cases}$$

In this formula the radicals and logarithms are positive, $0 \leq \arccos s \leq \pi/2$. The function $\chi(s)$ is regular in the vicinity of $s = 1$, and can be expanded into the series

$$\chi(s) = \sqrt[3]{2}(s - 1) + \frac{\sqrt[3]{2}}{20}(s - 1)^2 + \cdots .$$

The functions b_l and d_l that appear in the coefficients of asymptotic series (3.2.4) are also regular in the neighborhood of $a/r = 1$.

The expression

$$\frac{1}{i}\frac{4}{3}r\chi^{3/2}\left(\frac{a}{r}\right) = 2\left(a\arccos\frac{a}{r} - \sqrt{r^2 - a^2}\right)$$

has a simple geometric interpretation. Referring to Fig. 3.2, we denote the lengths of the ray segments between N and the points of tangency M and P by l_1 and l_2, while l_0 denotes the arc length of the circle $r = a$ between P and M. Then we have the obvious relation

$$\frac{1}{i}\frac{4}{3}r\chi^{3/2}\left(\frac{a}{r}\right) = l_0 - l_1 - l_2 \quad .$$

An analogous formula holds in the general case as well (Sect. 3.5) where, in contrast to the case considered here, we generally have $l_1 \neq l_2$. By introducing complex rays, we could also derive an analog of the formula for the shadow zone $r < a$ that was given above.

Equation (3.2.4) will be the pattern for constructing the field in the neighborhood of a caustic in the general case. However, before going on to these general constructions, we will have to look into the differential geometry of rays and wave fronts near a caustic in more detail. Let us proceed to carry out this (somewhat tedious) preparation.

3.3 The Ray Field and Eikonal in the Neighborhood of a Caustic

Let a given ray field have an envelope, i.e., a caustic. We shall consider the three-dimensional case, and the theory for caustics in the planar case will follow from our constructions by specialization. It will also be assumed that the rays and caustic always meet at osculation points of the first order.[1]

We first construct a curvilinear coordinate system (ζ, η) on the caustic which is based on the rays impinging tangentially onto the caustic. These rays form a field of directions on the caustic; we wish to lay out a family of curves on the caustic which has these directions as tangent vectors. This family of curves is indicated in Fig. 3.3 by the unbroken thin lines.

Points on a curve of this family can be characterized by the values of the eikonal there. Let the ray L_0 be tangent to a curve S of this family at a point M_0. We will take the value of the eikonal τ at this point on the ray for the parameter ζ along the curve (at the point M_0). Now let the parameter η characterize the curve. In this way we have introduced an orthogonal coordinate system on the caustic. That (ζ, η) is orthogonal follows from the fact that the lines $\zeta = $ const are also lines, where the eikonal is constant, i.e., are the intersections of wavefronts with the caustic; the lines $\eta = $ const, on the other hand, are tangent to rays on the caustic, and consequently perpendicular to the wavefronts and thus also to the lines $\zeta = $ const. Note that in the planar case when $c(M) = $ const, this choice of the parameter ζ corresponds to using arc length to parameterize the evolute (Sect. 3.1).

---

[1] E.g., in the planar case, the rays and the caustic have the same slope at the point of tangency, but different second derivatives [Translator's note].

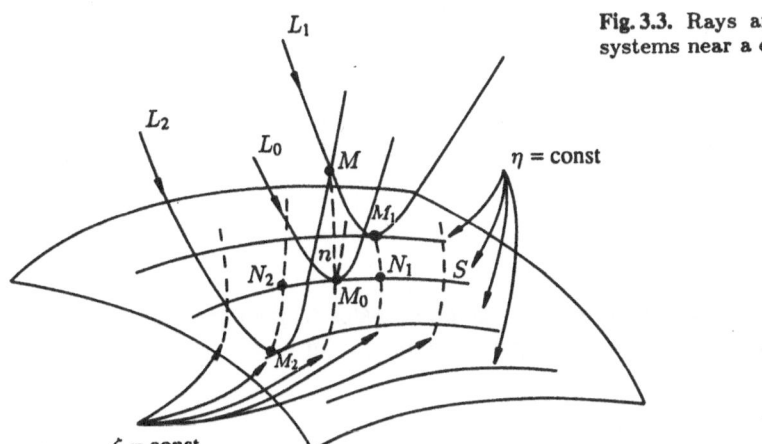

Fig. 3.3. Rays and coordinate systems near a caustic surface

The curvilinear coordinates (ζ, η) introduced on the caustic allow two separate coordinate systems to be defined in the neighborhood of the caustic: the ray coordinates (α, β, τ) and the orthogonal curvilinear coordinates (l, t, n), to which we will refer as the *caustic coordinate system*. The ray coordinates α and β for a given point M are identical to the parameters ζ and η at a point of tangency M_1 (or M_2) of the caustic to a ray passing through M. The location of M on the ray is specified by the value of the eikonal τ corresponding to this point. In the caustic coordinate system the coordinates l and t are identical to the parameters ζ and η at the point M_0 on the caustic which is at the base of the normal to the caustic which passes through M. The coordinate n is taken as the length of this same normal, i.e., the distance n of the point M to the caustic. We shall stipulate that $n>0$ on the illuminated side of the caustic (the side covered by the ray field). We put $n<0$ on the opposite side of the caustic.

It is evident from Fig. 3.3 that a point M in the vicinity of a caustic can be described in either of two sets of ray coordinates, associated with the rays L_1 or L_2. Therefore a point with the caustic coordinates (l, t, n), $n>0$, possesses two representations, $(\alpha_1, \beta_1, \tau_1)$ and $(\alpha_2, \beta_2, \tau_2)$, in ray coordinates.

In ray coordinates, the caustic is given by the equation $\tau = \alpha$, while in caustic coordinates it is $n = 0$. On the caustic itself, $\tau = \alpha = l = \zeta$, and $\beta = t = \eta$, i.e., the ray parameters (α, β), the caustic coordinates (l, t) and the coordinates (ζ, η) introduced on the caustic, all coincide.

The ray field that forms the caustic will be described in ray coordinates in the form

$$\vec{r} = \vec{r}(\alpha, \beta, \tau) \quad . \tag{3.3.1}$$

For fixed α and β we have in (3.3.1) a parametric equation for a ray tangent to the caustic at a point characteristized by the parameters $\zeta = \alpha$ and $\eta = \beta$. Putting $\alpha = \zeta$, $\beta = \eta$ and $\tau = \alpha(= \zeta)$ in (3.3.1), we evidently obtain a parametric equation for the caustic itself:

$$\vec{r} = \vec{R}(\zeta, \eta)$$

where $\vec{R}(\zeta, \eta) = \vec{r}(\zeta, \eta, \zeta)$. We will assume that the parameters (ζ, η) form a regular grid on the caustic. This assumption is equivalent to the condition that the vector product $\vec{R}_\zeta \times \vec{R}_\eta \neq 0$ everywhere. By virtue of the orthogonal nature of the coordinates (ζ, η) we have

$$\vec{R}_\zeta \cdot \vec{R}_\eta = 0$$

for their scalar product.

To evaluate $|\vec{R}_\zeta|$, note that the ray coordinates (α, β, τ) are orthogonal, and thus for arbitrary values of α, β, τ (and specifically for $\alpha = \zeta$, $\beta = \eta$, $\tau = \zeta$), that is, for points on the caustic, we have

$$\vec{r}_\alpha(\zeta, \eta, \zeta) \cdot \vec{r}_\tau(\zeta, \eta, \zeta) = 0 \quad . \tag{3.3.2}$$

Moreover, from (2.3.12), we have $|\vec{r}_\tau| = c(M)$. Then because of $\vec{R}_\zeta = \vec{r}_\alpha(\zeta, \eta, \zeta) + \vec{r}_\tau(\zeta, \eta, \zeta)$ and the fact that the vectors $\vec{R}_\zeta$ and $\vec{r}_\tau$ are directed along the ray, while $\vec{r}_\alpha$ is perpendicular to the ray, we obtain the following important relations:

$$\vec{r}_\alpha = 0 \ , \quad \vec{R}_\zeta = \vec{r}_\tau \ , \quad |\vec{R}_\zeta| = c_0(\zeta, \eta) \tag{3.3.3}$$

where $c_0(\zeta, \eta)$ is the value of the velocity at the point on the caustic with parameters ζ, η.

Next, let us find the equation $\vec{r} = \vec{r}(\alpha, \beta, \tau)$ of a ray tangent to the caustic at a point $M_0(\zeta = \alpha, \eta = \beta)$, using the (l, t, n) coordinates in the vicinity of the caustic. The eikonal equation [cf. (2.3.3) and Appendix A.2] in the coordinates

$$l = q_1 \ , \quad t = q_2 \ , \quad n = q_3$$

has the form

$$c^2(q_1, q_2, q_3) \sum_{i,j=1}^{3} g^{ij} \tau_{q_i} \tau_{q_j} - 1 = 0 \tag{3.3.4}$$

where $(g^{ij}) = (g_{ij})^{-1}$ and

$$g_{ij} = \vec{r}_{q_i}(q_1, q_2, q_3) \cdot \vec{r}_{q_j}(q_1, q_2, q_3) \tag{3.3.5}$$

are the components of the metric tensor.

Rays, or extremals of the Fermat functional, are also characteristics of the eikonal equation (3.3.4) and can be found as solutions of the Hamiltonian system:

$$\frac{dq_i}{d\sigma} = \frac{\partial}{\partial p_i} \mathcal{F}(q_i, p_j) \ , \tag{3.3.6}$$

$$\frac{dp_i}{d\sigma} = -\frac{\partial}{\partial q_i} \mathcal{F}(q_j, p_j) \ , \tag{3.3.7}$$

$$\mathcal{F}(q_j, p_j) = \frac{1}{2}\left(c^2 \sum_{i,j=1}^{3} g^{ij} p_i p_j - 1 \right)$$

that corresponds to (3.3.4).[2] The Hamiltonian system (3.3.6–7) is a first order system in the quantities $q_i = q_i(\sigma)$ and $p_i = p_i(\sigma)$. If a function $\tau(q_i)$ satisfies (3.3.4), then for a solution $q_i = q_i(\sigma)$, $p_i = p_i(\sigma)$ of (3.3.6–7), we have the relation [3.5]

$$\frac{\partial \tau}{\partial q_i} = p_i \, , \quad i = 1, 2, 3 \tag{3.3.8}$$

and consequently,

$$\frac{d\tau}{d\sigma} = \sum_i p_i \frac{dq_i}{d\sigma} = \sum_i p_i \frac{\partial}{\partial p_i} \mathcal{F}(q_i, p_j) \quad .$$

In our case,

$$\sum_i p_i \frac{\partial}{\partial p_i} \mathcal{F}(q_j, p_j) = c^2 \sum_{i,j} g^{ij} p_i p_j$$

and in view of (3.3.8) and (3.3.4), we have $d\tau/d\sigma = 1$. Thus we can choose the eikonal τ to be the parameter σ.

To evaluate the matrix elements g^{ij}, we express the position vector $\vec{r}(q_1, q_2, q_3)$ of a point near the caustic as

$$\vec{r}(q_1, q_2, q_3) = \vec{R}(q_1, q_2) + q_3 \vec{n}(q_1, q_2)$$

where $\vec{n}(q_1, q_2)$ is a unit vector normal to the caustic at the point (q_1, q_2), directed towards the illuminated side. Using (3.3.5) and evaluating the inverse of the matrix, we find that

$$g^{11} = \frac{1}{|\vec{R}_l(l, t)|^2} - 2 \frac{\vec{R}_l \cdot \vec{n}_l}{|\vec{R}_l|^4} n + O(n^2) \tag{3.3.9}$$

$$g^{22} = \frac{1}{|\vec{R}_t(l, t)|^2} + O(n) \; ; \quad g^{33} = 1 \; ,$$

$$g^{12} = g^{21} = O(n) \; ; \quad g^{13} = g^{31} = g^{23} = g^{32} = 0 \quad . \tag{3.3.10}$$

From differential geometry [3.6], we know that

$$\frac{\vec{R}_l \cdot \vec{n}_l}{|\vec{R}_l|^2} = \frac{1}{\varrho(l, t)}$$

whee $\varrho(l, t)$ is the radius of curvature of a normal cross-section of a surface [here, the caustic $\vec{r} = \vec{R}(l, t)$] taken tangentially to the coordinate line $t = \eta = $ const. The quantity $1/\varrho(l, t)$ is known as the *normal curvature* of the caustic surface along this coordinate line in an appropriate metric [3.7]. Moreover, since $|\vec{R}_l(l, t)|^2 = c_0^2(l, t)$ in view of (3.3.3), we can thus write (3.3.9) in the form

---

[2] The factor $\frac{1}{2}$ is introduced in order to simplify subsequent formulas and is related to the choice of the parameter σ.

$$g^{11} = \frac{1}{c_0^2(l,t)} - \frac{2}{c_0^2(l,t)} \cdot \frac{n}{\varrho(l,t)} + O(n^2) \quad . \tag{3.3.11}$$

Putting $\sigma = \tau$ in (3.3.6) and using (3.3.10) and (3.3.11), we obtain

$$\frac{dl}{d\tau} = p_1 + O(n)$$

$$\frac{dt}{d\tau} = p_2 \left[\frac{c_0^2(l,t)}{|\vec{R}_t(l,t)|^2} + O(n) \right]$$

$$\frac{dn}{d\tau} = p_3 [c_0^2(l,t) + O(n)] \quad . \tag{3.3.12}$$

In order to obtain the parametric equation for the ray tangent to the caustic at M_0, we integrate the system (3.3.12) of differential equations with initial data

$$l|_{\tau=\alpha} = \alpha \ , \quad t|_{\tau=\alpha} = \beta \ , \quad n|_{\tau=\alpha} = 0 \tag{3.3.13}$$

$$p_1|_{\tau=\alpha} = 1 \ , \quad p_2|_{\tau=\alpha} = p_3|_{\tau=\alpha} = 0 \quad . \tag{3.3.14}$$

The initial conditions (3.3.14) take the form indicated because of the choice of coordinates $l = \zeta$, $t = \eta$ on the caustic, as well as relation (3.3.8).

From (3.3.12) and (3.3.14) it follows that

$$\frac{dl}{d\tau}\bigg|_{\tau=\alpha} = 1 \ , \quad \frac{dt}{d\tau}\bigg|_{\tau=\alpha} = \frac{dn}{d\tau}\bigg|_{\tau=\alpha} = 0 \quad . \tag{3.3.15}$$

To evaluate $d^2n/d\tau^2|_{\tau=\alpha}$, differentiate the last equation of (3.3.12) with respect to τ. Putting $\tau = \alpha$, we get

$$\frac{d^2n}{d\tau^2}\bigg|_{\tau=\alpha} = c_0^2(l,t)\frac{dp_3}{d\tau}\bigg|_{\tau=\alpha} \quad .$$

The quantity $dp_3/d\tau|_{\tau=\alpha}$ can be determined from (3.3.7) for $i = 3$:

$$\frac{dp_3}{d\tau}\bigg|_{\tau=\alpha} = -\frac{1}{2}\frac{\partial}{\partial n}(c^2 g^{11})\bigg|_{\tau=\alpha} = -\frac{c_1(l,t)}{c_0(l,t)} + \frac{1}{\varrho(l,t)}$$

where $c_1(l,t) = \partial c/\partial n|_{\tau=\alpha}$, i.e., the normal derivative of the velocity at the point (l,t) on the caustic. Thus,

$$\frac{d^2n}{d\tau^2}\bigg|_{\tau=\alpha} = c_0^2\left(\frac{1}{\varrho} - \frac{c_1}{c_0}\right) = \frac{c_0^2}{P(l,t)} \ , \quad \frac{1}{P} = \frac{1}{\varrho} - \frac{c_1}{c_0} \quad . \tag{3.3.16}$$

The quantity P will be called the *effective radius of curvature* of a surface along the same direction as taken for the radius ϱ. In the present case, P is the effective radius of curvature of the caustic along a line $\eta = \text{const}$.

The effective radius of curvature plays an important role in the study of the behavior of a ray field in an inhomogeneous medium near several types of surfaces. As a result, we will encounter the quantity P again and again in the upcoming chapters of this book. In a mathematical context, P shows up as the normal radius of curvature of a line $\eta = $ const in the Riemannian metric given by the equations $g_{ik} = \delta_{ik}/c^2$.

The assumption made previously that the rays meet the caustic at an osculation point of the first order implies that the condition $d^2n/d\tau^2|_{\tau=\alpha} \neq 0$ is met, and (since the normal $\vec{n}$ points into the illuminated region) that $d^2n/d\tau^2 > 0$. It thus follows that for the case of first-order tangency which we have here, the effective radius of curvature of the caustic is always a positive quantity: $P > 0$.

The values of the derivatives (3.3.15, 16) that we have found along with the initial data (3.3.13) permit the equation for the ray in the neighborhood of the point M_0 to be written in the form

$$l = \alpha + (\tau - \alpha) + O[(\tau - \alpha)^2]$$
$$t = \beta + O[(\tau - \alpha)^2]$$
$$n = \frac{c_0^2(l,t)}{2P(l,t)}(\tau - \alpha)^2 + O[(\tau - \alpha)^3] \quad . \tag{3.3.17}$$

These equations can be taken as relating the caustic coordinates (l, t, n) of a point M to its ray coordinates (α, β, τ). In place of the coordinate n, let us introduce the coordinate

$$m = \pm \sqrt{n} \quad (n > 0) \quad .$$

The plus sign holds for points M on the portion of the ray *after* it has met the caustic, while the minus sign applies to points along the portion of the ray which has not yet reached the caustic.

Clearly,

$$\left. \frac{D(l,t,m)}{D(\alpha,\beta,\tau)} \right|_{\tau=\alpha} = \sqrt{\frac{c_0^2(l,t)}{P(l,t)}} \neq 0$$

and by continuity the determinant $D(l,t,m)/D(\alpha,\beta,\tau)$ differs from zero over some neighborhood of the caustic. It is therefore possible to express the ray coordinates (α,β,τ) in terms of (l,t,m) in a one-to-one fashion wherein the functions

$$\alpha = \alpha(l,t,m) \quad , \quad \beta = \beta(l,t,m) \quad , \quad \tau = \tau(l,t,m)$$

are smooth. From (3.3.17) it follows that

$$\alpha = l + O(m) \quad , \quad \beta = t + O(m^2) \quad , \quad \tau = l + O(m^2) \quad . \tag{3.3.18}$$

The accuracy of formulas (3.3.18) will, however, not suffice for what follows. We can obtain a more exact formula for τ by appealing directly to the eikonal equation. Because of the smoothness of the function $\tau(l,t,m)$, we can write it in the form

$$\tau(l,t,m) = l + \tau_2(l,t)m^2 + \tau_3(l,t)m^3 + O(m^4) \quad . \tag{3.3.19}$$

The coefficients $\tau_2(l,t)$ and $\tau_3(l,t)$ are found by plugging (3.3.19) into the eikonal equation (3.3.4), and then equating the coefficients of the terms involving n^0 and n^1 to zero. The calculation leads to the values

$$\tau_2(l, t) = 0$$

and

$$\tau_3^2 = \frac{4}{9} \frac{2}{c_0^2(l, t) P(l, t)} \quad .$$

Thus,[3]

$$\tau(l, t, m) = l + \frac{2}{3} \sqrt{\frac{2}{c_0^2(l, t) P(l, t)}} m^3 + O(m^4) \quad .$$

It is convenient to split the function $\tau(l, t, m)$ into its even and odd parts with respect to the variable m :

$$\tau_e(l, t, m) = \frac{\tau(l, t, m) + \tau(l, t, -m)}{2}$$

$$\tau_o(l, t, m) = \frac{\tau(l, t, m) - \tau(l, t, -m)}{2}$$

so that

$$\tau(l, t, m) = \tau_e(l, t, m) + \tau_o(l, t, m) \quad .$$

The even part τ_e is a smooth function of n, representable for small n as

$$\tau_e(l, t, \pm \sqrt{n}) = \xi(l, t, n) = l + O(n^2) \quad . \tag{3.3.20}$$

The odd part τ_o can clearly be expressed in the form

$$\tau_o(l, t, \pm \sqrt{n}) = \pm \frac{2}{3} \mu^{3/2}(l, t, n)$$

where, as $n \to 0$,

$$\mu(l, t, n) = \sqrt[3]{\frac{2}{c_0^2(l, t) P(l, t)}} n + O(n^2) \quad . \tag{3.3.21}$$

Finally, then, we have

$$\tau = \tau(l, t, \pm \sqrt{n}) = \xi(l, t, n) \pm \frac{2}{3} \mu^{3/2}(l, t, n) \quad . \tag{3.3.22}$$

If the propagation velocity, the ray field and the caustic are all infinitely differentiable functions, then the functions ξ and μ are representable [with an accuracy of $O(n^N)$, where N is as large as desired] by truncated Taylor series in powers of n, as $n \to 0$.[4] If the velocity, ray field and caustic are analytic, then ξ and μ will likewise be analytic functions of l, t and n, and in this case the Taylor series will be convergent.

---

[3] The choice of the positive sign for τ_3 is based on our having specified $m > 0$ for $\tau - l > 0$.

[4] When the propagation velocity, ray field or caustic has only a finite number of continuous derivatives, the number N is limited by the order of smoothness of those functions.

Formulas (3.3.20–22) allow us to clarify the geometrical character of the wave fronts near a caustic. Introduce a cartesian coordinate system (x, y, z) such that the xy-plane is tangent to the caustic at the point M_0 (Fig. 3.3). Denote the parameter ξ, η corresponding to M_0 by ξ_0, η_0. Direct the x-axis tangentially to the coordinate line $\eta = \eta_0$, in the sense of increasing τ, and the z-axis along the normal to the caustic at M_0, taking $z > 0$ when $n > 0$. Clearly, in the neighborhod of the caustic we have the relations

$$x = c_0(\zeta_0, \eta_0)(l - \zeta_0) + \ldots$$
$$y = t - \eta_0 + \ldots$$
$$z = n + \ldots \quad . \tag{3.3.23}$$

In the first approximation, a wave front passing through the origin can, as follows from (3.3.20–23), be represented by the equation

$$\frac{x}{c_0(\zeta_0, \eta_0)} \pm \frac{2}{3}\left(\frac{2}{c_0^2(\zeta_0, \eta_0)P(\zeta_0, \eta_0)}\right)^{1/2} z^{3/2} = 0$$

or,

$$z^3 = \tfrac{9}{8}P(\zeta_0, \eta_0)x^2$$

which is an equation for a semicubical parabola. The wave front therefore has an edge of regression on the caustic.[5]

In what follows, we will need a representation for the geometrical divergence

$$J = \frac{1}{c}\left|\frac{D(x, y, z)}{D(\alpha, \beta, \tau)}\right| \tag{3.3.24}$$

in (l, t, n) coordinates near the caustic. Rewriting (3.3.24) in the form

$$J = \frac{1}{c}\left|\frac{D(x, y, z)}{D(l, t, n)}\right|\left|\frac{D(l, t, n)}{D(\alpha, \beta, \tau)}\right|$$

and using (3.3.23) and (3.3.17), we get

$$J = \left|\frac{c_0^2(l, t)}{P(l, t)}(\tau - \alpha) + O[(\tau - \alpha)^2]\right|$$

or, in terms of l, t, m,

$$J = c_0(l, t)\sqrt{\frac{2}{P(l, t)}}|m|J_0(l, t, m) \tag{3.3.25}$$

---

[5] The physically-minded reader may wish to verify for himself that x, y, z all have units of length; this is not true either of the ray coordinates or of the caustic coordinates used in this chapter. For τ, α and l have units of time, while t, n and β have units of length. The metric tensor of the transformation between two of these coordinate systems will therefore have several elements which are not dimensionless quantities, but have units of velocity or inverse velocity. This should be kept in mind when reading the following section [Translator's note].

where $J_0(l, t, m) = 1 + O(m)$ is a smooth function of l, t and m. Replacing m by $-\sqrt{n}$ or $\sqrt{n}$ in (3.3.25) as appropriate, we obtain an expression for either the geometrical divergence J_1 of the ray field before its arrival at the caustic, or the geometrical divergence J_2 of the ray field after leaving the caustic.

3.4 Derivation of the Recurrence Relations

The objective of this chapter is to construct the field in the neighborhood of a caustic. In fact, what we will construct is a certain series, which formally satisfies the equation

$$\left(\Delta + \frac{\omega^2}{c^2(x, y, z)} \right) u = 0 \tag{3.4.1}$$

as $\omega \to \infty$. These formal solutions will correspond to a ray field of the caustic type, which we have studied in Sect. 3.3. Apparently, an arbitrary solution of (3.4.1), to which there corresponds a caustic ray field like the one considered in Sect. 3.3, will have a formal series of this type as its asymptotic expansion.

We seek the formal solution to (3.4.1) in a form suggested by the solution of the etalon problem of Sect. 3.2:

$$u = \left\{ \left[A_0(x, y, z) + \frac{A_1(x, y, z)}{(-i\omega)} + \frac{A_2(x, y, z)}{(-i\omega)^2} + \ldots \right] v(-\omega^{2/3}\mu) \right.$$
$$\left. + i \left[B_0(x, y, z) + \frac{B_1(x, y, z)}{(-i\omega)} + \ldots \right] v'1(-\omega^{2/3}\mu)\omega^{-1/3} \right\} e^{i\omega\xi} \omega^{-\nu} \,. \tag{3.4.2}$$

Here $\mu = \mu(x, y, z)$, $\xi = \xi(x, y, z)$, ν is a constant, and $v(t)$ is an Airy function. Expression (3.4.2) is analogous to (3.2.4).

It should be noted in (3.2.4) that the quantities corresponding to A_1, A_3, A_5, ..., and B_0, B_2, B_4, ..., were all zero. This is due to the symmetry of the etalon problem; the introduction of these coefficients is necessary in the general case. For $\mu > 0$, the asymptotic behavior of $v(t)$ (cf. Appendix A.1) gives

$$u \sim \frac{1}{2}\omega^{-\nu-1/6} \, e^{i\pi/4}(A_0\mu^{-1/4} - B_0\mu^{1/4}) \exp\left[i\omega\left(\xi - \frac{2}{3}\mu^{3/2} \right) \right]$$
$$+ \frac{1}{2}\omega^{-\nu-1/6} \, e^{i\pi/4}(A_0\mu^{-1/4} + B_0\mu^{1/4}) \exp\left[i\omega\left(\xi + \frac{2}{3}\mu^{3/2} \right) - \frac{\pi i}{2} \right] \tag{3.4.3}$$

as $\omega \to \infty$. To each point (x, y, z), therefore, belong the two eikonals

$$\tau_1 = \xi - \tfrac{2}{3}\mu^{3/2} \quad \text{and} \quad \tau_2 = \xi + \tfrac{2}{3}\mu^{3/2} \tag{3.4.4}$$

just as in the case of the field near a caustic (Sect. 3.3).

We assume that ξ and μ in (3.4.2,3) are the same functions which appear in (3.3.22). Expressions (3.4.4) will then be solutions of the eikonal equation

$$\left[\nabla \left(\xi \pm \frac{2}{3} \mu^{3/2} \right) \right]^2 = \frac{1}{c^2}$$

whence

$$(\nabla \xi)^2 + \mu (\nabla \mu)^2 = \frac{1}{c^2} \ , \quad \nabla \xi \cdot \nabla \mu = 0 \ . \tag{3.4.5}$$

The second relation in (3.4.5) indicates that the surfaces

$$\xi = \frac{\tau_1 + \tau_2}{2} = \text{const} \quad \text{and} \quad \mu = \left[\frac{3}{4}(\tau_2 - \tau_1) \right]^{2/3} = \text{const} \tag{3.4.6}$$

are mutually orthogonal. In the planar case, $\xi = \text{const}$ and $\mu = \text{const}$ form an orthogonal grid of lines in the neighborhood of the caustic. Equations (3.4.5) are important for what will follow.

Substituting (3.4.2) into (3.4.1), we obtain[6]

$$\left(\Delta + \frac{\omega^2}{c^2(x,y,z)} \right) u = v \, e^{i\omega\xi} \omega^{-\nu} \sum_{j=0}^{\infty} (-\mathcal{L}_2 A_j - \mathcal{L}_3 B_j + \Delta A_{j-1})(-i\omega)^{1-j}$$

$$+ i v' \, e^{i\omega\xi} \omega^{2/3-\nu} \sum_{j=0}^{\infty} (-\mathcal{L}_1 A_j - \mathcal{L}_2 B_j + \Delta B_{j-1})(-i\omega)^{-j} = 0 \tag{3.4.7}$$

$$A_{-1} \equiv 0 \ , \quad B_{-1} \equiv 0 \ .$$

Here

$$\mathcal{L}_1 f = 2 \nabla \mu \cdot \nabla f + f \Delta \mu$$
$$\mathcal{L}_2 f = 2 \nabla \xi \cdot \nabla f + f \Delta \xi$$
$$\mathcal{L}_3 f = 2 \mu \nabla \mu \cdot \nabla f + f \nabla \cdot (\mu \nabla \mu) \ .$$

We obtain from (3.4.7) a system of recurrence relations:

$$\mathcal{L}_2 A_j + \mathcal{L}_3 B_j = \Delta A_{j-1} \tag{3.4.8}$$

$$\mathcal{L}_1 A_j + \mathcal{L}_2 B_j = \Delta B_{j-1} \tag{3.4.9}$$

whose analysis will be taken up in the next section.

---

[6] Terms involving $\omega^2 v$ and $\omega^{5/3} v'$ are not present in (3.4.7) by virtue of (3.4.5). If μ and ξ in (3.4.2) were treated as unknown functions yet to be determined, then by equating the coefficients of $\omega^2 v$ and $\omega^{5/3} v'$ in (3.4.7) to zero, we would arrive at none other than equations (3.4.5) again for these functions to satisfy.

3.5 The Field in the Vicinity of a Caustic – First Approximation

Setting $j = 0$ in (3.4.8, 9), we have

$$\mathcal{L}_2 A_0 + \mathcal{L}_3 B_0 = 0$$
$$\mathcal{L}_1 A_0 + \mathcal{L}_2 B_0 = 0 \quad .$$

Multiplying the first equation by $\mu^{-1/4}$, the second by $\mu^{1/4}$, and adding and subtracting the results, we arrive without difficulty at the relations

$$2\nabla\tau_1 \cdot \nabla\Phi_0^{(1)} + \Phi_0^{(1)}\Delta\tau_1 = 0 \tag{3.5.1}$$

$$2\nabla\tau_2 \cdot \nabla\Phi_0^{(2)} + \Phi_0^{(2)}\Delta\tau_2 = 0 \tag{3.5.2}$$

where, as before,

$$\tau_1 = \xi - \tfrac{2}{3}\mu^{3/2} \ , \quad \tau_2 = \xi + \tfrac{2}{3}\mu^{3/2} \tag{3.5.3}$$

and

$$\Phi_0^{(1)} = A_0\mu^{-1/4} - B_0\mu^{1/4}$$
$$\Phi_0^{(2)} = A_0\mu^{-1/4} + B_0\mu^{1/4} \quad . \tag{3.5.4}$$

Equations (3.5.1 and 2) have been considered in detail in Chap. 2. Their solution, the formulas of the zeroth-order aproximation of the ray method, is

$$\Phi_0^{(1)} = \sqrt{\frac{c}{J_1(\alpha_1, \beta_1, \tau_1)}}\chi_0^{(1)}(\alpha_1, \beta_1)$$

$$\Phi_0^{(2)} = \sqrt{\frac{c}{J_2(\alpha_2, \beta_2, \tau_2)}}\chi_0^{(2)}(\alpha_2, \beta_2) \tag{3.5.5}$$

where J_1 is the divergence of the portion of the ray field which has not yet reached the caustic, while J_2 is the divergence for those rays which have already touched the caustic.

The functions $\chi_0^{(1)}(\alpha, \beta)$, $\chi_0^{(2)}(\alpha, \beta)$ depend only upon the parameters α, β which characterize the ray. We shall assume both to be sufficiently smooth functions of α and β.

From (3.5.5 and 4), it follows that

$$A_0 = \frac{1}{2}\left[\frac{\chi_0^{(1)}(\alpha_1, \beta_1)}{\sqrt{J_1(\alpha_1, \beta_1, \tau_1)}} + \frac{\chi_0^{(2)}(\alpha_2, \beta_2)}{\sqrt{J_2(\alpha_2, \beta_2, \tau_2)}}\right]\mu^{1/4}\sqrt{c(M)}$$

$$B_0 = \frac{1}{2}\left[\frac{\chi_0^{(2)}(\alpha_2, \beta_2)}{\sqrt{J_2(\alpha_2, \beta_2, \tau_2)}} - \frac{\chi_0^{(1)}(\alpha_1, \beta_1)}{\sqrt{J_1(\alpha_1, \beta_1, \tau_1)}}\right]\mu^{-1/4}\sqrt{c(M)} \quad . \tag{3.5.6}$$

Since $J_1 = J_2 = 0$ on the caustic, the denominators in (3.5.6) will vanish there.

Nevertheless, A_0 is remains bounded because of (3.3.21 and 25), while for B_0 to be bounded it is necessary and sufficient that

$$\chi_0^{(1)}(\alpha, \beta) = \chi_0^{(2)}(\alpha, \beta) = \chi_0(\alpha, \beta) \quad . \tag{3.5.7}$$

Indeed, we shall prove below that if $\chi_0(\alpha, \beta)$ is a smooth nonzero function, and if relation (3.5.7) holds, then the coefficients A_0 and B_0 are smooth functions of the caustic coordinates l, t, and n.

Consider the ratio $\chi_0(\alpha, \beta)/\sqrt{J(\alpha, \beta, \tau)}$. Since $\chi_0(\alpha, \beta)$ is a smooth function of α and β, we can express this ratio, using (3.3.18 and 25), in the form

$$\frac{\chi_0(\alpha, \beta)}{\sqrt{J(\alpha, \beta, \tau)}} = \chi_0(l, t) \sqrt[4]{\frac{P(l, t)}{2c_0^2(l, t)}} m^{-1/2} f_0(l, t, m) \tag{3.5.8}$$

where $f_0(l, t, m)$ is a smooth function of (l, t, m), with $f_0(l, t, m) = 1 + O(m)$. Splitting $f_0(l, t, m)$ into its even and odd parts:

$$\frac{1}{2}[f_0(l, t, m) + f_0(l, t, -m)] = L_0(l, t, m^2)$$

$$\frac{1}{2}[f_0(l, t, m) - f_0(l, t, -m)] = mK_0(l, t, m^2)$$

it is evident that L_0 and K_0 are smooth functions of (l, t, n), and that $L_0(l, t, n) = 1 + O(n)$. Now, to obtain the value of (3.5.8) on a ray which has not yet reached the caustic, i.e., for $\alpha = \alpha_1$ and $\beta = \beta_1$, the quantity m should be replaced by $-\sqrt{n}$, while on a ray leaving the caustic, i.e., for $\alpha = \alpha_2$ and $\beta = \beta_2$, we must put $\sqrt{n}$ in place of m. Consequently,

$$\frac{1}{2}\left(\frac{\chi_0(\alpha_1, \beta_1)}{\sqrt{J_1(\alpha_1, \beta_1, \tau_1)}} + \frac{\chi_0(\alpha_2, \beta_2)}{\sqrt{J_2(\alpha_2, \beta_2, \tau_2)}}\right)$$

$$= \chi_0(l, t) \sqrt[4]{\frac{P(l, t)}{2c_0^2(l, t)}} n^{-1/4} L_0(l, t, n) \tag{3.5.9}$$

$$\frac{1}{2}\left(\frac{\chi_0(\alpha_1, \beta_1)}{\sqrt{J_1(\alpha_1, \beta_1, \tau_1)}} - \frac{\chi_0(\alpha_2, \beta_2)}{\sqrt{J_2(\alpha_2, \beta_2, \tau_2)}}\right)$$

$$= \chi_0, (l, t) \sqrt[4]{\frac{P(l, t)}{2c_0^2(l, t)}} n^{1/4} K_0(l, t, n,) \quad . \tag{3.5.10}$$

Multiply the left sides of (3.5.9 and 10) by $\mu^{1/4}c^{1/2}(M)$ and $\mu^{-1/4}c^{1/2}(M)$ respectively. Then since $c(M)$ is a smooth nonvanishing function, and since by (3.3.21),

$$\mu^{1/4} = \left(\frac{2}{c_0^2(l, t)P(l, t)}\right)^{1/12} n^{1/4} \mu_1(l, t, n) \tag{3.5.11}$$

where $\mu_1(l, t, n) = 1 + O(n)$ is a smooth function, we arrive at the result that A_0 and B_0 are indeed smooth functions of the caustic coordinates (l, t, n), and that

$$A_0(l, t, n) = \chi_0(l, t) \left(\frac{P(l, t)}{2c_0(l, t)} \right)^{1/6} [1 + O(n)] \quad .$$

And so, the field $u(M)$ in the neighborhood of a caustic is, in the first approximation, described by the formula

$$u(M) = \frac{1}{2} \sqrt{c(M)} \left\{ \left[\mu^{1/4} \left(\frac{\chi_0(\alpha_1, \beta_1)}{\sqrt{J_1(\alpha_1, \beta_1, \tau_1)}} + \frac{\chi_0(\alpha_2, \beta_2)}{\sqrt{J_2(\alpha_2, \beta_2, \tau_2)}} + O\left(\frac{1}{\omega} \right) \right) \right] \right.$$
$$\times v(-\omega^{2/3} \mu) + \frac{i}{\omega^{1/3}} \left[\mu^{-1/4} \left(\frac{\chi_0(\alpha_2, \beta_2)}{\sqrt{J_2(\alpha_2, \beta_2, \tau_2)}} \right. \right.$$
$$\left. \left. - \frac{\chi_0(\alpha_1, \beta_1)}{\sqrt{J_1(\alpha_1, \beta_1, \tau_1)}} + O\left(\frac{1}{\omega} \right) \right) \right] v'(-\omega^{2/3} \mu) \right\} e^{i\omega \xi} \omega^{-\nu}. \quad (3.5.12)$$

A calculation using (3.5.12) would proceed in the following manner. Through each point M on the illuminated side of (and sufficiently close to) the caustic, there pass two rays: MM_1 with parameters (α_1, β_1) and $M_2 M$ with parameters (α_2, β_2) (Fig. 3.3). For each of these rays we find the corresponding divergences $J_1(\alpha_1, \beta_1, \tau_1)$ and $J_2(\alpha_2, \beta_2, \tau_2)$ from (3.3.24), and substitute these into (3.5.12). Next, the functions ξ and μ appearing in the exponent and in the arguments of the Airy functions are evaluated. At small distances from the caustic, this can be done using (3.3.20 and 21). At arbitrary distances, we must use more general formulas for ξ and μ which are derived below.

The values of the eikonals τ_1 and τ_2 at M, for the ray L_1, which has not yet arrived at the caustic and the ray L_2 which is leaving it, are given in terms of ξ and μ by (3.4.4); adding and subtracting, we get, see (3.4.6):

$$\xi = \frac{\tau_1 + \tau_2}{2} \quad \text{and} \quad \mu = \left[\frac{3}{4}(\tau_2 - \tau_1) \right]^{2/3} \quad . \tag{3.5.13}$$

To evaluate the difference $\tau_2 - \tau_1$ and mean $(\tau_1 + \tau_2)/2$ of τ_1 and τ_2, let $M_0(l, t)$ be the base of the normal to the caustic which passes through M (Fig. 3.3; for $n = 0$, take $l = \zeta, t = \eta$), and let $M_1(\alpha_1, \beta_1)$ and $M_2(\alpha_2, \beta_2)$ be the points of tangency of the rays L_1 and L_2 with the caustic. On the caustic, draw the coordinate lines $\eta = t, \zeta = \alpha_1, \zeta = \alpha_2$, and denote their intersections by N_1 and N_2 as shown.

For points *on* the caustic, the eikonal is a single-valued function. Let $\tau(M_1)$ and $\tau(M_2)$ be the values of the eikonal at M_1 and M_2, and introduce the notations

$$l_1 = \int_M^{M_1} \frac{ds}{c} \quad , \quad l_2 = \int_{M_2}^{M} \frac{ds}{c} \quad , \quad l_0 = \int_{N_2}^{N_1} \frac{ds}{c} \quad . \tag{3.5.14}$$

Since the eikonal is constant along any line $\zeta = \text{const}$,

$$\tau(N_2) = \tau(M_2) \quad \text{and} \quad \tau(N_2) = \tau(M_1) - l_0 \quad . \tag{3.5.15}$$

On the other hand,

$$\tau_1 = \tau(M_1) - l_1 \quad , \quad \tau_2 = \tau(M_2) + l_2 \quad . \tag{3.5.16}$$

From (3.5.15 and 16) it follows that

$$\tau_2 - \tau_1 = l_1 + l_2 - l_0 \quad . \tag{3.5.17}$$

For the mean value $(\tau_1 + \tau_2)/2$, it is not difficult to obtain the expression

$$\tfrac{1}{2}(\tau_1 + \tau_2) = \tau(M_0) + \tfrac{1}{2}[(l_2 - l_0'') - (l_1 - l_0')] \tag{3.5.18}$$

where $\tau(M_0)$ is the value of the eikonal at M_0, while

$$l_0' = \int\limits_{M_0}^{N_1} \frac{ds}{c} \; , \quad l_0'' = \int\limits_{N_2}^{M_0} \frac{ds}{c} \quad .$$

Expressions (3.5.17 and 18) are particularly convenient to use if $c \equiv 1$ (in which case l_0, l_1 and l_2 are simply the lengths of the respective lines) or if the points M_1 and M_2 lie on the same line $\eta = \text{const}$.

Having evaluated $\tau_2 - \tau_1$ and $(\tau_1 + \tau_2)/2$, we proceed to obtain the values of ξ and μ from (3.5.13). The function $\chi_0(\alpha, \beta)$ in (3.5.12) remains arbitrary, and must be determined from a matching condition between $u(M)$ and a given incident wave at the caustic (Sect. 3.7).

In the small region

$$0 \leq n < \text{const}\,\omega^{-2/3} \;, \quad \text{const} > 0 \tag{3.5.19}$$

near the caustic, (3.5.12) can be simplified considerably. In the region (3.5.19), the relations

$$\omega\xi = \omega l + O(\omega^{-1/3})$$

and

$$\omega^{2/3}\mu = \sqrt[3]{\frac{2}{c^2(l,t)P(l,t)}}\,\omega^{2/3}n + O(\omega^{-2/3})$$

hold. Therefore, the Airy function $v(-\omega^{2/3}\mu)$ and its derivative $v'(-\omega^{2/3}\mu)$ are of the same order, and in (3.5.12) the second term (containing the factor $\omega^{-1/3}$) can be neglected. Hence, the field in region (3.5.19) very near the caustic can be represented in the form

$$u(M) = \chi_0(l,t)\left(\frac{P(l,t)}{2c_0(l,t)}\right)^{1/6} v\left(-\sqrt[3]{\frac{2}{c_0^2(l,t)P(l,t)}}\,\omega^{2/3}n\right)$$

$$\times\, e^{i\omega l}\omega^{-\nu} + O(\omega^{-1/3-\nu}). \tag{3.5.20}$$

When things are analytic, expressions (3.5.12 and 20) can be extended into the shadow region $n < 0$ by analytic continuation in n. Near a caustic, waves going towards or away from the caustic *cannot be distinguished*. This caustic *boundary layer* appears wherever the Airy function cannot be replaced by its asymptotic expression, that is, whenever the argument of the Airy function v has order 1 or $\mu = O(\omega^{-2/3})$. Recalling that μ is of the same order of magnitude as the distance from the caustic as $\mu \to 0$, see (3.3.21), we conclude that the thickness of the boundary layer is likewise of the order $O(\omega^{-2/3})$. Thus, if n denotes the distance of an arbitrary point to the caustic, measured along a normal to the caustic, then

$$|n| = O(\omega^{-2/3}) \qquad (3.5.21)$$

for a point in the boundary layer.

In the shadow zone $\mu<0$, the Airy function decays exponentially outside of the boundary layer. The asymptotic form of the Airy function for $\mu<0$, $\omega\to\infty$ (Appendix A.1) leads to the estimate

$$u \sim O[\omega^{-\nu-1/6}\exp(-\tfrac{2}{3}|\mu|^{3/2}\omega)] \quad .$$

3.6 Determination of A_j and B_j for $j>0$

Now we come back to (3.4.8) and (3.4.9). Multiplying (3.4.8) by $\mu^{-1/4}$ and (3.4.9) by $\mu^{1/4}$, and adding and subtracting the results, we find

$$2\nabla\tau_1\cdot\nabla\Phi_j^{(1)} + \Phi_j^{(1)}\Delta\tau_1 = \mu^{-1/4}\Delta A_{j-1} - \mu^{1/4}\Delta B_{j-1}$$

$$2\nabla\tau_2\cdot\nabla\Phi_j^{(2)} + \Phi_j^{(2)}\Delta\tau_2 = \mu^{-1/4}\Delta A_{j-1} + \mu^{1/4}\Delta B_{j-1} \quad . \qquad (3.6.1)$$

Here,

$$\Phi_j^{(1)} = \mu^{-1/4}A_j - \mu^{1/4}B_j$$

$$\Phi_j^{(2)} = \mu^{-1/4}A_j + \mu^{1/4}B_j \quad . \qquad (3.6.2)$$

Exactly as in Chap. 2, it follows from (3.6.1) that

$$\Phi_j^{(1)} = \sqrt{\frac{c}{J_1}}\left[\chi_j^{(1)}(\alpha_1,\beta_1) + \int\limits_{\alpha_1}^{\tau_1}\sqrt{\frac{J_1}{c}}\frac{c^2}{2}(\mu^{-1/4}\Delta A_{j-1} - \mu^{1/4}\Delta B_{j-1})d\tau\right]$$

$$\Phi_j^{(2)} = \sqrt{\frac{c}{J_2}}\left[\chi_j^{(2)}(\alpha_2,\beta_2) + \int\limits_{\alpha_2}^{\tau_2}\sqrt{\frac{J_2}{c}}\frac{c^2}{2}(\mu^{-1/4}\Delta A_{j-1} + \mu^{1/4}\Delta B_{j-1})d\tau\right].$$

$$(3.6.3)$$

In (3.6.3) we have chosen the start of the integrals at $\tau_1 = \alpha_1$ and $\tau_2 = \alpha_2$, i.e., at the caustic.

From (3.6.2 and 3), we have

$$A_j = A_j^0 + A_j^1 \quad \text{and} \quad B_j = B_j^0 + B_j^1 \qquad (3.6.4)$$

where

$$\left.\begin{array}{c}A_j^0 \\ B_j^0\end{array}\right\} = \frac{1}{2}\left[\frac{\chi_j^{(2)}(\alpha_2,\beta_2)}{\sqrt{J_2(\alpha_2,\beta_2,\tau_2)}} \pm \frac{\chi_j^{(1)}(\alpha_1,\beta_1)}{\sqrt{J_1(\alpha_1,\beta_1,\tau_1)}}\right]\mu^{\pm1/4}\sqrt{c(M)} \qquad (3.6.5)$$

and

$$\left.\begin{array}{c} A_j^1 \\ B_j^1 \end{array}\right\} = \frac{1}{2}\left[\sqrt{\frac{c}{J_2}}\int_{\alpha_2}^{\tau_2}\sqrt{\frac{J_2}{c}}\frac{c^2}{2}(\mu^{-1/4}\Delta A_{j-1}+\mu^{1/4}\Delta B_{j-1})d\tau\right.$$

$$\left.\pm\sqrt{\frac{c}{J_1}}\int_{\alpha_1}^{\tau_1}\sqrt{\frac{J_1}{c}}\frac{c^2}{2}\left(\mu^{-1/4}\Delta A_{j-1}-\mu^{1/4}\Delta B_{j-1}\right)d\tau\right]\mu^{\pm 1/4}. \quad (3.6.6)$$

The expressions for A_j^0 and B_j^0 are completely analogous to (3.5.6) for A_0 and B_0, and consequently it is necessary and sufficient that

$$\chi_j^{(1)}(\alpha,\beta) = \chi_j^{(2)}(\alpha,\beta) = \chi_j(\alpha,\beta)$$

in order that they be bounded. If $\chi_j(\alpha,\beta)$ are smooth functions, then A_j^0 and B_j^0 will be smooth functions of the caustic coordinates (l,t,n) if the (non-vanishing) velocity, ray field and caustic are all smooth. We will show, finally, that A_j^1 and B_j^1 are likewise smooth functions of (l,t,n) if the coefficients A_{j-1} and B_{j-1} are smooth.

On the basis of (3.3.25, 21, and 17), the integral

$$\sqrt{\frac{c}{J}}\int_{\alpha}^{\tau}\sqrt{\frac{J}{c}}\frac{c^2}{2}(\mu^{-1/4}\Delta A_{j-1}\pm\mu^{1/4}\Delta B_{j-1})d\tau \quad (3.6.7)$$

can be written in the form

$$m^{3/2}[L_j(l,t,m^2)+mK_j(l,t,m^2)] \quad , \quad m = \pm\sqrt{n} \quad (3.6.8)$$

where $L_j(l,t,m^2)$ and $K_j(l,t,m^2)$ are smooth functions of l,t and m^2. To obtain the value of the integral (3.6.7) along a ray approaching the caustic, put $m = -\sqrt{n}$ in (3.6.8); to obtain it for a ray leaving the caustic, put $m = \sqrt{n}$. Thus

$$A_j^1 = nL_j(l,t,n)\left(\frac{\mu}{n}\right)^{1/4}$$

and

$$B_j^1 = nK_j(l,t,n)\left(\frac{n}{\mu}\right)^{1/4} \quad .$$

Since the ratio $(n/\mu)^{1/4}$ is smooth in l,t and n by virtue of (3.5.11), it follows then that A_j^1 and B_j^1 are likewise so. By the same token, the smoothness of the coeffcients A_j and B_j in expansion (3.4.2) has been proven for arbitrary j. If the velocity, ray field and caustic are analytic, then so also are A_j and B_j.

As far as the construction of A_j and B_j from formulas (3.6.4–6) is concerned, the same comments made at the end of Sect. 3.5 relative to the evaluation of A_0 and B_0 from (3.5.6) apply here as well.

We have only one thing left to clear up: how to choose the functions $\chi_j(\alpha,\beta)$. This will be taken up in the remaining section of this Chapter.

3.7 Determination of the χ_j

We frequently are faced with the problem of finding the field behind and near a caustic in a case where "everything is known" up to the caustic. For example, let a field be produced by an oscillating point source in an inhomogeneous medium, and let the corresponding central field of rays have an envelope, that is, a caustic. Then the complete ray expansion for the wave traveling toward the caustic is, in principle, known. The problem, then, is to find the caustic expansion (3.4.2) of the field. This boils down to the determination of the functions χ_j in formulas (3.5.12) and (3.6.5). A similar situation occurs in finding the shortwave asymptotics of a field resulting from the reflection of a known (e.g., plane) wave from a body whose shape is such that the reflected rays have an envelope. So if we known the ray expansion for the wave incident on the caustic, how do we find these functions χ_j?

Let the wave incident on the caustic be given by its ray expansion, see (2.1.1) and (2.4.8):

$$u^1 = \sum_{s=0}^{\infty} \frac{u_s^1(M)}{(-i\omega)^{s+\gamma}} e^{i\omega\tau_1}$$

$$\gamma = \text{const} \quad , \quad (-i)^\gamma = e^{-\pi i\gamma/2} \quad , \quad \tau_1 = \xi - \tfrac{2}{3}\mu^{3/2} \tag{3.7.1}$$

$$u_s^1(M) = \sqrt{\frac{c}{J_1}}\left[\psi_s(\alpha_1,\beta_1) + \int_{\tau_0(\alpha_1,\beta_1)}^{\tau} \sqrt{\frac{J_1}{c}}\,\frac{c^2}{2}\Delta u_{s-1}^1 d\tau\right] \quad . \tag{3.7.2}$$

Referring to (3.4.2), we have by virtue of (3.4.3)

$$\begin{aligned}
u &\sim e^{\pi i/4}\frac{\omega^{-1/6-\nu}}{2}(A_0\mu^{-1/4} - B_0\mu^{1/4})e^{i\omega(\xi-\frac{2}{3}\mu^{3/2})} \\
&+ e^{-\pi i/4}\frac{\omega^{-1/6-\nu}}{2}(A_0\mu^{-1/4} + B_0\mu^{1/4})e^{i\omega(\xi+\frac{2}{3}\mu^{3/2})} \\
&= e^{\pi i/4}\frac{\omega^{-1/6-\nu}}{2}\sqrt{\frac{c}{J_1}}\chi_0(\alpha_1,\beta_1)e^{i\omega\tau_1} \\
&+ e^{-\pi i/4}\frac{\omega^{-1/6-\nu}}{2}\sqrt{\frac{c}{J_2}}\chi_0(\alpha_2,\beta_2)e^{i\omega\tau_2}
\end{aligned} \tag{3.7.3}$$

in the first approximation as $\omega\to\infty$. Comparing the leading terms of (3.7.1 and 3), we obtain

$$e^{\pi i\gamma/2}\psi_0(\alpha_1,\beta_1)\omega^{-\gamma} = e^{\pi i/4}\frac{\omega^{-1/6-\nu}}{2}\chi_0(\alpha_1,\beta_1) \quad . \tag{3.7.4}$$

From (3.7.4) it follows that

$$\nu = \gamma + \tfrac{1}{6} \quad , \quad \chi_0(\alpha_1,\beta_1) = 2\exp\left[\frac{\pi i}{2}(\gamma - 1/2)\right]\psi_0(\alpha_1,\beta_1) \quad . \tag{3.7.5}$$

Note that we have found the zeroth order approximation (of the ray method) for the wave leaving the caustic in terms of the same order of approximation for the wave incident at the caustic. In the zeroth-order approximation, the wave incident on the caustic has the form

$$u^1 \sim \frac{\psi_0(\alpha_1, \beta_1)}{(-i\omega)^\gamma} \sqrt{\frac{c}{J_1}} e^{i\omega\tau_1} \quad .$$

The wave leaving the caustic [in view of (3.7.3–5)] then has the form

$$u^2 = \frac{\psi_0(\alpha_2, \beta_2)}{(-i\omega)^\gamma} \sqrt{\frac{c}{J_2}} e^{i\omega\tau_2 - \pi i/2} \quad .$$

These two expressions correspond completely, except that $\pi/2$ has been subtracted from the phase of u^2. This result is briefly summarized as follows: *there is a phase jump of $-\pi/2$ when a wave passes through a caustic.* The total field outside the shadow zone and the boundary layer near the caustic will, of course, be a superposition of u^1 and u^2.

Let us now turn to the higher-order approximations. Let $\psi_0, \psi_1, \ldots, \psi_j$ be given. Assume that $A_0, B_0, \ldots, A_{j-1}, B_{j-1}$ can be calculated from $\psi_0, \ldots, \psi_{j-1}$. Let us see how we can then find A_j and B_j or (what is the same) $\chi_j(\alpha, \beta)$. In (3.4.2), we replace v and v' by their asymptotic expansions (cf. Appendix A.1):

$$v(-t) \sim \frac{1}{2} t^{-1/4} \exp\left[i\frac{2}{3}t^{3/2} - i\frac{\pi}{4}\right]\left\{1 + \sum_{s=1}^{\infty} \frac{\Gamma(3s + \frac{1}{2})}{\sqrt{\pi}(2s)!}(9t^{3/2}i)^{-s}\right\}$$

$$+ \frac{1}{2} t^{-1/4} \exp\left[-i\frac{2}{3}t^{3/2} + i\frac{\pi}{4}\right]\left\{1 + \sum_{s=1}^{\infty} \frac{\Gamma(3s + \frac{1}{2})}{\sqrt{\pi}(2s)!}(-9t^{3/2}i)^{-s}\right\}$$

$$v'(-t) \sim -\frac{1}{2} t^{1/4} \exp\left[i\frac{2}{3}t^{3/2} + i\frac{\pi}{4}\right]\left\{1 + \sum_{s=1}^{\infty} \frac{\Gamma(3s + \frac{3}{2})}{\sqrt{\pi}(2s)!}\frac{(9t^{3/2}i)^{-s}}{\frac{1}{2} - 3s}\right\}$$

$$- \frac{1}{2} t^{1/4} \exp\left[-i\frac{2}{3}t^{3/2} - i\frac{\pi}{4}\right]\left\{1 + \sum_{s=1}^{\infty} \frac{\Gamma(3s + \frac{3}{2})}{\sqrt{\pi}(2s)!}\frac{(-9t^{3/2}i)^{-s}}{\frac{1}{2} - 3s}\right\}$$

The coefficient of $\omega^{-\nu-j-1/6} e^{i\omega\tau_1}$ in (3.4.2) ($\tau_1 = \xi - \frac{2}{3}\mu^{3/2}$) will then have the form

$$\frac{1}{2} \exp\left[\frac{\pi ij}{2} + \frac{\pi i}{4}\right](A_j\mu^{-1/4} - B_j\mu^{1/4})$$

$$+ \Xi(A_0, B_0, \ldots, A_{j-1}, B_{j-1}) \quad . \tag{3.7.6}$$

Here Ξ is an expression determined solely by the coefficient $A_0, B_0, \ldots, A_{j-1}, B_{j-1}$.

Equating (3.7.6) to the coefficient of $\omega^{-j-\gamma}$ the ray expansion (3.7.1) and using (3.7.2, 6.4, 6.5 and 6.6), we get

$$\frac{1}{2} \exp\left[\frac{\pi ij}{2} + \frac{\pi i}{4}\right]\sqrt{\frac{c}{J_1}}\left[\chi_j(\alpha_1, \beta_1)\right.$$

$$+ \int_{\alpha_1}^{\tau_1} \sqrt{\frac{J_1}{c}} \frac{c^2}{2} (\mu^{-1/4} \Delta A_{j-1} - \mu^{1/4} \Delta B_{j-1}) d\tau \bigg] + \Xi$$

$$= \sqrt{\frac{c}{J_1}} \bigg[\psi_j(\alpha_1, \beta_1) + \int_{\tau_0(\alpha_1, \beta_1)}^{\tau} \sqrt{\frac{J_1}{c}} \frac{c^2}{2} \Delta u_{j-1}^1 d\tau \bigg] \exp \bigg[\frac{\pi i}{2} (j + \gamma) \bigg] \ .$$

$$(3.7.7)$$

Note that the integrals in (3.7.7) and the function Ξ are determined by $\psi_0, \ldots,$ ψ_{j-1}. Thus, the function $\chi_j(\alpha, \beta)$ is uniquely determined in terms of $\psi_0, \ldots, \psi_{j-1},$ ψ_j. It is not difficult to see that, conversely, $\psi_0, \ldots, \psi_j$ are uniquely determined in terms of $\chi_0, \ldots, \chi_j$ for arbitrary j.

Let us close this section with the following comment. To determine A_l and B_l when $l < j$, we must construct the functions $\Phi_l^{(1)}$ and $\Phi_l^{(2)}$ (cf. Sect. 3.6), but (3.6.3) indicates that $\Phi_l^{(i)}$ is calculated by integrating along rays which leave the caustic. Thus, despite the deceptively simple appearance of (3.7.7), the dependence of $\chi_j(\alpha, \beta)$ on $\psi_0, \psi_1, \ldots, \psi_{j-1}$ is actually rather complicated.

3.8 Notes on the Literature

The classical theory of evolutes and involutes used in Sect. 3.1 can be found, e.g., in the text by *Fikhtengol'ts* [3.8] or in [3.9] or [3.6]. The derivation of the asymptotic formula (3.2.4) for the Bessel function can be found in the original papers or the book by *Olver* [3.2–4]. The Debye asymptotic expansion for the Bessel function is derived in the well-known treatise of *Watson* [3.1] and in the paper of *Petrashen'* et al. [3.10].

A large body of literature is devoted to the study of the field near a caustic. The Airy function itself was originally "invented" specifically to express the shortwave asymptotic behavior of the field near a caustic in a closed form. A more classical treatment of caustic fields is given in the monograph by *Landau* and *Lifšits* [Ref. 3.11, Chap. VII, Sect. 59].

The exposition given in this chapter is based on the work of *Gazaryan* [3.12] and *Kravtsov* [3.13] (primarily the latter); but the present development is much more detailed and mathematically rigorous than these papers. The refinements and additions of this chapter are due to *Babič*.

Ludwig [3.14] has obtained some interesting results in the theory of caustics which have much in common with those of *Kravtsov* [3.13]. The proof of (3.3.22) and of the smoothness of the A_j and B_j is closely analogous to the corresponding derivations in *Ludwig*'s paper. See also Sect. 3.2 of the book by *Babič* and *Kirpičnikova* [3.15].

The dominant term of the asymptotic expansion of the field near a caustic could also have been derived starting from a representation of the field in terms of *Maslov*'s canonical operator, see [3.16]. The succeeding terms of an expansion of the field in inverse powers of frequency could also be obtained in a similar way. *Buslaev* [3.17] not only indicated this possibility, but also constructed a generalized canonical operator. However, the evaluation of the resulting integral by the method of stationary phase (in the case of closely spaced saddle points) which must be done in this approach would be no simpler than the derivations of this chapter. The use of the canonical operator to represent the field (which is valid for quite general types of behavior of the corresponding ray field) does allow a number of rigorous results to be established. All the same, it seems to us that the method described in this chapter is preferable for obtaining the asymptotic expansion of a field near a caustic that has no singular points.

4. Derivation of Asymptotic Formulas for Eigenvalues and Eigenfunctions Using the Ray Method

In this chapter we present an important heuristic method for deriving asymptotic formulas for the eigenvalues and eigenfunctions of the equation

$$\left(\Delta + \frac{\omega^2}{c^2(M)}\right)u = 0 \quad (\omega^2 \text{ is an eigenvalue}) \quad .$$

We will call this method the *Keller-Rubinow method*.

4.1 Introductory Remarks

Since large values of ω correspond to eigenfunctions of high order, it is natural to expect that geometrical-optic considerations will play an important role in constructing asymptotic expansions of eigenfunctions. In fact, much of the work known to us on the asymptotic properties of eigenfunctions is based (more or less) on the ray method described in Chap. 2. The arguments of this chapter are also based on the ray method.

Suppose that we want to find asymptotic formulas (as $\omega \to \infty$) for the eigenfunctions and eigenvalues of the equation

$$\Delta u + \frac{\omega^2}{c^2(M)}u = 0 \quad (M \in \Omega) \tag{4.1.1}$$

subject to one of the following boundary conditions

$$u|_s = 0 \quad \text{or} \tag{4.1.2}$$

$$\left.\frac{\partial u}{\partial n}\right|_s = 0 \quad . \tag{4.1.3}$$

Here Ω is a bounded region within which equation (4.1.1) holds, S is the boundary of Ω, and M is a point of Ω.

In certain cases, we can construct an asymptotic expansion of the eigenfunctions in the form of a ray solution (Chap. 2). The field of rays corresponding to this ray solution has a very interesting property: if we follow the trajectory of one of the rays inside the region, then after a finite number of reflections the ray returns to its original trajectory. We say that the ray field transforms into itself. There will be as many terms in the asymptotic formula for $u(M)$ as there are rays passing through the point M. The asymptotic expansion of the eigenvalues is given by the condition that the ray solution be unique on this closed field of rays.

The general technique for applying the ray method to the theory of eigenvalues and eigenfunctions is considered in Sects. 4.2, 3. The later sections deal with examples which will be important in the sequel. The constructions of Sects. 4.2, 3 are independent of the number of spatial dimensions, although the presentation is geared to the three-dimensional case, with an indication each time of the adjustments required for the two-dimensional case.

4.2 Multi-Sheeted Covering Spaces

Let some family of rays, i.e. a family of extremals of the functional $\int(ds/c)$, be a *simple covering*[1] of some subregion Ω_1 of the region Ω, i.e. one and only one ray of the family passes through each point of this subregion.

Consider a continuously differentiable field of vectors $\vec{t}(\tau_1, \tau_2, \tau_3)$ (in the two-dimensional case, $\vec{t}(\tau_1, \tau_2)$)[2] tangent to the rays. Let us assume that the family of rays is such that[3]

$$\sum \tau_\mu^2 = \frac{1}{c^2} \quad , \quad \frac{\partial \tau_\mu}{\partial x_j} = \frac{\partial \tau_j}{\partial x_\mu} \quad .$$

In the neighborhood of each point M there then exists a function τ (the eikonal) whose differential $d\tau$ is equal to $\sum \tau_\mu dx_\mu$. The surfaces (lines) of constant τ are orthogonal to the rays and play the part of wave fronts. Let us call such a family of rays a *normal congruence*.

Let the rays of a normal congruence pass through a caustic. Assume further that the rays going away from the caustic (Fig. 4.1) form a second normal congruence that is likewise a simple covering of some subregion Ω_2 of Ω. Let us join subregion Ω_2 to subregion Ω_1 along the caustic, so that their union forms a manifold. Naturally, one and only one ray passes through each point of this resulting manifold. No point of this manifold covers any point of Ω in the

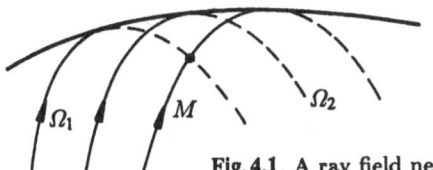

Fig. 4.1. A ray field near a caustic

---

[1] The term is borrowed from topology.

[2] In this chapter we use the notation $x = x_1$, $y = x_2$, $z = x_3$, $t_x = \tau_1, \ldots$

[3] In the terminology of the variational calculus, this family of rays forms a *Mayer family* of extremals of the functional $\int c^{-1} ds$.

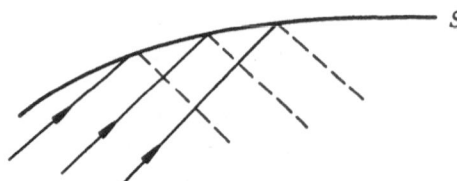

S   **Fig. 4.2.** A ray field near a reflecting surface

shadow region near the caustic; but in the illuminated region the manifold forms a two-sheeted covering of the points of Ω.

Consider next the rays of some normal congruence reflected from a surface *S*, and let the reflected rays also form a normal congruence (Fig. 4.2). The sub-regions that are covered by these normal congruences can be joined along *S* in a manner similar to the above, and then one and only one ray will pass through each point of the resulting manifold. Of particular interest is the case where the rays that touch a caustic are reflected from *S*, with the caustic intersecting *S* at a non-zero angle. In this case, the reflected rays will also have an envelope – a reflected caustic (Fig. 4.3). The incident rays cover the two subregions of Ω that are joined along the caustic *AO*, and are reflected from the section *OC* of the surface *S*, again covering two subregions of Ω that are joined along the reflected caustic *OB*. Let us make two more joins along *OC*, matching the two pairs of subregions together in such a way that the incident rays in one subregion are transformed into reflected rays in the other subregion which is attached to it.

Let us assume that *N* normal congruences which are simple coverings of the subregions Ω_1, Ω_2, ..., $\Omega_N \subset \Omega$, are joined in such a way that no free, unattached boundary remains. The boundaries of the regions Ω_i consist of caustics and reflecting boundaries S_i which are part of the boundary *S* of Ω. We call the resultant manifold a multi-sheeted covering space, analogous to Riemann surfaces in complex function theory and the sets of rays which weave together to form the normal congruences are called *closed congruences of rays*. No matter how far we might continue any ray belonging to a closed congruence of rays, reflecting it where it meets the surface *S* in accordance with the law of reflection, it will never leave the covering space, but only pass through caustics and reflecting surfaces from one normal congruence to another. This closed congruence of rays is the set of rays that is transformed into itself which was mentioned in the preceding section.

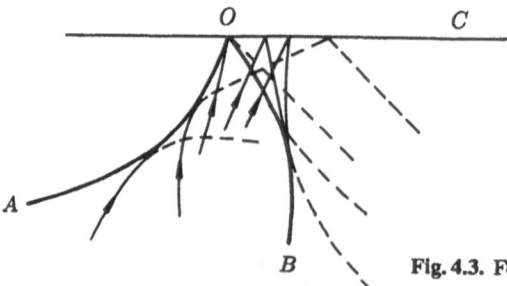

Fig. 4.3. Formation of a caustic by reflected rays

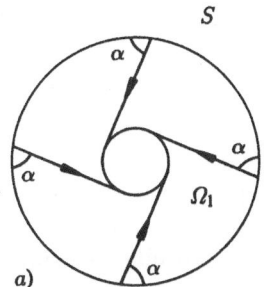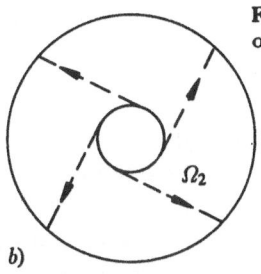

Fig. 4.4a,b. Normal congruences of rays for a circle

Example. Let the region Ω be the circle $r = \sqrt{x_1^2 + x_2^2} < a$, $c(M) = c(x_1, x_2)$ $\equiv 1$. Construct rays all at the same angle $\alpha(< \pi/2)$ to the circumference $r = a$ (Fig. 4.4a). These rays form a normal congruence bounded by the circle $S(r = a)$ and the caustic $r = \sqrt{x_1^2 + x_2^2} = a \cos \alpha$. Continuing the rays beyond the caustic, we get a second normal congruence (Fig. 4.4b). By reflecting the rays of the second normal congruence from S in accordance with the law of reflection, we again arrive at the first normal congruence. By joining regions Ω_1 and Ω_2 ($a \cos \alpha < r < a$) along the reflecting boundary $S(r = a)$ and the caustic $r = a \cos \alpha$, we get a multi-sheeted (in our case a two-sheeted) covering space "interwoven" by the rays of two normal congruences. This covering space enables us to find (see Sect. 4.4) the asymptotic expansion of the eigenvalues of a circle. Other examples of multi-sheeted covering spaces and closed congruences of rays will be found in Sect. 4.5.

Let us return once again to the general case. Let the ray solution $u_0 e^{i\omega\tau}$ be given on any normal congruence, where τ is the eikonal and u_0 satisfies the transport equation (2.4.1). Using formula (2.4.6) and the fact that the phase decreases by a jump of $\pi/2$ when the caustic is crossed, we can continue this ray solution throughout the covering space, except of course near the caustics, where we have seen that the ray solutions become infinite (see Sect. 2.8). At each point M the asymptotic expansion of $u(M)$ is expressed by the formula

$$u(M) \approx \sum u_{0j} \exp(i\omega\tau^{(j)}) \quad, \tag{4.2.1}$$

where the summation is carried out over all points of the covering space that have the same coordinate x_j as point M and lie on different normal congruences. In other words, the number of terms in (4.2.1) is equal to the number of rays of the closed congruence that pass through point M. In Fig. 4.1 for example, two rays pass through M, and corresponding to them are two terms in the sum (4.2.1).

4.3 Single-Valuedness of the Eigenfunctions and Quantization Conditions

The basic premise of the Keller-Rubinow method is that asymptotic formulas for eigenfunctions are given by ray solutions which are single-valued on certain closed congruences of rays. An arbitrary ray solution defined on some normal congruence (which is part of a closed congruence) does not, generally speaking, define a single-valued function throughout the entire covering space after continuation along the rays. When we continue a function along rays, we come back in the end onto the normal congruence on which this ray solution was originally defined. For single-valuedness of a ray solution it is necessary and sufficient that the modulus $|u_0 e^{i\omega\tau}| = |u_0|$ return to its original value, and that $\arg(u_0 e^{i\omega\tau})$ acquire an increment that is a multiple of 2π. As implied by (2.4.6), we have for $|u_0|$ the formula $|u_0| = |\psi_0|\sqrt{c/J}$, where ψ_0 is a function that depends only on the ray, c is the velocity of propagation and J is the divergence. If the closed congruence of rays is such that each ray is transformed into itself, then the function ψ_0 can be arbitrarily assigned, and $|u_0|$ will be single-valued. Otherwise a ray, in leaving some point A of wave front AB (Fig. 4.5) will, after being reflected several times and passing through caustics, return to some point B on the same wave front that is, generally speaking, different from point A. For $|u_0|$ to be single-valued, the equation

$$|u_0(B)| = |u_0(A)|\sqrt{\frac{c(B)}{c(A)}}\sqrt{\frac{|J(A)|}{|J(B)|}}$$

must be satisfied. This equality will hold if $\psi_0 = \text{const}$, which is what we will assume.

To ensure that $u_0 e^{i\omega\tau}$ is single-valued, it is necessary in addition that the increment of $\arg(u_0 e^{i\omega\tau})$ on any closed curve lying in a covering space be equal to a multiple of 2π. Let us return to the vector field $\vec{t}(\tau_1, \tau_2, \tau_3)$ [$\vec{t}(\tau_1, \tau_2)$ in the two-dimensional case] introduced at the beginning of Sect. 4.2. This vector field is locally a potential, i.e., in the vicinity of an arbitrary point there is a function $\tau(x_1, x_2, x_3)$ (the eikonal) such that $\nabla\tau = \vec{t}(\tau_1, \tau_2, \tau_3)$ [$(\nabla\tau = \vec{t}(\tau_1, \tau_2)$ in the two-dimensional case]. A ray passes through each point of the covering space; the gradient of the eikonal $\nabla\tau$ is directed along the ray, where $(\nabla\tau)^2 = 1/c^2$. Although the eikonal itself is not uniquely defined on a covering space, the components of its gradient are single-valued functions.

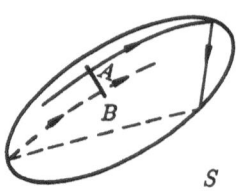

Fig. 4.5. Ray of a closed congruence passing through the original wavefront

Now we will derive a formula for the increment of $\arg(u_0\,e^{i\omega\tau})$ on an arbitrary contour Γ which will be important in what follows. We will assume that some direction of travel has been chosen on this contour[4]. To calculate the increment of $\arg(u_0\,e^{i\omega\tau})$ on Γ we will break Γ down into sufficiently small arcs, calculate the increments of $\arg(u_0\,e^{i\omega\tau})$ on these arcs, and add the results.

First consider the case where $AB(\subset\Gamma)$ is an arc lying completely on a normal congruence. Here, A is the beginning of the arc, and B is the end. Then the increment of $\arg(u_0\,e^{i\omega\tau})$ reduces to the increment of $\arg(e^{i\omega\tau})$, which in turn reduces to the integral

$$\omega\int_A^B d\tau = \omega\int_{AB}\sum\tau_\mu dx_\mu = \omega\int_A^B\sum\tau_\mu dx_\mu \quad . \tag{4.3.1}$$

(Recall that by virtue of the equalities $(\partial\tau_\mu/\partial x_j)=(\partial\tau_j/\partial x_\mu)$ this integral does not change if the contour joining A and B is deformed). Consider now another case where the points A and B belong to different normal congruences joined together along a caustic (Fig. 4.6). Draw the wave front K_1K_2 passing through A, and the ray passing through B. Let A_1 be the intersection point of this wave front and the ray passing through B. On the wave front K_1K_2, $\arg(u_0\,e^{i\omega\tau})=$ const. Along the ray, the phase decreases by $\pi/2$ on passing through the caustic (Sect. 2.8), and therefore

$$\arg(u_0\,e^{i\omega\tau})\Big|_A^B = \arg(u_0\,e^{i\omega\tau})\Big|_{A_1}^B = \omega\int_{A_1CB}\sum\tau_\mu dx_\mu - \frac{\pi}{2}\varepsilon_c$$

$$= \omega\int_{AA_1CB}\sum\tau_\mu dx_\mu - \frac{\pi}{2}\varepsilon_c = \omega\int_{ADB}\sum\tau_\mu dx_\mu - \frac{\pi}{2}\varepsilon_c \quad . \tag{4.3.2}$$

Here the integral over the ray A_1CB can be replaced by the integral over the arc $ADB\subset\Gamma$ by virtue of the fact that AA_1CB can be continuously deformed

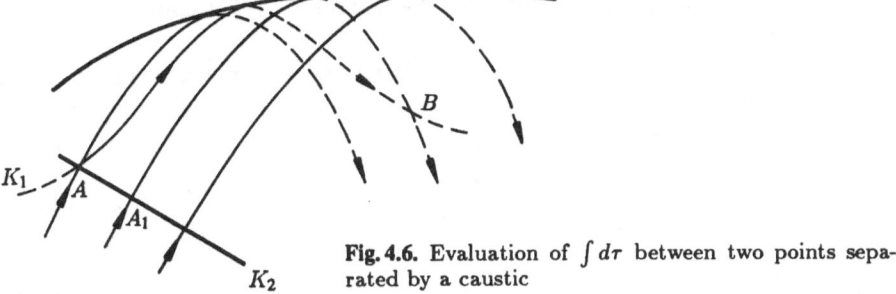

Fig. 4.6. Evaluation of $\int d\tau$ between two points separated by a caustic

---

[4] In other words, we will treat Γ as a *one-dimensional oriented closed chain* or *cycle* on the covering space (considered as a compact manifold) [4.1, 2].

into ADB. Obviously, the integral with respect to AA_1 is equal to zero: the vectors $d\vec{r}(dx_1, dx_2, dx_3)$ and $\vec{t}(\tau_1, \tau_2, \tau_3)$ are orthogonal on AA_1. The number $\varepsilon_c = 1$ if the rays of the normal congruence that contains point A are transformed to the normal congruence that contains B (Fig. 4.6). In this case we will say that the contour Γ passes through the caustic *along* the rays. In the case where the rays of a normal congruence (α) pass through the caustic into a normal congruence (β), the point A lies on the congruence (β), and B is on (α), we have $\varepsilon_c = -1$, and we will say that the contour Γ crosses the caustic *against* the rays. Let us turn now to the computation of $\arg(u_0\, e^{i\omega\tau})$ for rays reflected from a boundary S.

Let the boundary condition on S have the form (4.1.3). The ray solution $u_{0j} \exp(i\omega\tau^{(j)})$ that represents the incident wave will be reflected and transformed into the reflected wave $u_{0m} \exp(i\omega\tau^{(m)})$. The rays of the normal congruence that correspond to the incident wave will be reflected in accordance with the law of geometrical optics from S and converted to rays corresponding to the reflected wave. Boundary condition (4.1.3) implies

$$\tau^{(m)}\big|_S = \tau^{(j)}\big|_S \quad , \qquad u_{0j}\big|_S = u_{0m}\big|_S \quad , \tag{4.3.3}$$

whence

$$\arg(u_0\, e^{i\omega\tau})\bigg|_A^B = \omega \int_A^B \sum \tau_\mu \, dx_\mu \quad . \tag{4.3.4}$$

If the boundary condition on S has the form (4.1.2), then the formulas (4.3.3) are replaced by

$$\tau^{(m)}\big|_S = \tau^{(j)}\big|_S \quad , \qquad u_{0j}\big|_S = -u_{0m}\big|_S = u_{0m}\, e^{i\pi}\big|_S \quad . \tag{4.3.5}$$

By analogy with the case of a caustic, we have from (4.3.5)

$$\arg(u_0\, e^{i\omega\tau})\bigg|_A^B = \omega \int_A^B \sum \tau_\mu \, dx_\mu - \pi\varepsilon_r \quad , \tag{4.3.6}$$

where $\varepsilon_r = 1$ if the contour Γ or (what amounts to the same thing) the arc AB crosses the reflecting surface S along the rays, and $\varepsilon_r = -1$ if contour Γ crosses S against the rays.

Summing the increments of $\arg(u_0\, e^{i\omega\tau})$ over all arcs into which the contour Γ was broken up, and using (4.3.1, 2, 4, 6), we arrive at last at an expression for the increment of $\arg(u_0\, e^{i\omega\tau})$ over Γ:

$$\arg(u_0\, e^{i\omega\tau})\big|_\Gamma = \omega \int_\Gamma \sum \tau_\mu \, dx_\mu - \frac{\pi}{2} l_c - \pi l_r \quad , \tag{4.3.7}$$

where

$$l_c = l_c' - l_c'' \quad , \qquad l_r = l_r' - l_r'' \quad , \tag{4.3.8}$$

l_c' is the number of transitions of the contour Γ through the caustic along the rays, l_c'' is the number of transitions of Γ through the caustic against the rays, and

$l'_r (l''_r)$ is the number of transitions of Γ through the reflecting surface S along (against) the rays if the boundary condition takes form (4.1.2). If the boundary condition has form (4.1.3), then $l_r = 0$. We call l_c the *caustic index*, and l_r the *reflection index*.

Let Γ_1 and Γ_2 be two closed contours such that they can be transformed into one another by a continuous deformation. Neither the integral $\int \sum \tau_\mu dx_\mu$, nor the indices l_c, l_r change with deformation of the contour[5], therefore the increments of $\arg(u_0 e^{i\omega\tau})$ over Γ_1 and Γ_2 are the same:

$$\arg(u_0 e^{i\omega\tau})|_{\Gamma_1} = \arg(u_0 e^{i\omega\tau})|_{\Gamma_2} \quad . \tag{4.3.9}$$

There exist only a finite number of basis contours Γ_s, $s = 1, 2, \ldots, f$ such that any closed contour Γ is a linear combination of them with integer coefficients c_s[6]

$$\Gamma = \sum_{s=1}^{f} c_s \Gamma_s \quad . \tag{4.3.10}$$

Equation (4.3.10) is to be understood in the following sense: the contour Γ can be obtained by continuous deformation from the composite contour made up of contour Γ_1 repeated c_1 times, contour Γ_2 repeated c_2 times, ..., and contour Γ_f repeated c_f times. If $c_j < 0$, then we must also reverse the direction in which the contour Γ_j is traversed.

We readily get from (4.3.9) that

$$\arg(u_0 e^{i\omega\tau})|_\Gamma = \sum_{s=1}^{f} c_s \arg(u_0 e^{i\omega\tau})|_{\Gamma_s}. \tag{4.3.11}$$

In view of (4.3.7, 11), the requirement that for any contour Γ

$$\arg(u_0 e^{i\omega\tau})|_\Gamma = 2\pi n \quad (n = \text{ integer}) \quad ,$$

is equivalent to the conditions

$$\omega \int_{\Gamma_s} \sum \tau_\mu dx_\mu - \pi l_r^{(s)} - \frac{\pi}{2} l_c^{(s)} = 2\pi n_s \quad , (s = 1, 2, \ldots, f) \quad , \tag{4.3.12}$$

where n_s is a whole number, and Γ_s is a basis contour. Conditions (4.3.12) are called *quantization conditions*.

We shall assume that the basis contours Γ_s are linearly independent in the sense that none of them can be obtained by a continuous deformation of any linear combination of the others. Then the numbers n_s in (4.3.12) can be specified independently of each another. If there is given a fixed closed congruence of rays, then specifying one of the integers n_s will, as can be seen from (4.3.12),

---

[5] That the indices l_c and l_r remain constant as the contour is continuously deformed is implied by the general topological theorem on homologous invariance of the *intersection number*, which is discussed at the end of this section.

[6] This is implied by the fact that the fundamental group of a compact manifold has a finite number of generators [4.1].

determine ω, and the remaining $f-1$ quantization conditions will not generally be satisfied. Let us assume that rather than a single closed congruence of rays, we have an $(f-1)$-parameter family of such congruences. Furthermore, for each closed congruence of this family with arbitrary values of these parameters, let the number of linearly independent basis contours be equal to f. These basis contours, and hence the left-hand sides of (4.3.12) as well, will depend on the $f-1$ parameters that define the congruence. Hence, by arbitrarily setting the whole numbers n_s in expressions (4.3.12), we will be able to treat these relationships as a system of f equations in f unknowns: one of these unknowns is ω, and the others are the $f-1$ parameters that define the closed congruence. This is the case, for instance, with the closed congruence of the circle that was constructed in Sect. 4.2, where joining the two concentric annuli along their boundaries leads to a two-sheeted space, which can obviously be transformed to a torus by a continuous deformation. On the torus there are two linearly independent basis contours (Γ_1 and Γ_2 in Fig. 4.7), giving two quantization conditions. From the two equations we find ω and the single parameter (the angle α) that determines the congruence under investigation (for further details, see Sect. 4.4).

In closing this section, let us call attention to the relationship between the theory presented here and certain topological concepts[7]. First of all, in a neighborhood of every point (including those that belong to caustics and reflecting surfaces), a ray coordinate system (α, β, τ) can be introduced. (Here, as usual, the parameters α, β characterize rays, and the eikonal τ characterizes the points on a ray). In this way a multi-sheeted covering space can be treated as a differentiable manifold (obviously compact). Assuming that the hypotheses concerning caustics given in Sect. 3.3 are also valid here, we have on a caustic

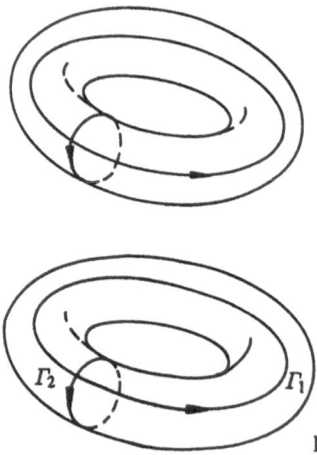

Fig. 4.7. Independent basis contours on the torus

---

[7] The reader who is unacquainted with the principles of topology may skip to the end of this section without loss of understanding of the rest of the development.

$$J = \frac{D(x_1, x_2, x_3)}{D(\alpha, \beta, \tau)} = 0 \quad , \qquad \frac{\partial J}{\partial \tau}\bigg|_{J=0} \neq 0 \quad .$$

Thus the set of points $J = 0$ may be treated as a two-dimensional closed chain embedded orientably in the covering space. The orientable nature of this embedding stems from the fact that the tangent vectors to the rays form a continuous vector field on the caustic, nowhere tangent to it. Let us recall that these considerations apply to the coordinate system α, β, τ, and that rays (α, β) = const are not tangent to the submanifold $J = 0$ by virtue of the condition $\partial J/\partial \tau \neq 0$. The side of the caustic towards which the tangents to the rays point can be taken as the "positive" side of the caustic. The caustic index $l_c(\Gamma)$, as directly implied by the definition, is a (Kronecker-Poincaré) intersection number of the one-dimensional oriented closed chain Γ and a caustic embedded orientably in the manifold. From the well known topological theorem on homologous invariance of the intersection number we get the equality

$$l_c(\Gamma) = l_c(\Gamma') \quad , \tag{4.3.13}$$

which holds when the chains Γ and Γ' are homologous. Moreover, (4.3.13) is valid when Γ and Γ' are homotopic, i.e. if they can be transformed into one another by continuous deformation. This entire discussion also applies to the reflection index $l_r(\Gamma)$, which is also an intersection number (of the chain Γ and a reflecting surface).

In all cases where the technique of Sects. 4.1–3 has been successfully applied, the multi-sheeted covering space has been a multidimensional torus, and as we know, homologous contours for a torus can be transformed into one another by continuous deformation. As shown by *Arnol'd* [4.3] in the case where the number of basis contours f is equal to the number of spatial dimensions, and closed congruences of rays form an $(f-1)$-parameter family, the covering space will, with certain refining assumptions, necessarily be a multidimensional torus. The conditions of Arnold's theorem have been satisfied for all problems which have been solved by this method so far.

4.4 Eigenvalues and Eigenfunctions of a Circle

Let the region Ω be a circle of radius a, and let the velocity c be equal to unity. Let us consider the eigenfunctions of the circle for the case of boundary condition (4.1.2). The case of boundary condition (4.1.3) is treated in a completely analogous fashion and we will only mention it briefly.

The closed congruence of rays that covers the annulus $a \cos \alpha \leq r \leq a$ (see Sect. 4.2) enables us to find the shortwave asymptotic expansion of the eigenfunctions of the circle. The covering space here consists of two copies of the annulus $a \cos \alpha \leq r \leq a$ joined along the circles $r = a$ and $r = a \cos \alpha$.

Let us now choose the basis contours Γ_1 and Γ_2. Let Γ_1 (Fig. 4.8) be the circle $r = a \cos \alpha = a_0$, the direction of travel on Γ_1 taken to be that induced

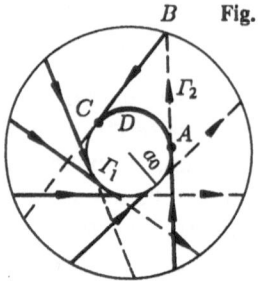

B Fig. 4.8. Independent basis contours on a covering space for a circle

on the contour by the rays. For Γ_2 we take the closed curve formed by the two rays AB and BC (which lie on different sheets) and by the arc CDA of circle a_0 that connects them. For Γ_1, both the caustic and reflection indices are equal to zero[8], and the differential form $\Sigma \tau_\mu dx_\mu$ is the same as the differential arc length. The first quantization condition gives

$$k \int_{\Gamma_1} \sum \tau_\mu dx_\mu = k \int_{\Gamma_1} ds = 2\pi a_0 k = 2\pi q$$
$$(q = 0, 1, 2, \ldots) \quad , \tag{4.4.1}$$

where $k = \omega/c = \omega$ is an eigenvalue.

For basis contour Γ_2, the caustic index l_c is equal to 1 (the contour crosses the caustic once along the rays in passing from one normal congruence to another). The reflection index l_r is obviously also equal to one.

The differential form $\Sigma \tau_\mu dx_\mu$ on the rays and the curve CDA is the same as the differential arc length ds (note, however, that while the eikonal increases while traversing the rays AB and BC, we are traversing the section CDA of the caustic "against the grain", and the eikonal decreases along this curve). The second quantization condition thus gives

$$k \int_{\Gamma_2} \sum \tau_\mu dx_\mu - \frac{\pi}{2} l_c - \pi l_r = 2\pi p \quad (p = 0, 1, \ldots) \quad , \quad \text{or}$$

$$k \left(\sqrt{a^2 - a_0^2} - a_0 \arccos \frac{a_0}{a} \right) = \pi \left(p + \frac{3}{4} \right) \quad . \tag{4.4.2}$$

Eliminating a_0 from equations (4.4.1 and 2), we get the equation for the eigenvalues k_{pq}

$$\sqrt{(k_{pq}a)^2 - q^2} - q \arccos \frac{q}{k_{pq}a} = \pi \left(p + \frac{3}{4} \right) \quad . \tag{4.4.3}$$

In the case of boudnary condition (4.1.3) the equation for the eigenvalues is

---

[8] The fact that the caustic index is equal to zero becomes obvious if Γ_1 is moved slightly from the caustic to a position internal to one of the rings.

76

$$\sqrt{(k_{pq}a)^2 - q^2} - q\arccos\frac{q}{k_{pq}a} = \pi\left(p + \frac{1}{4}\right) \ . \tag{4.4.4}$$

The transcendental equation (4.4.3) [or (4.4.4)] has simple approximate solutions in the two limiting case $p \gg 1$, $q = 0, 1, \ldots$ and $q \gg 1$, $p = 0, 1, \ldots$. Let us examine these limiting cases for equation (4.4.3).

1) When $p \gg 1$ and $q = 0, 1, 2, \ldots$, it is clear that $\pi p \sim k_{pq}a \gg 1$. Expanding $\sqrt{(k_{pq}a)^2 - q^2}$ and $\arccos(q/k_{pq}a)$ in powers of the ratio $q/k_{pq}a$ and limiting ourselves to the principal terms, we get

$$k_{pq}a = \pi(p + q/2 + 3/4) \ , \quad p \gg 1 \ , q = 0, 1, \ldots \ .$$

Determining k_{pq} from equation (4.4.1) we find

$$a_0 = \frac{a}{\pi}\frac{q}{p + q/2 + 3/4} \ .$$

This implies that the radius of the caustic a_0 that corresponds to the eigenvalues k_{pq} satisfies the inequality $a_0 \ll a$, and the closed congruence of rays covers almost the entire circle $r \leq a$[9].

2) When $q \gg 1$ and $p = 0, 1, \ldots$ the right-hand side of equation (4.4.3) is of the order of one or so, and we must assume that $k_{pq}a \sim q$. Let us expand the left-hand side of equation (4.4.3) in powers of the difference $1 - (q/k_{pq}a)$. Retaining the principal term of the expansion, we get

$$\frac{2^{3/2}}{3}k_{pq}a\left(1 - \frac{q}{k_{pq}a}\right)^{3/2} = \pi\left(p + \frac{3}{4}\right) \ ,$$

whence

$$k_{pq}a = q + \frac{q^{1/3}}{2^{1/3}}\left[\frac{3\pi}{2}\left(p + \frac{3}{4}\right)\right]^{2/3} \ , \tag{4.4.5}$$

and the corresponding value of a_0 is equal to

$$a_0 = a\left\{1 - \frac{1}{2^{1/3}}\left[\frac{3\pi}{2}\left(p + \frac{3}{4}\right)\right]^{2/3}\frac{1}{q^{2/3}}\right\} \ .$$

Thus the caustic $r = a_0$ is located in the neighborhood of the boundary $r = a$ of the circle, and the rays cover a narrow ring of width $\sim (a/2^{1/3})[(3\pi/2)(p + \frac{3}{4})]^{2/3}1/q^{2/3}$. It should be noted that on the caustic itself and in its vicinity the original representation of the eigenfunction in the form of the sum (4.2.1) does not hold. Since $a_0 \sim a$ in the case under consideration, representation (4.2.1) is also invalid on the circle $r = a$ proper, where the boundary conditions are satisfied. This situation results in the factor $[(3\pi/2)(p + \frac{3}{4})]^{2/3}$ in (4.4.5) not quite

---

[9] Strictly speaking, our constructions are not legitimate for small a_0 since as $a_0 \to 0$ the rays under consideration are focused in the center of the circle. However, an investigation of the explicit formulas for the eigenfunctions of a circle shows that the expressions derived here for a_0 and k_{pq} are still valid.

being accurate. As we shall see below (Sect. 7.4), this factor should actually be equal to the pth root t_p of the Airy function $v(-t)$. When p becomes large, we have the asymptotic behavior

$$t_p \approx \left[\frac{3\pi}{2} \left(p + \frac{3}{4} \right) \right]^{2/3} \quad ,$$

which shows that the accuracy of (4.4.5) increases with increasing p (so long as the condition $q \gg p$ is met).

Let us go on to construction of the eigenfunctions for the case of boundary condition (4.1.2). To do this, we first need to calculate $\tau^{(j)}$ and u_{0j}, $j = 1, 2$. Let us introduce the polar coordinates (r, ϕ) and assume that $\tau^{(1)}(a_0, 0) = 0$. Under this condition we calculate $\tau^{(1)}(r_0, \phi_0)$ and $\tau^{(2)}(r_0, \phi_0)$ at an aribitrary point $M(r_0, \phi_0)$ located in the ring $a_0 \leq r \leq a$. The calculation of $\tau^{(1)}$ reduces to the evaluation of the *optical distance* (the extremal value of the functional $\int ds$ of geometric optics) between the points $A(a_0, 0)$ and $M(r_0, \phi_0)$. This distance will be the sum of the arc lengths of the caustic AB and the segment BM tangent to the circle $r = a_0$ (Fig. 4.9). Obviously,

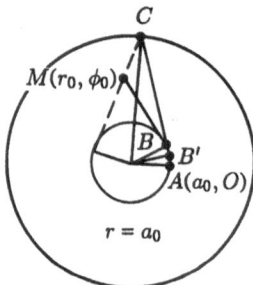

Fig. 4.9. Calculation of the optical distance for the ray field of a circle

$$\tau^{(1)}(r_0, \phi_0) = a_0 \left(\phi_0 - \arccos \frac{a_0}{r_0} \right) + \sqrt{r_0^2 - a_0^2} \quad .$$

To calculate $\tau^{(2)}(r_0, \phi_0)$ we must find the optical distance between the A and M for a ray reflected at the circle $r = a$. Obviously, $\tau^{(2)}(r_0, \phi_0) = AB' + B'C + CM$; whence

$$\tau^{(2)}(r_0, \phi_0) = a_0 \left[\phi_0 - \left(2 \arccos \frac{a_0}{a} - \arccos \frac{a_0}{r_0} \right) \right]$$
$$+ \sqrt{a^2 - a_0^2} + \left(\sqrt{a^2 - a_0^2} - \sqrt{r_0^2 - a_0^2} \right) \quad .$$

Using (4.4.2), we may write the result as

$$\tau^{(2)}(r_0, \phi_0) = a_0 \left(\phi_0 + \arccos \frac{a_0}{r_0} \right) - \sqrt{r_0^2 - a_0^2} + \frac{2\pi}{k} \left(p + \frac{3}{4} \right) \quad .$$

Clearly, the equations

$$\tau^{(1)}(r, q) = \tau^{(1)}(r_0, \phi_0) \quad ,$$
$$\tau^{(2)}(r, q) = \tau^{(2)}(r_0, \phi_0) \tag{4.4.6}$$

are equations for two involutes of the circle $r = a_0$ which intersect at the point r_0, ϕ_0. The radii of curvature of the curves (4.4.6) at (r_0, ϕ_0) are equal to each other and given by $\sqrt{r_0^2 - a_0^2}$. Since the amplitude u_0 varies as the inverse square root of the radius of curvature of the wave front (Sects. 2.3, 4), we have

$$u_{01}(r_0, \phi_0) = \frac{A_1}{\sqrt[4]{r_0^2 - a_0^2}} \quad ,$$

$$u_{02}(r_0, \phi_0) = \frac{A_2}{\sqrt[4]{r_0^2 - a_0^2}} \quad ,$$

where A_1, A_2 are some constants (note that a caustic is the locus of the centers of curvature for the wave fronts).

Boundary condition (4.1.2) gives $A_2 = A_1 e^{i\pi}$. Let us take $A_1 = (e^{-i\pi/4})/(2k^{1/2})$. Omitting the unimportant subscript "0" in the coordinates of an aribitrary point M, and considering the relation $k_{pq} a_0 = q$, we get

$$u_{01} e^{ik_{pq}\tau^{(1)}} + u_{02} e^{ik_{pq}\tau^{(2)}} = \frac{1}{\sqrt[4]{(k_{pq}r)^2 - q^2}}$$

$$\times \cos \left[\sqrt{(k_{pq}r)^2 - q^2} - q \arccos \frac{q}{k_{pq}r} - \frac{\pi}{4} \right] e^{iq\phi} \quad . \tag{4.4.7}$$

If we consider rays that coincide with the chords we have constructed in the ring $a_0 \leq r \leq a$ and extend them clockwise, then these rays lead us to a formula analogous to (4.4.7):

$$u_{03} e^{ik_{pq}\tau^{(3)}} + u_{04} e^{ik_{pq}\tau^{(4)}} = \frac{1}{\sqrt[4]{(k_{pq}r)^2 - q^2}}$$

$$\times \cos \left[\sqrt{(k_{pq}r)^2 - q^2} - q \arccos \frac{q}{k_{pq}r} - \frac{\pi}{4} \right] e^{-iq\phi} \quad . \tag{4.4.8}$$

Adding and subtracting (4.4.7 and 8), we arrive at a real expression for the eigenfunctions:

$$u_{pq}(r, \phi) = \frac{1}{\sqrt[4]{(k_{pq}r)^2 - q^2}}$$

$$\times \cos \left[\sqrt{(k_{pq}r)^2 - q^2} - q \arccos \frac{q}{k_{pq}r} - \frac{\pi}{4} \right] \begin{Bmatrix} \cos q\phi \\ \sin q\phi \end{Bmatrix} . \tag{4.4.9}$$

Equation (4.4.9) gives representations for the eigenfunctions in the ring $a_0 \leq r \leq a$ which is covered by rays, i.e., for $k_{pq}r > q$. To get an expression for the eigenfunctions behind the caustic (in the circle $r < a_0$) and in the vicinity of the caustic $r = a_0$ we must use the results of Chap. 3. In the case of boundary condition

(4.1.3) the expressions for the eigenfunctions are obtained in the same way. The asymptotic formulas (4.4.7–9) also hold for them, but the eigenvalues k_{pq} must be found from (4.4.4), rather than from the transcendental equation (4.4.3) as was the case for boundary condition (4.1.2).

Let us compare the resultant formulas with the exact values of the eigenfunctions. The solution of the equation

$$(\Delta + k^2)u = 0$$

in polar coordinates can be expressed in terms of functions having the form

$$u = \text{const } J_q(kr) \begin{cases} \cos q\phi, \\ \sin q\phi \end{cases}, \tag{4.4.10}$$

where J_q is a Bessel function of order q. The eigenvalues k_{pq} are determined from the equations

$$J_q(k_{pq}a) = 0 \quad,\text{or} \quad J'_q(k_{pq}a) = 0 \quad, \tag{4.4.11}$$

for boundary conditions (4.1.2 and 3), respectively.

When $kr \gg 1$ and $q = 0, 1, 2, \ldots < kr$, we have for the Bessel functions the asymptotic formula

$$J_q(kr) = \sqrt{\frac{2}{\pi}}[(kr)^2 - q^2]^{-1/4}\left\{\cos\left[\sqrt{(kr)^2 - q^2}\right.\right.$$

$$\left.\left. - q\arccos\frac{q}{kr} - \frac{\pi}{4}\right] + O\left(\frac{1}{\sqrt{(kr)^2 - q^2}}\right)\right\} . \tag{4.4.12}$$

Substituting (4.4.12) for the Bessel function into (4.4.11), we get the equations

$$\sqrt{(k_{pq}a)^2 - q^2} - q\arccos\frac{q}{k_{pq}a} = \pi\left(p + \frac{3}{4}\right) \quad,$$

$$\sqrt{(k_{pq}a)^2 - q^2} - q\arccos\frac{q}{k_{pq}a} = \pi\left(p + \frac{1}{4}\right) \quad,$$

for the eigenvalues k_{pq} in the first approximation. These equations agree completely with equations (4.4.3 and 4) which were obtained by the ray method. If we substitute (4.4.12) into (4.4.10) and substitute k_{pq} for k, we arrive at the asymptotic formula (4.4.9). Thus when $q < k_{pq}r$, we have complete agreement between the formulas found by the ray method and the asymptotic formulas resulting from the exact solution of the problem. However, comparing (4.4.9) at $q \sim k_{pq}r$ with the exact solution we find that the ray method is limited to some extent. When $q \sim k_{pq}r$, the asymptotic formula (4.4.12) for the Bessel function breaks down. In this case we must use an asymptotic formula for the Bessel function which employs an Airy function [4.4],

$$J_q(kr) \approx \frac{1}{\sqrt{\pi}}\left(\frac{2}{kr}\right)^{1/3} v(t) ,$$

$$t = \left(\frac{2}{kr}\right)^{1/3}(q - kr) .$$

To determine the eigenvalues k_{pq} we get (in the first approximation) the equations

$$\left(\frac{2}{k_{pq}a}\right)^{1/3}(q - k_{pq}a) = -t_p , \quad p = 0, 1, \ldots , \tag{4.4.13}$$

and

$$\left(\frac{2}{k_{pq}a}\right)^{1/3}(q - k_{pq}a) = -t'_p , \quad p = 0, 1, \ldots . \tag{4.4.14}$$

In these equations $-t_p$ and $-t'_p$ are the pth roots of the Airy function $v(t)$ and its derivative (see Appendix A.1).

For sufficiently large p, the asymptotic formulas

$$t_p = \left[\frac{3\pi}{2}\left(p + \frac{3}{4}\right)\right]^{2/3} + O\left(\frac{1}{p}\right) ,$$

$$t'_p = \left[\frac{3\pi}{2}\left(p + \frac{1}{4}\right)\right]^{2/3} + O\left(\frac{1}{p}\right)$$

apply. Solving equations (4.4.13 and 14) we get for the eigenvalues k_{pq} the following formulas:

$$k_{pq} = \frac{1}{a}\left[q + t_p\left(\frac{q}{2}\right)^{1/3} + \ldots\right]$$

$$k_{pq} = \frac{1}{a}\left[q + t'_p\left(\frac{q}{2}\right)^{1/3} + \ldots\right] .$$

If in the first of these t_p is replaced by its asymptotic expansion, we obtain formula (4.4.5) which was derived by the ray method. The asymptotic expansion of the eigenfunctions for this case, in contrast to (4.4.9), is represented in terms of the Airy function. For sufficiently large p, we can replace the Airy function by its asymptotic expression, and then we get the formula resulting from (4.4.9) under the condition that $k_{pq}a \sim q$. Thus, the formulas obtained by the ray method agree with the corresponding asymptotic formulas resulting from the exact solution only for sufficiently large p. The reason that formulas obtained using small p do not agree with the exact asymptotic formulas has already been pointed out.

4.5 Eigenvalues of an Ellipse

In this section we examine a second example of construction of eigenvalues by the ray method. We will get the eigenvalues of the Laplacian for an ellipse where one of the conditions (4.1.2 or 3) is satisfied on the boundary S. As in Sect. 4.4 we will take $c = 1$ and set $\omega = \omega/c = k$. We introduce the elliptical coordinate system (Fig. 4.10)

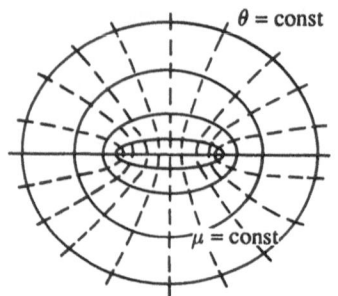

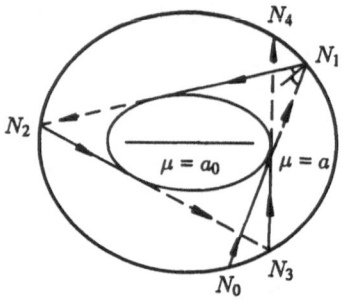

Fig. 4.10. Elliptical coordinate system   Fig. 4.11. Reflected rays in an ellipse

$$x_1 = h \cosh \mu \cos \theta \ , \quad x_2 = h \sinh \mu \sin \theta \ ,$$
$$0 \leq \mu < \infty \ , \qquad\qquad 0 \leq \theta \leq 2\pi \quad . \tag{4.5.1}$$

The coordinates lines are the ellipses

$$\mu = \text{const}$$

and hyperbolas

$$\theta = \text{const} \quad .$$

The square of an element of length is given by the formula

$$ds^2 = h^2 (\cosh^2 \mu - \cos^2 \theta)(d\theta^2 + d\mu^2). \tag{4.5.2}$$

Let us assume that $\mu = a$ defines the ellipse S bounding the region under consideration.

First of all, we will derive the eigenvalues that correspond to eigenfunctions concentrated in a strip adjacent to the boundary of the region. The constructions leading to the corresponding system of equations will be completely analogous to those of the preceding section.

Consider some chord $N_0 N_1$ of the ellipse $\mu = a$ (Fig. 4.11). Let this chord be tangent to an ellipse $\mu = a_0$ which is confocal with ellipse $\mu = a$. Take the chord $N_0 N_1$ as the initial ray. We now construct the chord $N_1 N_2$ in accordance with the law of reflection (the angles formed by the chords $N_0 N_1$ or $N_1 N_2$ and the normal to the ellipse at N_1 are equal). The chords $N_2 N_3, N_3 N_4, \ldots$, are constructed in the same fashion. Obviously the chords $N_1 N_2, N_2 N_3, \ldots$, describe rays that arise as a result of successive reflections of the initial ray $N_0 N_1$ at the points $N_1, N_2, \ldots$. In order to construct a closed congruence from the set of rays $N_i N_{i+1}$, we must prove that all rays $N_i N_{i+1}$ are tangent to the same ellipse $\mu = a_0$.

We will give a proof of this assertion[10] based on integration of the eikonal equation in elliptical coordinates (note that the technique used here plays an

---

[10] Other proofs can be found in the books by Salmon [4.5] and Vainshtein [4.6].

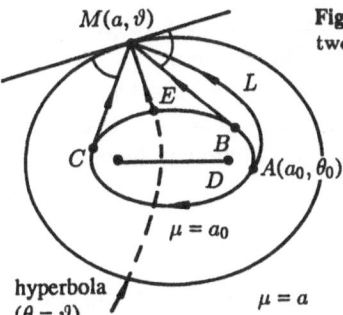

$M(a, \vartheta)$

L

E

B

C

$A(a_0, \theta_0)$

D

$\mu = a_0$

hyperbola
$(\theta = \vartheta)$

$\mu = a$

Fig. 4.12. Calculation of the increment of the eikonal between two points in the ellipse

important role in the study of the equation for saddle points corresponding to incident and reflected waves in the diffraction problem for an ellipse).

We will actually prove the following assertion which is equivalent to the one formulated above:

Theorem. *If* $M(a, \vartheta)$ *is an arbitrary point on the ellipse* $\mu = a$ *(Fig. 4.12) and* MB *and* MC *are tangents drawn from this point to the confocal ellipse* $\mu = a_0$, *then the angles that these tangents make with the ellipse* $\mu = a$ *are equal to each other.*

Let us first construct a solution of the eikonal equation $|\nabla \tau|^2 = 1$ that is real in the elliptical ring $a_0 \leq \mu \leq a$. In elliptical coordinates, the eikonal equation, as implied by (4.5.1, 2) takes the form

$$\left(\frac{\partial \tau}{\partial \theta} \right)^2 + \left(\frac{\partial \tau}{\partial \mu} \right)^2 = h^2 (\cosh^2 \mu - \cos^2 \theta)$$

(see Appendix A.2). Setting

$$\tau(\mu, \theta) = F(\mu) + \Phi(\theta) \quad , \tag{4.5.3}$$

we get

$$F'^2(\mu) + \Lambda^2 - h^2 \cosh^2 \mu = 0 \ ,$$
$$\Phi'^2(\theta) - \Lambda^2 + h^2 \cos^2 \theta = 0 \quad . \tag{4.5.4}$$

Here Λ is a separation constant. We can ensure that the solution (4.5.3) is real in $a_0 \leq \mu \leq a$ by setting

$$\Lambda = h \cosh a_0 \quad .$$

As solutions of (4.5.4) we take

$$F(\mu) = +h \int_{a_0}^{\mu} \sqrt{\cosh^2 \mu - \cosh^2 a_0} \, d\mu \ ,$$

$$\Phi(\theta) = +h \int_{\theta_0}^{\theta} \sqrt{\cosh^2 a_0 - \cos^2 \theta} \, d\theta. \tag{4.5.5}$$

The choices of sign in (4.5.5) corresponds to a family of rays tangent to the ellipse $\mu = a_0$ and going away from it towards the ellipse $\mu = a$.

Let us now calculate the increment of the eikonal τ as defined by (4.5.3 and 5) between a fixed point $A(a_0, \theta_0)$ located on the ellipse $\mu = a_0$ and the point $M(a, \vartheta)$. Denoting the unknown increment by $\tau|_A^M$ we obviously have

$$\tau|_A^M = \int_l (\nabla \tau \cdot d\vec{s}) \equiv \int_l (\nabla \tau \cdot \vec{s}) ds \quad ,$$

where l is a contour of integration joining A and M, and $\vec{s}$ is a unit vector tangent to l. The integral $\int_l (\nabla \tau \cdot d\vec{s})$, as we know (see Sect. 4.3), is invariant with respect to continuous deformations of l (with A and M fixed).

Let the contour l go from A to M such that the angle θ increases along l, but remains less than 2π. Let us combine the contour l (see Fig. 4.12) with the coordinate lines ABE ($\mu = a_0$) and EM ($\theta = \vartheta$). Then

$$\tau|_A^M = \int_{\theta_0}^{\vartheta} \tau_\theta \, d\theta + \int_{a_0}^{a} \tau_\mu \, d\mu$$

and by virtue of (4.5.3 and 5),

$$\tau|_A^M = h \int_{\theta_0}^{\vartheta} \sqrt{\cosh^2 a_0 - \cos^2 \theta} \, d\theta$$

$$+ h \int_{a_0}^{a} \sqrt{\cosh^2 \mu - \cosh^2 a_0} \, d\mu \quad .$$

On the other hand, we can combine the contour l with the arc AB of the ellipse $\mu = a_0$ and the tangent line BM. On this integration path $\nabla \tau = \vec{s}$, and consequently

$$\tau|_A^M \equiv \int_{ABM} (\vec{s} \cdot d\vec{s}) = |ABM| \quad ,$$

where $|ABM|$ is the length of the curve ABM. Comparing the last two equations we get

$$h \int_{\theta_0}^{\vartheta} \sqrt{\cosh^2 a_0 - \cos^2 \theta} \, d\theta + h \int_{a_0}^{a} \sqrt{\cosh^2 \mu - \cosh^2 a_0} \, d\mu = |ABM|.$$
$$(4.5.6)$$

In a similar manner, we can show that

$$-h \int_{2\pi + \theta_0}^{\vartheta} \sqrt{\cosh^2 a_0 - \cosh^2 \theta} \, d\theta + h \int_{a_0}^{a} \sqrt{\cosh^2 \mu - \cosh^2 a_0} \, d\mu$$

$$= |ADCM| \quad ,$$
$$(4.5.7)$$

where $|ADCM|$ is the length of the arc $ADCM$, consisting of the arc ADC of the ellipse $\mu = a_0$ and the tangent line CM.

Adding (4.5.6 and 7), we find that the length of the closed curve $ABMCDA$ is equal to

$$h \int_{\theta_0}^{2\pi+\theta_0} \sqrt{\cosh^2 a_0 - \cos^2 \theta} \, d\theta + 2h \int_{a_0}^{a} \sqrt{\cosh^2 \mu - \cosh^2 a_0} \, d\mu$$

$$= h \int_{0}^{2\pi} \sqrt{\cosh^2 a_0 - \cos^2 \theta} \, d\theta + 2h \int_{a_0}^{a} \sqrt{\cosh^2 \mu - \cosh^2 a_0} \, d\mu$$

and consequently is independent of ϑ (recall that ϑ is the coordinate of the point M on the ellipse $\mu = a$). Thus the length of the curve $ABMCDA$ will not change as point M moves along the ellipse $\mu = a$. Because of this, the variation of the integral $\int_{ABMCDA} ds$ is equal to zero, i.e.

$$\delta \int_{ABMCDA} ds = \frac{\partial}{\partial \vartheta} \int_{ABMCDA} \sqrt{\left(\frac{\partial x_1}{\partial \sigma}\right)^2 + \left(\frac{\partial x_2}{\partial \sigma}\right)^2} \, d\sigma = 0 \quad ,$$

where $x_1 = x_1(\sigma)$, $x_2 = x_2(\sigma)$ is a parametric equation of the curve $ABMCDA$. Calculating this variation according to the general rules of the calculus of variations [4.7], we find that at the point M

$$(\vec{s}_1 - \vec{s}_2) \cdot \delta \vec{r} = 0 \quad , \tag{4.5.8}$$

where $\delta \vec{r}$ is a vector tangent to the ellipse $\mu = a$ at M, $\vec{s}_1$ is a unit vector directed along the tangent line BM, and $\vec{s}_2$ is a unit vector directed along the tangent line MC. Relation (4.5.8) also proves that the angles that BM and MC make with the ellipse $\mu = a$ are equal.

Let us return to construction of the closed congruence of rays. The set of multiply reflected rays $N_i N_{i+1}$ tangent to the ellipse $\mu = a_0$ is easily broken down into two normal congruences.

One congruence is formed by segments tangent to the ellipse $\mu = a_0$ that originate at points on the ellipse $\mu = a$ and terminate at points on the ellipse $\mu = a_0$; the other is formed by segments of tangents that originate at points of the ellipse $\mu = a_0$, and terminate at points of the ellipse $\mu = a$ (Fig. 4.13a,b). One normal congruence (the solid lines) is transformed into the other one (the broken lines) as a result of reflection or by passing through the caustic. Thus we get a two-sheeted covering of the elliptical ring $a_0 \leq \mu \leq a$. Joining these two sheets along the caustic $\mu = a_0$ and the reflecting boundary $\mu = a$, we arrive at a two-sheeted covering space (Fig. 4.13c) that is homeomorphic to a torus. For the basis curves on this covering space we take the ellipse $\mu = a_0$, which is a caustic, and the closed curve consisting of an incident ray belonging to the second congruence, the corresponding reflected ray belonging to the first congruence, and the arc of the caustic $\mu = a_0$ that joins them, this arc being traversed in the direction opposite to that of the rays. For the sake of simplicity,

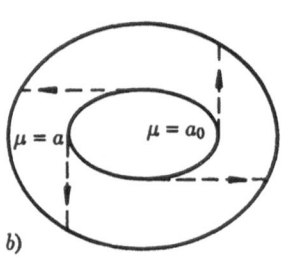

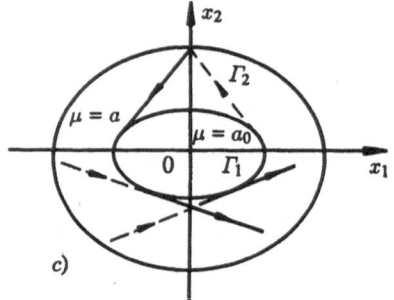

Fig. 4.13. (a, b) Normal congruences of rays for an ellipse. (c) Two-sheeted covering space for an ellipse

we will assume that the closed curve we have constructed is symmetric relative to the axis $0x_2$, so that the incident ray is transformed into the reflected ray at the point $\mu = a$, $\theta = \pi/2$, and on the arc of the caustic $\mu = a_0$, the angle θ varies over the range

$$\arcsin\frac{\sinh a_0}{\sinh a} < \theta < \pi - \arcsin\frac{\sinh a_0}{\sinh a} \quad .$$

Along these basis curves, the equalities

$$\sum \tau_\mu dx_\mu = \pm ds \quad ,$$

hold, where ds is an element of arc length on the basis curve. Therefore the quantization conditions (4.3.12) reduce to

$$kh \int_0^{2\pi} \sqrt{\cosh^2 a_0 - \cos^2 \theta}\, d\theta = 2\pi q \quad ,$$

$$2kh\left[\frac{\cosh a}{\sinh a}\sqrt{\sinh^2 a - \sinh^2 a_0} - \int_{\arcsin\frac{\sinh a_0}{\sinh a}}^{\pi/2} \sqrt{\cosh^2 a_0 - \cos^2 \theta}\, d\theta\right]$$

$$= \begin{cases} 2\pi p + 3\pi/2 \\ 2\pi p + \pi/2 \end{cases} \quad . \tag{4.5.9}$$

The right-hand side of the second equation is equal to $2\pi p + 3\pi/2$ in the case of condition (4.1.2), and to $2\pi p + \pi/2$ in the case of condition (4.1.3). The additional terms $3\pi/2$ and $\pi/2$ appear due to one intersection of the basis curve with the caustic, and one with the boundary of the region. Since the quantity k must be large, we must assume that $q \gg 1$ in the first of equations (4.5.9). The system of equations (4.5.9) determines the eigenvalues k_{pq} and the corresponding value of a_0.

Eliminating the unknown k from equations (4.5.9), we get

$$2q\left[\frac{\cosh a}{\sinh a}\sqrt{\sinh^2 a - \sinh^2 a_0} - \int\limits_{\arcsin\frac{\sinh a_0}{\sinh a}}^{\pi/2}\sqrt{\cosh^2 a_0 - \cos^2\theta}\,d\theta\right]$$

$$\times\left(\int\limits_0^{2\pi}\sqrt{\cosh^2 a_0 - \cosh^2\theta}\,d\theta\right)^{-1} = \begin{cases} p+3/4 \\ p+1/4 \end{cases}. \qquad (4.5.10)$$

The resulting equation for a_0 is not solvable for all p and q. The left-hand side of (4.5.10) takes on its maximum value of

$$\frac{q}{2}(\cosh a - 1)\ ,$$

when $a_0 = 0$. Thus, from equation (4.5.10) we can find that a_0 which satisfies the inequality $0 \le a_0 \le a$, and from that, a corresponding eigenvalue k_{pq} only if

$$0 \le p \le \frac{q}{2}(\cosh a - 1) - \begin{cases} 3/4 \\ 1/4 \end{cases}. \qquad (4.5.11)$$

However, there does exist a closed congruence of rays with corresponding eigenvalues k_{pq} for which the values of p and q obey

$$p > \frac{q}{2}(\cosh a - 1) - \begin{cases} 3/4 \\ 1/4 \end{cases}.$$

In fact, let the initial ray (the chord $N_0 N_1$ in Fig. 4.14) now intersect the segment between the foci F_1 and F_2 of the ellipse $\mu = a$. Such a ray will be tangent to some confocal hyperbola either inside or outside the ellipse $\mu = a$. Let the right and left branches of this hyperbola be described by the equations

$$\begin{cases} \theta = \theta_0, \\ \theta = 2\pi - \theta_0, \end{cases} \qquad \begin{cases} \theta = \pi - \theta_0, \\ \theta = \pi + \theta_0, \end{cases} \qquad (4.5.12)$$

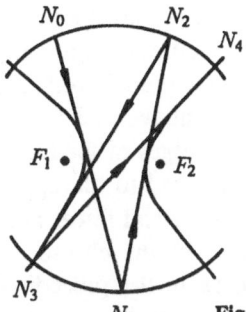

Fig. 4.14. Rays passing between the foci of the ellipse

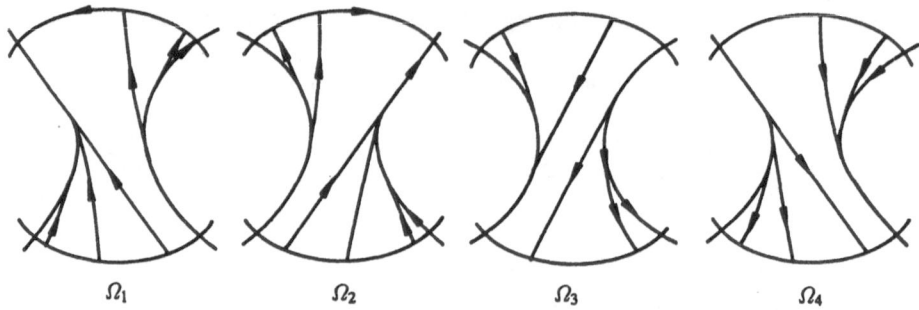

Fig. 4.15. Normal congruences whose rays all pass between the foci of an ellipse

respectively. The rays $N_1 N_2$, $N_2 N_3$, ... that arise as a result of multiple reflections of the initial ray from the ellipse $\mu = a$ (or possibly their extensions beyond $\mu = a$) will be tangent to branches of this same hyperbola. We have already proved in this section an analogous property for chords of the ellipse $\mu = a$ which touch the confocal ellipse $\mu = a_0$. The arguments are analogous for the case of a confocal hyperbola, and we omit them. Consider all possible tangents to the branches of the hyperbola defined by (4.5.12). The set of these tangents forms four normal congruences inside the ellipse (Fig. 4.15). The first congruence is formed by segments of these tangents which begin on the right branch of the hyperbola or on the lower arc of the ellipse $\mu = a$ and terminate on the left branch of (4.5.12) or on the upper arc of the $\mu = a$. The second congruence is formed by segments of tangents which originate on the left branch of (4.5.12) or on the lower arc of the ellipse $\mu = a$; the third by tangents originating on the right branch of the hyperbola or on the upper arc of the ellipse, and the fourth by segments of tangents starting on the left branch of (4.5.12) or on the upper arc of the ellipse $\mu = a$. Obviously the hyperbola (4.5.12) is a caustic for the congruences we have constructed. Rays of each of the four congruences are transformed into rays of two of the other congruences as a result of reflection and passage through caustics. For example, a ray of the first congruence is converted into a ray of the second congruence if it terminates on the left branch of the hyperbola (4.5.12), or into a ray of the third congruence if it ends on the upper arc of the ellipse $\mu = a$.

The four normal congruences that we have constructed cover four identical subregions Ω_1, Ω_2, Ω_3 and Ω_4, bounded by arcs of the ellipse $\mu = a$ and branches of the hyperbola (4.5.12). Let us superpose region Ω_1 on region Ω_2, region Ω_3 on region Ω_4, and join them together along the branches of the hyperbola (4.5.12) (Fig. 4.16). As a result, we get two "tubes" that are bounded above and below by arcs of the ellipse $\mu = a$. We now join the upper and lower edges of the tubes together; the upper and lower arcs of the ellipse on Ω_1 are joined to the upper and lower arcs of the ellipse on Ω_3 respectively, while the upper and lower arcs of the ellipse on Ω_2 are joined to the upper and lower arcs of the ellipse on Ω_4 respectively. Obviously the resultant closed manifold, as in the preceding case, is homeomorphic to a torus, and consequently there exist two basis curves

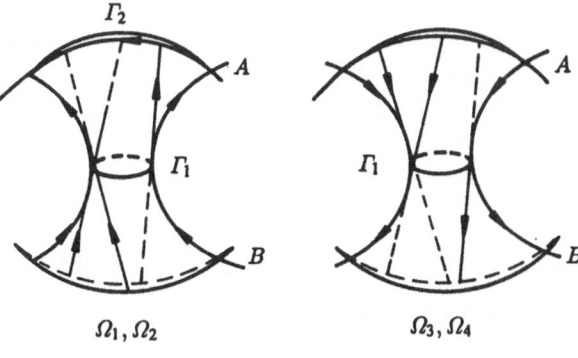

Fig. 4.16. Pairwise superposition of the normal congruences of Fig. 4.15

Ω_1, Ω_2 Ω_3, Ω_4

on the given manifold. As the first basis curve Γ_1, let us take the closed curve (one-dimensional closed chain) formed by the right-hand branch of (4.5.12) that belongs to Ω_1, and the left branch of (4.5.12) that belongs to Ω_3. As the second basis curve Γ_2, let us take the closed curve formed by the arcs of the ellipse $\mu = a$ belonging to Ω_1 and Ω_2. The arrows indicate the direction of traverse. On Γ_1, the direction of $\nabla \tau$ is the same as that of the tangent to Γ_1, and therefore the integral

$$\int_{\Gamma_1} \sum \tau_\mu dx_\mu$$

reduces to the length of Γ_1 :

$$\int_{\Gamma_1} \sum \tau_\mu dx_\mu = 4h \int_0^a \sqrt{\cosh^2 \mu - \cos^2 \theta_0}\, d\mu \quad . \tag{4.5.13}$$

Along the second basis curve, $\sum \tau_\mu dx_\mu = \tau_\theta d\theta$, and therefore we must first calculate τ_θ. We need to find expressions for the eikonals τ corresponding to the normal congruences of rays under consideration.

Using (4.5.3 and 4), and setting

$$\Lambda = h \cos \theta_0 \ , \qquad 0 \leq \theta_0 \leq \pi/2 \ ,$$

we find that $\Phi(\theta)$ is real if $|\cos \theta| < \cos \theta_0$, i.e., if $\theta_0 < \theta < \pi - \theta_0$ and $\pi + \theta_0 < \theta < 2\pi - \theta_0$. On the other hand, if $0 < \theta < \theta_0$, $\pi - \theta_0 < \theta < \pi + \theta_0$ and $2\pi - \theta_0 < \theta < 2\pi$ then the function $\Phi(\theta)$ is purely imaginary. The eikonal τ will be real in the region bounded by the arcs of the ellipse $\mu = a$ and the branches of the hyperbola (4.5.12), i.e., in precisely the region that was covered by the normal congruences of rays. Solving the differential equation (4.5.4) with $\Lambda = h \cos \theta_0$ we get

$$F(\mu) = \pm h \int \sqrt{\cosh^2 \mu - \cos^2 \theta_0}\, d\mu + C_1 \ ,$$

$$\Phi(\theta) = \pm h \int \sqrt{\cos^2 \theta_0 - \cos^2 \theta}\, d\theta + C_2 \quad .$$

The four different combinations of signs give four different values for $\nabla \tau$, corresponding to the four normal congruences that we have constructed.

By selecting the limits of integration and arbitrary constants in some specified manner, we can construct the function τ on a multi-sheeted covering space
so that this function remains continuous, for instance, on the left branch of hyperbola (4.5.12) and the upper arc of ellipse $\mu = a$.

Differentiating equation (4.5.3), we get for the first normal congruence covering subregion Ω_1,

$$\tau_\theta = h\sqrt{\cos^2 \theta_0 - \cos^2 \theta} \quad , \tag{4.5.14}$$

and for the second normal congruence covering Ω_2,

$$\tau_\theta = -h\sqrt{\cos^2 \theta_0 - \cos^2 \theta} \quad .$$

Having determined the values of τ_θ on Ω_1 and Ω_2, we return to the computation
of the integral over the second basis curve Γ_2. The formulas for τ_θ give us the
equation

$$\int_{\Gamma_2} \sum \tau_\mu dx_\mu = \int_{\Gamma_2} \tau_\theta d\theta$$

$$= h \int_{\theta_0}^{\pi - \theta_0} \sqrt{\cos^2 \theta_0 - \cos^2 \theta}\, d\theta$$

$$+ \left(-h \int_{\pi - \theta_0}^{\theta_0} \sqrt{\cos^2 \theta_0 - \cos^2 \theta}\, d\theta \right)$$

$$= 4h \int_{\theta_0}^{\pi/2} \sqrt{\cos^2 \theta_0 - \cos^2 \theta}\, d\theta \quad . \tag{4.5.15}$$

Substituting the integrals (4.5.13 and 15) into (4.3.12) we have

$$4kh \int_0^a \sqrt{\cosh^2 \mu - \cos^2 \theta_0}\, d\mu = \begin{cases} 2\pi p + 2\pi \,, \\ 2\pi p \,, \end{cases}$$

$$4kh \int_{\theta_0}^{\pi/2} \sqrt{\cos^2 \theta_0 - \cos^2 \theta}\, d\theta = 2\pi(q + 1/2) \quad . \tag{4.5.16}$$

The right-hand side of the first equation is equal to $2\pi p + 2\pi$ in the case of
boundary condition (4.1.2) and to $2\pi p$ in the case of condition (4.1.3), since Γ_1
intersects the boundary of the region twice. The additional term $2\pi \cdot (1/2)$ on the
right hand side of the second equation results from the fact that Γ_2 intersects the
caustic [the branches of the hyperbola (4.5.12)] twice.

For k to be large, we must take $p \gg 1$ in the first of equations (4.5.16).
Solving (4.5.16) for k and θ_0, we find the eigenvalues k_{pq} and the corresponding values of the parameter θ_0. Thus there will be a characteristic position for
the caustic corresponding to each eigenfunction. The equations (4.5.16) are not

consistent for all integers p and q. To show this, we eliminate the unknown k from (4.5.16):

$$
\int_{\theta_0}^{\pi/2} \sqrt{\cos^2\theta_0 - \cos^2\theta}\, d\theta \left(\int_0^a \sqrt{\cosh^2\mu - \cos^2\theta_0}\, d\mu \right)^{-1}
$$

$$
= \begin{cases} \dfrac{q+1/2}{p+1} \\ \dfrac{q+1/2}{p} \end{cases}. \tag{4.5.17}
$$

The left side of equation (4.5.17) takes on its maximum value of $(\cosh a - 1)^{-1}$ when $\theta_0 = 0$. Therefore, (4.5.17) has a solution only if the integers p and q satisfy the inequality

$$
0 \leq q + \tfrac{1}{2} \leq (\cosh a - 1)^{-1} \cdot \begin{cases} (p+1) \\ p \end{cases}. \tag{4.5.18}
$$

The eigenfunctions of the ellipse can be given explicitly in terms of Mathieu functions. These expressions reveal that any eigenfunction of an ellipse of sufficiently high indices p and/or q will have an asymptotic expansion corresponding to one of the two ray types that have been considered here.

The asymptotic expansion of the corresponding eigenvalue will also be expressed by the formulas we have found here. Thus the methods of Sect. 4.1–4.3 enable us, in the case of an ellipse, to find asymptotic expressions for all eigenvalues of sufficiently high order. As shown by actual computations [4.8], the asymptotic formulas found in this section give a good approximation even for eigenvalues with comparatively small values of p and q.

Summing up, we can say that for $q \gg 1$ and comparatively small values of p the eigenfunctions of the ellipse u_{pq} corresponding to the eigenvalues k_{pq} have an elliptic caustic and are concentrated in the vicinity of the boundary. The smaller p is, the smaller is the difference between a and a_0, see (4.5.10), i.e., the thinner will be the elliptical ring in which the eigenfunctions oscillate, and beyond which they decay exponentially. We will call eigenfunction such as these which are concentrated in the neighborhood of a boundary eigenfunctions of the *whispering gallery* type. If $p \gg 1$ and q is relatively small, then the eigenfunctions u_{pq} have a hyperbolic caustic. In this case, the smaller q is, the smaller will be the difference between $\pi/2$ and θ_0, see (4.5.17), i.e., the narrower will be the strip surrounding the minor axis of the ellipse within which the eigenfunctions oscillate, and outside of which they decay exponentially. Because of this, the eigenfunctions associated with $p \gg 1$ and $q = 0, 1, 2, \ldots$ can be called eigenfunctions of the *bouncing ball* type. In passing, we note that *no* eigenfunctions are concentrated around the major axis of the ellipse. This situation is due to the fact that the system of rays that arises as a result of multiple reflections (as we shall see later on) is stable in the vicinity of the minor axis of an ellipse, but unstable in the neighborhood of the major axis (see Chap. 5).

Equations (4.5.9 and 16) contain elliptic integrals, and therefore they cannot be directly solved in a general form. Below we will consider only two special cases:

1) Equations (4.5.9) when $a - a_0 \ll 1$;
2) Equations (4.5.16) when $\pi/2 - \theta_0 \ll 1$.

1) When $a - a_0 \ll 1$ the function $(\cosh a/\sinh a)\sqrt{\sinh^2 a - \sinh^2 a_0}$ and the elliptic integrals that enter into (4.5.9) can be represented by the first terms of their Taylor series expansions in $(a - a_0)$. This enables us to write (4.5.9) when $a - a_0 \ll 1$ in the form

$$
k\left\{ h \int_0^{2\pi} \sqrt{\cosh^2 a - \cos^2 \theta}\, d\theta - h \cosh a \sinh a \right.
$$

$$
\left. \times \int_0^{2\pi} \frac{d\theta}{\sqrt{\cosh^2 a - \cos^2 \theta}} (a - a_0) + O[(a - a_0)^2] \right\} = 2\pi q \quad , \quad (4.5.19)
$$

$$
2kh\left\{ \tfrac{2}{3}\sqrt{2 \cosh a \sinh a}(a - a_0)^{3/2} + O[(a - a_0)^{5/2}] \right\}
$$

$$
= \begin{cases} 2\pi(p + 3/4) \\ 2\pi(p + 1/4) \end{cases} \quad . \qquad\qquad\qquad (4.5.20)
$$

The first integral in equation (4.5.19) is evidently equal to the circumference of the ellipse $\mu = a$, which we will denote by L. The second integral in this equation can be converted to an integral that contains the radius of curvature ϱ of the ellipse $\mu = a$. Substituting the parametric equation of the ellipse $\mu = a$ into the formula for the radius of curvature of the curve

$$
\varrho = \frac{[(\partial x_1/\partial \sigma)^2 + (\partial x_2/\partial \sigma)^2]^{3/2}}{(\partial x_1/\partial \sigma)(\partial^2 x_2/\partial \sigma^2) - (\partial x_2/\partial \sigma)(\partial^2 x_1/\partial \sigma^2)}
$$

we get

$$
\varrho = h\frac{[\cosh^2 a - \cos^2 \theta]^{3/2}}{\cosh a \sinh a} \quad .
$$

Let us transform the integral in which we are interested into one with respect to the arc length s on the ellipse $\mu = a$ $(ds = h\sqrt{\cosh^2 a - \cos^2 \theta}\, d\theta)$, measured, say, from the point $x_1 = h \cosh a$, $x_2 = 0$ $(\theta = 0)$, and express the integrand in terms of the radius of curvature $\varrho(s)$:

$$
h \cosh a \sinh a \int_0^{2\pi} \frac{d\theta}{\sqrt{\cosh^2 a - \cos^2 \theta}}
$$

$$
= \cosh a \sinh a \int_0^{L} \frac{ds}{\cosh^2 a - \cos^2 \theta}
$$

$$= h^{2/3}(\cosh a \sinh a)^{1/3} \int_0^L \frac{ds}{\varrho^{2/3}(s)} \quad .$$

The resulting equation enables to write (4.5.19) in the form

$$kL - kh^{2/3}(\cosh a \sinh a)^{1/3} \int_0^L \frac{ds}{\varrho^{2/3}(s)}(a - a_0)$$

$$+ O[(a - a_0)^2] = 2\pi q \quad . \tag{4.5.21}$$

Eliminating the parameter a_0 of the caustic between (4.5.20 and 21), we get

$$kL - \left(\frac{k}{2}\right)^{1/3} \int_0^L \frac{ds}{\varrho^{2/3}(s)} \begin{cases} [\frac{3}{2}\pi(p + \frac{3}{4})]^{2/3} \\ [\frac{3}{2}\pi(p + \frac{1}{4})]^{2/3} \end{cases} + O(k^{-1/3}) = 2\pi q \quad ; \tag{4.5.22}$$

then solving (4.5.22) for k, we obtain the eigenvalues

$$k_{pq} = \frac{2\pi q}{L} + \left(\frac{\pi q}{L}\right)^{1/3} \frac{1}{L} \int_0^L \frac{ds}{\varrho^{2/3}(s)}$$

$$\times \begin{cases} [\frac{3}{2}\pi(p + \frac{3}{4})]^{2/3} \\ [\frac{3}{2}\pi(p + \frac{1}{4})]^{2/3} \end{cases} + O\left(\frac{p^{4/3}}{q^{1/3}}\right) \quad . \tag{4.5.23}$$

Finally, substituting k_{pq} into (4.5.20), we can determine the caustic parameter a_0 corresponding to the eigenfunction u_{pq}

$$a_0^{(pq)} = a - \frac{1}{(2\pi q)^{2/3}} \frac{L^{2/3}}{h^{2/3}} \frac{1}{(2 \cosh a \sinh a)^{1/3}}$$

$$\times \begin{cases} [\frac{3}{2}\pi(p + \frac{3}{4})]^{2/3} \\ [\frac{3}{2}\pi(p + \frac{1}{4})]^{2/3} \end{cases} + O\left[\left(\frac{p}{q}\right)^{4/3}\right] \quad . \tag{4.5.24}$$

[Keep in mind that in the last three equations the upper factor corresponds to boundary condition (4.1.2), and the lower to (4.1.3)].

Since we have assumed $kL \gg 1$ and $a - a_0 \ll 1$, formulas (4.5.23 and 24) are true when $2\pi q \gg 1$ and $(p/q)^{2/3} \ll 1$, i.e., for large values of q and fairly small values of p/q.

We can estimate the width of the strip in which the eigenfunctions are concentrated for small p and beyond which they decay exponentially. Let us denote the width of this strip by δ_{pq}. For small p the formula

$$\delta_{pq} = h \int_{a_0}^a \sqrt{\cosh^2 \mu - \cos^2 \theta} \, d\mu \approx h\sqrt{\cosh^2 a - \sin^2 \theta}(a - a_0)$$

$$\approx \left(\frac{\varrho}{2}\right)^{1/3} \left[\frac{3\pi L}{2}\left(p + \begin{cases} 3/4 \\ 1/4 \end{cases}\right)\right]^{2/3} \frac{1}{(2\pi q)^{2/3}}$$

$$\approx \left(\frac{\varrho}{2}\right)^{1/3} \left[\frac{3\pi}{2k_{pq}}\left(p + \left\{\begin{array}{c}3/4\\1/4\end{array}\right\}\right)\right]^{2/3}$$

is valid.

We call particular attention to the fact that for $p = 0, 1, 2, \ldots$ the elliptical caustic approaches the ellipse $\mu = a$ so closely that the ray representation of the field, which breaks down at the caustic, will be invalid throughout the entire elliptical ring $a_0 \leq \mu \leq a$ as well, including the boundary of the region, where we must satisfy the boundary conditions. Therefore it is also to be expected that the values of k_{pq} when $q \gg 1$ and $p = 0, 1, 2, \ldots$ as defined by formula (4.5.23) will differ from the true asymptotic values of k_{pq}. Later we will see that for small p the factor $[(3\pi/2)(p + \frac{3}{4})]^{2/3}$ must be replaced by the pth root t_p of the Airy function $v(-t)$, and the factor $[(3\pi/2)(p + \frac{1}{4})]^{2/3}$ must be replaced by the pth root t'_p of the derivative of the Airy function $v'(-t)$, the eigenfunctions also being expressed in terms of the Airy function.

2) Consider now equations (4.5.16) when $\pi/2 - \theta_0 \ll 1$. We expand the integrals entering into equations (4.5.16) in series with respect to powers of $\cos\theta_0$, retaining only the terms up to $\cos^2\theta_0$. Making the change of variables $x = (\cos\theta/\cos\theta_0)$, in the first integral, we get

$$\int_{\theta_0}^{\pi/2} \sqrt{\cos^2\theta_0 - \cos^2\theta}\, d\theta = \cos^2\theta_0 \int_0^1 \sqrt{\frac{1 - x^2}{1 - x^2\cos^2\theta_0}}\, dx$$

$$= \frac{\pi}{4}\cos^2\theta_0 + O(\cos^4\theta_0) \quad . \tag{4.5.25}$$

The expansion of the second integral is done directly:

$$\int_0^a \sqrt{\cosh^2\mu - \cos^2\theta_0}\, d\mu = \int_0^a \cosh\mu\, d\mu - \frac{1}{2}\cos^2\theta_0 \int_0^a \frac{d\mu}{\cosh\mu}$$

$$+ O(\cos^4\theta_0) = \sinh a - \frac{1}{2}\arcsin(\tanh a)\cos^2\theta_0 + O(\cos^4\theta_0) \quad .\tag{4.5.26}$$

Let us denote the minor semiaxis of the ellipse by $g/2 = h\sinh a$, and use the notation $\varrho_1 = h(\cosh^2 a/\sinh a)$ to designate the radius of curvature of the ellipse at the point $\theta = \pi/2$ where it intersects the minor axis. Then (4.5.26) can be rewritten in the form

$$\int_0^a \sqrt{\cosh^2\mu - \cos^2\theta_0}\, d\mu = \frac{g}{2h} - \frac{1}{2}\arcsin\sqrt{\frac{g}{2\varrho_1}}\cos^2\theta_0 + O(\cos^4\theta_0) \quad .$$
$$\tag{4.5.27}$$

Formulas (4.5.25 and 27) enable us to write (4.5.16) in the form

$$k\left\{\frac{g}{2} - \frac{h}{2}\arcsin\sqrt{\frac{g}{2\varrho_1}}\cos^2\theta_0 + O(\cos^4\theta_0)\right\} = \frac{\pi}{2}\left\{\begin{array}{c}(p+1)\\p\end{array}\right. \quad . \tag{4.5.28}$$

$$kh\pi\{\cos^2\theta_0 + O(\cos^4\theta_0)\} = 2\pi(q + 1/2) \quad .$$

Solving (4.5.28), we get for the eigenvalues k_{pq} and the parameters $\theta_0^{(pq)}$ of the hyperbolic caustic the following formulas:

$$k_{pq} = \frac{\pi}{g} \left\{ \left\{ \frac{p+1}{p} \right. + \frac{2q+1}{p} \arcsin \sqrt{\frac{g}{2\varrho_1}} + O\left[\left(\frac{q}{p} \right)^2 \right] \right\} \quad . \tag{4.5.29}$$

$$\cos^2 \theta_0^{(pq)} = 2(2q+1) \sinh a \left\{ \begin{array}{c} 1/(p+1) \\ 1/p \end{array} \right. + O\left[\left(\frac{q}{p} \right)^2 \right] \quad . \tag{4.5.30}$$

The factors $p+1$ and $1/(p+1)$ here apply to boundary condition (4.1.2), and the factors p and $1/p$ apply to (4.1.3). Formulas (4.5.29 and 30), obviously, are valid when $p \gg 1$ and $q \ll p$, i.e., at large values of p and fairly small values of q. If $q = 0, 1, 2, \ldots$, then the hyperbolic caustic [*branch* of hyperbola (4.5.12)] will be located close to the minor axis of the ellipse. The strip bounded by the caustic in which the eigenfunctions oscillate and beyond which they decay exponentially will have a width δ_{pq} defined by the formula

$$\delta_{pq} = 2h \int\limits_{\theta_0}^{\pi/2} \sqrt{\cosh^2 \mu - \cos^2 \theta}\, d\theta = 2h \left(\frac{\pi}{2} - \theta_0 \right) \cosh \mu$$

$$\approx 2h \sqrt{\frac{2q+1}{k_{pq}}} \quad .$$

The ray representation of the eigenfunction will be invalid within this narrow strip. We shall see later that the eigenfunction in this region must be expressed in terms of a parabolic cylinder function. Nonetheless, both formulas in (4.5.29) are valid, in contrast with the case of an elliptical caustic.

4.6 Notes on the Literature

We have basically followed the development of the paper of *Keller* and *Rubinow* [4.8], adding some refinements. Discussions with *Maslov* and *Lazutkin* were especially helpful in making these refinements.

The "quantization" concept introduced in Sect. 4.3 is closely related to actual quantization. It is therefore possible to trace the ideas underlying the constructions of Chap. 4 back to the seminal works of quantum mechanics (chief among which we should mention the paper by *Einstein* [4.9]).

An analog of the ray method (the so-called *quasi-classical* approximation) can also be developed for the Schrödinger equation. Here, the reciprocal of Planck's constant plays the role of the large parameter. The well-known Bohr-Sommerfeld quantization conditions for the bound states of a quantum-mechanical system (in other words, for the eigenfunctions of the Schrödinger equation) are then completely analogous to conditions (4.3.12).

We should also mention the close relationship between the methods of this chapter and the so-called *method of constructive interference*, which is often used by physicists to construct dispersion curves (cf., e.g., [4.10, 11]). Using an analogous ray method, *Vainshtein* [4.12] and *Bykov* [4.13] (see also [4.14]) have studied the eigenvalues and eigenfunctions of a triaxial ellipsoid.

In case no reflection takes place, the multi-sheeted covering space woven by the rays uniquely defines a closed lagrangian manifold, see [4.15]. In [4.15] as well as [4.16], a rigorous theory of a more general index than l_c, applicable to arbitrary lagrangian manifolds, is given. This general index is often referred to as the *Maslov index* or the *Maslov-Arnol'd index*.

As shown by *Lazutkin* (see [4.17] and references cited therein), the deeper development of the ideas underlying Chap. 4 involves the theory of dynamical systems, invariant tori, and the theories of *Kolmogorov*, *Mozer*, and *Arnol'd*.

5. The Ray Method "in the Small"

The Keller-Rubinow method, as described in the preceding chapter is applicable if there exists a closed congruence of rays, or more precisely an $(f - 1)$-parameter family of congruences. So far, such congruences have been successfully constructed only for the case where the variables in the corresponding eikonal equation are separable in suitable curvilinear coordinates. But if the variables in the eikonal equation *are* separable, then the problem of constructing the asymptotic expansion of the eigenfunctions can be reduced to finding asymptotic expansions of solutions of *ordinary* differential equations. Thus the constructions of Sects. 4.1–5 apparently only amount to an interesting geometric interpretation of asymptotic formulas that could be derived more easily by other means.

Now in point of fact, the method developed in the preceding chapter is applicable not only to actual closed congruences of rays, but also to closed congruences constructed from rays that satisfy Euler's equation and the law of reflection only in a first approximation. Such congruences will be called *approximate congruences*. Extending the Keller-Rubinow method to closed approximate congruences appreciably expands its range of application and permits the solution of problems not yielding to the separation of variables technique. The modification of the Keller-Rubinow method as applied to closed approximate congruences will be called the *ray method in the small*.

In implementing the ray method in the small, and in particular when constructing a closed approximate congruence, an especially important concept is that of stability of the system of rays in the first approximation. Therefore the question of the stability of the corresponding system of rays in the first approximation will be examined in detail throughout this entire chapter.

5.1 Eigenfunctions of the Whispering Gallery Type

In this section we will consider an important and very simple example of the application of the ray method in the small.

In Sect. 4.4 we found eigenvalues and eigenfunctions whose fields are concentrated near the boundary of a circle, and showed that the same kind of eigenfunctions exist for an ellipse. The problem of this section is to find eigenvalues corresponding to eigenfunctions concentrated near the boundary S of an arbitrary convex region Ω.

Let S have circumference L and be described by its "natural" equation $\varrho = \varrho(s)$, where ϱ is the radius of curvature of S, and s is the arc length measured along S from some fixed initial point. It is convenient to assume that s varies over the range $-\infty < s < \infty$, so that $\varrho(s)$ is a periodic function with period L. We will assume that $\varrho(s)$ has a continuous third derivative and that $0 < \varrho(s) < \infty$. (The latter inequality implies the strict convexity of S.)

A fundamental role is played by the system of rays that arises close to the boundary of region Ω as a result of numerous repeated reflections from S. It turns out that this system is stable in the first approximation. Let us give a precise definition of this concept, which will be important in what follows.

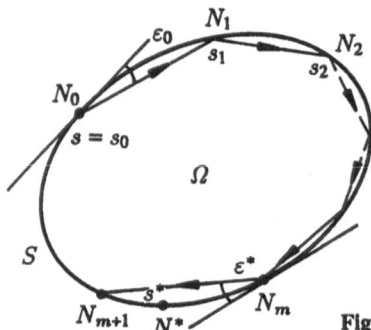

Fig. 5.1. Whispering-gallery rays within a convex region

Let N_0 (Fig. 5.1) be some arbitrarily chosen point on the boundary S for which $s = s_0$ and $0 \leq s_0 < L$. At N_0 construct a chord $N_0 N_1$ of the region Ω forming the glancing angle ε_0 $(0 < \varepsilon_0 < \pi/2)$ with the tangent to S at N_0. We take $N_0 N_1$ as the initial ray; as a result of successively reflecting this ray at the points $N_1, N_2, \ldots, N_m$ we get a system of multiply reflected rays $N_j N_{j+1}$, $j = 1, 2, \ldots, m$. We will assume that corresponding to the points of reflection N_j are values of the arc length $s = s_j$ $(s_{j-1} < s_j < \infty)$, $j = 1, 2, \ldots, m, \ldots$. Let s^* be some fixed value of the arc length s $(s_0 < s^* < \infty)$, and let the point N^* correspond to the value $s = s^*$ on S. We will let ε^* denote the glancing angle of the ray $N_m N_{m+1}$ for which $s_m \leq s^* < s_{m+1}$, i.e., the angle subtended by the arc $s_m s_{m+1}$ on which the point N^* is located. Obviously the number of reflections m which this ray has undergone is a function of ε_0 and s^*. For a smooth boundary S with fixed s^* and $\varepsilon_0 \to 0$, the number of reflections $m \to \infty$ and the glancing angle $\varepsilon^* \to 0$.

We will call a system of multiply reflected rays *stable in the first approximation* if a more rigorous condition is met: for each fixed value s^* there exists a $\delta(s^*)$ such that for all $\varepsilon_0 < \delta(s^*)$ the relative deflection of the ray $\varepsilon^*/\varepsilon_0$ at N^* satisfies the inequality $\varepsilon^*/\varepsilon_0 < K$, where the constant K is independent of the difference $s^* - s_0$ and is determined only by the properties of the contour S.

For a system of rays that arises as a result of repeated reflections near the contour S we shall prove an even stronger result – namely that

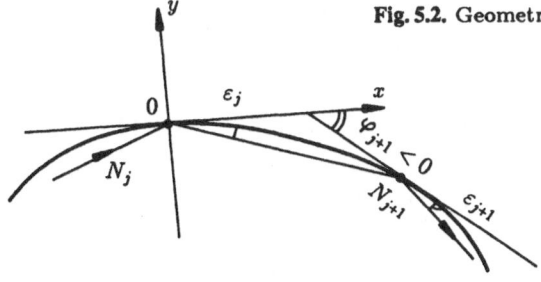

$$\varepsilon^* = E\varepsilon_0 + |s^* - s_0|O(\varepsilon_0^3) \quad . \tag{5.1.1}$$

Here E is a positive number, independent of ε_0 and uniformly bounded with respect to $|s^* - s_0|$ (i.e., for all $|s^* - s_0|$ the inequality $E < K_1$ is satisfied, where K_1 is a constant that does not depend on $|s^* - s_0|$). Stability in the first approximation follows easily from (5.1.1); in fact for fixed s^* one can always find a $\delta(s^*)$ such that $|s^* - s_0|O(\varepsilon_0^2) < 1$ when $\varepsilon_0 < \delta(s^*)$, and thus

$$\frac{\varepsilon_m}{\varepsilon_0} < K_1 + 1 = \text{const} \quad .$$

Let us now derive (5.1.1). Consider some intermediate reflection point N_j. Let the glancing angle of the ray at N_j be equal to ε_j, while at the next reflection point N_{j+1} it is equal to ε_{j+1}. We must find a relation between ε_{j+1} and ε_j. Introducing a local coordinate system $(x\,0\,y)$ with origin at the point $0 = N_j$ (Fig. 5.2), the parametric equation of S in these local coordinates can be written as

$$x = \int_{s_j}^{s} \cos\left(\int_{s_j}^{s'} \frac{d\xi}{\varrho(\xi)}\right) ds' = (s - s_j) - \frac{1}{6\varrho_j^2}(s - s_j)^3 + O[(s - s_j)^4], \tag{5.1.2}$$

$$y = \int_{s_j}^{s} \sin\left(\int_{s_j}^{s'} \frac{d\xi}{\varrho(\xi)}\right) ds' = -\frac{1}{2\varrho_j}(s - s_j)^2 + \frac{\varrho_j'}{6\varrho_j^2}(s - s_j)^3$$

$$+ \frac{1}{24}\left[\left(\frac{\varrho'}{\varrho^2}\right)_j' + \frac{1}{\varrho_j^3}\right](s - s_j)^4 + O[(s - s_j)^5] \quad , \tag{5.1.3}$$

where

$$\varrho_j = \varrho(s_j) \quad , \quad \varrho_j' = \left.\frac{d\varrho(s)}{ds}\right|_{s=s_j} \quad , \quad \left(\frac{\varrho'}{\varrho^2}\right)_j = \left.\frac{d}{ds}\left(\frac{\varrho'(s)}{\varrho^2(s)}\right)\right|_{s=s_j} \quad .$$

The arc length s_{j+1} corresponding to the point N_{j+1} must satisfy the equation

$$y(s_{j+1}) = -x(s_{j+1})\tan \varepsilon_j$$

which can be rewritten with the use of (5.1.2 and 3) in the form

$$-\frac{1}{2\varrho_j}(s_{j+1} - s_j)^2 + \frac{\varrho_j'}{6\varrho_j^2}(s_{j+1} - s_j)^3$$

$$+\frac{1}{24}\left[\left(\frac{\varrho'}{\varrho^2}\right)_j + \frac{1}{\varrho_j^3}\right](s_{j+1} - s_j)^4 + O[(s_{j+1} - s_j)^5]$$

$$= -\tan\varepsilon_j\left\{(s_{j+1} - s_j) - \frac{1}{6\varrho_j^2}(s_{j+1} - s_j)^3 + O[(s_{j+1} - s_j)^4]\right\}. \quad (5.1.4)$$

Solving equation (5.1.4) for $(s_{j+1} - s_j)$ and expanding $\tan\varepsilon_j$ in powers of ε_j we find that

$$s_{j+1} - s_j = 2\varrho_j\varepsilon_j + \frac{4}{3}\varrho_j\varrho_j'\varepsilon_j^2 + \left(\frac{4}{9}\varrho_j\varrho_j'^2 + \frac{2}{3}\varrho_j^2\varrho_j''\right)\varepsilon_j^3 + O(\varepsilon_j^4) \quad . \quad (5.1.5)$$

The angle φ_{j+1} formed by the tangent to S at N_{j+1} and the OX axis is

$$\varphi_{j+1} = -\int\limits_{s_j}^{s_{j+1}}\frac{ds}{\varrho(s)} = -\frac{1}{\varrho_j}(s_{j+1} - s_j) + \frac{\varrho_j'}{2\varrho_j^2}(s_{j+1} - s_j)^2$$

$$+\frac{1}{6}\left(\frac{\varrho'}{\varrho^2}\right)_j(s_{j+1} - s_j)^3 + O[(s_{j+1} - s_j)^4]$$

or, expressing $s_{j+1} - s_j$ in terms of ε_j as given by (5.1.5), we get

$$\varphi_{j+1} = -2\varepsilon_j + \frac{2}{3}\varrho_j'\varepsilon_j^2 - \left(\frac{4}{9}\varrho_j'^2 - \frac{2}{3}\varrho_j\varrho_j''\right)\varepsilon_j^3 + O(\varepsilon_j^4) \quad .$$

Since $\varepsilon_{j+1} = -\phi_{j+1} - \varepsilon_j$,

$$\varepsilon_{j+1} = \varepsilon_j - \frac{2}{3}\varrho_j'\varepsilon_j^2 + \left(\frac{4}{9}\varrho_j'^2 - \frac{2}{3}\varrho_j\varrho_j''\right)\varepsilon_j^3 + O(\varepsilon_j^4) \quad . \quad (5.1.6)$$

To find the dependence of $\varepsilon^* \equiv \varepsilon_m$ on ε_0 construct the ratio

$$\frac{\varepsilon_m}{\varepsilon_0} = \frac{\varepsilon_1}{\varepsilon_0}\cdot\frac{\varepsilon_2}{\varepsilon_1}\cdots\frac{\varepsilon_m}{\varepsilon_{m-1}}$$

$$= \prod_{j=0}^{m-1}\left[1 - \frac{2}{3}\varrho_j'\varepsilon_j + \left(\frac{4}{9}\varrho_j'^2 - \frac{2}{3}\varrho_j\varrho_j''\right)\varepsilon_j^2 + O(\varepsilon_j^3)\right] \quad ,$$

whence

$$\frac{\varepsilon_m}{\varepsilon_0} = \exp\left\{\sum_{j=0}^{m-1}\ln\left[1 - \frac{2}{3}\varrho_j'\varepsilon_j + \left(\frac{4}{9}\varrho_j'^2 - \frac{2}{3}\varrho_j\varrho_j''\right)\varepsilon_j^2 + O(\varepsilon_j^3)\right]\right\} \quad .$$

Expressing ε_j in terms of $s_{j+1} - s_j$ and expanding the logarithm as a power series, we get

$$\frac{\varepsilon_m}{\varepsilon_0} = \exp\left\{-\frac{1}{3}\sum_{j=0}^{m-1}\left[\frac{\varrho_j'}{\varrho_j} + \left(\frac{\varrho'}{\varrho}\right)_j'\frac{s_{j+1} - s_j}{2} + O[(s_{j+1} - s_j)^2]\right]\right.$$

$$\left.\times(s_{j+1} - s_j)\right\} \quad .$$

The latter can be rewritten as

$$\frac{\varepsilon_m}{\varepsilon_0} = \exp\left\{-\frac{1}{3}\sum_{j=0}^{m-1}\left[\frac{\varrho'(s)}{\varrho(s)}\bigg|_{s=s_j+\frac{1}{2}(s_{j+1}-s_j)}\right.\right.$$
$$\left.\left.\times (s_{j+1}-s_j)+O[(s_{j+1}-s_j)^3]\right]\right\} \quad . \tag{5.1.7}$$

Now, the sum

$$\sum_{j=0}^{m-1}\frac{\varrho'(s)}{\varrho(s)}\bigg|_{s=s_j+\frac{1}{2}(s_{j+1}-s_j)}\cdot(s_{j+1}-s_j)$$

is a Riemann sum approximating the value of the integral $\int_0^{s_m}[(\varrho'(s)/\varrho(s)]ds$ [5.1]), and therefore

$$-\frac{1}{3}\sum_{j=0}^{m-1}\frac{\varrho'(s)}{\varrho(s)}\bigg|_{s=s_j+\frac{1}{2}(s_{j+1}-s_j)}\cdot(s_{j+1}-s)$$

$$= \ln\sqrt[3]{\frac{\varrho_0}{\varrho_m}}+|s_m-s_0|O[(\max_j|s_{j+1}-s_j|)^2] \quad .$$

The arc length s_m corresponding to N_m can be determined from ε_0 and s^*. However, the exact value of s_m is not essential in what follows; it is important only that $|s^*-s_m|<L$. Moreover, since

$$\sum_{j=0}^{m-1}O[(s_{j+1}-s_j)^3] = \sum_{j=0}^{m-1}O[(s_{j+1}-s_j)^2](s_{j+1}-s_j)$$

$$= |s^*-s_0|O[(\max_j|s_{j+1}-s_j|)^2] \quad ,$$

(5.1.7) imply that

$$\varepsilon_m = \varepsilon_0\sqrt[3]{\frac{\varrho_0}{\varrho_m}}\left\{1+|s^*-s_0|O[(\max_j|s_{j+1}-s_j|)^2]\right\} \quad .$$

On the basis of this, we can state that for fixed s^* the quantity ε_j is $O(\varepsilon_0)$, and consequently by virtue of (5.1.5),

$$\max_j|s_{j+1}-s_j| = O(\varepsilon_0) \quad .$$

Thus the relative deflection of a ray is

$$\frac{\varepsilon^*}{\varepsilon_0} = \sqrt[3]{\frac{\varrho_0}{\varrho_m}}+|s^*-s_0|O(\varepsilon_0^2) \quad . \tag{5.1.8}$$

Since we assume that $0<\varrho(s)<\infty$, the quantity $E = \sqrt[3]{\varrho_0/\varrho_m}$ remains bounded, from which we infer stability in the first approximation[1].

---

[1] To prove stability of the system of rays in the first approximation it would only have been necessary to assume that $\varrho''(s)$ is continuous and to show that

$$\frac{\varepsilon^*}{\varepsilon_0} = \sqrt[3]{\frac{\varrho_0}{\varrho_m}}+|s^*-s_0|O(\varepsilon_0) \quad . \qquad \text{(continued on p. 102)}$$

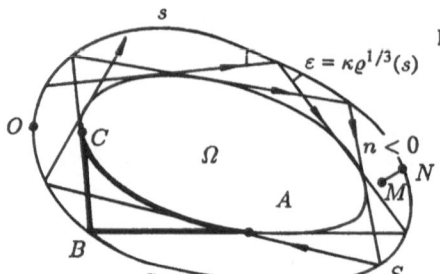

Fig. 5.3. Invariant set of whispering-gallery rays

Let us now construct a closed approximate congruence of rays. We shall construct this congruence from a system of rays which have been multiply reflected from S. We will require this system to be stable in the first approximation, but the law of reflection need only be satisfied to first order. Our analysis will be carried out by analogy between the problem under consideration and the corresponding problem for a circle.

Consider a set of rays (Fig. 5.3) that fills some neighborhood of the boundary S, with glancing angles defined by the formula

$$\varepsilon(s) = \kappa \varrho^{-1/3}(s) \quad , \tag{5.1.9}$$

where κ is a constant which is sufficiently small (e.g., $\kappa \ll \min|\varrho^{1/3}/\varrho'|$, as implied in the preceding derivation). In the first approximation the rays of this set are invariant relative to reflections from S, i.e. when any ray of this set is reflected the resulting ray belongs to the same set. Actually, to first order, (5.1.8) implies that the product $\varepsilon_m \varrho_m^{1/3}$ is independent of m. After constructing the envelope of this set of rays, we introduce the coordinates s and n for a point M, where n is the length of the normal MN from M to some point N on S, and s is the arc length along S measured from some initial point to the point N – the base of normal. We will assume that $n<0$ for points in Ω. The coordinate system (s, n) is regular in a strip $|n|<\varrho_{\min}$ containing S.

The equation of a ray that intersects S at $s = s_j$ at a glancing angle $\varepsilon = \kappa \varrho^{-1/3}(s_j)$, takes the form

$$y = -x \tan \varepsilon \quad (\varepsilon = \varepsilon_j) \quad , \tag{5.1.10}$$

in the local (x, y) coordinate system (Fig. 5.2).

The coordinates (x, y) and (n, s) are related by the simple expressions

$$x = x(s) - ny'(s) \quad ,$$
$$y = y(s) + nx'(s) \quad , \tag{5.1.11}$$

where, as before, $x = x(s)$ and $y = y(s)$ are the parametric equations of the curve S. Equations (5.1.11) enable us to write the equation of the ray (5.1.10) in (n, s) coordinates as:

However, the latter estimate would have been insufficient for the calculations of Chap. 12, which is why we have imposed a more rigorous smoothness condition on S, and proved (5.1.8).

$$n = \frac{y(s) + x(s)\tan\varepsilon}{y'(s)\tan\varepsilon - x'(s)} \quad .$$

Using (5.1.2 and 3), and considering that $\tan\varepsilon = \kappa\varrho^{-1/3}(s_j) + O(\kappa^3)$, we get[2]

$$n = -\kappa\varrho^{-1/3}(s_j)(s - s_j) + \frac{1}{2\varrho(s_j)}(s - s_j)^2 - \frac{\varrho'(s_j)}{6\varrho^2(s_j)}(s - s_j)^3$$

$$+ O[(s - s_j)^4 \; , \; \kappa(s - s_j)^3 \; , \; \kappa^2(s - s_j)^2 \; , \; \kappa^3(s - s_j)] \quad .$$

Replacing $\varrho(s_j)$ with the first two terms of its expansion in powers of $(s_j - s)$

$$\varrho(s_j) = \varrho(s) + \varrho'(s)(s_j - s) + O[(s_j - s)^2] \quad ,$$

we have. to the same accuracy as before.
The coefficients $a_m(s, \nu)$ and $b_m(s, \nu)$ here contain the partial derivatives with respect to s and ν of the desired polynomials $\alpha_n(s, \nu)$ and $\beta_n(s, \nu)$. Equations (8.2.17) will be identically satisfied in ω if and only if

$$a_m(s, \nu) = 0 \quad , \quad m = -4, \; -3, \; -2, \ldots \quad ,$$
$$b_m(s, \nu) = 0 \quad , \quad m = -2, \; -1, \; 0, \ldots \quad . \tag{8.2.19}$$

Equations (8.2.19) are a recurrent system of partial differential equations that must be satisfied by the unknown polynomials $\alpha_n(s, \nu)$ and $\beta_n(s, \nu)$. Let us find the principal terms $\alpha_{-2}(s, \nu)$, $\alpha_{-1}(s, \nu)$, $\alpha_0(s, \nu)$ and $\beta_0(s, \nu)$ of the formal series. This will permit the equations of (8.2.19) for $m \geq 1$ to be substantially simplified. After making the necessary calculations we get

$$a_{-4}(s, \nu) \equiv -\left(\frac{\partial\alpha_{-2}}{\partial\nu}\right)^2 = 0 \quad , \tag{8.2.20}$$

$$a_{-3}(s, \nu) \equiv -2\frac{\partial\alpha_{-2}}{\partial\nu}\frac{\partial\alpha_{-1}}{\partial\nu} - \left(\frac{\partial\alpha_{-2}}{\partial\nu}\right)^2\frac{\nu}{\varrho} = 0 \quad , \tag{8.2.21}$$

$$a_{-2}(s, \nu) \equiv -\left(\frac{\partial\alpha_{-1}}{\partial\nu}\right)^2 - 2\frac{\partial\alpha_{-2}}{\partial\nu}\frac{\partial\alpha_0}{\partial\nu} - 2\frac{\partial\alpha_{-2}}{\partial\nu}\frac{\partial\alpha_{-1}}{\partial\nu}\frac{\nu}{\varrho}$$
$$+ i\frac{\partial^2\alpha_{-2}}{\partial\nu^2} - \left(\frac{\partial\alpha_{-2}}{\partial s}\right)^2 + \frac{1}{c_0^2(s)} = 0 \quad , \tag{8.2.22}$$

$$a_{-1}(s, \nu) \equiv -2\frac{\partial\alpha_{-1}}{\partial\nu}\frac{\partial\alpha_0}{\partial\nu} - \left(\frac{\partial\alpha_{-1}}{\partial\nu}\right)^2\frac{\nu}{\varrho} + i\frac{\partial^2\alpha_{-1}}{\partial\nu^2}$$
$$- 2\frac{\partial\alpha_{-2}}{\partial s}\frac{\partial\alpha_{-1}}{\partial s} + \left(\frac{\partial\alpha_{-2}}{\partial s}\right)^2\frac{\nu}{\varrho} + e_1(s)\frac{\nu}{\varrho} = 0 \quad , \tag{8.2.23}$$

$$b_{-2}(s, \nu) \equiv 2i\frac{\partial\alpha_{-2}}{\partial\nu}\frac{\partial\beta_0}{\partial\nu} = 0 \quad , \tag{8.2.24}$$

by our closed approximate congruence, we select two basis curves as we did for the circle. Then the system of equations (4.3.12) for boundary condition $u|_S = 0$ will now take the form

$$\begin{cases} k\Lambda = 2\pi q, \\ k\eta = 2\pi \left(p + \frac{3}{4}\right), \end{cases}$$

where Λ is the circumference of the caustic, η is the sum of the lengths of a pair of incident and reflected rays minus the arc length of the segment of caustic between points of tangency ($AB + BC - CDA$, for example as shown in Fig. 5.3), while p and q are integers.

An element ds_k of arc length from the caustic is equal to

$$ds_k = \frac{ds}{\varrho(s)}[\varrho(s) + n(s) + O(\kappa^4)] = \left[1 - \frac{1}{2}\varrho^{-2/3}(s)\kappa^2 + O(\kappa^4)\right] ds .$$

Therefore we have

$$\Lambda = \int_0^L ds_k = L - \frac{\kappa^2}{2} \int_0^L \frac{d\tau}{\varrho^{2/3}(\tau)} + O(\kappa^4) ,$$

where L is the circumference of the contour S. By virtue of the fact that terms proportional to κ^3 are absent from (5.1.14), the quantity η can be calculated to leading order terms as if the boundary S and its caustic were concentric circles whose radii were $\varrho(s)$ and $\varrho(s) + n(s)$, respectively. It turns out to be equal to:

$$\eta = \frac{2}{3}\kappa^3 + O(\kappa^5) .$$

Thus we get a system of equations which determines the eigenvalues:

$$k\left[L - \frac{\kappa^2}{2} \int_0^L \frac{d\tau}{\varrho^{2/3}(\tau)} + O(\kappa^4)\right] = 2\pi q ,$$

$$k\left[\frac{2}{3}\kappa^3 + O(\kappa^5)\right] = 2\pi\left(p + \frac{3}{4}\right) , \tag{5.1.15}$$

whence we obtain:

$$k_{p,q} = \frac{2\pi q}{L} + \frac{[\frac{3\pi}{2}(p + \frac{3}{4})]^{2/3}}{2^{1/3}}\left(\frac{2\pi q}{L}\right)^{1/3}\frac{1}{L}\int_0^L \frac{d\tau}{\varrho^{2/3}(\tau)} + \cdots . \tag{5.1.16}$$

Since $k_{p,q}L$ must be large, in this formula we need $q \gg 1$ and $p = 0, 1, 2, 3, \ldots$; the correction terms are $O(p^{4/3}q^{-1/3})$. After determining the quantity $k_{p,q}$ we can find the corresponding value of κ from the second of equations (5.1.15), and by substituting it into (5.1.14) we get an equation for the caustic:

$$n = -\frac{1}{2}\varrho^{1/3}(s)\frac{[3\pi(p + 3/4)]^{2/3}}{(k_{p,q})^{2/3}} + O\left[\left(\frac{p}{q}\right)^{4/3}\right] .$$

The pq-eigenfunctions are concentrated near and oscillatory only in the region between this caustic and S, a layer whose thickness is proportional to $\varrho^{1/3}(s)k_{p,q}^{-2/3}$.

We have seen that in the case of a circle, the eigenvalues obtained from asymptotic formulas for the Bessel functions contain the root t_p of the Airy function $v(-t)$ rather than the factor $[(3\pi/2)(p + \frac{3}{4})]^{2/3}$. Replacing $[(3\pi/2)(p + \frac{3}{4})]^{2/3}$ by t_p in (5.1.16), we get a more exact expression for the eigenvalues

$$k_{p,q} = \frac{2\pi q}{L} + \frac{t_p}{2^{1/3}}\left(\frac{2\pi q}{L}\right)^{4/3}\frac{1}{L}\int_0^L \frac{d\tau}{\varrho^{2/3}(\tau)} + \dots . \qquad (5.1.17)$$

In Chap. 7 we will justify this replacement and find further terms in expansion (5.1.17).

5.2 Eigenvalues of the Bouncing Ball Type

In studying the eigenfunctions of the ellipse we found that the ellipse has a subset of eigenfunctions concentrated in the neighborhood of the minor axis. In the first approximation, the eigenvalues corresponding to these eigenfunctions were completely determined by the length of the minor axis of the ellipse and its radii of curvature at the intersections with the minor axis.

This suggests that these eigenfunctions are local in nature and that their asymptotic behavior would not change if the ellipse were modified in some smooth fashion outside of the vicinity of its intersections with the minor axis. It is natural therefore to expect that a subset of eigenfunctions with similar properties will exist for arbitrary regions as well.

In this section we shall assume that such a subset of eigenfunctions does in fact exist, and we shall construct asymptotic expansions for the corresponding subset of eigenvalues, using the ray method in the small. The fact that the ray method in the small does not encounter any fundamental difficulties and that, even in this case, it can be carried through completely, can be taken as further support for the existence of such subsets of concentrated eigenfunctions for arbitrary regions[3]. Consider the plane region Ω bounded by a sufficiently smooth curve Σ. Our first goal is to construct a system of multiply reflected rays in Ω that is stable in the first approximation and analogous to the system of rays of an ellipse which is bounded by confocal hyperbolas.

An *extremal diameter* of the region Ω will be any chord that forms right angles with tangents to Σ at the ends of the chord. A ray of light that is initially coincident with some extremal diameter CD of Ω will, after any number of reflections from Σ at C or D, still coincide with CD. Now let us assume that the initial ray N_0N_1 does not coincide with the extremal diameter CD, but is

---

[3] The asymptotic formulas for eigenvalues obtained to a first approximation by the ray method in the small can be proved rigorously (Sect. 7.6 and Chap. 8).

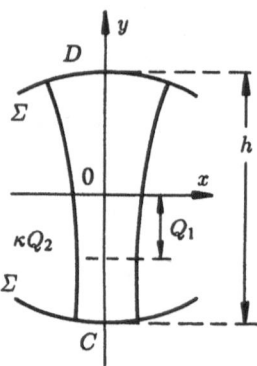

Fig. 5.4. The caustics in the neighborhood of the stable diameter of the domain

located fairly close to it. By reflection of the ray N_0N_1 from Σ at N_1, a new ray N_1N_2 results. Due to subsequent reflections, further rays N_2N_3, N_3N_4, ..., N_mN_{m+1}, ... appear. Let us examine the system of rays

$$N_0N_1, \; N_1N_2, \; \ldots, \; N_mN_{m+1}, \; \ldots \tag{5.2.1}$$

for stability. Let h (Fig. 5.4) be the length of an extremal diameter, and let r_1 and r_2 be the radii of curvature of Σ at C and D respectively. We locate the origin of a cartesian coordinate system xOy at the midpoint of CD, and let the Oy axis coincide with the diameter itself. We will assume that the rays N_jN_{j+1}, $j = 0, 1, \ldots, m, \ldots$ of the system (5.2.1) are described by the equation

$$x = \alpha_j y + \beta_j \frac{h}{2} \; . \tag{5.2.2}$$

With each reflection, the parameters α_j and β_j of the ray take on new values. We assume that the parameters α_0 and β_0 of the initial ray are small quantities:

$$\varepsilon_0 \equiv \sqrt{\alpha_0^2 + \beta_0^2} \ll 1 \; ,$$

i.e. we assume that the initial ray is near the diameter CD. We introduce the two-dimensional vectors $\vec{V}_j = (\alpha_j, \beta_j)$. Then, to order ε_0^2, the vectors $\vec{V}_j$ will undergo a linear transformation with each reflection. In the following, we will use the symbol $\vec{O}(\varepsilon_0^2)$ to denote vectors whose components are of order ε_0^2. After two reflections, we get the vector $\vec{V}_2$, for which

$$\vec{V}_2 = A\vec{V}_0 + \vec{O}(\varepsilon_0^2) \; , \quad A = \begin{pmatrix} a_{11}, & a_{12} \\ a_{21}, & a_{22} \end{pmatrix} \; .$$

The matrix elements a_{ik} are expressed in terms of the radii of curvature r_1 and r_2 and the length h of the extremal diameter by the formulas

$$a_{11} = \frac{1}{r_1 r_2}[(h - r_1)(h - r_2) + h(h - 2r_2)] \; ,$$

$$a_{12} = \frac{h}{r_1 r_2}(2h - r_1 - r_2) \; ,$$

$$a_{21} = \frac{1}{r_1 r_2}[(h - r_1)(h - 2r_2) + (h - r_2)(h - 2r_1)] \; ,$$

$$a_{22} = \frac{1}{r_1 r_2}[(h - r_1)(h - r_2) + h(h - 2r_1)] \quad . \tag{5.2.3}$$

We can readily verify that Det $A = 1$. After $m = 2l$ reflections, we will have the vector $\vec{V}_m$

$$\vec{V}_m = A^l \vec{V}_0 + m\vec{O}(\varepsilon_0^2) \quad . \tag{5.2.4}$$

Obviously if m is fixed, then $|\vec{V}_m| \to 0$ as $\varepsilon_0 \to 0$. If, in addition, for each fixed m a $\delta(m) > 0$ can be found such that when $\varepsilon_0 < \delta(m)$ the relative deflection of the ray $|\vec{V}_m|$ is less than $K\varepsilon_0$, where the constant K is independent of m, we will call the system of rays (5.2.1) *stable in the first approximation*.

To determine the conditions for which this will occur, consider the transformation (5.2.4) in the linear approximation

$$\vec{V}_m = A^l \vec{V}_0 \quad . \tag{5.2.5}$$

Let λ_1 and λ_2 be the eigenvalues of A. We first assume that $\lambda_1 \neq \lambda_2$. Then there exists a matrix U (det $U \neq 0$) which, by the similarity transformation, diagonalizes the matrix A

$$UAU^{-1} = \begin{pmatrix} \lambda_1 & 0 \\ 0 & \lambda_2 \end{pmatrix} \quad ,$$

whence

$$A = U^{-1} \begin{pmatrix} \lambda_1 & 0 \\ 0 & \lambda_2 \end{pmatrix} U \quad \text{and} \quad A^l = U^{-1} \begin{pmatrix} \lambda_1^l & 0 \\ 0 & \lambda_2^l \end{pmatrix} U \quad .$$

Thus, in the linear approximation

$$\begin{pmatrix} \alpha_m \\ \beta_m \end{pmatrix} = U^{-1} \begin{pmatrix} \lambda_1^l & 0 \\ 0 & \lambda_2^l \end{pmatrix} U \begin{pmatrix} \alpha_0 \\ \beta_0 \end{pmatrix} \quad . \tag{5.2.6}$$

Formula (5.2.6) implies that system of rays (5.2.1) is stable in the first approximation if the eigenvalues λ_1 and λ_2 satisfy the conditions

$$|\lambda_1| \leq 1 \; , \quad |\lambda_2| \leq 1 \quad (\lambda_1 \neq \lambda_2). \tag{5.2.7}$$

Since $\lambda_1 \lambda_2 = \text{Det } A = 1$, (5.2.7) reduces to the requirement that

$$|\lambda_1| = |\lambda_2| = 1 \; , \quad \lambda_1 \neq \lambda_2 \quad . \tag{5.2.8}$$

The characteristic equation for λ_1 and λ_2 takes the form

$$\lambda^2 - \frac{2}{r_1 r_2}[2h^2 - 2h(r_1 + r_2) + r_1 r_2]\lambda + 1 = 0 \quad . \tag{5.2.9}$$

Equation (5.2.8) will be satisfied if the discriminant d of (5.2.9) is less than zero:

$$d = 4h(h - r_1 - r_2)(h - r_1)(h - r_2)\frac{1}{r_1^2 r_2^2} < 0 \quad . \tag{5.2.10}$$

Inequality (5.2.10) is equivalent to one of two systems of inequalities:

1. $h < r_1 + r_2$, $h > r_1$, $h > r_2$.
2. $h < r_1 + r_2$, $h < r_1$, $h < r_2$.

If now $\lambda_1 = \lambda_2$, then $d = 0$ and $\lambda_1 = \lambda_2 = \pm 1$.

When its eigenvalues are equal, then we can only reduce the matrix A to the Jordan form

$$\begin{pmatrix} \lambda & \gamma \\ 0 & \lambda \end{pmatrix}$$

and its l-th power will be equal to

$$\begin{pmatrix} \lambda^l & l\gamma\lambda^{l-1} \\ 0 & \lambda^l \end{pmatrix} .$$

If $|\lambda| = 1$ and $\gamma \neq 0$, the element $l\gamma\lambda^{l-1}$ of the latter matrix increases without limit as l increases, and therefore system of rays (5.2.1) will be unstable in the first approximation.

Thus when $d = 0$ the system of rays is stable in the first approximation only in the case $\gamma = 0$, $\lambda_1 = \lambda_2 = \pm 1$. The matrix A in this case will be represented in the form

$$A = U^{-1} \begin{pmatrix} \lambda_1 & 0 \\ 0 & \lambda_2 \end{pmatrix} U = \pm U^{-1} \begin{pmatrix} 1 & 0 \\ 0 & 1 \end{pmatrix} U ,$$

whence it follows that $A = \pm I$, where I is the identity matrix. The equality $A = \pm I$ is possible only if $r_1 = r_2 = h$, i.e. when the radii of curvature and the diameter h are equal.

Note that if the diameter CD is locally minimum, then $h < r_1 + r_2$, while if CD is locally maximum, then $h > r_1 + r_2$. Thus a system of rays that is stable in the first approximation cannot exist in the neighborhood of a locally maximum diameter of a region, and can exist in the neighborhood of a locally minimum diameter only when one of the supplementary conditions

1) $h > r_1$, $h > r_2$; 2) $h < r_1$, $h < r_2$; 3) $r_1 = r_2 = h$. $\qquad$ (5.2.11)

is satisfied as well.

In Sect. 4.5, it was found that the eigenfunctions of an ellipse can concentrate only in the vicinity of the minor axis of an ellipse, and do not concentrate in the neighborhood of the major axis. It is natural to conclude that this is a result of the fact that no stable system of rays exists in the vicinity of the major axis of an ellipse.

From now on we will consider only systems of rays that are stable in the first approximation, and only the corresponding subsets of eigenfunctions. Eigenvalues of A that are complex-conjugate and equal to one in magnitude can be written as

$$\lambda_1 = e^{i\phi} , \quad \lambda_2 = e^{-i\phi}, \qquad\qquad (5.2.12)$$

where

$$\phi = 2 \arccos \sqrt{\left(1 - \frac{h}{r_1}\right)\left(1 - \frac{h}{r_2}\right)} \,. \tag{5.2.13}$$

Let us now construct the closed ray congruences. The basis for our constructions will be the set of rays (5.2.2). From now on we will assume that the parameters α_j and β_j of the rays undergo linear transformations (5.2.5) upon each reflection. Thus the closed congruences that we will construct will be made up of approximate rays rather than from true reflected rays.

First of all, from the entire family of rays (5.2.2) we must isolate the set of rays that is invariant to reflections. Corresponding to each matrix A^l, $l = 0$, ± 1, ± 2, ..., is some transformation in the linear space of vectors $\binom{\alpha}{\beta}$. We will have constructed a set of vectors that is invariant to reflections if we can isolated a subset of vectors in this linear space that is invariant relative to the group of linear transformations consisting of the matrices A^l, $l=0$, ± 1, ± 2, ... Passing to a new basis

$$\binom{\gamma}{\delta} = U \binom{\alpha}{\beta}$$

in this space is equivalent to substitution of the matrix $U A^l U^{-1}$ for A^l. Let U be the matrix that converts A to diagonal form:

$$UAU^{-1} = \begin{pmatrix} \lambda_1 & 0 \\ 0 & \lambda_2 \end{pmatrix} = \begin{pmatrix} e^{i\phi} & 0 \\ 0 & e^{-i\phi} \end{pmatrix} \,.$$

It can be easily shown that the elements of the second row of U will be complex-conjugate to the elements of the first row ($U_{2k} = U_{1k}^*$; $k = 1$, 2), and consequently corresponding to real α and β will be complex-conjugate values of γ and δ, so that $\delta = \gamma^*$. In the new basis, the components γ_m and δ_m of a ray that has undergone $m = 2l$ reflections are related to the initial components γ_0 and δ_0 by the transformation

$$\binom{\gamma_m}{\delta_m} = \begin{pmatrix} e^{i\phi} & 0 \\ 0 & e^{-i\phi} \end{pmatrix}^l \binom{\gamma_0}{\delta_0} \,.$$

In the real space $a = (\gamma + \delta)/2$, $b = (\gamma - \delta)/2i$, this transformation reduces to a rotation through the angle ϕ. When $\phi \neq \pi(q_1/q_2)$, where q_1, q_2 are integers, the only invariant to the rotation is the length of a vector, and the invariant subsets are sets of vectors all of which have the same length[1]. Since the length $\sqrt{a^2 + b^2}$ of the vector (a, b) is given by the expression

$$\left(\frac{\gamma + \delta}{2}\right)^2 + \left(\frac{\gamma - \delta}{2i}\right)^2 = \gamma\delta, \tag{5.2.14}$$

---

[1] We will not consider here the case where $\phi = \pi(q_1/q_2)$ and q_1, q_2 are integers, since then the group of linear transformations has other invariants in addition to vector length. We note only that the examination of such invariants leads to a number of interesting consequences relating to the properties of extremal diameters.

which, by virtue of γ and δ being complex conjugates, is also equal to $|\gamma|^2 = |\delta|^2$, the invariant subset of vectors in the space $\binom{\gamma}{\delta}$ is described by the formulas

$$\gamma = \frac{\kappa}{2}e^{i\mu} \ , \quad \delta = \frac{\kappa}{2}e^{-i\mu} \ , \tag{5.2.15}$$

where κ is fixed and μ belongs to the interval $0, 2\pi$.

Operating on the set of vectors (5.2.15) by U^{-1}, we get an invariant subset in the space of real vectors $\vec{V}(\alpha, \beta)$:

$$\binom{\alpha}{\beta} = U^{-1}\binom{\frac{\kappa}{2}e^{i\mu}}{\frac{\kappa}{2}e^{-i\mu}} = \binom{\kappa \cos \mu}{\kappa a_{12}^{-1}[- a_{11} \cos \mu + \cos(\phi + \mu)]} . \tag{5.2.16}$$

Thus the set of rays that is invariant relative to reflections has the form

$$x = \kappa y \cos \mu + \kappa a_{12}^{-1}[- a_{11} \cos \mu + \cos(\phi + \mu)]\frac{h}{2} \ . \tag{5.2.17}$$

Now we must break this set of rays down into a finite number of normal congruences and then form a closed congruence of rays. Let us first construct the envelope of the family of rays (5.2.17). Differentiating (5.2.17) with respect to the parameter μ and then eliminating μ, we get an envelope equation in the form

$$\frac{x^2}{\kappa^2} - (y + Q_1)^2 = Q_2^2 \ , \tag{5.2.18}$$

where

$$Q_1 = \frac{\cos \phi - a_{11}}{a_{12}}\frac{h}{2} \ , \quad Q_2 = \frac{\sin \phi}{a_{12}}\frac{h}{2} \ . \tag{5.2.19}$$

Equation (5.2.18) is a hyperbola with the center displaced relative to the midpoint of the diameter by an amount $-Q_1$, with transverse axis $|\kappa Q_2|$ and conjugate axis $|Q_2|$ (Fig. 5.4). If the radii of curvature of the boundary at C and D are equal to each other, i.e., if $r_1 = r_2 = r$, then $\cos \phi = a_{11}$ and $Q_1 = 0$. In this case the center of the hyperbola coincides with the midpoint of the diameter.

Hence the invariant set of rays is bounded by two caustics: the left and right branches of the hyperbola (5.2.18). For small κ the invariant set of rays covers a narrow region between the branches of the hyperbola. Obviously, the invariant set of rays (5.2.17) contains two normal congruences. The first is formed by rays originating on the lower part of the boundary of the region and on the right arc of the hyperbola and terminating on the upper part of the boundary and the left arc of the hyperbola. The second is formed by rays originating on the lower part of the boundary and the left branch of the hyperbola. The rays of the second congruence combined with the rays of the first congruence make up the whole of the set (5.2.17). If we also consider rays reflected by the upper part of the boundary and going from the top down, we get two more normal congruences − a third and fourth − for which the hyperbola (5.2.18) is also a caustic. These four normal congruences cover four identical subregions completely analogous to subregions Ω_1, Ω_2, Ω_3 and Ω_4 that we considered in constructing the eigenfunctions of the ellipse (Sect. 4.5). By joining these subregions together in exactly the same way as Ω_j were joined in the case of the ellipse, we get a closed manifold that is homeomorphic to a torus. Having access to explicit equations for the rays that form the four normal congruences, we can easily write down expressions for the components of the gradient of the eikonal τ on each subregion. We do this by eliminating μ from the three relations

$$x/\kappa = (y + Q_1)\cos\mu - Q_2\sin\mu \ ,$$
$$\tau_1/\tau_2 = \kappa\cos\mu \ , \quad \tau_1^2 + \tau_2^2 = 1 \ ,$$

that follow from (5.2.17 and 19), the fact that the vector (τ_1, τ_2) is parallel to the ray (5.2.17), and the eikonal equation $((\nabla\tau)^2 = 1)$[5]. Solving the remaining equations for τ_1 and τ_2, we get

$$\tau_1 = \tau_1(x, y, \kappa) \ , \quad \tau_2 = \tau_2(x, y, \kappa) \ . \tag{5.2.20}$$

In (5.2.20) it is natural to limit ourselves to the lowest-order terms with respect to κ since we have only satisfied the law of reflection in the first approximation in our construction of the closed congruence. Then,

$$\tau_1 = \pm \frac{x(y + Q_1) \mp \kappa Q_2^2 \sqrt{1 + \frac{(y+Q_1)^2}{Q_2^2} - \frac{x^2}{\kappa^2 Q_2^2}}}{(y + Q_1)^2 + Q_2^2} + O(\kappa^2) \ , \tag{5.2.21}$$

$$\tau_2 = \pm \left[1 - \frac{1}{2}\left(\frac{x(y + Q_1) \mp \kappa Q_2^2 \sqrt{1 + \frac{(y+Q_1)^2}{Q_2^2} - \frac{x^2}{\kappa^2 Q_2^2}}}{(y + Q_1)^2 + Q_2^2} \right)^2 \right] + O(\kappa^3). \tag{5.2.22}$$

The four different combinations of signs in (5.2.21) and the corresponding four combinations of signs in (5.2.22) give the values of τ_1 and τ_2 on all four subregions of the closed manifold.

Let us now select the basis curves on the multi-sheeted covering space. As basis curve Γ_1 (Fig. 5.5) let us take the diameter CD of the region located on the

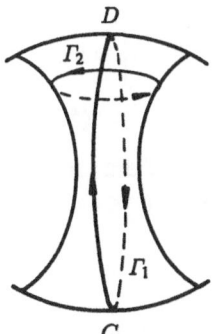

Fig. 5.5. Basis curves on the covering space for bounding-ball rays

---

[5] Recall that we set $c = 1$.

first and third sheet of the space; as basis curve Γ_2 we take the double segment $y = y_0$, $|x| \le |\kappa| \sqrt{(y_0 + Q_1)^2 + Q_2^2}$, traversed on the first and second sheets of the space. (The directions of travel of the basis curves are indicated in Fig. 5.5 by the arrows). For points on the first basis curve,

$$\tau_2 = \pm \left(1 - \frac{1}{2} \frac{\kappa^2 Q_2^2}{(y + Q_1)^2 + Q_2^2} \right) + O(\kappa^3) \quad . \tag{5.2.23}$$

For points of the second,

$$\tau_1 = \frac{y_0 + Q_1}{(y_0 + Q_1)^2 + Q_2^2} x \mp \kappa |Q_2|$$

$$\times \sqrt{1 - \frac{\kappa^{-2} x^2}{(y_0 + Q_1)^2 + Q_2^2}} \frac{1}{\sqrt{(y_0 + Q_1)^2 + Q_2^2}} + O(\kappa^2) \quad . \tag{5.2.24}$$

Formulas (5.2.23 and 24), together with the directions of travel along the basis curves which we have chosen enable us to write (4.3.12) for boundary condition $\partial u / \partial n |_\Sigma = 0$ in the form

$$k \left\{ 2 \int_{-h/2}^{h/2} \left[1 - \frac{1}{2} \frac{\kappa^2 Q_2^2}{(y_0 + Q_1)^2 + Q_2^2} \right] dy + O(\kappa^3) \right\} = 2\pi p \ ,$$

$$k \left\{ 2\kappa |Q_2| \int_{-\kappa \sqrt{(y_0 + Q_1)^2 + Q_2^2}}^{\kappa \sqrt{(y_0 + Q_1)^2 + Q_2^2}} \sqrt{1 - \frac{\kappa^{-2} x^2}{(y_0 + Q_1)^2 + Q_2^2}} \right.$$

$$\times \left. \frac{dx}{\sqrt{(y_0 + Q_1)^2 + Q_2^2}} + O(\kappa^3) \right\} = 2\pi (q + 1/2) \quad . \tag{5.2.25}$$

The right-hand side of the second equation takes consideration of the fact that basis curve Γ_2 intersects the caustic twice. (For the boundary condition $u |_\Sigma = 0$, $p+1$ should be substituted for p in the first equation.) Let us calculate the integrals in the left-hand sides of (5.2.25)

$$k \left[h + \frac{1}{2} |Q_2| \left(\arctan \frac{h/2 + Q_1}{|Q_2|} - \arctan \frac{-h/2 + Q_1}{|Q_2|} \right) \kappa^2 + O(\kappa^3) \right] = \pi p \ ,$$

$$k[|Q_2| \kappa^2 + O(\kappa^3)] = 2q + 1 \quad . \tag{5.2.26}$$

Equations (5.2.26) are easily solved for k and κ. The difference of arc tangents is conveniently expressed in terms of the arc cosine. After the appropriate calculation we get

$$k_{p,q} = \frac{\pi p}{h} + \frac{q + 1/2}{h}$$

$$\times \arccos \left\{ \Lambda \left[1 + \left(\frac{h Q_2}{Q_1^2 + Q_2^2 - h^2/4} \right)^2 \right]^{-1/2} \right\} + O \left(\frac{q^{3/2}}{p^{1/2}} \right) \ , \tag{5.2.27}$$

$$\kappa_{p,q} = \left(\frac{h}{|Q_2|}\frac{2q+1}{\pi p}\right)^{1/2} + O\left(\frac{q}{p}\right) , \tag{5.2.28}$$

where

$$\Lambda = \text{sign}(Q_1^2 + Q_2^2 - h^2/4) .$$

The form of the remainder term in formulas (5.2.27 and 28) shows that these formulas are applicable if

$$p \gg 1 , \quad q = 0, 1, 2, \ldots = O(1) . \tag{5.2.29}$$

When condition (5.2.29) is satisfied, $\kappa_{p,q} \ll 1$, and consequently the envelope of the closed congruence of rays is located close to the diameter of the region, and the rays themselves are concentrated in a small neighborhood of the diameter. This conclusion agrees with out initial assumption that the rays undergo a linear transformation when they are reflected which was made in the construction of the closed congruence of rays.

We see immediately from (5.2.18) for the caustics that the width of the strip that encloses the diameter of the region within which the eigenfunctions under consideration are concentrated is of order κ. If we assume that $q = O(1)$, then we see from (5.2.27 and 28) that $\kappa = O(1/\sqrt{k})$. The width of the strip will have this same order.

Formula (5.2.27) for the eigenvalues can be put in a more instructive form. To do this, we express the quantities Q_1 and Q_2 in terms of the radii of curvature r_1 and r_2 of the boundary of Σ at C and D. First converting from Q_1 and Q_2 to elements of A using (5.2.19 and 13), and then using (5.2.3), we get

$$k_{p,q} = \frac{\pi p}{h} + \frac{q + 1/2}{h}\arccos\left[\Lambda\sqrt{\left(1 - \frac{h}{r_1}\right)\left(1 - \frac{h}{r_2}\right)}\right] + O\left(\frac{q^{3/2}}{p^{1/2}}\right) , \tag{5.2.30}$$

where

$$\Lambda = \text{sign}(r_1 + r_2 - 2h) . \tag{5.2.31}$$

When $r_1 = r_2 = r$ this formula reduces to (4.5.29) for the eigenvalues of an ellipse. Recalling (5.2.13) for the angle ϕ — the argument of the eigenvalue λ_1 of A — we rewrite (5.2.30) in the form

$$k_{p,q} = \frac{\pi p}{h} + \frac{q + 1/2}{h}\left[\frac{\Lambda\phi}{2} + (1 - \Lambda)\frac{\pi}{2}\right] + O\left(\frac{q^{3/2}}{p^{1/2}}\right) .$$

Formula (5.2.30) is well known in the theory of open resonators. In fact, for the case of an open region, the eigenvalues must be complex. However, for natural oscillations confined to the vicinity of caustics, and consequently concentrated in a small neighborhood of the resonator axis, radiation losses can usually be disregarded. This disregard of radiation losses can be thought of as solving the problem for a closed region whose properties in the vicinity of C and D are the same as those in the actual (open) region.

113

In Chap. 8 we shall see how (5.2.30) can be derived in another way. This other way will enable us to give a rigorous justification for (5.2.30) at the same time.

5.3 Eigenvalues of the Whispering Gallery Type for a Nonconstant Wave Velocity

To study the behavior of rays in the neighborhood of the boundary of a region we will first have to derive the Euler equation that describes rays which are close to a curve S. In this section, S will be used to denote the boundary of Ω.

Let us introduce a coordinate system (s, n) in the neighborhood of S where n is the length of the normal dropped from a point M to S, and s is the arc length along S measured from some origin to the base of the normal. These coordinates, already used in Sect. 5.1, will be of frequent use to us in what follows.

Let the magnitude of the normal be positive for points M located to the left of S, viewed as we look along S in the direction of increasing s (Fig. 5.6). We will also assume that the radius of curvature $\varrho(s)$ of S is positive if the center of curvature lies to the right of S, i.e. among negative values of n, and that the radius of curvature is negative otherwise. If we introduce a unit vector $\vec{t}_0$ tangent to S and pointed in the direction of increasing arc length s, and a unit vector $\vec{n}_0$ normal to S and directed toward positive values of n, then in accordance with our sign conventions the formulas

$$\frac{d\vec{t}_0}{ds} = -\frac{1}{\varrho}\vec{n}_0 \ , \qquad \frac{d\vec{n}_0}{ds} = \frac{1}{\varrho}\vec{t}_0 \ , \tag{5.3.1}$$

will be true. Clearly, the coordinates (s, n) are orthogonal.

Let $d\sigma$ be a differential arc length of an arbitrary curve near S. Applying the Pythagorean theorem to the infinitesimal triangle ABC (Fig. 5.7), we get

$$(d\sigma)^2 = \left[\frac{\varrho + n}{\varrho}ds\right]^2 + dn^2 = \left(1 + \frac{n}{\varrho}\right)^2 ds^2 + dn^2 \ . \tag{5.3.2}$$

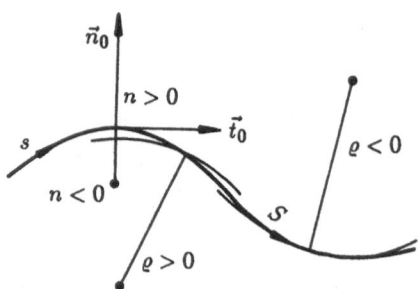

Fig. 5.6. The (s, n) coordinates near S

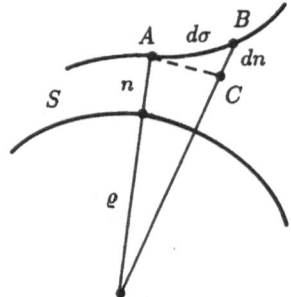

Fig. 5.7. Geometry of a curve near S

114

Thus the metric coefficients h_s and h_n in the (s, n) coordinate system (Appendix A.2) will be equal to

$$h_s = 1 + \frac{n}{\varrho} \, , \quad h_n = 1 \, , \tag{5.3.3}$$

respectively. The propagation velocity of waves in (s, n) coordinates will be denoted as before by $c(s, n)$. We will further assume that the function $c(s, n)$ is differentiable a sufficient number of times with respect to both variables. In (s, n) coordinates the basic (Fermat) functional of geometrical optics

$$T = \int \frac{d\sigma}{c(x, y)} \tag{5.3.4}$$

takes the form

$$T = \int \frac{\sqrt{\left(1 + \frac{n}{\varrho(s)}\right)^2 + \left(\frac{dn}{ds}\right)^2}}{c(s, n)} ds \, . \tag{5.3.5}$$

We wish to obtain the Euler equation for the rays, i.e. for the extremals of (5.3.5) in (s, n) coordinates, assuming that the deviation of the rays from S is small both in terms of distance and of the angle of inclination of the tangent.

Let $n = n(s)$ be the equation of a ray. Then our assumption means that the quantities $|n|$ and $|dn/ds|$ are small quantities[6]. Let us expand the integrand

$$\Phi\left(s, n, \frac{dn}{ds}\right) = \sqrt{\left(1 + \frac{n}{\varrho(s)}\right)^2 + \left(\frac{dn}{ds}\right)^2} \, c^{-1}(s, n)$$

of (5.3.5) in powers of n and dn/ds, limiting ourselves to quadratic terms:

$$\begin{aligned}
\Phi\left(s, n, \frac{dn}{ds}\right) = \frac{1}{c_0(s)} &\left\{ 1 + \left(\frac{1}{\varrho(s)} - \frac{c_1(s)}{c_0(s)}\right) n \right. \\
&+ \left(\frac{c_1^2(s)}{c_0^2(s)} - \frac{c_2(s)}{c_0(s)} - \frac{1}{\varrho(s)}\frac{c_1(s)}{c_0(s)}\right) n^2 \\
&+ \left. \frac{1}{2}\left(\frac{dn}{ds}\right)^2 + O\left[n^3, \, n\left(\frac{dn}{ds}\right)^2\right] \right\} \, , \tag{5.3.6}
\end{aligned}$$

where

$$c_0(s) = c(s, 0), \quad c_1(s) = \left. \frac{\partial c(s, n)}{\partial n}\right|_{n=0} \, ,$$

$$c_2(s) = \left. \frac{1}{2}\frac{\partial^2 c(s, n)}{\partial n^2}\right|_{n=0} \, .$$

---

[6] More precisely, we reckon the dimensionless quantities

$$\frac{n}{\varrho} \quad \text{and} \quad \left.\frac{1}{c}\frac{\partial^j c}{\partial n^j}\right|_{n=0} n^j \, ,$$

to be small.

Expansion (5.3.6) gives an Euler equation in the form

$$\frac{d^2n}{ds^2} - \frac{c_0'(s)}{c_0(s)}\frac{dn}{ds} - 2\left(\frac{c_1^2(s)}{c_0^2(s)} - \frac{1}{\varrho(s)}\frac{c_1(s)}{c_0(s)} - \frac{c_2(s)}{c_0(s)}\right)n$$

$$-\left(\frac{1}{\varrho(s)} - \frac{c_1(s)}{c_0(s)}\right) + O\left[n^2, \; n\frac{dn}{ds} \; , \; n\frac{d^2n}{ds^2} \; , \; \left(\frac{dn}{ds}\right)^2\right] = 0. \quad (5.3.7)$$

Equation (5.3.7) implies that S is itself a ray if and only if the equality

$$\frac{1}{\varrho(s)} - \frac{c_1(s)}{c_0(s)} = 0$$

is satisfied along S. The quantity

$$P(s) = \left(\frac{1}{\varrho(s)} - \frac{c_1(s)}{c_0(s)}\right)^{-1} \qquad\qquad (5.3.8)$$

which plays an important part in what follows will be called the *effective radius of curvature* of the curve S.

We can now find the condition under which superposition of rays multiply reflected by the boundary S of region Ω occurs in the neighborhood of the boundary. In other words, we will determine the conditions near the boundary under which the whispering gallery effect will arise. Obviously these conditions are local in nature, and so in deriving them we do not have to assume that Ω is bounded nor that S is closed.

Let Ω be located to the right of S, i.e. in $n<0$. The angle between a ray and the tangent to S pointed along the direction of increasing arc length will be called the glancing angle. We will adopt the convention of always taking the glancing angle to be positive. We say that superposition of waves multiply reflected by S occurs if a ray reflected at a point $N_j = N(s_j)$ at a glancing angle ε_j meets S at the point $N_{j+1} = N(s_{j+1})$, where corresponding to small ε_j are small $\Delta s_j = s_{j+1} - s_j > 0$. The condition thus formulated for superposition to occur can be written in the form

$$\lim_{\varepsilon_j \to 0} \frac{\Delta s_j}{\varepsilon_j} > 0 \; , \quad j = 0, 1, 2, \dots . \qquad\qquad (5.3.9)$$

We can express condition (5.3.9) in terms of the characteristics of S and the velocity $c(s, n)$. Let $n = n_j(s)$ be the equation of a ray reflected at N_j at a glancing angle ε_j. For the way in which Ω is located relative to S, we have

$$\left.\frac{dn_j}{ds}\right|_{s=s_j} = -\tan\varepsilon_j \; . \qquad\qquad (5.3.10)$$

Setting $s = s_j$, $n = 0$ in (5.3.7), and taking the limit as $\varepsilon_j \to 0$, we get

$$\lim_{\varepsilon_j \to 0} \left.\frac{d^2n_j}{ds^2}\right|_{s=s_j} = \frac{1}{\varrho(s_j)} - \frac{c_1(s_j)}{c_0(s_j)} = \frac{1}{P(s_j)} \; .$$

The point N_{j+1} where ray $n = n_j(s)$ meets S is determined from the equation $n_j(s_{j+1}) = 0$. Representing the function $n_j(s)$ by its Taylor series

$$n_j(s) = n'_j(s_j)(s - s_j) + \tfrac{1}{2}n''_j(s_j)(s - s_j)^2$$
$$+ \tfrac{1}{3!}n'''_j(s_j)(s - s_j)^3 + O[(s - s_j)^4] \tag{5.3.11}$$

and taking (5.3.10) into consideration, we can write the equation $n_j(s_{j+1}) = 0$ in the form

$$- \tan \varepsilon_j + \tfrac{1}{2}n''_j(s_j)(s_{j+1} - s_j)$$
$$+ \tfrac{1}{3!}n'''_j(s_j)(s_{j+1} - s_j)^2 + O[(s_{j+1} - s_j)^3] = 0 \quad .$$

If the effective radius of curvature $P(s)$, (5.3.8) takes on finite, non-zero values at all points of the boundary S, then $\lim\limits_{\varepsilon_j \to 0} n''_j(s_j) \neq 0$. From the last equation we obtain

$$\Delta s_j = s_{j+1} - s_j = \frac{2}{n''_j(s_j)}\varepsilon_j - \frac{4}{3}\frac{n'''_j(s_j)}{[n''_j(s_j)]^3}\varepsilon_j^2 + O(\varepsilon_j^3) \tag{5.3.12}$$

and

$$\lim_{\varepsilon_j \to 0} \frac{\Delta s_j}{\varepsilon_j} = \frac{2}{\lim\limits_{\varepsilon_j \to 0} n''_j(s_j)} = 2P(s_j) \quad . \tag{5.3.13}$$

Thus for superposition of multiply reflected waves to occur in Ω, i.e. for the whispering gallery effect to arise, it is sufficient for the inequality

$$\frac{1}{\varrho(s)} - \frac{c_1(s)}{c_0(s)} = \frac{1}{P(s)} > 0 \tag{5.3.14}$$

to be satisfied for all points on the curve S. If Ω is located to the left of S, then

$$\left.\frac{dn_j}{ds}\right|_{s=s_j} = \tan \varepsilon_j$$

and superposition of multiply reflected waves close to the boundary of Ω occurs when

$$\frac{1}{P(s)} < 0 \quad .$$

It follows from this that the whispering gallery effect can take place only on one side of S. Thus for instance when $P(s) > 0$ the effect arises to the right of S where $n < 0$, and does not arise to the left of $S(n > 0)$.

The geometric significance of the effective radius of curvature $P(s)$ can be drawn from (5.3.13). Replacing the infinitesimals Δs and ε by other equivalent infinitesimals, we get for $P(s)$

$$P(s) = \lim_{h \to 0} \frac{|NM|^2}{8h} \quad ,$$

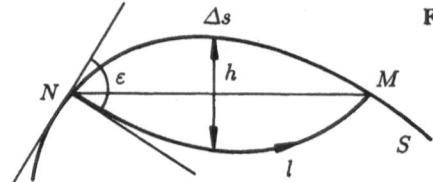

Fig. 5.8. On the geometrical significance of $P(s)$

in which $|NM|$ is the length of chord NM of the lune (Fig. 5.8) formed by the boundary and the ray NlM, and h is its thickness (the maximum length of any of its chords perpendicular to NM).

On the other hand, the effective radius of curvature is (within a factor of c_0) the radius of curvature in a Riemannian space with a suitably selected metric. In fact, consider the (x, y) plane as a Riemannian manifold with metric defined by

$$d\chi^2 = \frac{1}{c^2(x, y)}(dx^2 + dy^2) \quad .$$

Its geodesics will be extremals of the integral $\int d\chi$, i.e. rays. The Riemannian radius of curvature ϱ_R of the curve $\vec{x} = \vec{x}(\chi)$ [$\vec{x} = (x, y)$ and χ is the Riemannian arc length] is defined as in the Euclidean case by the equation

$$\frac{D^2\vec{x}}{d\chi^2} = -\frac{1}{\varrho_R}\vec{n}_R \quad ,$$

where $\vec{n}_R$ is a unit normal (in the sense of the Riemannian metric) directed to the left from the curve $\vec{x} = \vec{x}(\chi)$. The symbol $D/d\chi$ denotes covariant differentiation. It can be easily shown by calculation that at any point of the contour S,

$$\varrho_R = \frac{1}{c_0}P \quad ,$$

where P is the effective radius of curvature.

If the velocity of propagation is constant, it was shown in Sect. 5.1 [see (5.1.6)] that the glancing angle ε_{j+1} at a reflection point N_{j+1} is related to the glancing angle ε_j at the preceding reflection point N_j by the expression

$$\varepsilon_{j+1} = \varepsilon_j - \tfrac{2}{3}\varrho'_j\varepsilon_j^2 + O(\varepsilon_j^3), \tag{5.3.15}$$

where ϱ'_j is the derivative of the radius of curvature of boundary S at point N_j. We will show that in the case of a nonhomogeneous medium, where the rays are curvilinear (Fig. 5.9), the derivative of the radius of curvature ϱ'_j must be replaced by

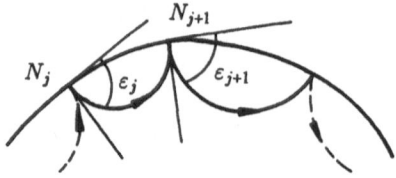

Fig. 5.9. Whispering-gallery rays in an inhomogeneous medium

$$c_0(s_j)\frac{d}{ds}[P(s)c_0^{-1}(s)]\Big|_{s=s_j} \quad .$$

It is obvious that

$$\tan \varepsilon_{j+1} = n_j'(s_{j+1}) \quad \text{or} \quad \varepsilon_{j+1} = n_j'(s_{j+1}) + O(\varepsilon_{j+1}^3) \quad . \tag{5.3.16}$$

Differentiating (5.3.11) and substituting for $(s_{j+1} - s_j)$ its expansion (5.3.12), we get

$$n_j'(s_{j+1}) = n_j'(s_j) + 2\varepsilon_j + \frac{2}{3}\frac{n_j'''(s_j)}{[n_j''(s_j)]^2}\varepsilon_j^2 + O(\varepsilon_j^3) \quad .$$

Thus, (5.3.16) can be written as

$$\varepsilon_{j+1} = \varepsilon_j + \frac{2}{3}\lim_{\varepsilon_j \to 0}\frac{n_j'''(s_j)}{[n_j''(s_j)]^2}\varepsilon_j^2 + O(\varepsilon_j^3) \quad . \tag{5.3.17}$$

Let us calculate $\lim_{\varepsilon_j \to 0} n_j'''(s_j)$. To do this, we differentiate (5.3.7), set $s = s_j$ and take the limit as $\varepsilon_j \to 0$, obtaining

$$\lim_{\varepsilon_j \to 0} n_j'''(s_j) = -\frac{d}{ds}[P(s)c_0^{-1}(s)]\Big|_{s=s_j} c_0(s_j)P^{-2}(s_j) \quad .$$

This enables us to rewrite (5.3.17) as[7]

$$\varepsilon_{j+1} = \varepsilon_j - \frac{2}{3}c_0(s_j)\frac{d}{ds}[P(s)c_0^{-1}(s)]\Big|_{s=s_j}\varepsilon_j^2 + O(\varepsilon_j^3) \quad . \tag{5.3.18}$$

Using (5.3.18), we can readily find, just as in Sect. 5.1, the relation between the glancing angle ε_m after m reflections and the initial glancing angle ε_0

$$\frac{\varepsilon_m}{\varepsilon_0} = \sqrt[3]{\frac{c_0(s_m)P(s_0)}{c_0(s_0)P(s_m)}} + O(\varepsilon_0) \quad . \tag{5.3.19}$$

Formula (5.3.19) implies that when (5.3.14) holds, any ray that has met the contour S at a small enough angle will remain within a small neighborhood of S after a correspondingly large number of subsequent reflections, i.e., the system of multiply reflected rays will be stable in the first approximation.

If account is taken of terms of order ε_j^3 in (5.3.12 and 17), then it will be found that the correction term in (5.3.19), as in the case of $c(s, n) \equiv 1$, will have order ε_0^2.

Let us now derive a formula for the eigenvalues of the whispering gallery type oscillations for the problem

---

[7] If we introduce (see above) the Riemannian radius of curvature ϱ_R, then (5.3.18) will have the same form as the analogous formula in the case $c \equiv 1$ (Sect. 5.1):

$$\varepsilon_{j+1} = \varepsilon_j - \frac{2}{3}\varepsilon_j^2\frac{d\varrho_R}{d\chi}$$

[$d/d\chi$ is differentiation with respect to the Riemannian arc length: $d\chi^2 = (dx^2 + dy^2)/c^2$].

$$\Delta u + \frac{\omega^2}{c^2(s,n)} u = 0 \; , \quad \frac{\partial u}{\partial n}\bigg|_s = 0 \; ,$$

where S is a closed curve that bounds the region Ω. Let us assume that $n<0$ within Ω and that the inequality

$$0<P(s)<\infty \tag{5.3.20}$$

is satisfied on S. In this case, as we have just found, superposition of multiply reflected waves occurs in Ω near the boundary S, and the system of multiply reflected rays is stable in the first approximation.

For a set of rays that is invariant in the first approximation relative to reflections at boundary S, the glancing angle $\varepsilon(s)$ is defined by

$$\varepsilon(s) = \kappa \sqrt[3]{\frac{c_0(s)}{P(s)}} \; .$$

The envelope or caustic of such a set is described by

$$n = -\tfrac{1}{2}[P(s)c_0^2(s)]^{1/3}\kappa^2 + O(\kappa^3). \tag{5.3.21}$$

To get the natural frequencies, the invariant set of rays must be broken down into two normal congruences. A closed congruence will be formed by curvilinear rays going from boundary S to the envelope, and by rays going from the envelope to the boundary. As in the case of constant velocity in Sect. 5.1, we shall take as basis curves the caustic $\mathcal{L}_k$ and the curvilinear triangle l_k formed by two rays having a common point on S and the segment of the caustic between their points of tangency. Taking into consideration that along these basis curves we have

$$\sum \tau_j \, dx_j = \pm \frac{d\sigma}{c(s,n)} = \pm \frac{1}{c_0(s)}\left[1 - \frac{c_1(s)}{c_0(s)}n + O(n^2)\right] d\sigma \; ,$$

where $d\sigma$ is an element of length of the curve, we get the two equations

$$\omega \left[\int_0^L \frac{ds}{c_0(s)} - \frac{1}{2}\kappa^2 \int_0^L \frac{ds}{c_0^{1/3}(s)P^{2/3}(s)} + O(\kappa^3) \right] = 2\pi q \; ,$$

$$\omega \left[\tfrac{2}{3}\kappa^3 + O(\kappa^4) \right] = 2\pi \left(p + \tfrac{1}{4} \right) \; , \tag{5.3.22}$$

where L is the circumference of S. (For the boundary condition $u|_s = 0$, the term $p + 1/4$ must be replaced by $p + 3/4$.) Solving the system of equations (5.3.22), we find the natural frequencies

$$\omega_{p,q} = \frac{1}{\displaystyle\int_0^L \frac{ds}{c_0(s)}} \left\{ 2\pi q + \left[\frac{3\pi(p + 1/4)}{2} \right]^{2/3} \right.$$

$$\times \left[\pi q \bigg/ \int_0^L \frac{ds}{c_0(s)} \right]^{1/3} \int_0^L \frac{ds}{c_0^{1/3}(s)P^{2/3}(s)} + \ldots \Bigg\} \ ,$$

The coefficients $a_m(s, \nu)$ and $b_m(s, \nu)$ here contain the partial derivatives with respect to s and ν of the desired polynomials $\alpha_n(s, \nu)$ and $\beta_n(s, \nu)$. Equations (8.2.17) will be identically satisfied in ω if and only if

$$a_m(s, \nu) = 0 \quad , \quad m = -4, \, -3, \, -2, \ldots \quad ,$$
$$b_m(s, \nu) = 0 \quad , \quad m = -2, \, -1, \, 0, \ldots \quad . \tag{8.2.19}$$

Equations (8.2.19) are a recurrent system of partial differential equations that must be satisfied by the unknown polynomials $\alpha_n(s, \nu)$ and $\beta_n(s, \nu)$. Let us find the principal terms $\alpha_{-2}(s, \nu)$, $\alpha_{-1}(s, \nu)$, $\alpha_0(s, \nu)$ and $\beta_0(s, \nu)$ of the formal series. This will permit the equations of (8.2.19) for $m \geq 1$ to be substantially simplified. After making the necessary calculations we get

$$a_{-4}(s, \nu) \equiv -\left(\frac{\partial \alpha_{-2}}{\partial \nu}\right)^2 = 0 \quad , \tag{8.2.20}$$

$$a_{-3}(s, \nu) \equiv -2\frac{\partial \alpha_{-2}}{\partial \nu}\frac{\partial \alpha_{-1}}{\partial \nu} - \left(\frac{\partial \alpha_{-2}}{\partial \nu}\right)^2\frac{\nu}{\varrho} = 0 \quad , \tag{8.2.21}$$

$$a_{-2}(s, \nu) \equiv -\left(\frac{\partial \alpha_{-1}}{\partial \nu}\right)^2 - 2\frac{\partial \alpha_{-2}}{\partial \nu}\frac{\partial \alpha_0}{\partial \nu} - 2\frac{\partial \alpha_{-2}}{\partial \nu}\frac{\partial \alpha_{-1}}{\partial \nu}\frac{\nu}{\varrho}$$
$$+ i\frac{\partial^2 \alpha_{-2}}{\partial \nu^2} - \left(\frac{\partial \alpha_{-2}}{\partial s}\right)^2 + \frac{1}{c_0^2(s)} = 0 \quad , \tag{8.2.22}$$

$$a_{-1}(s, \nu) \equiv -2\frac{\partial \alpha_{-1}}{\partial \nu}\frac{\partial \alpha_0}{\partial \nu} - \left(\frac{\partial \alpha_{-1}}{\partial \nu}\right)^2\frac{\nu}{\varrho} + i\frac{\partial^2 \alpha_{-1}}{\partial \nu^2}$$
$$- 2\frac{\partial \alpha_{-2}}{\partial s}\frac{\partial \alpha_{-1}}{\partial s} + \left(\frac{\partial \alpha_{-2}}{\partial s}\right)^2\frac{\nu}{\varrho} + e_1(s)\frac{\nu}{\varrho} = 0 \quad , \tag{8.2.23}$$

$$b_{-2}(s, \nu) \equiv 2i\frac{\partial \alpha_{-2}}{\partial \nu}\frac{\partial \beta_0}{\partial \nu} = 0 \quad , \tag{8.2.24}$$

that some eigenfunctions of $\mathscr{A}$ will be concentrated near a stable ray $\mathscr{D}$.

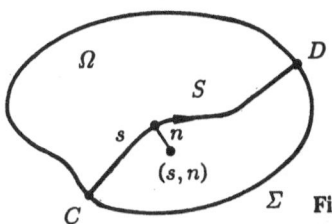

Σ **Fig. 5.10.** An extremal ray of an inhomogeneous region

This section will be concerned with finding the asymptotic expansions of the corresponding eigenvalues and the derivation of an analytical condition for the stability of an extremal ray.

As in the preceding section, let us associate coordinates s and n with S. We take the length of an extremal ray S to be equal to h, and measure the arc length s along the ray from the point C. Since S in the present case is a ray, the condition $1/P(s) = 0$ is satisfied along S, and for rays close to S we have (Sect. 5.3) the equation

$$\frac{d^2n}{ds^2} - \frac{c_0'(s)}{c_0(s)}\frac{dn}{ds} + 2\frac{c_2(s)}{c_0(s)}n + (\text{quadratic terms}) = 0 \quad . \tag{5.4.1}$$

We eliminate the term containing the first derivative in (5.4.1) by passing to a new unknown function

$$y(s) = \sqrt{\frac{c_0(0)}{c_0(s)}}\, n(s) \quad . \tag{5.4.2}$$

In the linear approximation, we get the equation

$$y''(s) + K(s)y(s) = 0 \quad , \tag{5.4.3}$$

for $y(s)$, where

$$K(s) = \frac{1}{2}\frac{c_0''(s)}{c_0(s)} - \frac{3}{4}\left(\frac{c_0'(s)}{c_0(s)}\right)^2 + 2\frac{c_2(s)}{c_0(s)} \quad . \tag{5.4.4}$$

Let an initial ray going in the direction of increasing arc length along S be described by the equation $n = n_0(s)$. Let the reflected ray going in the opposite direction be described by the equation $n = n_1(s)$, and in general let $n = n_l(s)$ be the equation of the ray arising as a result of l reflections of the initial ray. Let us derive conditions which in the first approximation will relate the displacement $n_{m-1}(0)$, $m = 2l$ and the angular coefficient $n_{m-1}'(0)$ of a ray incident on Σ in the vicinity of C, to the displacement $n_m(0)$ and angular coefficient $n_m'(0)$ of the reflected ray. Since we are interested in the first approximation only, we can replace Σ near C by the arc of a circle with radius equal to the radius of curvature r_1 of Σ at C. We see easily from Fig. 5.11 that

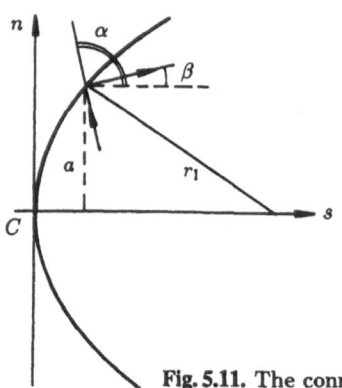

Fig. 5.11. The connection between parameters of incident and reflected rays

$$\pi - \alpha - \arcsin\frac{a}{r_1} = \beta + \arcsin\frac{a}{r_1} \quad . \tag{5.4.5}$$

Since in the first approximation

$$n_m(0) = n_{m-1}(0) = a \; ,$$

$$n'_m(a) = \beta, \quad n'_{m-1}(0) = \alpha - \pi, \quad \arcsin\frac{a}{r_1} = \frac{a}{r_1} \quad , \tag{5.4.6}$$

then (5.4.5) can be rewritten as

$$n'_m(0) + \frac{1}{r_1}n_m(0) = -\left[n'_{m-1}(0) + \frac{1}{r_1}n_{m-1}(0)\right] \quad . \tag{5.4.7}$$

For the functions $y_m(s)$ and $y_{m-1}(s)$ [which are related to $n_m(s)$ and $n_{m-1}(s)$ by (5.4.2)] we get

$$y_m(0) = y_{m-1}(0) \; ,$$

$$y'_m(0) + q(0)y_m(0) = -(y'_{m-1}(0) + q(0)y_{m-1}(0)) \quad , \tag{5.4.8}$$

in place of (5.4.6 and 7), where

$$q(0) = \frac{1}{2}\frac{c'_0(0)}{c_0(0)} + \frac{1}{r_1} \quad .$$

In a similar way we can find the conditions that relate $y_m(h)$, $y'_m(h)$ to $y_{m+1}(h)$, $y'_{m+1}(h)$. These conditions take the form

$$y_{m+1}(h) = y_m(h) \; ,$$

$$y'_{m+1}(h) + q(h)y_{m+1}(h) = -[y'_m(h) + q(h)y_m(h)] \quad , \tag{5.4.9}$$

where $q(h) = \frac{1}{2}(c'_0(h)/c_0(h)) - (1/r_2)$ and r_2 is the radius of curvature of Σ at D.

The initial ray is specified by the conditions

$$y_0(0) = \alpha_0, \quad y'_0(0) + q(0)y_0(0) = \beta_0 \; ,$$

$$\sqrt{[q(0)\alpha_0]^2 + \beta_0^2} \ll 1 \quad . \tag{5.4.10}$$

For a ray which has been reflected j times we put

$$y_j(0) = \alpha_j, \quad y'_j(0) + q(0)y_j(0) = \beta_j \quad .$$

We express α_m, β_m, $m = 2l$, in terms of α_0, β_0 by constructing two linearly independent solutions $f_1(s)$ and $f_2(s)$ of (5.4.3) which satisfy the conditions

$$f_1(0) = a_1 > 0 \; , \quad f'_1(0) + q(0)f_1(0) = 0$$

$$f_2(h) = a_2 > 0 \; , \quad f'_2(h) + q(h)f_2(h) = 0 \tag{5.4.11}$$

and we introduce the following notation:

$$f_1(h) = b_1 \; , \quad f'_1(h) + q(h)f_1(h) = c_1$$

$$f_2(0) = b_2 \; , \quad f'_2(0) + q(0)f_2(0) = c_2 \tag{5.4.12}$$

and assume that $c_2 \neq 0$. Then the equation of the initial ray $y = y_0(s)$ can be written as

$$y_0(s) = \left(\frac{\alpha_0}{a_1} - \frac{b_2}{a_1 c_2} \beta_0 \right) f_1(s) + \frac{1}{c_2} \beta_0 f_2(s) \quad . \tag{5.4.13}$$

Obviously the parameters α_j and β_j which define the ray undergo a linear transformation with each reflection (in the first approximation). The parameters α_m and β_m that correspond to the ray arising after $2l$ reflections are expressed in terms of α_0 and β_0 by the transformation

$$\binom{\alpha_m}{\beta_m} = A^l \binom{\alpha_0}{\beta_0} \quad ,$$

where the elements a_{ik} of the matrix A can be expressed in terms of a_1, b_1, c_1 and a_2, b_2, c_2. It can be readily calculated that

$$a_{11} = 2 \frac{b_1 b_2}{a_1 a_2} - 1 \ , \quad a_{12} = 2 \left(\frac{b_2}{c_2} - \frac{b_1 b_2^2}{a_1 a_2 c_2} \right) \ ,$$

$$a_{21} = -2 \frac{b_1 c_2}{a_1 a_2} \ , \quad a_{22} = 2 \frac{b_1 b_2}{a_1 a_2} - 1 \ . \tag{5.4.14}$$

As in the case of a homogeneous medium, the system of multiply reflected curvilinear rays $n = n_j(s)$ will be stable in the first approximation if the eigenvalues λ_1 and λ_2 of A are distinct and have magnitude equal to one. The characteristic equation for A has the form

$$\lambda^2 - 2 \left(2 \frac{b_1 b_2}{a_1 a_2} - 1 \right) \lambda + 1 = 0 \quad ,$$

and consequently

$$\lambda_{1,2} = 2 \frac{b_1 b_2}{a_1 a_2} - 1 \pm \sqrt{ \left(2 \frac{b_1 b_2}{a_1 a_2} - 1 \right)^2 - 1 } \quad .$$

Thus the inequality

$$\left| 2 \frac{b_1 b_2}{a_1 a_2} - 1 \right| < 1 \quad \text{or}$$

$$0 < \frac{f_1(h) f_2(0)}{f_1(0) f_2(h)} < 1 \tag{5.4.15}$$

gives the condition for stability of a system of rays $n_j(s)$, $j = 0, 1, 2, \dots$. If

$$\left| 2 \frac{b_1 b_2}{a_1 a_2} - 1 \right| = 1 \quad , \tag{5.4.16}$$

then $\lambda_1 = \lambda_2$, and in order for the system of rays $n_j(s)$ to be stable, we must stipulate in addition that $a_{12} = a_{21}$ (see Sect. 5.2), i.e.

$$2 \left(\frac{b_2}{c_2} - \frac{b_1 b_2^2}{a_1 a_2 c_2} \right) = -2 \frac{b_1 c_2}{a_1 a_2} \quad . \tag{5.4.17}$$

Equations (5.4.16, 17) are simultaneously satisfied only if $b_1 = b_2 = 0$, i.e., if

$$f_1(h) = f_2(0) = 0 \quad . \tag{5.4.18}$$

Finally, consider the case where (5.4.3) has a non-zero solution that satisfies the homogeneous boundary conditions

$$y'(0) + q(0)y(0) = 0 \ , \quad y'(h) + q(h)y(h) = 0 \ ,$$

and consequently the solutions $f_1(s)$ and $f_2(s)$ are no longer linearly independent. In this case, we may take for the second solution $f_2(s)$ any solution of (5.4.3) that is linearly independent of $f_1(s)$. Now the elements of A will be equal to

$$a_{11} = a_{22} = 1 \ , \quad a_{12} = 0 \ , \quad a_{21} = 2(a_1 a_2 - b_1 b_2)b_1^{-1}c_2^{-1} \ ,$$

so that $\lambda_1 = \lambda_2 = 1$, and the stability condition is written as $a_{21} = 0$, or

$$f_1(h)f_2(0) = f_1(0)f_2(h) \quad . \tag{5.4.19}$$

Let us denote the argument of the complex number

$$2\frac{f_1(h)f_2(0)}{f_1(0)f_2(h)} - 1 + i\sqrt{1 - \left(2\frac{f_1(h)f_2(0)}{f_1(0)f_2(h)} - 1\right)^2}$$

by ϕ, and assume that $0 \leq \phi \leq \pi$. We easily calculate that

$$\phi = 2\arctan\sqrt{\frac{f_1(0)f_2(h)}{f_1(h)f_2(0)} - 1} = 2\arccos\sqrt{\frac{f_1(h)f_2(0)}{f_1(0)f_2(0)}} \tag{5.4.20}$$

[compare (5.2.13)]. The eigenvalues of A can be expressed in terms of ϕ as

$$\lambda_1 = e^{i\phi} \ , \quad \lambda_2 = e^{-i\phi} \quad .$$

We can now determine the eigenvalues of the problem

$$\Delta u + \frac{\omega^2}{c^2(s,n)}u = 0 \ , \quad \left.\frac{\partial u}{\partial \nu}\right|_{\Sigma} = 0 \quad (\nu \text{ is the normal to } \Sigma) \ ,$$

that correspond to eigenfunctions concentrated in the vicinity of an extremal ray CD of Ω. We will assume that stability condition (5.4.15) is satisfied for an extremal ray CD, and consequently that there exists a system of multiply reflected rays in the neighborhood of CD that is stable in the first approximation.

To find the eigenvalues we must first construct a set of rays close to the extremal ray that is invariant to reflections. Just as in Sect. 5.2, this problem reduces to finding subsets of the vector space $\binom{\alpha}{\beta}$ that are transformed into themselves when operated on by the matrices A^l.

From (5.2.16) we see that we get an invariant set of rays if we choose their parameters (α, β) in the form

$$\alpha = \kappa \cos \mu \ ,$$
$$\beta = \kappa a_{12}^{-1}[-a_{11} \cos \mu + \cos(\phi + \mu)] \ ,$$
$$\kappa \ll 1 \ , \quad 0 \leq \mu \leq 2\pi \ , \tag{5.4.21}$$

where a_{11} and a_{12} are expressed in terms of f_1 and f_2 by (5.4.14, 11, 12). Taking into consideration that $\cos \phi = a_{11}$, we get

$$\beta = -\kappa \frac{\sin \phi}{a_{12}} \sin \mu \quad .$$

If we introduce the notation

$$\frac{b_2 \sin \phi}{c_2 a_{12}} = Q \quad ,$$

then from (5.4.14) we can easily calculate that

$$Q^2 = \frac{b_1 b_2}{a_1 a_2 - b_1 b_2} \quad . \tag{5.4.22}$$

Using Q, the invariant set of rays can now be written as

$$y = \kappa \left[\frac{1}{a_1} (\cos \mu + Q \sin \mu) f_1(s) - \frac{1}{b_2} (Q \sin \mu) f_2(s) \right] \quad . \tag{5.4.23}$$

By the usual method we find the envelope of the set of rays (5.4.23):

$$\frac{y^2}{\kappa^2} = \frac{Q^2}{a_1 b_2} \left[\frac{a_2}{b_1} f_1^2(s) - 2 f_1(s) f_2(s) + \frac{a_1}{b_2} f_2^2(s) \right] \quad . \tag{5.4.24}$$

Obviously (5.4.24) has two branches, so that the invariant set of rays (5.4.23) is bounded by two caustics, for one of which $y>0$, and for the other $y<0$. Just as in Sect. 5.2, the invariant set of rays is splits up into two normal congruences (the only difference being that we now have curvilinear rather than straight rays). We form the two other normal congruences of rays from the set of rays which arises from the reflection of the first set and propagates in the direction from D to C. Joining these normal congruences together along the caustics and sections of the boundary, we get a closed congruence that covers a four-sheeted covering space homeomorphic to a torus. Using (5.4.23) which describes the invariant set of rays, we can find the components of $\nabla \tau$ on each sheet of the space. Carrying out calculations analogous to those of Sect. 5.2, we get

$$\tau_n = \pm \left[\frac{\frac{1}{2} \frac{d}{ds} n^2(s)}{c_0(s) n^2(s)} n \mp \kappa^2 \frac{Q}{a_1 b_2} W \frac{1}{c_0(0) n(s)} \sqrt{1 - \frac{n^2}{n^2(s)}} \right] + O(\kappa^2) \quad ,$$

$$\tau_s = \pm \frac{1}{c(s,n)} \left[1 - \frac{c^2(s,n)}{2} \tau_n^2 + O(\kappa^3) \right] \quad ,$$

where s and n are coordinates of a point in space,

$$n(s) = \sqrt{\frac{c_0(s)}{c_0(0)}} y(s) = \kappa \sqrt{\frac{c_0(s)}{c_0(0)}} \sqrt{\frac{Q^2}{a_1 b_2}}$$

$$\times \left[\frac{a_2}{b_1} f_1^2(s) - 2 f_1(s) f_2(s) + \frac{a_1}{b_2} f_2^2(s) \right]^{1/2} \tag{5.4.25}$$

is the right-hand side of the equation of the envelope (5.4.24) in these coordinates, and $W = f_1 f_2' - f_1' f_2$ is the Wronskian of f_1 and f_2.

The basis curves are also chosen as before: Γ_1 is the extremal ray traversed on the first and third sheets in opposite directions, and Γ_2 is the double segment $s = s_0$ and $-n(s_0) < n < n(s_0)$ that belongs to the first and second sheets, with s_0 some fixed value of the arc length s.

Since $n = 0$ along Γ_1, the first equation of (4.3.12) used in determining the natural frequencies is written as

$$2\omega \left[\int_0^h \frac{ds}{c_0(s)} - \frac{\kappa^2}{2} \frac{Q^2}{a_1^2 b_2^2} W^2 \frac{1}{c_0^2} \int_0^h \frac{c_0(s)}{n^2(s)} ds \right] = 2\pi p \quad .$$

When we integrate τ_n along Γ_1 the integral of the first term (which is an odd function of n) vanishes, and therefore the second equation of (4.3.12) takes the form

$$2\omega\kappa^2 \frac{Q}{a_1 b_2} \frac{W}{c_0(0)n(s_0)} \int_{-n(s_0)}^{n(s_0)} \sqrt{1 - \frac{n^2}{n^2(s_0)}} \, dn = 2\pi \left(q + \frac{1}{2} \right) \quad . \tag{5.4.26}$$

Using (5.4.25) and evaluating the integral in (5.4.26) we get the following system of equations:

$$\omega \left(\int_0^h \frac{ds}{c_0(s)} - \frac{\kappa^2}{2} \frac{W^2}{a_1 b_2} \frac{1}{c_0(0)} \int_0^h \frac{ds}{\frac{a_2}{b_1} f_1^2(s) - 2 f_1(s) f_2(s) + \frac{a_1}{b_2} f_2^2(s)} \right) = \pi p \quad ,$$

$$\omega\kappa^2 \frac{Q}{a_1 b_2} W \frac{1}{c_0(0)} = 2q + 1 \quad ,$$

from which we can readily find the natural frequencies

$$\omega_{p,q} = \left(\int_0^h \frac{ds}{c_0(s)} \right)^{-1} \left(\pi p + \frac{2q+1}{2} \frac{W}{Q} \int_0^h \frac{ds}{\frac{a_2}{b_1} f_1^2(s) - 2 f_1(s) f_2(s) + \frac{a_1}{b_2} f_2^2(s)} \right)$$

$$p \gg 1 \, , \quad q = 0, 1, 2, \ldots \, , \tag{5.4.27}$$

and the values of κ^2 :

$$\kappa_{p,q}^2 = \frac{2q+1}{\omega_{p,q}} \frac{a_1 b_2}{Q} \frac{c_0(0)}{W} \quad .$$

Substituting the values found for $\kappa_{p,q}$ into (5.4.24), we get the equation for the caustic:

$$n^2 = \frac{2q+1}{\omega_{p,q}} \frac{Q}{W} c_0(s) \left[\frac{a_2}{b_1} f_1^2(s) - 2 f_1(s) f_2(s) + \frac{a_1}{b_2} f_2^2(s) \right] \quad , \tag{5.4.28}$$

which bounds the strip within which the corresponding eigenfunction is concentrated. It can be easily seen that when $q = O(1)$ the width of this strip is of order $\omega_{p,q}^{-1/2}$. This conclusion is important in the analyses of Chaps. 6 and 8. In calculating Q, we must take the square root of (5.4.22). The sign of the root must be selected so that the right side of (5.4.28) is positive for all values of s. From this requirement we come to the conclusion that

$$Q = \frac{\text{sign}\left(\frac{a_1}{b_2}W\right)}{\sqrt{\frac{a_1 a_2}{b_1 b_2} - 1}} \quad . \tag{5.4.29}$$

Recalling the way that a_1, a_2, b_1 and b_2 are related to the values of solutions $f_1(s)$ and $f_2(s)$ at 0 and h, we obtain

$$n^2 = \frac{2q+1}{\omega_{p,q}} \frac{Q}{W} \left[\frac{f_2(h)}{f_1(h)} f_1^2(s) - 2f_1(s)f_2(s) + \frac{f_1(0)}{f_2(0)} f_2^2(s)\right]$$

and

$$Q = \frac{\text{sign}\left(\frac{f_1(0)}{f_2(0)}W\right)}{\sqrt{\frac{f_1(0)f_2(h)}{f_1(h)f_2(0)} - 1}} \quad ,$$

in place of (5.4.28 and 29).

It remains for us to compute the integral in formula (5.4.27) for the eigenvalues. Introducing the complex number

$$Z = f_1(s) + iQ\left[\frac{a_1}{b_2} f_2(s) - f_1(s)\right] \quad ,$$

the argument of which we denote by $\Psi(s)$, we see that if $s = 0$, obviously $Z = f_1(0) = a_1 > 0$, and we take $\Psi(0) = 0$.

Let us calculate the derivative of $\Psi(s)$:

$$\Psi'(s) = \frac{W}{Q}\left[\frac{a_2}{b_1} f_1^2(s) - 2f_1(s)f_2(s) + \frac{a_1}{b_2} f_2^2(s)\right]^{-1} \quad . \tag{5.4.30}$$

By virtue of the choice of the sign in (5.4.29), $\Psi'(s) > 0$, and consequently the function $\Psi(s)$ increases monotonically with increasing s. Let us denote the positive roots of $f_1(s)$ by s_j, $j = 1, 2, 3, \ldots$. Since the function $\Psi(s)$ increases monotonically, it can be stated that if

$$s_j \leq s \leq s_{j+1} \quad ,$$

then

$$\pi(j - 1/2) \leq \Psi(s) \leq \pi(j + 1/2)$$

for all $j \geq 1$.

Using (5.4.30), we arrive at the equation

$$\frac{W}{Q}\int_0^h \left[\frac{a_2}{b_1} f_1^2(s) - 2f_1(s)f_2(s) + \frac{a_1}{b_2} f_2^2(s)\right]^{-1} ds = \int_0^h d\Psi(s) = \Psi(h) \quad .$$

If the function $f_1(s)$ has N roots in the interval $(0, h)$, then $\Psi(h) > \pi(N - 1/2)$ and consequently

$$\Psi(h) = \arctan Q\left(\frac{a_1}{b_2}\frac{f_2(h)}{f_1(h)} - 1\right) + \pi N \quad . \tag{5.4.31}$$

Here the principal branch of the arc tangent is used, whose range is $(-\pi/2, \pi/2)$. Substituting the values of Q, a_1 and b_1, (5.4.31) can be rewritten as

$$\Psi(h) = \text{sign}\left(\frac{f_1(0)}{f_2(0)}W\right)\arctan\sqrt{\frac{f_1(0)f_2(h)}{f_1(h)f_2(0)} - 1} + \pi N$$

or, recalling (5.4.20), as

$$\Psi(h) = \text{sign}\left(\frac{f_1(0)}{f_2(0)}W\right)\frac{\phi}{2} + \pi N \quad ,$$

where ϕ is the phase of the eigenvalue λ_1 of A. Thus we arrive at last at the following formula for the eigenvalues:

$$\omega_{p,q} = \left(\int_0^h \frac{ds}{c_0(s)}\right)^{-1}$$

$$\times \left\{\pi p + \left(q + \frac{1}{2}\right)\left[\text{sign}\left(\frac{f_1(0)}{f_2(0)}W\right)\frac{\phi}{2} + \pi N\right]\right\} \quad . \tag{5.4.32}$$

We have derived this formula for the eigenvalues considering only the case where the stability condition (5.4.15) is satisfied. For conditions (5.4.18 and 5.4.19), the formula for $\omega_{p,q}$ can be derived from (5.4.32) by taking the limit.

In the case of condition (5.4.18), taking the limit as $f_1(h) \to 0$, $f_2(0) \to 0$ leads to the formula

$$\omega_{p,q} = \left(\int_0^h \frac{ds}{c_0(s)}\right)^{-1}\left[\pi p + \left(q + \frac{1}{2}\right)\left(\frac{\pi}{2} + \pi N\right)\right] \quad , \tag{5.4.33}$$

and in the case of condition (5.4.19), passage to the limit $[f_1(0)f_2(h)/f_1(h) f_2(0)] \to 1$ gives the formula

$$\omega_{p,q} = \left(\int_0^h \frac{ds}{c_0(s)}\right)^{-1}\left[\pi p + \left(q + \frac{1}{2}\right)\pi N\right] \quad . \tag{5.4.34}$$

5.5 Notes on the Literature

Only two-dimensional problems were considered in this chapter. However, similar three- and multi-dimensional problems can be studied by the same sorts of methods.

The possibility of using the ray method to obtain eigenvalues and eigenfunctions which are concentrated either near the boundary or near the minimal diameter of an arbitrary convex region was first pointed out by *Keller* and *Rubinow* [5.2]. The *Keller-Rubinow* method (Chap. 4) relies heavily on the assumption that a closed congruence of true rays exists which depends continuously on a large enough number of parameters. It is now known, however, that stable systems of multiply reflected rays bounded by caustics which depend continuously on a *single* parameter do not exist in general, even for problems in two dimensions (this follows from the work of *Arnol'd* [5.3], *Moser* [5.4] and *Mogilevskii* [5.5]).

That there is a subset of eigenfunctions concentrated in the vicinity of a closed chain (extremal diameter or ray) which is stable in the first approximation was first suggested by *Buldyrev* [5.6]. He is also responsible for the major results of Chap. 5.

The ray method in the small has been used to determine the eigenfrequencies of multiple-mirror resonators by *Buldyrev* and *Popov* [5.7] and by *Popov* [5.8]. Paraxial bundles of rays were used by *Bykov* [5.9] to find the resonant frequencies of open resonators (see also [5.10]). *Bykov*'s main idea was to replace a given open resonator by a reflecting ellipsoid for which a paraxial bundle of rays near the mirror axis behaves the same as it does in the open resonator.

6. The Parabolic Equation Method

In this chapter the parabolic equation method will be used to obtain eigenfunctions of the whispering gallery type and to find solutions of the wave equation concentrated in the neighborhood of a ray.

6.1 Introductory Remarks

In the preceding chapter the ray method in the small was used to find the leading terms in the asymptotic expansions of whispering gallery and bouncingball eigenvalues. Despite the heuristic nature of the ray method in the small, the constructions of Chap. 5 have great value. We shall undertake to treat these problems more exactly in this chapter, using some of the derivations from Chap. 5 to guide us. In particular, an important factor will be the result of Sect. 5.3 that eigenfunctions of the whispering gallery type are concentrated in a layer whose thickness is of order $O(\omega^{-2/3})$ when $p = O(1)$.

The thickness of the layer surrounding an extremal ray of a region in which eigenfunctions of the bouncing-ball type are concentrated is of order $O(\omega^{-1/2})$. This follows immediately from equations (5.4.28) for the caustics that bound this layer [assuming that $q = O(1)$ in this equation].

Thus a region where an eigenfunction of either of these types is concentrated has the nature of a boundary layer whose thickness approaches zero as $\omega \to \infty$.

It is natural to derive a corresponding boundary layer equation for the desired functions and to find their asymptotic properties by solving this equation. The method of boundary layer equations in diffraction problems is well established. This technique is traditionally called the *parabolic equation method* (although the boundary layer equation of diffraction theory is strictly speaking an equation of the Schrödinger type).

The parabolic equation method enables us to find asymptotic expansions for the eigenvalues and eigenfunctions[1] related both to an extremal ray, and to the whispering gallery phenomenon (and this does not begin to exhaust its range of application).

---

[1] Here (and in subsequent chapters) we will deal only with the formal construction of asymptotic expansions of eigenvalues and eigenfunctions. For a rigorous justification, see Sect. 7.5.

The parabolic equation method as presented here gives only the dominant terms of the asymptotic expansions of diffraction problems[2]. Use of this method is complicated still further by the fact that in deriving the "parabolic equation" it is not always possible to restrict ourselves to the principal terms of the equation and its boundary conditions. *Ivanov* [6.1] in a very interesting paper was the first to encounter this situation.

6.2 Derivation of the Parabolic Equation for Eigenfunctions of the Whispering Gallery Type

In this section and the following one, we will use the parabolic equation method to study the same problem that we considered in Sect. 5.3, namely the asymptotic expansion of solutions to the boundary value problem

$$\left(\Delta + \frac{\omega^2}{c^2(x,y)}\right)u = 0 \ , \quad x, y \in \Omega \ , \quad u \not\equiv 0 \ , \tag{6.2.1}$$

$$u|_S = 0 \quad \text{or} \quad \left.\frac{\partial u}{\partial n}\right|_S = 0 \ , \tag{6.2.2}$$

which behave as waves of the whispering gallery type.

We further assume that the effective radius of curvature $P(s)$ of boundary S is positive (Sect. 5.3), i.e., that the inequality

$$\frac{1}{P(s)} \equiv \frac{1}{\varrho(s)} - \frac{c_1}{c_0} > 0 \tag{6.2.3}$$

is valid (the notation follows that of Chap. 5).

Condition (6.2.3) ensures a whispering gallery effect in region Ω close to S. If we do not require that (6.2.3) be satisfied, then, as might be expected, we will not be able to construct an asymptotic expansion for the eigenfunctions by the parabolic equation method. If indeed $P(s)$ is negative instead of positive, then we will not be able to construct a solution of the parabolic equation which decays rapidly with distance from the boundary S of Ω.

Near S we introduce, as in Sect. 5.3, the coordinates s, n (s is arc length, and n is the normal distance from S; for points in Ω, $n<0$). Using the classical expression for the Laplacian in arbitrary orthogonal coordinates (Appendix A.2), (6.2.1) can be written in the form

---

[2] It can, however, be refined so as to yield higher-order terms as well (see the references to this chapter). The method presented in Chaps. 7-9 will also do this and can be used to find the complete asymptotic expansions for solutions of diffraction problems, and in some cases to rigorously justify these expansions as well. Let us also note that the method of Chap. 7 can be used to investigate problems that have not so far been solvable using the parabolic equation method.

$$\frac{1}{h_1 h_2}\left[\frac{\partial}{\partial q_1}\left(\frac{h_2}{h_1}\frac{\partial u}{\partial q_1}\right)+\frac{\partial}{\partial q_2}\left(\frac{h_1}{h_2}\frac{\partial u}{\partial q_2}\right)\right]+\frac{\omega^2}{c^2}u$$

$$\equiv\left(1+\frac{n}{\varrho}\right)^{-1}\left[\frac{\partial}{\partial s}\left(1+\frac{n}{\varrho}\right)^{-1}\frac{\partial u}{\partial s}+\frac{\partial}{\partial n}\left(1+\frac{n}{\varrho}\right)\frac{\partial u}{\partial n}\right]+\frac{\omega^2}{c^2}u=0\quad.$$

(6.2.4)

[Recall that (s, n) are orthogonal coordinates, and that the metric coefficients h_n and h_s are given by (5.3.3)].

We will look for solutions of these equations that have the nature of waves propagating along S and tending rapidly to zero outside of a boundary layer. It is natural to set

$$u = U(n, s, \omega)\exp\left(i\omega\int_{s_0}^{s}\frac{ds}{c(s, 0)}\right)\quad,$$

(6.2.5)

where s_0 is some fixed value of arc length s, $U(n, s, \omega)$ is a function which, for large ω, varies more slowly than the phase factor $\exp[i\omega\int_{s_0}^{s}ds/c(s, 0)]$. Following Fock's terminology [6.2], we will call this function an *attenuation function*, and derive a parabolic equation for it.

Substituting (6.2.5) into (6.2.4), we arrive at

$$\left(1+\frac{n}{\varrho(s)}\right)U_{nn}+\left(1+\frac{n}{\varrho(s)}\right)^{-1}U_{ss}+\frac{1}{\varrho(s)}U_n+\frac{2i\omega}{c(s, 0)}\left(1+\frac{n}{\varrho(s)}\right)^{-1}U_s$$

$$+U_s\frac{\partial}{\partial s}\left(\frac{1}{1+\frac{n}{\varrho(s)}}\right)+i\omega U\frac{\partial}{\partial s}\frac{1}{c(s, 0)\left(1+\frac{n}{\varrho(s)}\right)}$$

$$+\omega^2\left[\left(1+\frac{n}{\varrho(s)}\right)\frac{1}{c^2(s, n)}-\frac{1}{1+\frac{n}{\varrho(s)}}\frac{1}{c^2(s, 0)}\right]U=0\quad.$$

(6.2.6)

We get a parabolic equation for U by disregarding certain terms in (6.2.6). To clarify which terms in equation (6.2.6) are the principal terms we refer to Sects. 5.1, 3, where it was shown that eigenfunctions of the whispering gallery type are concentrated in a boundary layer whose thickness is of order $O(\omega^{-2/3})$. Therefore it is natural to assume that the attenuation factor U is appreciably different from zero only at values of n in the region

$$0\leq-n\leq O(\omega^{-2/3})\quad,$$

and it is only for such values of n that we will consider (6.2.6). We will further assume that

$$U = O(1)\ ,\quad\frac{\partial U}{\partial n}=O(\omega^{2/3})\ ,\quad\frac{\partial^2 U}{\partial n^2}=O(\omega^{4/3})\quad.$$

(6.2.7)

The calculations of Sects. 5.1, 3 also show that from the standpoint of geometrical optics the phase of u will not exactly be equal to $\omega\int_{s_0}^{s}ds/c(s, 0)$, since the integral is taken along the boundary which, although it is close to the rays, is *not* itself a ray.

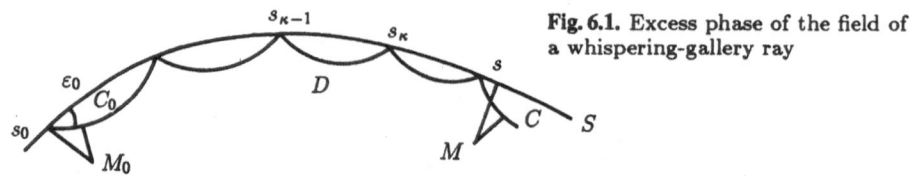

Fig. 6.1. Excess phase of the field of a whispering-gallery ray

We can calculate the order of the difference between the phase of u as calculated by the ray method in the small, and the integral $\omega \int_{s_0}^{s} ds/c(s,0)$. Let M_0 and M (Fig. 6.1) be two points with coordinates (s_0, n_0) and (s, n) located in the boundary layer, i.e.

$$0 \leq (-n, -n_0) \leq O(\omega^{-2/3}) \quad,$$

let $M_0 C_0$ be a wave front passing through M_0, let C_0 be the point of intersection of the wave front and the ray departing from the point $s = s_0$, $n = 0$ at a glancing angle ε_0. Similarly, let MC be a wave front passing through M, let C be the point of intersection of the wave front with the repeatedly rereflected ray leaving the point $s = s_0$, $n = 0$. For rays departing from the point $s = s_0$, $n = 0$ and belonging entirely to the boundary layer, the glancing angle ε_0 does not exceed the order of $O(\omega^{-1/3})$. This conclusion follows from the formula for the glancing angle (Sect. 5.3)

$$\varepsilon(s) = \kappa \sqrt[3]{c(s,0)P(s)}$$

and from the equation for the envelope of the multiply reflected rays

$$n = -\tfrac{1}{2}[P(s)c^2(s,0)]^{1/3}\kappa^2 + O(\kappa^3) \quad.$$

Actually, eliminating κ from these equations and taking into consideration that in the boundary layer $-n$ does not exceed $O(\omega^{-2/3})$, we get the inequality $\varepsilon(s) \leq O(\omega^{-1/3})$. If we now evaluate the length of the side $s_0 C_0$ of the infinitesimal right triangle $s_0 C_0 M_0$:

$$|s_0 C_0| = n_0 \sin \angle s_0 M_0 C_0 = n_0 \sin \varepsilon_0$$
$$= O(\omega^{-2/3})O(\omega^{-1/3}) = O\left(\frac{1}{\omega}\right) \quad,$$

we see that the phase of u at M_0 differs from that at s_0 by an amount of the order of

$$\omega \int_{s_0 C_0} \frac{ds}{c} \sim \omega c^{-1}(s_0, 0)|s_0 C_0| = O(1) \quad.$$

Similarly, the phases at M and s differ by no more than $O(1)$. Thus the change in phase of U in moving from M_0 to M is equal to

$$\omega \int_{s_0 s} \frac{ds}{c(s,n)} - \omega \int_{s_0}^{s} \frac{ds}{c(s,0)} + O(1)$$

where the first integral is calculated along the arcs of the rays (such as $s_{k-1}Ds_k$) reflected from S at the points s_j.

To calculate the first integral it is convenient to use (5.3.6), and we thus obtain

$$\omega \int_{s_0 s}^{s} \frac{ds}{c(s,n)} - \omega \int_{s_0}^{s} \frac{ds}{c(s,0)} = \omega \int_{s_0}^{s} \left[\Phi\left(s, n, \frac{dn}{ds}\right) - \frac{1}{c(s,0)} \right] ds$$

$$\approx \int_{s_0}^{s} \frac{1}{c_0(s)} \left(\frac{1}{\varrho(s)} - \frac{c_1(s)}{c_0(s)} \right) \omega n \, ds = \int_{s_0}^{s} O(\omega^{1/3}) \, ds \quad ,$$

since on the rays $n = O(\omega^{-2/3})$.

Thus, a geometrical optics analysis shows that the phase factor of U should take the form $\exp[i \int_{s_0}^{s} O(\omega^{1/3}) ds]$. Therefore, we will assume that the derivatives of U with respect to arc length can be estimated by

$$\frac{\partial U}{\partial s} = O(\omega^{1/3}) \; , \quad \frac{\partial^2 U}{\partial s^2} = O(\omega^{2/3}) \quad . \tag{6.2.8}$$

Using (6.2.7 and 8) we can isolate the principal terms in (6.2.6) [they will be of order $O(\omega^{4/3})$ and $O(\omega)$]. Deleting the remaining terms which have at most the order of $O(\omega^{2/3})$, we get the parabolic equation for this problem:

$$\frac{\partial^2 U}{\partial n^2} + \frac{2i\omega}{c(s)} \frac{\partial U}{\partial s} + i\omega U \frac{\partial}{\partial s} \frac{1}{c(s)} + \omega^2 n \frac{2}{c^2(s)} \frac{U}{P(s)} = 0 \quad , \tag{6.2.9}$$

where $P(s)$ is the effective radius of curvature, see (6.2.3), and $c(s) \equiv c(s,0) = c_0(s)$.

The last term in (6.2.9) precludes the use of separation of variables to solve the equation. The next section will deal chiefly with the transformation of (6.2.9) into a form which does allow separation of variables.

6.3 Solution of the Parabolic Equation (6.2.9); Asymptotic Expansion of Eigenfunctions of the Whispering Gallery Type

In the process of manipulating (6.2.9), we will disregard terms of order $O(\omega^{2/3})$. Since (6.2.9) differs from the original equation (6.2.6) by terms of just this order, this will not introduce any additional error. Let us introduce the new variable

$$\nu = n\omega^{2/3}\eta(s) \quad , \tag{6.3.1}$$

where $\eta(s)$ is a function to be determined. In contrast to the infinitesimal variable $n = O(\omega^{-2/3})$ as $\omega \to \infty$, the variable ν is finite, $\nu = O(1)$. Equation (6.2.9) can be rewritten in the form

$$
\omega^{4/3}\left(\frac{\partial^2 U}{\partial \nu^2} + \frac{2\nu}{c^2(s)\eta^3(s)P(s)}U\right) + \frac{i\omega}{c(s)\eta^2(s)}
$$

$$
\times \left(2\frac{\partial U}{\partial s} + 2\nu\frac{\eta'(s)}{\eta(s)}\frac{\partial U}{\partial \nu} - \frac{c'(s)}{c(s)}U\right) = 0 \quad .
$$

If we set

$$
\eta(s) = \sqrt[3]{\frac{2}{c^2(s)P(s)}} \quad , \tag{6.3.2}
$$

then the leading terms [of order $O(\omega^{4/3})$] will comprise an operator whose variables are separable. The only term that will interfere with the separation of variables is $(i\omega/c\eta^2)2\nu(\eta'/\eta)(\partial U/\partial \nu)$ of order $O(\omega)$. To eliminate this term, we make the substitution

$$
U = \exp[i\nu^2\omega^{-1/3}\phi(s)]V(\nu, s, \omega) \quad , \tag{6.3.3}
$$

where the function $\phi(s)$ is to be determined.

Substituting (6.3.3) for U, and disregarding terms of order $O(\omega^{2/3})$, we get

$$
\omega^{4/3}\left(\frac{\partial^2 V}{\partial \nu^2} + 4i\nu\omega^{-1/3}\phi\frac{\partial V}{\partial \nu} + 2i\omega^{-1/3}\phi V + \nu V\right)
$$

$$
+ \frac{i\omega}{c}\left(\frac{c^2 P}{2}\right)^{2/3}\left(2\frac{\partial V}{\partial s} + 2\nu\frac{\partial V}{\partial \nu}\frac{(c^{2/3}P^{1/3})'}{c^{2/3}P^{1/3}} - \frac{c'}{c}V\right) = 0 \quad .
$$

We choose $\phi(s)$ in such a way that the equation does not contain terms of the form $\nu(\partial V/\partial \nu)$. Obviously we must set

$$
\phi = \frac{1}{4c}\frac{d}{ds}\left(\frac{c^2 P}{2}\right)^{2/3} \quad . \tag{6.3.4}
$$

The resultant equation

$$
\omega^{4/3}(V_{\nu\nu} + \nu V) + \frac{2i\omega}{c}\left(\frac{c^2 P}{2}\right)^{2/3}\frac{\partial V}{\partial s} + \frac{i\omega}{3c}\left(\frac{c^2 P}{2}\right)^{2/3}
$$

$$
\times V\frac{d}{ds}\ln\left(\frac{P}{c}\right) = 0 \tag{6.3.5}
$$

will now permit separation of variables. Using the classical method for this separation, we arrive at the expression

$$
V = w(-\gamma - \nu)\exp\left[-\gamma i\left(\frac{\omega}{2}\right)^{1/3}\int_{s_0}^{s}\frac{ds}{c^{1/3}P^{2/3}}\right]\left(\frac{c}{P}\right)^{1/6} \quad . \tag{6.3.6}
$$

Here w is an arbitrary solution of Airy's equation $w''_{tt} = tw$, γ is a separation constant and s_0 is some fixed value of arc length s; in the following we will set $s_0 = 0$.

Combining expressions (6.2.5 and 6.3.3), we get the approximate formula

$$u = V \exp[i\nu^2 \omega^{-1/3} \phi(s)] \exp\left(i\omega \int_0^s \frac{ds}{c}\right) \quad , \tag{6.3.7}$$

as a solution of the original equation (6.2.4) in which V, ν and ϕ are determined by (6.3.6, and 6.3.1, 4) respectively. The function u must be concentrated in the boundary strip as $\omega \to \infty$, and therefore we must take our solution of Airy's equation to be $v(-\gamma - \nu)$ (Appendix A.1).

In fact, from (6.3.1, 2) and the fact that $P(s) > 0$ we see that when $\gamma = O(1)$ and $-n\omega^{2/3} \to +\infty$ the argument of the Airy function $-\gamma - \nu$ will also become infinite (recall that $n < 0$). It follows from the asymptotic formula for v that $v(-\gamma - \nu) \to 0$ when $-\gamma - \nu \to \infty$, and therefore $u \to 0$ as well.

If we require that u satisfy the boundary conditions (6.2.2), we find that in the case of the boundary condition $u|_S = 0$ the parameter γ must be a root of $v(-t)$, and in the case of the boundary condition $\frac{\partial u}{\partial n}\big|_S = 0$ it must be a root of the derivative $v'(-t)$ (Appendix A.1), i.e. in (6.3.6) we must set

$$\gamma = \gamma_p = t_p \ , \quad p = 0, 1, 2, \dots \ , \quad \text{when} \quad u|_S = 0 \tag{6.3.8}$$

and

$$\gamma = \gamma_p = t'_p \ , \quad p = 0, 1, 2, \dots \ , \quad \text{when} \quad \frac{\partial u}{\partial n}\bigg|_S = 0 \ . \tag{6.3.9}$$

At last we get

$$u = \left(\frac{c}{P}\right)^{1/6} \exp\left(i\omega \int_0^s \frac{ds}{c}\right) v(-\gamma_p - \nu)$$

$$\times \exp\left[-\gamma_p i \left(\frac{\omega}{2}\right)^{1/3} \int_0^s \frac{ds}{c^{1/3} P^{2/3}}\right] \exp[i\nu^2 \omega^{-1/3} \phi(s)] \ . \tag{6.3.10}$$

There is one thing worth noting about this solution. The Airy equation $v''(t) - tv(t) = 0$ has $t = 0$ as a turning point: for $t > 0$ the function $v(t)$ monotonically approaches zero, but when $t < 0$, $v(t)$ is oscillatory. Thus when $-\nu < \gamma_p$, (6.3.10) is oscillatory, and when $-\nu > \gamma_p$ it damps out rapidly. Using (6.3.1) and (6.3.2), the equation of the curve $-\nu = \gamma_p$ can be written as

$$n = -\omega^{-2/3} \sqrt[3]{\frac{c^2 P(s)}{2}} \gamma_p \ . \tag{6.3.11}$$

If in (6.3.11) we (formally) replace γ_p by the asymptotic expansion of the p-th root of function $v'(t)$:

$$\gamma_p \approx \left[\frac{3\pi}{2}\left(p + \frac{1}{4}\right)\right]^{2/3} \ ,$$

then (6.3.11) becomes the same as (5.3.21 and 24) [in (5.3.21) we must replace κ by an expression for it found from the second quantization condition (5.3.22)].

Thus we can consider (6.3.11) to be a more precise equation for the caustic, which of course was to be expected since the caustic, like (6.3.11), separates the wave zone from the shadow zone where the solution rapidly damps out.

Now let S be a boundary of a finite[3] region Ω. The function u must be a single-valued function near S. If L is the circumference of S, then obviously the requirement that u be single-valued means that $u(s, n)$ must be a periodic function with period $L : u(s + L, n) \equiv u(s, n)$. By requiring that u as given by (6.3.10) be periodic we easily get the desired formulas for the eigenvalues and eigenfunctions.

Since the factor $\exp[i\nu^2 \omega^{-1/3}\phi(s)]$ in (6.3.10) is already periodic because of the properties of $\phi(s)$, we require only that the remaining expression be periodic, whence

$$\omega_{pq} \int_0^L \frac{ds}{c} - \gamma_p \left(\frac{\omega_{pq}}{2}\right)^{1/3} \int_0^L \frac{ds}{c^{2/3}P^{2/3}} = 2\pi q \quad (q \text{ is an integer} \gg 1) \ ,$$

(6.3.12)

and thus for large ω_{pq},

$$\int_0^L \frac{ds}{c}\omega_{pq} = 2\pi q + \gamma_p \int_0^L \frac{ds}{c^{1/3}P^{2/3}} \left(\pi q \Big/ \int_0^L \frac{ds}{c}\right)^{1/3} + \dots \ ,$$

(6.3.13)

where $\gamma_p = t_p$ or $\gamma_p = t'_p$ [see (6.3.8 and 3.9)]. The ray method in the small leads to an analogous formula, (5.3.23). Substituting (6.3.13) into (6.3.10), we get an asymptotic expression for the eigenfunction that corresponds to the eigenvalue ω_{pq}. (Further terms of the asymptotic expansions of the eigenvalues ω_{pq} and eigenfunctions u_{pq} can be found by the methods of Chap. 7).

6.4 Derivation of the Basic Parabolic Equation for the Case Where S Is a Ray

Now let the curve S have an effective radius of curvature that is infinite, i.e. let S be a ray.

We first consider the problem of finding the asymptotic expansions of solutions u of the equation

$$\left(\Delta + \frac{\omega^2}{c^2(x, y)}\right)u = 0 \ ,$$

(6.4.1)

which differ appreciably from zero only in the vicinity of S and decay rapidly with increasing distance from this curve. In contrast to Sects. 6.2 and 6.3 we

---

[3] The finiteness of Ω has not been used anywhere up to this point.

will not take S to be the boundary of a region. The neighborhood of S is now an internal boundary layer, and no boundary conditions will be assigned on it. We have already encountered such a problem in the preceding chapter when we examined the eigenfunctions concentrated in the neighborhood of an extremal ray CD of a region Ω (Sect. 5.4). So, let S be a ray, and let a solution u of (6.4.1) be concentrated in the neighborhood of S for large ω. In exactly the same way as in the preceding sections we will carry out all calculations in (s, n) coordinates, where s is the arc length along S measured from some fixed point, and n is the normal distance from S (taking $n > 0$ on one side of S and $n < 0$ on the other).

An examination of the eigenfunctions concentrated close to an extremal ray of a region (Sect. 5.4) suggests that u will be concentrated in a layer

$$
|n| \leq O\left(\frac{1}{\sqrt{\omega}}\right) , \tag{6.4.2}
$$

or, more precisely, u must be oscillatory in the layer between the caustics, which are defined by (5.4.28):

$$
|n| = n(s) = \frac{\sqrt{2q+1}}{\sqrt{\omega}}\left|\frac{Q}{W}\right|\left|\sqrt{c_0(s)}\left|\frac{a_2}{b_1}f_1^2(s) - 2f_1(s)f_2(s) + \frac{a_1}{b_2}f_2^2(s)\right|^{1/2} , \tag{6.4.3}
$$

where $f_1(s)$ and $f_2(s)$ are linearly independent solutions of (5.4.3) and

$$
Q = \left(\frac{a_1 a_2}{b_1 b_2} - 1\right)^{-1/2} ,
$$

$$
W = f_1 f_2' - f_2 f_1' = \text{const} , \quad a_j = \text{const} , \quad b_j = \text{const} , \quad \frac{a_1 a_2}{b_1 b_2} > 1 .
$$

When $|n| > n(s)$ the function u must change from an oscillatory function to one which is exponentially damped as $\omega \to \infty$. Isolating the phase factor, we seek u in the form

$$
u = U(n, s, \omega)\exp\left(i\omega \int_0^s \frac{ds}{c(s, 0)}\right), \tag{6.4.4}
$$

where U is an attenuation function.

If we assume that

$$
U = O(1) , \tag{6.4.5}
$$

then in view of (6.4.2) it is natural to assume that

$$
\frac{\partial U}{\partial n} = O(\sqrt{\omega}) , \quad \frac{\partial^2 U}{\partial n^2} = O(\omega) . \tag{6.4.6}
$$

Since the curve S is a ray in the present case, the phase of u, at least on S itself, is completely taken care of by the term $\omega \int_0^s c^{-1} ds$ appearing in the exponent in (6.4.4) (which is the essential difference between this case and the whispering gallery case). It is natural therefore to assume that the attenuation function behaves smoothly along S, and that

$$\frac{\partial U}{\partial s} = O(1) , \quad \frac{\partial^2 U}{\partial s^2} = O(1) .$$

(6.4.7)

Substituting the form (6.4.4) for u into (6.4.1), we arrive once again at equation (6.2.6). In contrast to the whispering gallery case, the coefficient of U will be of order $O(n^2)$ as n approaches zero.

Relations (6.4.5–7) show that when (6.4.2) is satisfied, the dominant terms in (6.2.6) will be of order $O(\omega)$ followed by terms of order $O(\sqrt{\omega})$, $O(1)$, $O(1/\sqrt{\omega})$ and so forth. Disregarding terms of order $O(\sqrt{\omega})$ and lower, we arrive after some simple manipulation at the basic parabolic equation for the problem:

$$\frac{\partial^2 U}{\partial n^2} + \frac{2i\omega}{c(s)} \frac{\partial U}{\partial s} + i\omega U \frac{\partial}{\partial s} \frac{1}{c(s)} - \omega^2 n^2 \varPhi U = 0 ,$$

$$\left(\varPhi = \frac{c_{nn}(s)}{c^3(s)} , \quad c(s) = c(s,0) , \quad c_{nn}(s) = \left. \frac{\partial^2 c(s,n)}{\partial n^2} \right|_{n=0} \right) .$$

(6.4.8)

6.5 Solution of the Parabolic Equation (6.4.8)

The function U that satisfies (6.4.8) can be expressed in terms of solutions of ordinary linear differential equations.

By analogy with Sect. 6.3, we introduce a new independent variable

$$\nu = n\sqrt{\omega}\eta(s) ,$$

(6.5.1)

where $\eta(s)$ is a function yet to be specified.

Equation (6.4.8) takes the form

$$\omega \left[U_{\nu\nu} + \frac{2i\eta'}{c\eta^3} \nu U_\nu + \frac{2i}{c\eta^2} U_s - \left(\nu^2 \frac{\varPhi}{\eta^4} + \frac{ic'}{c^2\eta^2} \right) U \right] = 0.$$

(6.5.2)

The use of separation of variables in this equation is precluded by the terms involving νU_ν and $\nu^2 U$. We cannot reduce this equation to a form that permits separation of variables simply by an appropriate choice of $\eta(s)$. Let

$$U(n, s, \omega) = V(\nu, s, \omega) e^{i\nu^2 h(s)} ,$$

(6.5.3)

where the function $h(s)$ is to be specified. Substituting (6.5.3) into (6.5.2), we get an equation for V :

$$V_{\nu\nu} + 4i\nu \left(h + \frac{\eta'}{2c(s)\eta^3} \right) V_\nu + \frac{2i}{c(s)\eta^2} V_s$$

$$- \nu^2 \left[4h^2 + \frac{4h\eta'}{\eta^3 c(s)} + \frac{2h'}{\eta^2 c(s)} + \frac{\varPhi}{\eta^4} \right] V + i \left(2h - \frac{c'(s)}{c(s)\eta^2} \right) V = 0. \quad (6.5.4)$$

In order for this equation to admit separation of variables, it will be enough to relate η and h by:

$$h + \frac{\eta'}{2c\eta^3} = 0 ,$$

(6.5.5)

140

$$4h^2 + \frac{4h\eta'}{c\eta^3} + \frac{2h'}{c\eta^2} + \frac{1}{\eta^4}\Phi = 1 \quad . \tag{6.5.6}$$

Eliminating $h(s)$ from equation (6.5.5) and substituting into (6.5.6), we obtain a nonlinear second-order differential equation for $\eta(s)$. This can be converted to a somewhat simpler differential equation by introducing a new function $\sigma(s)$ by means of

$$\eta(s) = \frac{1}{\sqrt{c(s)}\sigma(s)} \quad , \tag{6.5.7}$$

for which we get

$$\sigma''(s) + K(s)\sigma(s) = \frac{1}{\sigma^3(s)} \quad , \tag{6.5.8}$$

where

$$K(s) = c^2\Phi + \frac{1}{2}\frac{c''}{c} - \frac{3}{4}\frac{c'^2}{c^2} \quad , \tag{6.5.9}$$

and

$$\Phi = \frac{c_{nn}(s)}{c^3(s)} \quad , \quad c' = \left.\frac{\partial c(s,n)}{\partial s}\right|_{n=0} \quad , \quad c'' = \left.\frac{\partial^2 c(s,n)}{\partial s^2}\right|_{n=0} \quad .$$

If there is a real non-zero function σ that satisfies (6.5.8), then in accordance with (6.5.5 and 6.5.7) we have

$$h(s) = \frac{\sigma^2}{2}\left(\frac{1}{2}\frac{c'}{c} + \frac{\sigma'}{\sigma}\right) \quad , \tag{6.5.10}$$

and equation (6.5.4) can be rewritten in the form

$$V_{\nu\nu} - \nu^2 V + 2i\sigma^2(s)V_s + i\sigma^2(s)\left(\frac{\sigma'(s)}{\sigma(s)} + \frac{1}{2}\frac{c'}{c}\right)V = 0 \quad . \tag{6.5.11}$$

Equation (6.5.11) does permit separation of variables. Its solutions (within a constant multiplier) have the form

$$V(\nu, s, \omega) = \frac{c^{1/4}(s)}{\sqrt{\sigma(s)}}\exp\left(\frac{\lambda}{2i}\int_0^s \frac{ds}{\sigma^2(s)}\right)D_{(\lambda-1)/2}(\sqrt{2}\nu), \tag{6.5.12}$$

where λ is a separation factor, and $D_q(z)$ is a parabolic cylinder function, which satisfies the equation

$$\frac{d^2 D_q(z)}{dz^2} + \left(q + \frac{1}{2} - \frac{z^2}{4}\right)D_q(z) = 0 \quad . \tag{6.5.13}$$

Substituting (6.5.3, 10, and 12) into (6.4.4), we get

$$u = \frac{c^{1/4}(s)}{\sqrt{\sigma(s)}}\exp\left[i\omega\int_0^s \frac{ds}{c(s)} + \frac{1}{2}i\nu^2\sigma^2(s)\left(\frac{1}{2}\frac{c'(s)}{c(s)} + \frac{\sigma'(s)}{\sigma(s)}\right)\right.$$

$$\left. + \frac{\lambda}{2i}\int_0^s \frac{ds}{\sigma^2(s)}\right]D_{(\lambda-1)/2}(\sqrt{2}\nu) \quad . \tag{6.5.14}$$

We can now find an asymptotic expansion (as $\omega \to \infty$) for the solutions that vanish as $|\nu| \to \infty$ (on both sides of S). Equation (6.5.13) has solutions that approach zero both when $z \to \infty$, and when $z \to -\infty$ for $q = 0, 1, 2, \ldots$ and only for these q. The corresponding solutions take the form

$$D_q(z) = 2^{-q/2} e^{-z^2/4} H_q\left(\frac{z}{\sqrt{2}}\right) \quad ,$$

where $H_q(z)$ is a Hermite polynomial [6.3,4]. Therefore it is necessary to set $\lambda = 2q + 1$ in (6.5.14). As a result we get asymptotic formulas for the solutions of (6.4.1) that are concentrated in the neighborhood of the given ray S and approach zero with increasing distance from the ray on both sides:

$$u_q = C_q \frac{c^{1/4}(s)}{\sqrt{\sigma(s)}} \exp\left[i\omega \int_0^s \frac{ds}{c(s)} + \frac{i\nu^2 \sigma^2(s)}{2}\left(\frac{1}{2}\frac{c'(s)}{c(s)} + \frac{\sigma'(s)}{\sigma(s)}\right)\right.$$

$$\left. - i\left(q + \frac{1}{2}\right) \int_0^s \frac{ds}{\sigma^2(s)}\right] D_q(\sqrt{2}\nu) \quad . \tag{6.5.15}$$

In (6.5.15), C_q is an arbitrary constant, the quantity ν is defined by (6.5.1,7), and function $\sigma(s)$ is a solution of the nonlinear equation (6.5.8).

The function $\sigma(s)$ can be expressed in terms of two linearly independent solutions of the linear equation

$$y''(s) + K(s)y(s) = 0 \quad . \tag{6.5.16}$$

When this is done, the solution of the nonlinear equation (6.5.8) is reduced to the solution of the simpler linear equation (6.5.16). We can establish the relationship between the function $\sigma(s)$ and solutions of (6.5.16) on the basis of geometrical-optics considerations, as follows.

The function $\psi(\nu) = D_q(\sqrt{2}\nu)$ satisfies the equation

$$\psi_{\nu\nu}'' + (2q + 1 - \nu^2)\psi = 0 \quad ,$$

which has two turning points,

$$\nu = \pm\sqrt{2q + 1} \quad . \tag{6.5.17}$$

When $|\nu| < \sqrt{2q + 1}$, the function $\psi(\nu)$ is oscillatory, and when $|\nu| > \sqrt{2q + 1}$ $|\psi(\nu)|$ decreases monotonically with increasing $|\nu|$, approaching zero as $\nu \to \pm\infty$. Replacing ν in (6.5.17) with its value as given by (6.5.1,7), we come to the conclusion that with increasing distance from S, i.e. with increasing $|n|$, the function u oscillates at first, and beginning at

$$|n| = \frac{\sigma(s)}{\sqrt{\omega}}\sqrt{c(s)}\sqrt{2q + 1} \tag{6.5.18}$$

it becomes a decaying function.

We now return to the ray treatment of this problem (Sect. 5.4). The caustics (6.4.3) are curves between which the corresponding function u oscillates, while

142

outside of the strip bounded by these curves the function u approaches zero (as $\omega \to \infty$). Note that the curves derived from (6.4.3) and (6.5.18) are identical. This immediately leads to the relation

$$\sigma(s) = \left(\frac{a_2}{b_1} f_1^2 - 2f_1f_2 + \frac{a_1}{b_2} f_2^2 \right)^{1/2} \frac{1}{W\sqrt{\frac{a_1a_2}{b_1b_2} - 1}} \quad , \tag{6.5.19}$$

where

$$\frac{a_1a_2}{b_1b_2} > 1 \quad , \quad W = W(f_1, f_2) = f_1f_2' - f_1'f_2 = \text{const}$$

and f_j, $j = 1, 2$ are the specially chosen linearly independent solutions of (6.5.16). It can be shown by direct substitution that (6.5.19) is actually a solution of (6.5.8).

The generalization of this result is

Theorem 1. If the fundamental matrix $\|a_{ik}\|$, $i, k = 1, 2$ and the fundamental system of solutions y_1, y_2 of (6.5.16) are related by

$$\det\|a_{ik}\| W^2(y_1, y_2) = 1 \quad ; \quad W = y_1y_2' - y_2y_1' = \text{const} \quad , \tag{6.5.20}$$

then the function

$$\sigma(s) = \sqrt{\sum_{i,k=1}^{2} a_{ik}y_iy_k} \tag{6.5.21}$$

is a solution of (6.5.8).

We also have

Theorem 2. No matter what the initial conditons

$$\sigma(s)\big|_{s=s_0} = c_1 > 0 \quad , \quad \sigma'(s)\big|_{s=s_0} = c_2 \tag{6.5.22}$$

are, there exists a matrix $\|a_{ik}\|$, satisfying (6.5.20) such that the function $\sigma(s)$ defined by (6.5.21) is a solution of (6.5.8) which obeys (6.5.22).

The proof of these theorems is elementary and is given in Appendix A.3.

6.6 Notes on the Literature

The method of parabolic equations for solving diffraction problems was first proposed by *Leontovič* [6.5] and *Fock* [6.6] (see also [6.7]) in the 1940s. *Fock*'s work has been collected in [6.2].

Ivanov [6.1] applied the parabolic equation method to the diffraction of short waves by a smooth convex cylinder, and discovered the presence of rapidly oscillating phase factors in the asymptotic expansion of the solution. *Buslaev* [6.8] was the first to find the amplitude factor within the framework of the parabolic equation method.

The asymptotic expansions for whispering-gallery-type eigenfunctions were obtained by *Buldyrev* [6.9]. *Molotkov* [6.10] found the asymptotic expansion for the field in the shadow zone in terms of the Airy function $w_1(t)$ for a medium which is inhomogeneous in two spatial coordinates.

The parabolic equation method has been used by *Vainshtein* [6.11], *Buldyrev* [6.9] and *Lazutkin* [6.12] to find bouncing-ball eigenfunctions in the case $c(x,y) \equiv 1$. When the velocity $c(x,y)$ is allowed to be nonconstant, additional difficulties arise, and these were overcome by *Babič* and *Lazutkin* [6.13].

In a paper by *Levey* and *Felsen* [6.14], a method resembling that of the parabolic equation made it possible to describe the field near a caustic which is partially shadowed by an opaque object. The parabolic equation method can also be used in problems involving objects with edges. In particular, the classical problem of diffraction by a wedge has been solved by *Ufimtsev* [6.15] using this approach.

By now, the parabolic equation method has been widely applied to various concrete problems of wave propagation and diffraction; a comprehensive bibliography on the subject would be inordinately lengthy. A brief historical sketch and a list of the fundamental papers on the parabolic equation method can be found in the article by *Tappert* [6.16], which deals with applications of the method to the propagation of underwater sound waves in the ocean. Higher-order terms in an asymptotic expansion for which the solution of the parabolic equation is the leading term can be found in a systematic fashion. This has been done, for example in the book by *Babič* and *Kirpičnikova* [Ref. 6.17, Chaps. 5 and 6], which also contains further references to the literature.

7. Asymptotic Expansions of Eigenfunctions Concentrated Close to the Boundary of a Region

In Sects. 5.1, 3 and 6.3, eigenvalues and eigenfunctions concentrated in the neighborhood of the boundary of a plane convex region have been constructed by the ray method in the small and by the parabolic equation method. Such eigenfunctions were called eigenfunctions of the *whispering gallery type*. It was shown that eigenvalues of the whispering gallery type oscillate in a layer whose thickness is of order $\omega^{-2/3}$, and are exponentially damped outside this layer. In this chapter we will consider this same problem by another method which will enable us not only to find the complete asymptotic expansion of the eigenvalues and eigenfunctions, but also to derive some rigorous results. Moreover, asymptotic expansions will also be constructed for the eigenvalues and eigenfunctions of the *external region* problem.

7.1 Introductory Remarks

Let a plane region Ω be bounded by a reasonably smooth curve S. We will assume that the velocity of wave propagation in Ω is a function $c(x, y)$ which is differentiable a sufficient number of times.

We wish to construct the eigenfunctions of the whispering gallery type on Ω, i.e. non-zero solutions of the equation

$$\frac{\partial^2 u}{\partial x^2} + \frac{\partial^2 u}{\partial y^2} + \frac{\omega^2}{c^2(x, y)} u = 0 \quad , \tag{7.1.1}$$

that satisfy the condition

$$u|_S = 0 \tag{7.1.2}$$

and are concentrated in a sufficiently narrow region near the boundary. Cases where the Neumann boundary condition

$$\left. \frac{\partial u}{\partial n} \right|_S = 0 \tag{7.1.3}$$

or the mixed boundary condition

$$\frac{\partial u}{\partial n} - i\omega g(s)u|_S = 0 \tag{7.1.4}$$

with a variable coefficient $g(s)$ are satisfied on S are treated in a similar fashion. Therefore these cases will be considered in less detail in this chapter.

The method that we will use to construct the eigenfunctions of the whispering gallery type can be called the *etalon (or model) problem method*. The basic idea of the method in the case of Helmholtz equation (a partial differential equation) is similar to the method of comparison equations in the theory of ordinary differential equations.

The starting point of the etalon problem method is the study of a ray field – the extremals of the geometrical optics functional. The next step is to choose the simplest *etalon problem* that permits an exact solution (for instance, by separation of variables) in which the field of rays has the same singularities as in the original problem. An analysis of the solution of the etalon problem enables us to choose a particular form for the desired expansion of the solution of the original problem. Substituting this expansion into the equations and boundary conditions of the original problem and requiring that they be (formally) satisfied to all orders, we can obtain a series of relations among the coefficients of the expansion. These resultant relations enable us to find the unknown functions which enter into these coefficients.

By determining a sufficiently large number of terms in the expansion, we can ensure that the solution satisfies the equation and boundary conditions up to some predetermined accuracy. As has already been noted, we can, in some cases, rigorously demonstrate that the expansion for the eigenvalues derived in this way is asymptotic as $\omega \to \infty$.

By virtue of the fact that the parabolic equation method also gives the asymptotic expansion of the solutions of diffraction problems (although only the principal term was obtained by the method of Chap. 6), we may think of the method presented here as a further development of the parabolic equation method. Indeed, the form of the *etalon functions*, i.e. the functions entering into the desired expansion, might have been established on the basis of solution of a parabolic equation. However, the solution of the etalon problem, in contrast to the solution of this parabolic equation, indicates not only the form of the standard functions, but also the way that further approximations depend on the observation point.

7.2 Eigenfunctions of the Circle for the Case $c = \text{const}$

In order for the whispering gallery effect to arise in the neighborhood $n < 0$ of the boundary S of Ω, the effective radius of curvature $P(s)$ of this region must be positive:

$$P(s) = \left(\frac{1}{\varrho(s)} - \frac{c_1(s)}{c_0(s)} \right)^{-1} > 0 \quad . \tag{7.2.1}$$

This result was found in Sect. 5.3 from ray considerations. Of the problems mentioned in Sect. 7.1, for which condition (7.2.1) is satisfied, the simplest is (7.1.1 and 2) for a circle, with the propagation velocity constant. Problem (7.1.1 and 2) for the circle $r \leq \varrho = $ const $(n = r - \varrho)$ can be regarded as an etalon problem for the whole class of problems wherein condition (7.2.1) is satisfied. In this section we will construct solutions of the Helmholtz equation that are concentrated close to the boundary of the circle, and their corresponding eigenfunctions of the whispering gallery type (in the remainder of this section, the quantity ϱ is a constant). Taking this simplest problem as a guide, we will "guess" the form of the asymptotic expansions of eigenfunctions of the whispering gallery type for the general case.

In polar coordinates r, ϕ the Helmholtz equation takes the form

$$\frac{1}{r} \frac{\partial}{\partial r} \left(r \frac{\partial u(r, \phi)}{\partial r} \right) + \frac{1}{r^2} \frac{\partial^2 u(r, \phi)}{\partial \phi^2} + \frac{\omega^2}{c^2} u(r, \phi) = 0 \quad .$$

Using separation of variables, we obtain the following class of solutions to this equation:

$$u(r, \phi) = J_\zeta \left(\frac{\omega}{c} r \right) e^{i\zeta\phi} \quad , \tag{7.2.2}$$

where $J_\zeta(\omega r/c)$ is the Bessel function of the first kind, and ζ is a separation constant. At this point, it is customary to require that the solution be periodic in ϕ,

$$u(r, \phi + 2\pi) = u(r, \phi) \quad ,$$

which results in integer values for ζ.

However, we shall defer the imposition of periodicity on the solution until the final step of the construction of the eigenfunctions.

We will assume that $\phi \in (-\infty, \infty)$, and consider the infinitely-sheeted surface $r \leq \varrho$, $-\infty < \phi < \infty$. This surface can be obtained by removing from the ordinary Riemann surface of the function $\ln(x + iy) = \ln(r\, e^{i\phi})$ those points whose distance from the origin is greater than ϱ.

Let condition (7.1.2) be satisfied on this infinitely-sheeted surface. This condition leads to the equation

$$J_\zeta(k\varrho) = 0 \quad , \quad k = \omega/c \quad . \tag{7.2.3}$$

Since we are interested in the asymptotic expansions of the eigenfunctions as $\omega \to \infty$, we will replace the Bessel function in (7.2.3) by its asymptotic expansion.

We write the uniform asymptotic expansion of the Bessel function as[1]

$$J_\zeta(kr) = \sqrt{\frac{2}{\pi}} \left[\frac{T_r}{\zeta^2 - (kr)^2} \right]^{1/4} \left\{ \frac{d}{dz} \left[z + \frac{p_1(z)}{(kr)^2} + \dots + \frac{p_n(z)}{(kr)^{2n}} \right. \right.$$
$$\left. \left. + O\left((kr)^{-2n-2} \right) \right] \right\}^{-1/2}$$

---

[1] This expansion was first derived in a somewhat different form by *Cherry* [7.1].

$$\times v\left[T_r + \frac{p_1(z)}{(kr)^{4/3}} + \dots + \frac{p_n(z)}{(kr)^{2n-2/3}} + O\left((kr)^{-2n-4/3}\right)\right] \ ,$$

$$(7.2.4)$$

where T_r and z are defined by

$$T_r = \zeta^{2/3}z = \left[\frac{3}{2}ikr\int_1^{\zeta/kr} \arccos x \, dx\right]^{2/3}$$

$$= 2\left(\frac{kr}{2}\right)^{2/3}\left(\frac{\zeta}{kr}-1\right)\left[1 - \frac{1}{30}\left(\frac{\zeta}{kr}-1\right)+\dots\right] \ ;$$

$v(t)$ is the Airy function which decays as $t \to \infty$ and $p_1(z)$, $p_2(z)$, ..., $p_n(z)$ are functions of z, regular at $z = 0$ and determined from a certain system of recurrence relations. Formula (7.2.4) is true when

$$|\arg \zeta| < \frac{\pi}{2} \ , \quad 0 < \varepsilon < \left|\frac{kr}{\zeta}\right| < M < \infty \ , \quad \left|\arg\frac{kr}{\zeta}\right| < \frac{\pi}{2} \ , \quad |kr| \to \infty$$

(ε and M are fixed constants).

Substituting (7.2.4) into (7.2.3), we get for determination of the constant ζ the equation

$$T_\varrho + \frac{1}{(k\varrho)^{4/3}}p_1(\zeta^{-2/3}T_\varrho) + \dots$$

$$+\frac{1}{(k\varrho)^{2n-2/3}}p_n(\zeta^{-2/3}T_\varrho) + O\left((k\varrho)^{-2n-4/3}\right) = -t_p \ , \qquad (7.2.5)$$

$$T_\varrho = 2\left(\frac{k\varrho}{2}\right)^{2/3}\left(\frac{\zeta}{k\varrho}-1\right)[1 - \dots] \ ,$$

where $(-t_p)$ is the pth zero of $v(t)$, $p = 0, 1, 2, \dots$. Solving (7.2.5) for $\zeta/k\varrho - 1$, we find

$$\zeta = \zeta_p = k\varrho\left\{1 + \left(\frac{-t_p}{2}\right)\left(\frac{2}{k\varrho}\right)^{2/3} + \frac{1}{30}\left(\frac{-t_p}{2}\right)^2\left(\frac{2}{k\varrho}\right)^{4/3}\right.$$

$$\left.+ O\left[t_p^3\left(\frac{1}{k\varrho}\right)^2\right]\right\} \ .$$

We substitute the resultant value of ζ_p into (7.2.2) and replace the Bessel function by its asymptotic expansion:

$$u(r, \phi) = \text{const}\left(\frac{2}{k\varrho}\right)^{1/3}\exp\left\{ik\left[s + \frac{-t_p}{2}\left(\frac{2}{k\varrho}\right)^{2/3}s + \frac{(-t_p)^2}{120}\left(\frac{2}{k\varrho}\right)^{4/3}s\right.\right.$$

$$\left.+ i\left(\frac{-t_p}{30}\varrho + \frac{1}{20}\varrho\tilde{\nu}\right)\left(\frac{2}{k\varrho}\right)^{5/3} + O\left(t_p^3\frac{4}{k^2\varrho^2}\right)\right]\right\}$$

$$\times v\left\{-t_p - \tilde{\nu} + \left(\frac{2}{15}t_p\tilde{\nu} + \frac{3}{20}\tilde{\nu}^2\right)\left(\frac{2}{k\varrho}\right)^{2/3} + O\left[t_p^2\left(\frac{2}{k\varrho}\right)^{4/3}\right]\right\} \ ,$$

$$(7.2.6)$$

148

where $s = \phi\varrho$ is the arc length along the circle, and

$$\tilde{\nu} = 2\left(\frac{k\varrho}{2}\right)^{2/3}\frac{r-\varrho}{\varrho} < 0$$

is a normalized distance between the observation point (r, ϕ) and the boundary of the circle.

Now, the Airy function $v(t)$ damps out exponentially when $t > 0$, and oscillates when $t < 0$. The point $t = 0$ is a turning point for this function. A solution of (7.2.6) will oscillate when

$$-t_p + O\left[\left(\frac{2}{k\varrho}\right)^{2/3}\right] < \tilde{\nu} < 0$$

and damp out exponentially when

$$\tilde{\nu} < -t_p + O\left[\left(\frac{2}{k\varrho}\right)^{2/3}\right] \quad .$$

Thus we say of our solution (7.2.6) that it is concentrated in the boundary layer

$$-t_p \leq \tilde{\nu} \leq 0.$$

In terms of r this layer can be written as

$$\left[1 - \frac{t_p}{2}\left(\frac{2}{k\varrho}\right)^{2/3}\right]\varrho < r \leq \varrho \quad .$$

The thickness of the layer in which the solutions (7.2.6) are concentrated has order $(k\varrho)^{-2/3}$, which agrees completely with what we found in Chaps. 5 and 6. A solution of the form (7.2.6) is a wave that travels along the boundary of the circle.

To get the eigenfunction (standing wave), it remains for us to satisfy the condition of periodicity with respect to ϕ. This condition will be satisfied if we require that (7.2.6) be the same when $s = 2\pi\varrho$ as it is when $s = 0$, i.e., if

$$2\pi\varrho k\left[1 - \frac{t_p}{2}\left(\frac{2}{k\varrho}\right)^{2/3} + \frac{t_p^2}{120}\left(\frac{2}{k\varrho}\right)^{4/3} + \ldots\right] = 2\pi q, \qquad (7.2.7)$$

where q is an integer. Equation (7.2.7) defines the natural frequencies

$$k_{p,q} = \frac{q}{\varrho}\left\{1 + \frac{t_p}{2}\left(\frac{2}{q}\right)^{2/3} + 3\frac{t_p^2}{40}\left(\frac{2}{q}\right)^{4/3} + O\left[t_p^3\left(\frac{2}{q}\right)^2\right]\right\} \quad . \qquad (7.2.8)$$

As implied by the correction term, formula (7.2.8) is valid for $q \gg 1$ and values of p which are not too large. If the values found for $k_{p,q}$ are substituted into (7.2.6), we get the asymptotic expansion of the eigenfunctions $u_{p,q}(r, \phi)$. Thus, the asymptotic expansions of solutions of the Helmholtz equation which are concentrated in a boundary layer and of the corresponding eigenfunctions take the form of an exponential term multiplied by an Airy function. The arguments of the exponential function and the Airy function are power series in $k^{-1/3}$.

The coefficients of these series are polynomials with respect to the normalized distance $\tilde{\nu}$.

Having determined the structure of the asymptotic formulas in the etalon problem, let us now go on to investigate the general problem.

7.3 Construction of Solutions of the Helmholtz Equation in a Boundary Layer

In this section, we will construct solutions of (7.1.1) which are concentrated in the neighborhood of a curve S (where S is not necessarily closed), and satisfy (7.1.2 or 1.3) on S.

As in Sect. 5.3, we introduce the coordinates s and n. We will consider the effective radius of curvature $P(s)$ of S to be positive, see (7.2.1). Let us construct solutions of equation (7.1.1) on the side of S where $n<0$. By virtue of (7.2.1), rays multiply reflected by the boundary overlap with each other on this side. In (s, n) variables, (7.1.1) takes the form (see Appendix A.2):

$$\frac{1}{h_1 h_2}\left[\frac{\partial}{\partial q_1}\left(\frac{h_2}{h_1}\frac{\partial u}{\partial q_1}\right) + \frac{\partial}{\partial q_2}\left(\frac{h_1}{h_2}\frac{\partial u}{\partial q_2}\right)\right] + \frac{\omega^2}{c^2(s,n)}u$$

$$= \left(1 + \frac{n}{\varrho(s)}\right)^{-1}\left\{\frac{\partial}{\partial n}\left[\left(1 + \frac{n}{\varrho(s)}\right)\frac{\partial u}{\partial n}\right]\right.$$

$$\left. + \frac{\partial}{\partial s}\left[\left(1 + \frac{n}{\varrho(s)}\right)^{-1}\frac{\partial u}{\partial s}\right]\right\} + \frac{\omega^2}{c^2(s,n)}u = 0 \quad . \tag{7.3.1}$$

We will assume that the propagation velocity $c(s, n)$ in the boundary layer can be represented by a truncated Taylor series with a large enough number of terms

$$c(s, n) = c_0(s) + c_1(s)n + \ldots + c_{N-1}(s)n^{N-1} + O(n^N) \quad . \tag{7.3.2}$$

The coefficients c_j $(j = 0, 1, 2, \ldots)$ of (7.3.2) and the radius of curvature $\varrho = \varrho(s)$ of curve S will be assumed to be sufficiently smooth functions of the arc length s.

Since we are interested in solutions that are concentrated in a boundary layer of thickness of the order of $\omega^{-2/3}$, we will convert to a normalized distance — a reduced normal — ν in (7.3.1), setting

$$\nu = n\omega^{2/3} \quad .$$

Writing (7.3.1) in s and ν gives

$$\omega^{4/3}A\frac{\partial^2 u}{\partial \nu^2} + \omega^{2/3}B\frac{\partial u}{\partial \nu} + C\frac{\partial^2 u}{\partial s^2} + \omega^{-2/3}D\frac{\partial u}{\partial s} + \omega^2 Eu = 0 \quad , \tag{7.3.3}$$

where

$$A = 1 + \omega^{-2/3}\frac{\nu}{\varrho(s)} \quad , \quad B = \frac{1}{\varrho(s)} \quad , \quad C = \left(1 + \omega^{-2/3}\frac{\nu}{\varrho(s)}\right)^{-1} \quad ,$$

$$D = \frac{\nu}{\varrho^2(s)}\varrho'(s)\left(1 + \omega^{-2/3}\frac{\nu}{\varrho(s)}\right)^{-2} \quad ,$$

$$E = \frac{1}{c(s, \nu\omega^{-2/3})}\left(1 + \omega^{-2/3}\frac{\nu}{\varrho(s)}\right).$$

The coefficients C, D and E are represented by truncated Taylor series:

$$C = \sum_{m=0}^{[\frac{M+1}{2}]-1}(-1)^m\left(\frac{\nu}{\varrho}\right)^m\omega^{-2m/3} + O(\omega^{-M/3}) \quad ,$$

$$D = \frac{\varrho'}{\varrho}\sum_{m=0}^{[\frac{M+1}{2}]-1}(-1)^m(m+1)\left(\frac{\nu}{\varrho}\right)^{m+1}\omega^{-2m/3} + O(\omega^{-M/3}) \quad ,$$

$$E = \sum_{m=0}^{[\frac{M+1}{2}]-1}e_m(s)\left(\frac{\nu}{\varrho}\right)^m\omega^{-2m/3} + O(\omega^{-M/3}) \quad ,$$

where $[(M+1)/2] = M/2$ for even M, and $[(M+1)/2] = (M+1)/2$ for odd M. In the expression for E, the functions $e_m(s)$ depend on $\varrho(s)$ and the coefficients $c_j(s)$, $j = 0, 1, 2, \ldots$ appearing in the velocity expansion (7.3.2). The first few coefficients $e_m(s)$ are equal to

$$e_0(s) = \frac{1}{c_0^2(s)} \quad , \quad e_1(s) = \frac{1}{c_0^2(s)} - 2\varrho(s)\frac{c_1(s)}{c_0^3(s)} \quad ,$$

$$e_2(s) = -2\varrho(s)\frac{c_1(s)}{c_0^3(s)} + \varrho^2(s)\frac{3c_1^2(s) - 2c_2(s)c_0(s)}{c_0^4(s)} \quad .$$

We will seek the solutions of (7.3.3) in a form analogous to (7.2.6):

$$u(s, \nu) = \text{const}\exp\left\{i\sum_{m=-3}^{M-1}\alpha_m(s, \nu)\omega^{-m/3} + O(\omega^{-M/3})\right\}$$

$$\times v\left\{\sum_{m=0}^{M-1}\beta_m(s, \nu)\omega^{-m/3} + O(\omega^{-M/3})\right\} \quad , \tag{7.3.4}$$

where $\alpha_m(s, \nu)$ and $\beta_m(s, \nu)$ are polynomials in ν with coefficients depending on s which must be determined, and $v(Z)$ is the Airy function. Our first task is to get a recurrent system of differential equations from which we could successively determine all the polynomials $\alpha_m(s, \nu)$ and $\beta_m(s, \nu)$. We substitute expression (7.3.4) into (7.3.3), and taking the Airy equation into consideration we replace the second derivative $v''(Z)$ by the product $Zv(Z)$, where

$$Z = \sum_{m=0}^{M-1}\beta_m(s, \nu)\omega^{-m/3} + O(\omega^{-M/3})$$

is the argument of the Airy function. After cancelling the exponential function, (7.3.3) takes the form

$$a(\alpha_j, \beta_j; \omega^{-1/3})v(Z) + b(\alpha_j, \beta_j; \omega^{-1/3})v'(Z) = 0 \ ,$$

where

$$a(\alpha_j, \beta_j; \omega^{-1/3}) = \omega^{4/3}\left[\sum_{m=-6}^{M-1} a_m(\alpha_j, \beta_j)\omega^{-m/3} + O(\omega^{-M/3})\right] \ ,$$

$$(7.3.5)$$

$$b(\alpha_j, \beta_j; \omega^{-1/3}) = \omega^{4/3}\left[\sum_{m=-3}^{M-1} b_m(\alpha_j, \beta_j)\omega^{-m/3} + O(\omega^{-M/3})\right]$$

$$(7.3.6)$$

and $a_m(\alpha_j, \beta_j)$ and $b_m(\alpha_j, \beta_j)$ are differential operators applied to the polynomials $\alpha_j(s, \nu)$ and $\beta_j(s, \nu)$. Since the functions $v(Z)$ and $v'(Z)$ are linearly independent, the expressions $a(\alpha_j, \beta_j; \omega^{-1/3})$ and $b(\alpha_j, \beta_j; \omega^{-1/3})$ must be independently equal to zero identically in ω. Setting the coefficients of expansions (7.3.5 and 7.3.6) equal to zero, we get the sought recurrence system of equations

$$a_m(\alpha_j, \beta_j) = 0 \ , \quad m \geq -6 \ ,$$
$$b_m(\alpha_j, \beta_j) = 0 \ , \quad m \geq -3 \ . \tag{7.3.7}$$

The system of equations (7.3.7) enables a step-by-step construction of all the unknown polynomials α_j and β_j. First we must analyze the first few equations of the system separately. This will enable us to write the later equations in a simpler form. Setting $m = -6, -5, -4$ in the first equation of the system, we get

$$a_{-6}(\alpha_j, \beta_j) \equiv \left(\frac{\partial\alpha_{-3}}{\partial\nu}\right)^2 = 0, \tag{7.3.8}$$

$$a_{-5}(\alpha_j, \beta_j) \equiv 2\frac{\partial\alpha_{-3}}{\partial\nu}\frac{\partial\alpha_{-2}}{\partial\nu} = 0 \ , \tag{7.3.9}$$

$$a_{-4}(\alpha_j, \beta_j) \equiv \frac{\nu}{\varrho}\left(\frac{\partial\alpha_{-3}}{\partial\nu}\right)^2 + \left(\frac{\partial\alpha_{-2}}{\partial\nu}\right)^2$$
$$+ 2\frac{\partial\alpha_{-3}}{\partial\nu}\frac{\partial\alpha_{-2}}{\partial\nu} = 0 \ . \tag{7.3.10}$$

From these equations we see that

$$\frac{\partial\alpha_{-3}}{\partial\nu} = \frac{\partial\alpha_{-2}}{\partial\nu} = 0 \ . \tag{7.3.11}$$

In view of (7.3.11), the next six equations of the system ($m = -3, -2, -1$) can be written as

$$a_{-3}(\alpha_j, \beta_j) \equiv 0 \ ,$$

$$a_{-2}(\alpha_j, \beta_j) \equiv -\left(\frac{\partial \alpha_{-1}}{\partial \nu}\right)^2 - \left(\frac{\partial \alpha_{-3}}{\partial s}\right)^2 + e_0(s) = 0 \ , \tag{7.3.12}$$

$$a_{-1}(\alpha_j, \beta_j) \equiv -2\frac{\partial \alpha_{-3}}{\partial s}\frac{\partial \alpha_{-2}}{\partial s} = 0 \ , \tag{7.3.13}$$

$$b_{-3}(\alpha_j, \beta_j) \equiv 0 \ ,$$
$$b_{-2}(\alpha_j, \beta_j) \equiv 0 \ ,$$
$$b_{-1}(\alpha_j, \beta_j) \equiv 2\frac{\partial \alpha_{-1}}{\partial \nu}\frac{\partial \beta_0}{\partial \nu} = 0 \ .$$

The last of these means that $\partial \alpha_{-1}/\partial \nu = 0$ or $\partial \beta_0/\partial \nu = 0$. If we assume that $\partial \beta_0/\partial \nu = 0$, then we arrive at a contradiction with the known solution (7.2.6) for a circle, in which $\partial \beta_0/\partial \nu = (2/\varrho)^{1/3}c_0^{-2/3} \neq 0$. Therefore we set

$$\frac{\partial \alpha_{-1}}{\partial \nu} = 0 \ ,$$

and then from (7.3.12, 13) we get

$$\frac{\partial \alpha_{-3}}{\partial s} = \pm\sqrt{e_0(s)} = \pm\frac{1}{c_0(s)} \neq 0 \tag{7.3.14}$$

and

$$\frac{\partial \alpha_{-2}}{\partial s} = 0 \ . \tag{7.3.15}$$

Equation (7.3.14) is easy to integrate. Taking into consideration that $\alpha_{-3}(s,\nu)$ does not depend on ν, we get

$$\alpha_{-3}(s,\nu) \equiv \alpha_{-30}(s) = \pm \int_{d_{-3}}^{s} \frac{d\tau}{c_0(\tau)} \ , \tag{7.3.16}$$

where d_{-3} is some arbitrary constant. The two signs in the formula lead to two solutions describing waves propagating in opposite directions along the S.

We will retain only the $+$ sign in (7.3.16). Equations (7.3.11, 15) mean that

$$\alpha_{-2}(s,\nu) = d_{-2} \ , \tag{7.3.17}$$

where d_{-2} is an arbitrary constant that does not depend on s or ν.

Turning now to the next two equations of (7.3.7) for $m = 0$, and taking into consideration not only (7.3.11), but also (7.3.14, 15, 17), we get

$$a_0(\alpha_j, \beta_j) \equiv -\left(\frac{\partial \alpha_0}{\partial \nu}\right)^2 + \mathrm{i}\frac{\partial^2 \alpha_0}{\partial \nu^2} + \beta_0\left(\frac{\partial \beta_0}{\partial \nu}\right)^2$$

$$+ \frac{\nu}{\varrho}\frac{1}{c_0^2(s)} - 2\frac{1}{c_0(s)}\frac{\partial \alpha_{-1}}{\partial s} + e_1(s)\frac{\nu}{\varrho} = 0 \ , \tag{7.3.18}$$

153

$$b_0(\alpha_j, \beta_j) \equiv 2\frac{\partial \alpha_0}{\partial \nu}\frac{\partial \beta_0}{\partial \nu} + \frac{\partial^2 \beta_0}{\partial \nu^2} = 0 \quad . \tag{7.3.19}$$

Eliminating the derivatives $\partial \alpha_0/\partial \nu$ and $\partial^2 \alpha_0/\partial \nu^2$ between (7.3.18 and 19) gives

$$-2i\frac{\partial^3 \beta_0}{\partial \nu^3}\frac{\partial \beta_0}{\partial \nu} - (1-2i)\left(\frac{\partial^2 \beta_0}{\partial \nu^2}\right)^2 + 4\beta_0\left(\frac{\partial \beta_0}{\partial \nu}\right)^4$$

$$+4\left[\frac{1}{c_0^2(s)} + e_1(s)\right]\frac{\nu}{\varrho(s)}\left(\frac{\partial \beta_0}{\partial \nu}\right)^2 - 8\frac{1}{c_0(s)}\frac{\partial \alpha_{-1}}{\partial s}\left(\frac{\partial \beta_0}{\partial \nu}\right)^2 = 0 \quad . \tag{7.3.20}$$

Now suppose that $\beta_0(s, \nu)$ were a polynomial of degree l:

$$\beta_0(s, \nu) = \beta_{0l}(s)\nu^l + \ldots + \beta_{00}(s) \quad .$$

Then if $l \geq 2$, the left side of (7.3.20) would be a polynomial of degree $5l - 4$:

$$4l^4\beta_{0l}^5(s)\nu^{5l-4} + \ldots = 0 \quad . \tag{7.3.21}$$

Since (7.3.21) must be satisfied identically in ν

$$\beta_{0l}(s) = 0$$

and consequently $\beta_0(s, \nu)$ can only be a polynomial of the first degree:

$$\beta_0(s, \nu) = \beta_{01}(s)\nu + \beta_{00}(s) \quad . \tag{7.3.22}$$

Substituting (7.3.22) into (7.3.20), we get

$$\left\{\beta_{01}^3 + \left[\frac{1}{c_0^2(s)} + e_1(s)\right]\frac{1}{\varrho(s)}\right\}\nu + \left(\beta_{01}^2\beta_{00} - 2\frac{1}{c_0(s)}\frac{\partial \alpha_{-1}}{\partial s}\right) = 0 \quad .$$

$$\tag{7.2.23}$$

whence

$$\beta_{01}(s) = -\sqrt[3]{\left[\frac{1}{c_0^2(s)} + e_1(s)\right]\frac{1}{\varrho(s)}} = -\frac{2^{1/3}}{c_0^{2/3}(s)P^{1/3}(s)} \quad .$$

The function $\beta_{00}(s)$ is as yet arbitrary. Later on this function will be determined from the boundary conditions. Setting the constant term in (7.3.23) equal to zero and integrating, we find

$$\alpha_{-1}(s, \nu) = \alpha_{-10}(s) = \frac{1}{2^{1/3}}\int_{d_{-1}}^{s}\frac{\beta_{00}(\tau)}{c_0^{1/3}(\tau)P^{2/3}(\tau)}d\tau. \tag{7.3.24}$$

where d_{-1} is an arbitrary constant.

Since $\beta_0(s, \nu)$ is a polynomial of first degree, (7.3.19) implies that $\alpha_0(s, \nu)$ is a polynomial of zero degree:

$$\alpha_0(s, \nu) = \alpha_{00}(s) \quad . \tag{7.3.25}$$

To find the polynomials $\alpha_1(s, \nu)$ and $\beta_1(s, \nu)$ and also to determine $\alpha_{00}(s)$, we use the equations of (7.3.7) for $m = 1$. Taking account of the values of the polynomials already found, and leaving the still unknown $\alpha_1(s, \nu)$, $\beta_1(s, \nu)$, $\alpha_{00}(s)$ in the left sides of the equations, we get

$$i\frac{\partial^2 \alpha_1}{\partial \nu^2} + 2(\beta_{01}\nu + \beta_{00})\beta_{01}\frac{\partial \beta_1}{\partial \nu} + \beta_{01}^2 \cdot \beta_1$$

$$- \frac{2}{c_0}\frac{d\alpha_{00}}{ds} = i\frac{c_0'}{c_0^2}, \tag{7.3.26}$$

$$2i\beta_{01}\frac{\partial \alpha_1}{\partial \nu} + \frac{\partial^2 \beta_1}{\partial \nu^2} = -\frac{2i}{c_0}\left(\frac{d\beta_{01}}{ds}\nu + \frac{d\beta_{00}}{ds}\right) . \tag{7.3.27}$$

Eliminating the function $\alpha_1(s,\nu)$ from (7.3.26, 27), we get

$$-\frac{\partial^3 \beta_1}{\partial \nu^3} + 4\beta_{01}^2(\beta_{01}\nu + \beta_{00})\frac{\partial \beta_1}{\partial \nu} + 2\beta_{01}^3 \beta_1 - \frac{4}{c_0}\beta_{01}\frac{d\alpha_{00}}{ds}$$

$$= 2i\frac{c_0'}{c_0^2} + 2i\frac{1}{c_0}\frac{d\beta_{01}}{ds} . \tag{7.3.28}$$

In the same way as before, we find that $\beta_1(s,\nu)$ can be only a polynomial of zero degree:

$$\beta_1(s,\nu) = \beta_{10}(s) . \tag{7.3.29}$$

The function $\beta_{10}(s)$ remains arbitrary; it is determined like $\beta_{00}(s)$ from the boundary conditions. Since $\beta_1(s,\nu)$ is independent of ν, the left and right sides of equation (7.3.28) contain only the variable s. Integrating this equation, we determine the function $\alpha_{00}(s)$:

$$\alpha_{00}(s) = i\frac{1}{2}\ln\frac{\beta_{01}(d_0)c_0(d_0)}{\beta_{01}(s)c_0(s)} + \frac{1}{2}\int_{d_0}^{s} c_0(\tau)\beta_{01}^2(\tau)\beta_{10}(\tau)d\tau , \tag{7.3.30}$$

where d_0 is an arbitrary constant.

Substituting (7.3.29) into (7.3.27) and then integrating this equation with respect to ν, we find $\alpha_1(s,\nu)$, which is a polynomial of second degree:

$$\alpha_1(s,\nu) = \alpha_{12}(s)\nu^2 + \alpha_{11}(s)\nu + \alpha_{10}(s)$$

$$= -\frac{1}{2}\frac{1}{c_0(s)\beta_{01}(s)}\frac{d\beta_{01}}{ds}\nu^2 - \frac{1}{c_0(s)\beta_{01}(s)}\frac{d\beta_{00}}{ds}\nu + \alpha_{10}(s). \tag{7.3.31}$$

Thus we have advanced one more step in determining the unknown coefficients of expansion (7.3.4).

To take the next step, i.e. to find the polynomials $\beta_2(s,\nu)$ and $\alpha_2(s,\nu)$, and also to determine $\alpha_{10}(s)$, we must refer to the next pair of equations from (7.3.7), corresponding to $m = 2$. By solving these equations, we construct the polynomials $\beta_2(s,\nu)$ and $\alpha_2(s,\nu)$ (except for their constant terms $\beta_{20}(s)$ and $\alpha_{20}(s)$), and also find the function $\alpha_{10}(s)$ expressed in terms of $\beta_{20}(s)$. The function $\beta_{20}(s)$ remains arbitrary and will be found later from the boundary conditions.

By using the indicated scheme we can find all subsequent polynomials $\alpha_m(s,\nu)$ and $\beta_m(s,\nu)$. To prove this, let us assume that we have solved the system of equations (7.3.7) up to the pair of equations corresponding to $m = r-1$.

In other words we will assume that the polynomials $\alpha_j(s, \nu)$ for $j \leq r - 2$ and the polynomials $\beta_j(s, \nu)$ for $j \leq r - 2$ are completely known, and that $\alpha_{r-1}(s, \nu)$ is known with the exception of its constant term $\alpha_{r-1,0}(s)$. To construct $\alpha_r(s, \nu)$ and $\beta_r(s, \nu)$, and also to determine the function $\alpha_{r-1,0}(s)$, we write out the equations of (7.3.7) that correspond to $m = r$. On the left sides of the equations we put only the unknown polynomials $\alpha_r(s, \nu)$, $\beta_r(s, \nu)$ and the function $\alpha_{r-1,0}(s)$:

$$i\frac{\partial^2 \alpha_r}{\partial \nu^2} + 2(\beta_{01}\nu + \beta_{00})\beta_{01}\frac{\partial \beta_r}{\partial \nu} + \beta_{01}^2 \beta_r$$
$$- \frac{2}{c_0}\frac{d\alpha_{r-1,0}}{ds} = \Phi_r^{(1)}(s, \nu) \quad , \tag{7.3.32}$$

$$2i\beta_{01}\frac{\partial \alpha_r}{\partial \nu} + \frac{\partial^2 \beta_r}{\partial \nu^2} = \Phi_r^{(2)}(s, \nu) \quad . \tag{7.3.33}$$

The right sides of the equations $\Phi_r^{(1)(2)}(s, \nu)$ depend on the polynomials already constructed ($\alpha_j(s, \nu)$ and $\beta_j(s, \nu)$, $j \leq r - 1$), and can be expanded in powers of ν.

We now prove that the degrees of the polynomials $\alpha_r(s, \nu)$ and $\beta_r(s, \nu)$ for even r are equal to $r/2$ and $r/2 + 1$ respectively, and for odd r are equal to $(r + 3)/2$ and $(r - 1)/2$. The proof will be by mathematical induction.

Let us assume that the degrees of $\alpha_j(s, \nu)$ and $\beta_j(s, \nu)$ when $0 \leq j \leq r - 1$ are equal respectively to $(j + 3)/2$ and $(j - 1)/2$ if j is odd, and are equal to $j/2$ and $j/2 + 1$ if j is even. For $j = 0, 1$, this assumption is satisfied in accordance with (7.3.25 and 22).

It can be shown that when this assumption is made, $\Phi_r^{(1)}(s, \nu)$ and $\Phi_r^{(2)}(s, \nu)$ are polynomials of degree $r/2 + 1$ and $r/2 - 1$ for even r, and polynomials of degree $(r - 1)/2$ and $(r + 1)/2$ for odd r. Using this result, we find the degrees of $\alpha_r(s, \nu)$ and $\beta_r(s, \nu)$. Eliminating $\alpha_r(s, \nu)$ between (7.3.32 and 33):

$$-\frac{\partial^3 \beta_r}{\partial \nu^3} + 4(\beta_{01}\nu + \beta_{00})\beta_{01}^2\frac{\partial \beta_r}{\partial \nu} + 2\beta_{01}^3 \beta_r - \frac{4}{c_0}\beta_{01} - \frac{d\alpha_{r-1,0}}{ds}$$
$$= 2\Phi_r^{(1)}(s, \nu)\beta_{01} - \frac{\partial \Phi_r^{(2)}(s, \nu)}{\partial \nu} \quad . \tag{7.3.34}$$

Equation (7.3.34) implies that $\beta_r(s, \nu)$ is a polynomial of degree $r/2 + 1$ for even r, and a polynomial of degree $(r - 1)/2$ for odd r. Thus we have proved our statement as to the degree of the polynomial $\beta_r(s, \nu)$. After the degree of $\beta_r(s, \nu)$ has been established, the degree of $\alpha_r(s, \nu)$ is easily found from (7.3.33), and is found to be equal to $r/2$ for even r and to $(r + 3)/2$ for odd r. This is just the result we should have expected.

The coefficients of $\beta_r(s, \nu)$ are determined as follows. Into (7.3.34) we substitute $\beta_r(s, \nu)$ expanded in powers of ν with undetermined coefficients depending on s. We then equate the coefficients of different powers (≥ 1) of ν on the left and right sides of (7.3.34). This brings us to a system of algebraic equations from which all coefficients of polynomial $\beta_r(s, \nu)$ are found with the exception of its constant term $\beta_{r0}(s)$ which is determined later on from the boundary conditions.

Finally, we set the constant terms equal in (7.3.34). Integrating the result with respect to s, we find the function $\alpha_{r-1,0}(s)$.

After $\beta_r(s, \nu)$ has been found, we find $\alpha_r(s, \nu)$ by integrating (7.3.33) with respect to ν. The constant term of $\alpha_r(s, \nu)$ is determined by integrating the next pair of equations from (7.3.7) and is expressed in terms of $\beta_{r+1,0}(s)$.

The results of the calculations made in this manner are given below for the case $r = 2$. The functions $\Phi_r^{(1)}(s, \nu)$ and $\Phi_r^{(2)}(s, \nu)$ when $r = 2$ are

$$\Phi_2^{(1)}(s, \nu) = \left(\frac{\partial \alpha_1}{\partial \nu}\right)^2 + \frac{2}{c_0(s)} \frac{\partial (\alpha_1 - \alpha_{10})}{\partial s} + \left(\frac{\partial \alpha_{-1}}{\partial s}\right)^2$$
$$- \frac{2}{c_0(s)} \frac{\partial \alpha_{-1}}{\partial s} \frac{\nu}{\varrho(s)} - \beta_0 \left(\frac{\partial \beta_0}{\partial \nu}\right)^2 \frac{\nu}{\varrho(s)}$$
$$+ \frac{\nu^2}{c_0^2(s)\varrho^2(s)} - e_2 \frac{\nu^2}{\varrho^2(s)} \quad .$$

(7.3.35)

$$\Phi_2^{(2)}(s, \nu) = -\frac{1}{\varrho(s)} \frac{\partial \beta_0}{\partial \nu} - 2i \frac{1}{c_0(s)} \frac{\partial \beta_1}{\partial s} \quad .$$

The second-degree polynomial $\beta_2(s, \nu)$ is

$$\beta_2(s, \nu) \equiv \beta_{22}(s)\nu^2 + \beta_{21}(s)\nu + \beta_{20}(s) \quad , \tag{7.3.36}$$

where

$$\beta_{22}(s) = \frac{2^{1/3}}{10} \frac{P^{2/3}(s)}{c_0^{2/3}(s)} Q(s) \quad ,$$

$$\beta_{21}(s) = \frac{1}{15} \beta_{00}(s) \left(2P(s)Q(s) - 10\frac{1}{\varrho(s)}\right)$$

and

$$Q(s) = \frac{1}{\varrho^2(s)} + \frac{2}{\varrho(s)P(s)} + \frac{1}{9}\left\{\frac{d}{ds} \ln \left[c_0^2(s)P(s)\right]\right\}^2$$
$$+ \frac{1}{3}c_0(s)\frac{d}{ds}\left\{\frac{1}{c_0(s)}\frac{d}{ds} \ln \left[c_0^2(s)P(s)\right]\right\}$$
$$+ \frac{2c_1(s)}{c_0(s)\varrho(s)} - \frac{3c_1^2(s) - 2c_2(s)c_0(s)}{c_0^2(s)} \quad . \tag{7.3.37}$$

The function $\beta_{20}(s)$ remains undetermined. The constant term of the polynomial $\alpha_1(s, \omega)$ is equal to

$$\alpha_{10}(s) = 2^{-1/3} \int_{d_1}^{s}\left\{\frac{\beta_{20}(\tau)}{c_0^{1/3}(\tau)P^{2/3}(\tau)} - \frac{\beta_{00}^2(\tau)}{2^{4/3}P^{4/3}(\tau)}\right\}$$

$$\times c_0^{1/3}(\tau)\left[1 + \frac{4}{3}P(\tau)\left(\frac{2}{5}P(\tau)Q(\tau) - \frac{2}{\varrho(\tau)}\right)\right]\Bigg\}d\tau \quad , \qquad (7.3.38)$$

where d_1 is an arbitrary constant.

The polynomial $\alpha_2(s, \nu)$ is of first-degree:

$$\alpha_2(s, \nu) \equiv \alpha_{21}(s)\nu + \alpha_{20}(s)$$

$$= \left[-\frac{i}{10}P(s)Q(s) + \frac{i}{2\varrho(s)} + \frac{P^{1/3}(s)}{2^{1/3}c_0^{1/3}(s)}\frac{d\beta_{10}}{ds}\right]\nu + \alpha_{20}(s) \quad . \qquad (7.3.39)$$

The function $\alpha_{20}(s)$ is determined in the next step, along with the polynomials $\alpha_3(s, \nu)$ and $\beta_3(s, \nu)$. Omitting the calculations involved in finding α_3 and β_3, we will present only the value of $\alpha_{20}(s)$:

$$\alpha_{20}(s) = \int\limits_{d_2}^{s}\Bigg\{\frac{d}{d\tau}\left[\frac{i}{2^{1/3}}c_0^{2/3}P^{1/3}\left(\frac{1}{2}P^{-1} - \varrho^{-1} + \frac{1}{5}PQ\right)\beta_{00}\right.$$

$$\left. - \frac{2^{1/3}}{3}c_0^{1/3}P^{-2/3}\beta_{10}'\right] + \frac{i}{2^{1/3}}c_0^{2/3}P^{1/3}$$

$$\times\left(\frac{1}{5}PQ - \frac{3}{2\varrho} - \frac{1}{2^{1/3}}c_0^{-1/3}P^{-1/3}\beta_{10}'\right)\beta_{00}'$$

$$+ 2^{1/3}c_0^{-2/3}P^{-1/3}\left(\frac{1}{\varrho} - \frac{4}{15}PQ\right)\beta_{00}\beta_{10}$$

$$+ \frac{1}{2^{1/3}}c_0^{-1/3}P^{-2/3}\beta_{30}\Bigg\}d\tau \quad . \qquad (7.3.40)$$

Here d_2 is an arbitrary constant, and the function $\beta_{30}(s)$, like β_{00}, β_{10} and β_{20}, remains undetermined and will be determined later from the boundary conditions.

Thus the polynomials $\alpha_m(s, \nu)$, $m = -3, -2, -1, 0, 1, 2$ appearing in (7.3.4) are determined by (7.3.16, 17, 24, 25, 31, 38, 39, and 40), and the polynomials $\beta_m(s, \nu)$, $m = 0, 1, 2$ are determined by (7.3.22, 29, and 36). The solution (7.3.4) of (7.1.1) can be written as

$$u(s, \nu) = \text{const}\sqrt[6]{\frac{c_0(s)}{c_0(d_0)}\frac{P(d_0)}{P(s)}}\exp\Bigg\{i\omega\int\limits_{d_{-3}}^{s}\frac{d\tau}{c_0(\tau)} + i\omega^{2/3}d_{-2}$$

$$+ i\left(\frac{\omega}{2}\right)^{1/3}\int\limits_{d_{-1}}^{s}\frac{\beta_{00}(\tau)d\tau}{c_0^{1/3}(\tau)P^{2/3}(\tau)} + \frac{i}{2}\int\limits_{d_0}^{s}c_0(\tau)\beta_{01}^2(\tau)\beta_{10}(\tau)d\tau$$

$$+ \frac{i}{\omega^{1/3}}\left[-\frac{\beta_{01}'(s)}{2c_0(s)\beta_{01}(s)}\nu^2 - \frac{\beta_{00}'(s)}{c_0(s)\beta_{01}(s)}\nu + \alpha_{10}(s)\right]$$

$$+ \frac{i}{\omega^{2/3}}\alpha_2(s, \nu) + O(\omega^{-1})\Bigg\}\nu\Bigg\{\left[-\frac{2^{1/3}}{c_0^{2/3}(s)P^{1/3}(s)}\nu + \beta_{00}(s)\right]$$

$$+ \frac{\beta_{10}(s)}{\omega^{1/3}} + \frac{\beta_2(s, \nu)}{\omega^{2/3}} + O\left(\frac{1}{\omega}\right)\Bigg\} \quad . \qquad (7.3.41)$$

Now, assume that $\alpha_m(s, \nu)$ and $\beta_m(s, \nu)$ are determined up to $m = M - 1$ inclusive, and that they are used to construct the function

$$u^{(M)}(s, \nu) = \text{const} \cdot \exp\left[i \sum_{m=-3}^{M-1} \alpha_m(s, \nu)\omega^{-m/3}\right]$$

$$\times v\left[\sum_{m=0}^{M-1} \beta_m(s, \nu)\omega^{-m/3}\right] . \qquad (7.3.42)$$

Let us evaluate the remainder R_M (the uncompensated terms) in the Helmholtz equation (7.3.3) when $u^{(M)}(s, \nu)$ is substituted into the equation. We will estimate R_M not only in the boundary layer $-n = O(\omega^{-2/3})$, where the reduced normal remains bounded but also in the wider strip $-n = O(\omega^{-\varepsilon})(\varepsilon > 0)$, for which $-\Lambda\omega^{2/3-\varepsilon} < \nu \leq 0$, $\Lambda = O(1) > 0$. We obtain the estimate

$$|R_M| \leq C\omega^{4/3+(3/4)\varepsilon_1+(\varepsilon_1/2-1/3)M} \exp(-\Delta\omega^{(3/2)\varepsilon_1})$$
$$\{C, \Delta = \text{const} > 0\} \qquad (7.3.43)$$

which is true for $-\nu = O(\omega^{\varepsilon_1})$, $0 \leq \varepsilon_1 < 2/3$. [The estimate (7.3.43) is crucial to Sect. 7.6, and also to Chaps. 11 and 12 for the investigation of Green's functions.]

To derive (7.3.43), note that (7.3.5 and 6) imply

$$|R_M| < \text{const } \omega^{-(M-4)/3}$$

$$\times \left\{\left|\Phi_M^{(1)}(s, \nu)\exp\left[i \sum_{m=-3}^{M-1} \alpha_m(s, \nu)\omega^{-m/3}\right] v\left[\sum_{m=0}^{M-1} \beta_m(s, \nu)\omega^{-m/3}\right]\right|\right.$$

$$\left. + \left|\Phi_M^{(2)}(s, \nu)\exp\left[i \sum_{m=-3}^{M-1} \alpha_m(s, \nu)\omega^{-m/3}\right] v'\left[\sum_{m=0}^{M-1} \beta_m(s, \nu)\omega^{-m/3}\right]\right|\right\} ,$$

$$(7.3.44)$$

where $\Phi_M^{(1)}(s, \nu)$ and $\Phi_M^{(2)}(s, \nu)$ are the polynomials from the right sides of (7.3.32, 33) when $r = M$. If $|\nu| = O(1)$, the estimate (7.3.43) is an obvious consequence of (7.3.44).

Let us now set $\nu = \omega^{\varepsilon_1}\Lambda$, where $\varepsilon_1 > 0$ and $|\Lambda| < \text{const} < \infty$. We want to find those values of ε_1 for which (7.3.43) holds as before. To do this, we must evaluate $\Phi_M^{(1)}, \Phi_M^{(2)}$, the exponential function, and the Airy function in accordance with (7.3.44).

Since $\Phi_M^{(1)}(s, \nu)$ is a polynomial in ν of degree no greater than $M/2 + 1$, and $\Phi_M^{(2)}(s, \nu)$ is a polynomial in ν of degree no greater than $(M + 1)/2$, the inequalities

$$\omega^{-(M-4)/3}|\Phi_M^{(1)}(s, \nu)| < C_1\omega^{4/3+\varepsilon_1+(\varepsilon_1/2-1/3)M}$$

and

$$\omega^{-(M-4)/3}|\Phi_M^{(2)}(s, \nu)| < C_2\omega^{4/3+(1/2)\varepsilon_1+(\varepsilon_1/2-1/3)M}$$

are satisfied, where C_1 and C_2 are positive constants.

Let us now evaluate the behavior of the Airy function and its derivative as $\omega \to \infty$. Since the degree of the polynomials $\beta_m(s, \nu)$ does not exceed $(m+2)/2$,

$$|\beta_m(s, \nu)\omega^{-m/3}| < C_3\omega^{[(m+2)/2]\varepsilon_1 - m/3} \quad ,$$

where C_3 is a positive constant, and in the sum

$$\sum_{m=0}^{M-1} \beta_m(s, \nu)\omega^{-m/3}$$

the first term $\beta_0(s, \nu) \sim \omega^{\varepsilon_1}$ will be the main term as $\omega \to \infty$ if

$$\omega^{[(m+2)/2]\varepsilon_1 - m/3 - \varepsilon_1} \to 0 \ , \quad m = 1, 2, \ldots,$$

as $\omega \to \infty$, i.e. if

$$0 \le \varepsilon_1 < 2/3 \quad .$$

Thus when $0 < \varepsilon_1 < 2/3$ and $\omega \to \infty$, the arguments of the Airy function and its derivative increase as ω^{ε_1}. Then from the asymptotic formulas for the Airy function (Appendix A.1) we get the estimates

$$\left| v\left[\sum_{m=0}^{M-1} \beta_m(s, \nu)\omega^{-m/3} \right] \right| = O\left[\omega^{-\varepsilon_1/4} \exp\left(-\Delta_1 \omega^{3\varepsilon_1/2} \right) \right] \ , \quad \Delta_1 > 0,$$

$$\left| v'\left[\sum_{m=0}^{M-1} \beta_m(s, \nu)\omega^{-m/3} \right] \right| = O\left[\omega^{\varepsilon_1/4} \exp\left(-\Delta_1 \omega^{3\varepsilon_1/2} \right) \right] \ , \quad \Delta_1 > 0 \ .$$

$$(7.3.45)$$

Finally, turning to the evaluation of the terms in the sum appearing in the argument of the exponential function, since the degree of polynomial $\alpha_m(s, \nu)$ does not exceed $(m + 3)/2$, we have that

$$|\alpha_m(s, \nu)| \omega^{-m/3} < C_4 \omega^{[(m+3)/2]\varepsilon_1 - m/3}$$

and consequently when $0 \le \varepsilon_1 \le 2/3 - \varepsilon$, $\varepsilon > 0$, we have

$$\left| \exp\left[i \sum_{m=-3}^{M-1} \alpha_m(s, \nu)\omega^{-m/3} \right] \right| = \left| \exp\left[i \sum_{m=1}^{M-1} \alpha_m(s, \nu)\omega^{-m/3} \right] \right|$$

$$= O\left[\exp\left(\Delta_2 \omega^{(3/2)\varepsilon_1 - (\varepsilon/2)} \right) \right], \quad \Delta_2 > 0.$$

Combining these estimates, we get the following inequality for values of ε_1 in the range $0 \le \varepsilon_1 \le 2/3 - \varepsilon$:

$$|R_M| < C_5 \omega^{4/3 + (3/4)\varepsilon_1 + [(\varepsilon_1/2) - (1/3)]M} O\left[\exp\left(-\Delta_3 \omega^{(3/2)\varepsilon_1} \right) \right]$$

$$\{C_5 > 0, \ \Delta_3 > 0\} \quad . \tag{7.3.46}$$

We can therefore state that (7.3.43) is uniformly satisfied in the boundary layer

$$|\nu| = O(\omega^{2/3-\epsilon}) \ , \quad \epsilon > 0 \ , \quad \text{or} \tag{7.3.47}$$

$$|n| = O(\omega^{-\epsilon}) \ , \quad \epsilon > 0 \ . \tag{7.3.48}$$

7.4 Eigenfunctions of the Whispering Gallery Type

In this section, expressions (7.3.4) will be used to get asymptotic formulas for the eigenvalues of certain boundary value problems, and approximate formulas will be written out for the corresponding eigenfunctions. Let us first consider the eigenvalues and eigenfunctions of the Helmholtz equation (7.1.1) which satisfy (7.1.2) on the boundary S of Ω. We will assume that (7.2.1) is satisfied at points of S. We will be concerned only with those eigenfunctions which are concentrated near the boundary of the region.

Substituting (7.3.4) into (7.1.2), and cancelling the exponential function, we get

$$v \left[\sum_{m=0}^{M-1} \beta_m(s,0)\omega^{-m/3} + O\left(\omega^{-M/3}\right) \right] = 0 \ . \tag{7.4.1}$$

Equation (7.4.1) must be satisfied identically in ω. It determines the remaining constant terms $\beta_m(s,0) = \beta_{m0}(s)$ of the polynomials $\beta_m(s,\nu)$, which had been arbitrary up to this point.

As a result, we get

$$\begin{aligned} \beta_{00}(s) &= -t_p \ , \quad p = 0, 1, 2, \ldots, \\ \beta_{m0}(s) &= 0 \ , \quad m \geq 1 \ . \end{aligned} \tag{7.4.2}$$

The numbers $t_0, t_1, t_2, \ldots$ are zeros of the Airy function $v(-t)$. The constant terms $\alpha_{m0}(s)$ of the polynomials $\alpha_m(s,\nu)$, $m \geq 1$, are expressed as integrals with variable upper limit and fixed lower limit. The integrands contain $\beta_{m0}(s)$, $m = 0, 1, 2, \ldots$, which have been determined by (7.4.2). By a suitable choice of the arbitrary constant multiplier in (7.3.4 and 41), we can set the fixed lower limits of integration d_m in these integrals equal to zero, i.e., we can assume that

$$\alpha_m(0,0) = 0 \ , \quad m \geq -3 \ . \tag{7.4.3}$$

Using (7.3.16, 17, 24, 30, 38 and 40) for α_{m0}, $m = -3, -2, -1, 0, \ldots$, and taking conditions (7.4.2 and 3) into account, we get

$$\alpha_{-30} = \int_0^s \frac{d\tau}{c_0(\tau)} \ ,$$

$$\alpha_{-20} = 0 \ ,$$

$$\alpha_{-10} = \frac{t_p}{2^{1/3}} \int_0^s \frac{d\tau}{c_0^{1/3}(\tau)P^{2/3}(\tau)} \ ,$$

$$\alpha_{00} = \frac{i}{2} \ln \frac{P^{1/3}(s) \, c_0^{1/3}(0)}{c_0^{1/3}(s) \, P^{1/3}(0)} \quad ,$$

$$\alpha_{10} = -\frac{t_p^2}{2^{5/3}} \int_0^s \frac{c_0^{1/3}(\tau)}{P^{4/3}(\tau)} \left\{ 1 + \frac{4}{3} P(\tau) \left[\frac{2}{5} P(\tau) Q(\tau) - \frac{2}{\varrho(\tau)} \right] \right\} d\tau \quad ,$$

$$\alpha_{20} = -i \frac{t_p}{2^{1/3}} \left\{ c_0^{2/3}(s) P^{1/3}(s) \left[\frac{1}{2} P^{-1}(s) - \varrho^{-1}(s) + \frac{1}{5} P(s) Q(s) \right] \right.$$
$$\left. - c_0^{2/3}(0) P^{1/3}(0) \left[\frac{1}{2} P^{-1}(0) - \varrho^{-1}(0) + \frac{1}{5} P(0) Q(0) \right] \right\} \quad ,$$

$$p = 0, 1, 2, \ldots \quad . \tag{7.4.4}$$

The solution of equation (7.1.1) can be written in the first approximation as

$$u(s, \nu) = u_p(s, \nu) = \text{const} \sqrt[6]{\frac{c_0(s)}{P(s)} \frac{P(0)}{c_0(0)}}$$

$$\times \exp \left\{ i \left[\omega \int_0^s \frac{d\tau}{c_0(\tau)} - \omega^{1/3} \frac{t_p}{2^{1/3}} \int_0^s \frac{d\tau}{c_0^{1/3}(\tau) P^{2/3}(\tau)} \right] \right\}$$

$$\times v \left[-2^{1/3} \frac{1}{c_0^{2/3}(s) P^{1/3}(s)} \nu - t_p \right] \quad p = 0, 1, 2, 3, \ldots \quad . \tag{7.4.5}$$

The Airy function $v(Z)$, as we know, oscillates when $Z < 0$, and is exponentially damped when $Z > 0$. Consequently, the functions $u_p(s, \nu)$, $p = 0, 1, 2, \ldots$, will oscillate in the strip

$$-\frac{t_p}{2^{1/3}} c_0^{2/3}(s) P^{1/3}(s) < \nu \leq 0$$

and damp out exponentially when

$$\nu < -\frac{t_p}{2^{1/3}} c_0^{2/3}(s) P^{1/3}(s) \quad .$$

Thus also if $P(s) > 0$, the solutions $u_p(s, \nu)$ are concentrated in a strip located where the normal is negative, i.e., to the right of S, precisely where the multiply-reflected waves overlap with each other. To the left of S we cannot construct such concentrated solutions, no matter how we choose the solution of the Airy equation in formula (7.4.5) instead of $v(Z)$.

In the case of a bounded region Ω the desired eigenfunctions $u_p(s, \nu)$ must also satisfy the obvious conditions of periodicity with respect to the variable s:

$$u_p(s, \nu) = u_p(s + L, \nu) \quad , \tag{7.4.6}$$

where L is the perimeter of S.

The effective radius of curvature $P(s)$ as well as the coefficients of the expansion of $c(s, \nu)$ in powers of the normal ν, (i.e., the functions $c_m(s)$, $m \geq 0$) are periodic functions with period L when Ω is bounded. Since the coefficients of the polynomial $\beta_m(s, \nu)$ are expressed directly in terms of $P(s)$ and $c_m(s)$, $m \geq 0$, all the $\beta_m(s, \nu)$ will be periodic functions of s. In exactly the same way, the coefficients of $\alpha_m(s, \nu)$ are periodic functions, with the exception of their constant terms $\alpha_{m0}(s)$, $m \geq -3$. Substituting (7.3.4) into the periodicity conditions (7.4.6), and cancelling the obviously periodic factors in the result, we arrive at the equation

$$\exp\left\{i \sum_{m=-3}^{M-1} [\alpha_{m0}(s + L) - \alpha_{m0}(s)]\omega^{-m/3} + O\left(\omega^{-M/3}\right)\right\} = 1 \quad,$$

which by virtue of (7.4.3) can be rewritten as

$$\sum_{m=-3}^{M-1} \alpha_{m0}(L)\omega^{-m/3} + O\left(\omega^{-M/3}\right) = 2\pi q \quad, \tag{7.4.7}$$

where q is a positive integer.

Equation (7.4.7) serves to determine the eigenvalues $\omega_{p,q}$ of the problem at hand. In the first approximation,

$$\omega_{p,q} \approx \omega_q = \frac{2\pi q}{\alpha_{-30}(L)} = 2\pi q \Big/ \int_0^L \frac{d\tau}{c_0(\tau)} \quad. \tag{7.4.8}$$

Since all our calculations relate to the case $\omega \gg 1$, we must assume that the integer $q \gg 1$. Solving (7.4.7) by the method of successive approximations, we get an expansion in fractional powers of ω_q for the eigenvalues $\omega_{p,q}$:

$$\omega_{p,q} = \omega_q[1 + a_1(\omega_q)^{-1/3} + a_2(\omega_q)^{-2/3} + a_3(\omega_q)^{-1} + a_4(\omega_q)^{-4/3} + \ldots] \,,$$
$$p = 0, 1, 2, 3, \ldots, \quad q \gg 1 \quad, \tag{7.4.9}$$

in which the coefficients a_k are expressed in terms of $\alpha_{m0}(L)$. Using (7.4.4), we find that

$$a_1 = 0 \,, \quad a_2 = \frac{t_p}{2^{1/3}} \int_0^L \frac{d\tau}{c_0^{1/3}(\tau)P^{2/3}(\tau)} \left(\int_0^L \frac{d\tau}{c_0(\tau)}\right)^{-1} \,,$$

$$a_3 = 0 \,, \quad a_4 = \frac{a_2^2}{3} + \frac{t_p^2}{2^{5/3}} \int_0^L \frac{c_0^{1/3}(\tau)}{P^{4/3}(\tau)} \left[1 + \frac{4}{3}P(\tau)\right]$$

$$\times \left(\frac{2}{5}P(\tau)Q(\tau) - \frac{2}{\varrho(\tau)}\right)\right] d\tau \left(\int_0^L \frac{d\tau}{c_0(\tau)}\right)^{-1} \quad. \tag{7.4.10}$$

It can in fact be shown that all the odd coefficients of expansion (7.4.9) are equal to zero. Formulas (7.4.9 and 10) determine the asymptotic expansions of the eigenvalues $\omega_{p,q}$ for $q \gg 1$. Let us note that the second subscript p enters into the expression for the eigenvalues $\omega_{p,q}$ only via $\beta_{00} = -t_p$ – the root of the Airy function $\nu(t)$.

163

Substituting the values found for the natural frequencies $\omega_{p,q}$ into (7.3.41) and considering (7.4.2,4), we get asymptotic formulas for the corresponding eigenfunctions:

$$
u_{p,q} = A_{p,q} \sqrt[6]{\frac{c_0(s)}{P(s)}} \exp\left\{ i\omega_q \int_0^s \frac{d\tau}{c_0(\tau)} \right.
$$

$$
+ i\omega_q^{1/3} \left[a_2 \int_0^s \frac{d\tau}{c_0(\tau)} - \frac{t_p}{2^{1/3}} \int_0^s \frac{d\tau}{c_0^{1/3}(\tau)P^{2/3}(\tau)} \right]
$$

$$
+ \frac{i}{\omega_q^{1/3}} \left[a_4 \int_0^s \frac{d\tau}{c_0(\tau)} - a_2 \frac{t_p}{3\cdot 2^{1/3}} \int_0^s \frac{d\tau}{c^{1/3}(\tau)P^{2/3}(\tau)} \right.
$$

$$
- \frac{t_p^2}{2^{5/3}} \int_0^s \frac{c_0^{1/3}(\tau)}{P^{4/3}(\tau)} \left[1 + \frac{4}{3}P(\tau)\left(\frac{2}{5}P(\tau)Q(\tau) - \frac{2}{\varrho(\tau)} \right) \right] d\tau
$$

$$
+ \frac{1}{6c_0(s)} \left(\frac{2c_0'(s)}{c_0(s)} + \frac{P'(s)}{P(s)} \right) \nu^2 \right]
$$

$$
+ O(\omega^{-2/3}) \right\} v\left(-t_p - \frac{2^{1/3}}{c_0^{2/3}(s)P^{1/3}(s)} \nu + O(\omega_q^{-2/3}) \right) \quad . \quad (7.4.11)
$$

Here $A_{p,q}$ does not depend on s or ν, and the quantities ω_q, $P(s)$, $Q(s)$, a_2, a_4 are given in (7.4.8), (7.2.1), (7.3.37) and (7.4.10).

If the region considered is a circle of radius ϱ, then $c_0 = 1$, $c_k(s) = 0$ for $k \geq 1$, and $P(\tau) = \varrho$, so consequently

$$
a_2 = \frac{t_p}{2^{1/3}} \frac{1}{\varrho^{2/3}} , \qquad a_4 = \frac{t_p^2}{2^{2/3}} \frac{1}{\varrho^{4/3}} \frac{3}{10} \quad .
$$

These values agree with the corresponding coefficients in formula (7.2.8) for the eigenvalues of a circle.

Let us now obtain an asymptotic formula for eigenvalues and eigenfunctions in the case where condition (7.1.3) is satisfied on the boundary of region Ω:

$$
\left. \frac{\partial u}{\partial n} \right|_S = 0 \quad .
$$

We differentiate (7.3.4) with respect to the normal and substitute into condition (7.1.3). After cancelling the exponential function, we arrive at

$$
i \left[\sum_{m=1}^{M-1} \left. \frac{\partial \alpha_m}{\partial \nu} \right|_{\nu=0} \omega^{-m/3} + O\left(\omega^{-M/3} \right) \right]
$$

$$
\times v \left[\sum_{m=0}^{M-1} \beta_m(s,0)\omega^{-m/3} + O\left(\omega^{-M/3} \right) \right]
$$

$$
+ \left[\sum_{m=0}^{M-1} \left. \frac{\partial \beta_m}{\partial \nu} \right|_{\nu=0} \omega^{-m/3} + O\left(\omega^{-M/3} \right) \right]
$$

164

$$\times v' \left[\sum_{m=0}^{M-1} \beta_m(s,0)\omega^{-m/3} + O\left(\omega^{-M/3}\right) \right] = 0 \quad . \tag{7.4.12}$$

The first summation starts at $m = 1$ because

$$\frac{\partial \alpha_{-3}}{\partial v} = \frac{\partial \alpha_{-2}}{\partial v} = \frac{\partial \alpha_{-1}}{\partial v} = \frac{\partial \alpha_0}{\partial v} = 0 \quad .$$

In the first approximation, (7.4.12) takes the form

$$\beta_{01}(s) \cdot v' \{\beta_{00}(s) + O(\omega^{-1/3})\} + O(\omega^{-1/3}) = 0 \quad .$$

Since $\beta_{01} \neq 0$,

$$\beta_{00}(s) = -t'_p \quad , \quad p = 0, 1, 2, 3, \dots \quad ,$$

where the t'_p are zeros of the derivative of the Airy function $v'(-t)$ (see Appendix A.1).

We now expand the Airy function and its derivative in (7.4.12) in a Taylor series in powers of the expression $\sum_{m=1}^{M-1} \beta_m(s,0)\omega^{-m/3} + O(\omega^{-M/3})$. Then expanding the left side of equalaity (7.4.12) in powers of $\omega^{-1/3}$ we get

$$\sum_{m=1}^{M-1} f_m(\alpha_j, \beta_j)\omega^{-m/3} + O(\omega^{-M/3}) = 0 \quad . \tag{7.4.13}$$

The coefficients $f_m(\alpha_j, \beta_j)$ depend on $\beta_j(s,0) = \beta_{j0}(s)$,

$$\left. \frac{\partial \beta_j}{\partial v} \right|_{v=0} = \beta_{j1}(s) \quad , \quad \text{and} \quad \left. \frac{\partial \alpha_j}{\partial v} \right|_{v=0} = \alpha_{j1}(s) \quad , \quad j \leq m \quad .$$

Equation (7.4.13) must hold identically in ω. Consequently, all coefficients $f_m(\alpha_j, \beta_j)$ must be equal to zero. The equations

$$f_m(\alpha_j, \beta_j) = 0 \quad , \quad m = 1, 2, 3, \dots \quad , \tag{7.4.14}$$

serve to determine the functions $\beta_{m0}(s)$. We readily find that

$$f_1 = [i\alpha_{11}(s) - t'_p \beta_{01}(s)\beta_{10}(s)]v(-t'_p) \quad .$$

Since $\beta_{00} = \text{const}$, we have $\alpha_{11}(s) = 0$, see (7.3.31), and f_1 vanishes when

$$\beta_{10}(s) = 0 \quad . \tag{7.4.15}$$

Taking (7.4.15) into consideration, we find

$$f_2 = [i\alpha_{21}(s) - t'_p \beta_{01}(s)\beta_{20}(s)]v(-t'_p) \quad .$$

Setting f_2 equal to zero, we get

$$\beta_{20}(s) = \frac{i\alpha_{21}(s)}{t'_p \beta_{01}(s)} = -\frac{1}{2^{4/3} t'_p} c_0^{2/3}(s) P^{1/3}(s) \left[\frac{1}{5} P(s)Q(s) - \frac{1}{\varrho(s)} \right] \quad .$$

Having determined the functions $\beta_{00}(s)$, $\beta_{10}(s)$ and $\beta_{20}(s)$, we can completely determine the constant terms of the polynomials α_{-1}, α_0 and α_1:

$$\alpha_{-10}(s) = -\frac{t'_p}{2^{1/3}} \int\limits_0^s \frac{d\tau}{c_0^{1/3}(\tau)P^{2/3}(\tau)} \ ,$$

$$\alpha_{00}(s) = \frac{i}{2}\ln\frac{P^{1/3}(s)c_0^{1/3}(0)}{c_0^{1/3}(s)P^{1/3}(0)} \ ,$$

$$\alpha_{10}(s) = -2^{-5/3} \int\limits_0^s \frac{c_0^{1/3}(\tau)}{P^{1/3}(\tau)} \left[\frac{t'^2_p}{P(\tau)} + \left(\frac{1}{t'_p} + \frac{8}{3}t'^2_p\right) \right.$$

$$\left. \times \left(\frac{1}{5}P(\tau)Q(\tau) - \frac{1}{\varrho(\tau)}\right)\right] d\tau \ .$$

In these formulas we have taken the lower limit of integration equal to zero in order to satisfy (7.4.3).

The subsequent equations of (7.4.14) enable us to determine all the remaining functions $\beta_{m0}(s)$. The functions $\beta_{m0}(s)$ in turn enable us to completely determine the constant terms of the polynomials α_m, i.e., the functions $\alpha_{m0}(s)$.

Substituting the values found for $\alpha_{m0}(s)$ into the periodicity condition (7.4.6), we get an equation for determining the natural frequencies $\omega_{p,q}$ corresponding to the Neumann problem. Once again, we can expand the solution of this equation in negative powers of $\omega_q = 2\pi q \big/ \int_0^L (d\tau/c_0(\tau))$ and we have

$$\omega_{p,q} = \omega_q[1 + a'_1\omega_q^{-1/3} + a'_2\omega_q^{-2/3} + a'_3\omega_q^{-1} + a'_4\omega_q^{-4/3} + \ldots] \ ,$$
$$q \gg 1 \ , \quad p = 0, 1, 2, 3, \ldots \ ,$$

where

$$a'_1 = 0 \ , \quad a'_2 = \frac{t'_p}{2^{1/3}} \int\limits_0^L \frac{d\tau}{c_0^{1/3}(\tau)P^{2/3}(\tau)} \left(\int\limits_0^L \frac{d\tau}{c_0(\tau)}\right)^{-1} \ , \quad a'_3 = 0 \ ,$$

$$a'_4 = \frac{a'^2_2}{3} + 2^{-5/3} \int\limits_0^L \frac{c_0^{1/3}(\tau)}{P^{1/3}(\tau)} \left[\frac{t'^2_p}{P(\tau)} + \left(\frac{1}{t'_p} + \frac{8}{3}t'^2_p\right)\right.$$

$$\left. \times \left(\frac{1}{5}P(\tau)Q(\tau) - \frac{1}{\varrho(\tau)}\right)\right] d\tau \left(\int\limits_0^L \frac{d\tau}{c_0(\tau)}\right)^{-1} \ , \ldots \ .$$

To get approximate expressions for the eigenfunctions of the Neumann problem, the values found for $\omega_{p,q}$ must be substituted into (7.3.41).

In physical problems, in addition to the Dirichlet and Neumann conditions at the boundary of a region, a condition of the third kind (7.1.4) is also frequently encountered:

$$\left.\frac{\partial u}{\partial n} - i\omega g(s)u\right|_S = 0 \ .$$

Here $g(s)$ is a positive function which is related to the absorption of waves by the boundary of the region. The dimensionless quantity $1/c_0(s)g(s)$ is usually called the *normal impedance of the boundary*.

Using the class of solutions for the Helmholtz equation constructed above, we can easily get asymptotic formulas for the eigenvalues for the case of condition (7.1.4) as well. Substituting expression (7.3.4) into (7.1.4), expanding the Airy function and its derivative in a Taylor series in powers of $\sum_{m=1}^{M-1} \beta_{m0}(s)$ $\times \omega^{-m/3} + O(\omega^{-M/3})$, and setting the coefficients of different powers of $\omega^{-1/3}$ equal to zero, i.e., proceeding just as in the previous case, we get a system of equations analogous to system (7.4.14), from which the functions $\beta_{m0}(s)$ are determined. The calculations lead to the following results:

$$\beta_{00}(s) = -t_p , \quad p = 0, 1, 2, 3, \dots ,$$

$$\beta_{10}(s) = i2^{1/3} \frac{1}{c_0^{2/3}(s)P^{1/3}(s)g(s)} ,$$

$$\beta_{20}(s) = 0 .$$

Since the values of $\beta_{00}(s)$ and $\beta_{20}(s)$ coincide with the values of these functions in the case of condition $u|_S = 0$, the functions $\alpha_{-10}(s)$ and $\alpha_{10}(s)$ will be given as before by (7.4.4). For $\alpha_{00}(s)$ we get

$$\alpha_{00}(s) = i\frac{1}{2}\ln\frac{P^{1/3}(s)c_0^{1/3}(0)}{c_0^{1/3}(s)P^{1/3}(0)} + i\int_0^s \frac{d\tau}{c_0(\tau)P(\tau)g(\tau)} ,$$

in place of (7.4.4).

The integral term in the expression for $\alpha_{00}(s)$ shifts the eigenvalues $\omega_{p,q}$ from the real axis into the lower half-plane:

$$\omega_{p,q} = \omega_q \left[1 + \frac{t_p}{2^{1/3}} \int_0^L \frac{d\tau}{c^{1/3}(\tau)P^{2/3}(\tau)} \left(\int_0^L \frac{d\tau}{c_0(\tau)} \right)^{-1} \omega_q^{-2/3} \right.$$
$$\left. - i\int_0^L \frac{d\tau}{c_0(\tau)P(\tau)g(\tau)} \left(\int_0^L \frac{d\tau}{c_0(\tau)} \right)^{-1} \omega_q^{-1} + \dots \right] .$$

The magnitude of this shift, which is given in the first approximation by the integral

$$\int_0^L \frac{d\tau}{c_0(\tau)P(\tau)g(\tau)} \left(\int_0^L \frac{d\tau}{c_0(\tau)} \right)^{-1} ,$$

characterizes what is called in quantum mechanics the half-life of this quasi-stationary state due to absorption of the energy of the natural oscillation by the boundary of the region.[2]

---

[2] In electromagnetics and acoustics, this quantity is related to the so-called *quality factor* of the resonant mode [Translator's note].

7.5 Eigenfunctions of the Region Exterior to Ω

Consider the eigenfunctions of the region $C\Omega$ complementary to Ω. First, we should note that we are dealing with eigenfunctions that are a little unusual. For the moment, let the velocity $c \equiv 1$. The proof of the uniqueness theorem for the exterior Dirichlet and Neumann problems for the Helmholtz equation [7.2] implies that any solution of the equation

$$(\Delta + k^2)u(M) = 0 , \quad \Delta = \frac{\partial^2}{\partial x^2} + \frac{\partial^2}{\partial y^2} ,$$

$$M(x, y) \in C\Omega , \tag{7.5.1}$$

that satisfies the boundary condition

$$u|_S = 0 \quad \text{or} \quad \frac{\partial u}{\partial n}\bigg|_S = 0 \tag{7.5.2}$$

and is quadratically integrable over the whole exterior of Ω, is identically zero. The proof can be easily extended to the case of complex k. Thus, if there exist nontrivial solutions of (7.5.1) with condition (7.5.2), they cannot, in contrast to ordinary eigenfunctions, be quadratically integrable. Let us deal with this question more precisely.

Let us denote the Green's function for the exterior of Ω by $G(M_0, M, k)$, where:

$$(\Delta + k^2)G = -\delta(M - M_0) , \quad k > 0 ,$$

$$\lim_{r \to \infty} \sqrt{r}\left(\frac{\partial G}{\partial r} - ikG\right) = 0 ,$$

$$G|_S = 0 \quad \text{or} \quad \frac{\partial G}{\partial n}\bigg|_S = 0 . \tag{7.5.3}$$

We know that the Green's function can be analytically continued from real values of k into the upper half of the complex k-plane, and that it will be regular there (as a function of k). When $G(M_0, M, k)$ is analytically continued into the lower half-plane (Im $k < 0$), we may encounter poles. Close to such a pole $k = k_0$,

$$G(M_0, M, k) = \frac{U(M_0, M)}{(k - k_0)^\alpha} + \text{ less singular terms } . \tag{7.5.4}$$

We know [7.3] that the function $U(M_0, M)$ is a linear combination of functions $u_j(M)$ satisfying (7.5.1 and 2) at $k = k_0$. These functions $u_j(M)$ we will call *eigenfunctions of the exterior of* Ω. Obviously an analogous definition of eigenfunctions is applicable (with certain additional restrictions) in the case of variable $c = c(x, y)$ as well.

The eigenfunctions of the exterior of a circle (they can easily be found in explicit form by the method of separation of variables) increase exponentially as $r = \sqrt{x^2 + y^2} \to \infty$. It is possible that this circumstance holds also in the general case where the circle is replaced by an arbitrary convex finite region Ω.

The eigenfunctions of the whispering gallery type constructed in the preceding sections do not exhaust the entire set of eigenfunctions with large indices. The eigenfunctions of the exterior of Ω that will be investigated below also represent only a fraction of the eigenfunctions, and the corresponding eigenvalues are a subsequence of eigenvalues which are closest to the real axis of all eigenvalues with large index. This can be demonstrated for the exterior of a circle by direct calculations which are elementary, though cumbersome.

To construct the eigenfunctions of the exterior of Ω, we must first find a certain class of asymptotic solutions of equation (7.1.1). As before we introduce the coordinates s and n and assume that $c = c(s, n)$. In contrast to Sect. 7.3, we now assume that $n > 0$. As before, $P(s) > 0$. We seek a solution of (7.3.1) (in which now $n > 0$) in the form

$$
u(s, \nu) = \text{const} \, \exp\left[i \sum_{m=-3}^{M-1} \alpha_m(s, \nu)\omega^{-m/3} + O(\omega^{-M/3}) \right]
$$

$$
\times w_1 \left[\sum_{m=0}^{M-1} \beta_m(s, \nu)\omega^{-m/3} + O(\omega^{-m/3}) \right] , \quad \nu = \omega^{2/3}n ,
$$

(7.5.5)

which is suggested by the asymptotic expansion of the exact solution of the external problem of diffraction on a circle. Here $w_1(Z)$ is the Airy function of the first kind (see Appendix A.1), while $\alpha_m(s, \nu)$ and $\beta_m(s, \nu)$ are unknown polynomials in ν with coefficients that depend on s. Functions of the form (7.5.5), which we will go on to construct later, will, as we shall see, satisfy the principle of limiting absorption in a certain sense.

The substitution of the Airy function $w_1(Z)$ for the Airy function $v(Z)$ does not affect the calculations of Sect. 7.3 as both these functions satisfy the same equation. Therefore, as long as we are dealing only with equation (7.3.1), formulas for the polynomials α_m and β_m in (7.5.5) can be found by extending the formulas of Sect. 7.3 for the analogous polynomials in (7.3.4) to values of ν which are greater than zero. However, these formulas will contain arbitrary functions which must be determined from the boundary conditions.

Now let us seek solutions of (7.1.1) in $C\Omega$ (the exterior of Ω) which satisfy condition (7.1.2). Substituting (7.5.5) into (7.1.2) gives

$$
w_1 \left(\sum_{m=0}^{M-1} \beta_m(s, 0)\omega^{-m/3} + O(\omega^{-M/3}) \right) = 0 .
$$

(7.5.6)

Equation (7.5.6) defines the constant terms $\beta_m(s, 0) = \beta_{m0}(s)$ of the polynomials $\beta_m(s, \nu)$ which had remained arbitrary when (7.1.1) was satisfied:

$$
\beta_{00} = \xi_p , \quad p = 0, 1, 2, \dots ,
$$

$$
\beta_{m0} = 0 , \quad m = 1, 2, \dots .
$$

(7.5.7)

The complex numbers $\xi_p = t_p \, e^{i(\pi/3)}$ are the zeros of $w_1(\xi)$. The new formulas (7.5.7) for the constant terms of $\beta_m(s, \nu)$ lead to new formulas for the constant

169

terms of the polynomials $\alpha_m(s, \nu)$, as compared to Sect. 7.4. For these terms, we get expressions that differ from (7.4.4) in the substitution of ξ_p for $(-t_p)$.

Substituting the expressions for α_m and β_m from Sect. 7.3 into (7.5.5), and taking (7.5.7) into consideration, we get

$$u(s, \nu) = u_p(s, \nu) \equiv \text{const} \sqrt[6]{\frac{c_0(s)}{P(s)}} \exp\left\{ i\omega \int_0^s \frac{d\tau}{c_0(\tau)} \right.$$

$$+ \frac{i\xi_p \omega^{1/3}}{2^{1/3}} \int_0^s \frac{d\tau}{c_0^{1/3}(\tau) P^{2/3}(\tau)} + \frac{i}{\omega^{1/3}} \frac{\nu^2}{6 c_0(s)} \left(\frac{2 c_0'(s)}{c_0(s)} + \frac{P'(s)}{P(s)} \right)$$

$$- \frac{i\xi_p^2}{\omega^{1/3} 2^{5/3}} \int_0^s \frac{c_0^{1/3}(\tau)}{P^{4/3}(\tau)} \left[1 + \frac{4}{3} P(\tau) \left(\frac{2}{5} P(\tau) Q(\tau) - \frac{2}{\varrho(\tau)} \right) \right] d\tau$$

$$\left. + O(\omega^{-2/3}) \right\} w_1 \left(\xi_p - \frac{2^{1/3} \nu}{c_0^{2/3}(s) P^{1/3}(s)} + O(\omega^{-2/3}) \right) ,$$

$$p = 0, 1, 2, 3, \ldots \quad . \tag{7.5.8}$$

The functions $c_0(s)$, $P(s)$ and $Q(s)$ in (7.5.8) are defined in formulas (7.3.2, 2.1, and 3.37). The first term in the exponent describes a wave propagating along the boundary of the region with velocity $c_0(s)$. The second term, by virtue of the fact that $\arg \xi_p = \pi/3$, $p = 0, 1, 2, \ldots$, has a negative real part whose absolute value increases with increasing s. Because of this, the intensity of the wave decreases exponentially as it propagates along the boundary. Thus, solutions (7.5.8) represent waves creeping along the boundary which continuously shed energy into the surrounding space. In the what follows we will call solutions of the type (7.5.8) *creeping waves* (see also Sect. 11.2).

We stated above that the solutions (7.5.5) satisfy the principle of limiting absorption "in a certain sense". We can now state more precisely what was meant. Assume that the medium is absorbing, which can be accomplished by letting $\text{Im}(\omega) > 0$. Then $\arg(-\omega^{2/3} n) = \arg(-\nu)$ will be between $-\pi/3$ and $-\pi$ $(n > 0)$. As implied by the asymptotic expansion, $w_1(t) \to 0$ where $-\pi < \arg t < -\pi/3$ and $|t| \to \infty$ (see Appendix A.1, and in particular Fig. A.1.3). If we begin to move away from S along the normal $(|\nu| \to \infty)$, then if $-\pi < \arg(-\nu) < -\pi/3$,

$$w_1 \left(\xi_p - \frac{2^{1/3} \nu}{c_0^{2/3}(s) P^{1/3}(s)} \right) \to 0 ,$$

i.e. (7.5.8) decays rapidly with increasing distance from S when $\text{Im}\,\omega > 0$. This property of the formal solution (7.5.5) is what we mean by the principle of limiting absorption[3].

---

[3] At the beginning of this section we noted that there are no eigenfunctions on the exterior of Ω that decay rapidly enough with increasing distance from S to be of finite energy. There is no contradiction here, since the condition of L-periodicity with respect to s (L being the perimeter of S), which is necessary for (7.5.5) to be an eigenfunction, has yet to be imposed.

Formulas (7.5.8) define on the exterior of Ω the asymptotic expansions of solutions of (7.1.1) that satisfy the homogeneous boundary condition $u|_S = 0$ in the case where the boundary S extends to infinity. On the other hand, if S has a finite perimeter L, then just as in the case of the interior problem the additional condition of periodicity (7.4.6) must be imposed. Substitution of (7.5.8) into (7.4.6) leads once more to condition (7.4.7) in which, however, the $\alpha_m(s, \nu)$ are now different functions, those appearing in (7.5.5) and defined by (7.5.8).

In the first approximation for the natural frequencies $\omega_{p,q}$, (7.4.8) is true as before, since the expressions for α_{-3} in the interior and exterior problems are the same. In higher approximations [if we seek $\omega_{p,q}$ in the form (7.4.9)] a difference shows up, amounting to the necessity of substituting for a_2 and a_4 the coefficients b_2 and b_4 obtained from a_2 and a_4 by substituting ξ_p for $-t_p$. Obviously, this difference also applies to (7.4.11) for the eigenfunctions $u_{p,q}$ as well. Finally we have

$$
\omega_{p,q} = \omega_q \left[1 + \frac{b_2}{\omega_q^{2/3}} + \frac{b_4}{\omega_q^{4/3}} + O\left(\frac{1}{\omega_q^2}\right) \right] \quad , \tag{7.5.9}
$$

$$
\begin{aligned}
u_{p,q}(s, \nu) = B_{p,q} \sqrt[6]{\frac{c_0(s)}{P(s)}} \exp\Bigg\{ & i\omega_q \int_0^s \frac{d\tau}{c_0(\tau)} \\
& + i\omega_q^{1/3} \left[b_2 \int_0^s \frac{d\tau}{c_0(\tau)} + \frac{\xi_p}{2^{1/3}} \int_0^s \frac{d\tau}{c_0^{1/3}(\tau) P^{2/3}(\tau)} \right] \\
& + \frac{i}{\omega_q^{1/3}} \left[\frac{\nu^2}{6 c_0(s)} \left(\frac{2 c_0'(s)}{c_0(s)} + \frac{P'(s)}{P(s)} \right) \right. \\
& + b_4 \int_0^s \frac{d\tau}{c_0(\tau)} + \frac{b_2 \xi_p}{3 \cdot 2^{1/3}} \int_0^s \frac{d\tau}{c_0^{1/3}(\tau) P^{2/3}(\tau)} \\
& \left. - \frac{\xi_p^2}{2^{5/3}} \int_0^s \frac{c_0^{1/3}(\tau)}{P^{4/3}(\tau)} \left[1 + \frac{4}{3} P(\tau) \left(\frac{2}{5} P(\tau) Q(\tau) - \frac{2}{\varrho(\tau)} \right) \right] d\tau \right] \\
& + O\left(\omega_q^{-2/3}\right) \Bigg\} w_1 \left(\xi_p - \frac{2^{1/3} \nu}{c_0^{2/3}(s) P^{1/3}(s)} + O\left(\omega_q^{-2/3}\right) \right) \\
& (p = 0, 1, 2, \ldots; \quad q \gg 1) \quad .
\end{aligned} \tag{7.5.10}
$$

The coefficients B_{pq} do not depend on the coordinates s and ν, and the numbers b_2, b_4 are defined by

$$
b_2 = -\frac{\xi_p}{2^{1/3}} \int_0^L \frac{d\tau}{c_0^{1/3}(\tau) P^{2/3}(\tau)} \left(\int_0^L \frac{d\tau}{c_0(\tau)} \right)^{-1} ,
$$

171

$$b_4 = \frac{b_2^2}{3} + \frac{\xi_p^2}{2^{5/3}} \int_0^L \frac{c_0^{1/3}(\tau)}{P^{4/3}(\tau)} \left[1 + \frac{4}{3} P(\tau) \left(\frac{2}{5} P(\tau) Q(\tau) - \frac{2}{\varrho(\tau)} \right) \right] d\tau$$

$$\times \left(\int_0^L \frac{d\tau}{c_0(\tau)} \right)^{-1} . \tag{7.5.11}$$

From formula (7.5.9) and the expression for b_2 in (7.5.11) we see that the eigenvalues $\omega_{p,q}$ in constrast to the case of the interior problem, are shifted into the lower half of the complex ω-plane. The way that formulas (7.5.9 and 10) are related to (7.5.4) will be established in Chap. 11 after construction of the asymptotic expansion of the Green's function.

Solutions of (7.1.1) for the exterior $C\Omega$ of Ω under other boundary conditions can be found in similar fashion. We will write out the final result for the case of boundary condition (7.1.4). For simplicity, we assume that the propagation velocity c is constant.

If the boundary S of Ω extends to infinity, we have the following set of asymptotic solutions to (7.1.1, 4):

$$u_p(s, \nu) = \frac{C_p}{\varrho^{1/6}(s)} \exp \left[iks + i\xi_p \left(\frac{k}{2} \right)^{1/3} \int_0^s \frac{d\tau}{\varrho^{2/3}(\tau)} + \int_0^s \frac{d\tau}{\varrho(\tau)g(\tau)} \right.$$

$$+ \frac{i\varrho'(s)\nu^2}{6k^{1/3}\varrho(s)} + \frac{i\xi_p^2 \lambda(s)}{(k/2)^{1/3}} + O\left(\frac{1}{k^{2/3}} \right) \right] w_1 \left[\xi_p - \nu \left(\frac{2}{\varrho(s)} \right)^{1/3} \right.$$

$$\left. - \frac{i}{(k/2)^{1/3} \varrho^{1/3}(s)g(s)} + O\left(\frac{1}{k^{2/3}} \right) \right] , \quad p = 0, 1, 2, \ldots .$$

$$\tag{7.5.12}$$

Here C_p does not depend on the coordinates, $k = \omega c^{-1}$, and

$$\lambda(s) = \int_0^s \varrho^{-4/3}(\tau) \left[\frac{1}{60} + \frac{4}{135} \varrho'^2(\tau) - \frac{2}{45} \varrho(\tau)\varrho''(\tau) \right] d\tau . \tag{7.5.13}$$

If the boundary S has a finite perimeter L, the additional condition of periodicity leads to eigenvalues and eigenfunctions like those of (7.5.9, 10).

Finally, we note that, in addition to all the asymptotic solutions given above, we also have the asymptotic solutions

$$u_p^-(s, \nu) , \quad p = 0, 1, 2, \ldots,$$

corresponding to selection of the lower sign in formula (7.3.16) when the eikonal equation is solved.

7.6 Justification of the Asymptotic Formulas[4]

Up to this point we have constructed expressions which satisfy the equations and boundary conditions approximately as $\omega \to \infty$. Corresponding to these "quasi-eigenfunctions" are "quasi-eigenvalues" (Sect. 7.4). The question naturally arises as to whether these quasi-eigenfunctions and quasi-eigenvalues are at all close to any true eigenfunctions and eigenvalues of Ω.

Using certain simple theorems from the theory of self-adjoint operators in Hilbert space, we can prove that the *quasi-eigenvalues* that we have constructed are close to elements of a certain subset of the *true* eigenvalues.

An analogous assertion cannot be made with respect to the *eigenfunctions*. Neverthelss, the theory of self-adjoint operators does enable us to prove a certain theorem on a relationship between quasi-eigenfunctions and true eigenfunctions. These considerations are based on the following theorem.

Theorem 7.1. *Let A be a self-adjoint operator with discrete spectrum on a Hilbert space. Furthermore, let $\lambda_1, \lambda_2, \ldots, \lambda_n, \ldots$ be a sequence of nonnegative real numbers approaching infinity, and $u_1, u_2, \ldots, u_n, \ldots$ be a sequence of elements of the Hilbert space with a norm of unity such that*
  1) u_n belongs to the domain of definition of A, and

$$2) \; \|(A - \lambda_n)u_n\| \leq C\lambda_n^{-\zeta} \; , \tag{7.6.1}$$

where C and ζ are constants which do not depend on n. Then
  1) there exists a subsequence of eigenvalues of A:

$$\lambda_{\kappa_1}^* \; , \; \lambda_{\kappa_2}^* \; , \; \lambda_{\kappa_3}^*, \ldots$$

such that

$$|\lambda_n - \lambda_{\kappa_n}^*| < C\lambda_n^{-\zeta} \; , \tag{7.6.2}$$

$$2) \; \|u_n - P_{\Delta_n}u_n\| \leq \frac{C}{C_1}\lambda_n^{-(\zeta-\zeta_1)} \; , \tag{7.6.3}$$

where P_{Δ_n} is a projection operator onto the part of the spectrum of A which lies in the interval

$$\Delta_n = \left(\lambda_n - C_1\lambda_n^{-\zeta_1}, \; \lambda_n + C_1\lambda_n^{-\zeta_1}\right) \; , \tag{7.6.4}$$

and C_1 and ζ_1 are arbitrary positive constants.

It is shorter to prove the theorem than to state it. In fact, let d_λ be the distance from a point λ to the spectrum of operator A, let R_λ be the resolvent of A $(R_\lambda = (\lambda - A)^{-1})$, and let φ_κ and λ_κ^* be the eigenelements and eigenvalues of A.

---

[4] Understanding of this section requires that the reader be acquainted with the principles of the theory of self-adjoint operators in Hilbert space. This section need not be read for understanding of subsequent chapters of the book.

The first assertion of the theorem follows immediately from the inequalities

$$\frac{1}{d_{\lambda_n}} = \|R_{\lambda_n}\| \geq \frac{\|R_{\lambda_n}(\lambda_n - A)u_n\|}{\|(\lambda_n - A)u_n\|} = \frac{\|u_n\|}{\|(\lambda_n - A)u_n\|} \geq \frac{1}{C\lambda_n^{-\zeta}} \quad . \tag{7.6.5}$$

The second assertion is a consequence of the chain of inequalities

$$C^2\lambda_n^{-2\zeta} \geq \|(A - \lambda_n)u_n\|^2 = \sum_\kappa |\lambda_\kappa^* - \lambda_n|^2 |(u_n, \varphi_\kappa)|^2$$

$$\geq \sum_{\lambda_\kappa^* \notin \Delta_n} |\lambda_\kappa^* - \lambda_n|^2 |(u_n, \varphi_\kappa)|^2 \geq C_1^2\lambda_n^{-2\zeta_1} \|P_{\Delta_n}u_n - u_n\|^2 \quad . \tag{7.6.6}$$

Problems (7.1.1, 2), or (7.1.1, 3) can be taken as those of finding the eigenelements of self-adjoint operators in some Hilbert space.

Let us introduce the Hilbert space L_{2c} of functions $\{u\}$ defined on Ω, with the scalar product being defined by

$$(u_1, u_2)_{L_{2c}} = \int_\Omega u_1(x, y)\overline{u_2(x, y)}\frac{1}{c^2}d\Omega \quad .$$

It is easily seen by integration by parts that for any two twice-continuously differentiable functions u_1 and u_2 satisfying (7.1.2 or 3), the equality

$$\int_\Omega (-c^2\Delta u_1)\overline{u}_2\frac{1}{c^2}d\Omega = \int_\Omega u_1\overline{(-c^2\Delta u_2)}\frac{1}{c^2}d\Omega$$

holds, or in other words

$$(-c^2\Delta u_1, u_2)_{L_{2c}} = (u_1, \ -c^2\Delta u_2)_{L_{2c}} \quad ,$$

i.e. the operator $-c^2\Delta$ is symmetric over such functions.

If the equality $-c^2\Delta u = f(f \in L_{2c})$ and the satisfaction of the boundary condition (7.1.2) [or (7.1.3)] are understood in a generalized sense[5], then the operator $-c^2\Delta$ will be a self-adjoint operator in L_{2c} and the theorem proved above will be applicable to this operator. Thus the generalized operator $-c^2\Delta$ will be self-adjoint in L_{2c} and will have a discrete spectrum. For a sufficiently smooth contour S and function $c(x, y) > 0$, the eigenfunctions of the operator $-c^2\Delta$ with boundary condition (7.1.2 or 3) will be sufficiently smooth functions in the closure of Ω. These statements can be easily proved by the methods of the monograph by *Ladyzhenskaya* [7.4]. To use Theorem 1, we must construct a sequence of functions that would play the part of sequence $u_n(x, y)$. This will be

---

[5] We say that the equality $-c^2\Delta u = f(f \in L_{2c})$ holds for a function $u \in L_{2c}$ in a generalized sense, and boundary condition (7.1.2 or 3) is satisfied in a generalized sense, if for any function v which is twice-continuously differentiable in the closed region Ω and which satisfies the boundary condition (7.1.2 or 3), the equality

$$\int_\Omega u(-c^2\Delta v)\frac{1}{c^2}d\Omega = \int_\Omega fv\frac{1}{c^2}d\Omega$$

holds.

our next task. Let us turn first to the case of boundary condition (7.1.2): $u|_S = 0$. We set

$$u_{p,q}^{(M)} = \exp\left(i \sum_{m=-3}^{M-1} \alpha_m(s,v)\omega^{-m/3} \right)$$

$$\times v \left(\sum_{m=0}^{M-1} \beta_m(s,v)\omega^{-m/3} \right), \tag{7.6.7}$$

where the β_m satisfy (7.4.2), and ω is determined from the equation

$$\sum_{m=-3}^{M-1} \alpha_{m0}(L)\omega^{-m/3} = 2\pi q \quad , \tag{7.6.8}$$

where q is a positive integer. The function $u_{p,q}^{(M)}$ exactly satisfies the periodicity condition and boundary condition (7.1.2).

To be able to construct the functions α_m and β_m, $m \leq M-1$ that appear in (7.6.7), and to ensure that (7.6.7) satisfies (7.1.1) with the accuracy indicated at the end of Sect. 7.3, it is sufficient that velocity $c(x,y)$ have $M+1$ derivatives, and that the radius of curvature have M derivatives. We note that the function $u_{p,q}^{(M)}$ is defined in the boundary strip $0 \geq n \geq \text{const} < 0$ and that in order to use Theorem 1 we must construct functions u_n defined throughout the entire region Ω.

To construct the desired sequence of functions, we must multiply $u_{p,q}^{(M)}$ by an appropriately selected neutralizer function $\eta(n)$ equal to one in the neighborhood of the point $n = 0$, and equal to zero outside the boundary strip. Let $\varepsilon_1 > \varepsilon_2 > 0$ be two fixed positive numbers, where $\varepsilon_1 < 2/3$.

By means of the numbers ε_1 and ε_2, we define two boundary strips

$$-C_1\omega^{-2/3+\varepsilon_1} \equiv n_1 \leq n \leq 0 \ , \quad C_1 = \text{const} > 0 \quad , \tag{7.6.9}$$

and

$$-C_2\omega^{-2/3+\varepsilon_2} \equiv n_2 \leq n \leq 0 \ , \quad C_2 = \text{const} > 0 \quad . \tag{7.6.10}$$

Since $\varepsilon_1 > \varepsilon_2$, the strip (7.6.10) will be narrower for sufficiently large ω and will be completely contained within the strip (7.6.9). Let us define the function $\eta(n)$ in the interval $n_1 \leq n \leq 0$ so that $\eta(n) \equiv 1$ when $n_2 \leq n \leq 0$, $\eta(n) \equiv 0$ in the neighborhood of $n = n_1$, and we will assume that $\eta(n)$ has derivatives of all orders. We note that in the strip

$$n_1 \leq n \leq n_2 \tag{7.6.11}$$

the function $u_{p,q}^{(M)}(s,n)$, by virtue of (7.3.45), decays exponentially in ω together with all its derivatives.

Consider now the function $V_{p,q}(x,y)$ defined in Ω by

$$V_{p,q}(x,y) = \begin{cases} u_{p,q}^{(M)}\eta(n) & (x,y) \in \{n_1 \leq n \leq 0\} \ , \\ 0 \ , & (x,y) \notin \{n_1 \leq n \leq 0\} \ . \end{cases}$$

The functions $V_{p,q}(x, y)$ satisfy the boundary condition (7.1.2) exactly. In the strip (7.6.11), by virtue of the properties of the factor $u_{p,q}^{(M)}$, the function $V_{p,q}(M)$ and all its derivatives decay exponentially in ω, or more precisely we have the estimates

$$V_{p,q} = O[\exp(-\Delta\omega^{3\varepsilon_2/2})] \ , \quad \Delta > 0 \ ,$$

and

$$\frac{\partial^{l_1+l_2} V_{p,q}}{\partial s^{l_1} \partial n^{l_2}} = O\left[\omega^{l_1+2l_2/3} \exp(-\Delta\omega^{3\varepsilon_2/2})\right] \ , \quad \Delta > 0 \ ,$$

see (7.3.43–46).

Normalizing $V_{p,q}(x, y)$ we get the function

$$\tilde{V}_{p,q}(x, y) = \frac{V_{p,q}(x, y)}{\sqrt{\iint\limits_{\Omega} |V_{p,q}|^2 \frac{1}{c^2} d\Omega}} \ . \tag{7.6.12}$$

On the basis of our estimate (7.3.43), we can state that the function $\tilde{V}_{p,q}(x, y)$ will satisfy the equation

$$(-c^2\Delta - \omega_{p,q}^2)\tilde{V}_{p,q}(x, y) = \frac{O(\omega_{p,q}^{-(M-4)/3})}{\sqrt{\iint\limits_{\Omega} |V_{p,q}|^2 \frac{1}{c^2} d\Omega}} \ .$$

The denominator in the latter formula can be easily estimated from below. In fact, in a boundary strip with thickness of order $O(\omega^{-2/3})$, $V_{p,q} = O(1)$, and outside of this strip it decays exponentially, and therefore

$$\iint\limits_{\Omega} |V_{p,q}(x, y)|^2 \frac{1}{c^2} d\Omega = O(\omega^{-2/3})$$

and

$$(-c^2\Delta - \omega_{p,q}^2)\tilde{V}_{p,q}(x, y) = O\left(\omega_{p,q}^{-(M-5)/3}\right) \ . \tag{7.6.13}$$

The functions $\tilde{V}_{p,q}(x, y)$ (p fixed, $q \to \infty$) can serve as the sequence u_n in Theorem 1.

In the case of boundary condition (7.1.3): $(\partial u/\partial n)|_S = 0$ the functions $\tilde{V}_{p,q}$ are constructed somewhat differently. If we take the function $u_{p,q}^{(M)}$, defined by (7.6.7), with the condition that polynomials α_m and β_m satisfy conditions (7.6.8 and 4.14), then such a function, although periodic in s, will only satisfy the boundary condition approximately with accuracy $O\left(\omega^{(-2/3)[(M/2)-1]}\right)$, since

$$\left.\frac{\partial u_{p,q}^{(M)}}{\partial n}\right|_{n=0} = O\left(\omega^{(-2/3)[M/2-1]}\right) \ . \tag{7.6.14}$$

Estimate (7.6.14) follows from formulas (7.4.12–14) and the relation $\nu = n\omega^{2/3}$.

Therefore in the case of boundary condition (7.1.3) we have to define the function $V_{p,q}$ by

$$V_{p,q}(x,y) = u_{p,q}^{(M)}\eta(n)\exp\left(-\left.\frac{\partial u_{p,q}^{(M)}/\partial n}{u_{p,q}^{(M)}}\right|_{n=0}n\right) \tag{7.6.15}$$

in the strip $n_1 \leq n \leq 0$.[6] Outside this strip let $V_{p,q} = 0$. The function (7.6.15) will satisfy condition (7.1.3) exactly. The normalized function

$$\tilde{V}_{p,q}(x,y) = \frac{V_{p,q}(x,y)}{\sqrt{\iint_{\Omega}|V_{p,q}|^2 d\Omega}} \tag{7.6.16}$$

will satisfy boundary condition (7.1.3) and equation (7.6.13). Obviously, the functions (7.6.16) for fixed p and $q \to \infty$ can serve as the sequence u_n in Theorem 1 for the case of boundary condition (7.1.3).

Thus the necessary sequence of functions has been constructed for both boundary conditions. Consequently, by virtue of Theorem 1 there exist subsequences of eigenvalues $(\omega_{p,q}^*)^2$ of the operator $-c^2\Delta$ corresponding to boundary condition (7.1.2 or 3) such that

$$|\omega_{p,q}^2 - (\omega_{p,q}^*)^2| \leq C\omega_{p,q}^{-(M-5)/3}, \quad p \text{ fixed}, \quad q = 1, 2, 3, \dots$$

where $\omega_{p,q}$ is a solution of equation (7.6.8), whence for $M > 2$

$$|\omega_{p,q} - \omega_{p,q}^*| \leq C\omega_{p,q}^{-(M-2)/3}. \tag{7.6.17}$$

In the first approximation (Sect. 7.4), $\omega_{p,q} \sim 2\pi q \left[\int_0^L (ds/c)\right]^{-1}$, and therefore the right side of (7.6.17) can be replaced by $O(q^{-(M-2)/3})$.

Replacing $\omega_{p,q}$ in (7.6.17) by its asymptotic expansion (7.4.9) in the case of boundary condition (7.1.2) or by the analogous expansion in the case of boundary condition (7.1.3), we finally get the following result.

Theorem 2. *If the velocity $c(x,y)$ and radius of curvature of contour S have smoothness sufficient for the constructions of Sects. 7.3, 4, then for each $p = 0, 1, 2, \dots$, there exists a subsequence of natural frequencies $\omega_{p,q}^*$, $q = 1, 2, \dots$, such that*

$$\omega_q\left(1 + \sum_{j=1}^{M} a_{jp}\omega_q^{-j/3}\right) = \omega_{p,q}^* + O(q^{-(M-2)/3}). \tag{7.6.18}$$

The coefficients a_{jp} in (7.6.18) in the case of boundary condition (7.1.2) are the coefficients a_j in formula (7.4.9), and are the analogous coefficients a_j' in the case of boundary condition (7.1.3).

---

[6] The function $u_{p,q}^{(M)}|_{n=0} \neq 0$ and moreover $(u_{p,q}^{(M)})^{-1}|_{n=0} = O(1)$. This is implied by the fact that the Airy function does not have multiple roots.

As we have already pointed out, in order for the calculations of Sects. 7.3 and 7.4 to be possible, it is sufficient that the velocity c have $M + 1$ derivatives, and that the radius of curvature have M derivatives.

7.7 Notes on the Literature

Lord Rayleigh [7.5], in considering the Helmholtz equation in a circle (in connection with the whispering gallery effect), called attention to the fact that at high frequencies there are solutions of this equation that oscillate in a boundary strip and are rapidly damped outside the bounds of this strip. Further investigation of such solutions was done in papers by *Krasnushkin* [7.6], and by *Krasnushkin* and *Mustel'* [7.7]. In these papers, the authors do not confine themselves to the acoustic case as did *Rayleigh*, but examine electromagnetic oscillations as well. They called the corresponding solutions normal waves that cling to a waveguide wall.

The construction of the surface-layer type of solutions to the wave equation as infinite series in inverse powers of frequency is related to constructions of boundary-layer theory [7.8]; however, the reversal of sign in the term involving a small parameter multiplying the highest derivative leads to the Airy function and entirely different calculations.

The form of the solution (7.3.4), an exponential function multiplied by an Airy function, with arguments that are infinite series in powers of $\omega^{-1/3}$, was taken from the paper by *Buldyrev* [7.9], as were the basic calculations of the first four sections of the chapter. In the asymptotic theory of ordinary differential equations, the prototype of the series (7.3.4) was suggested by *Cherry* [7.1]. Besides *Cherry*'s asymptotic expansion, there are also asymptotic expansions in *Olver*'s form [7.10] (a sum of two asymptotic series, one of which is multiplied by an Airy function, the other by its derivative). The *Olver* form was used by *Lewis* et al. [7.11] to get some interesting asymptotic expansions from which some of the formulas of Sect. 7.5 can be derived as special cases. In many respects the constructions of this paper are analogous to those of Chap. 3. Other applications of *Olver*'s method can be found in papers by *Mukhina* and *Molotkov* [7.12], and *Kirpičnikova* [7.13], which are devoted to the theory of elastic surface waves.

The idea of seeking the terms of the asymptotic series in the form of polynomials in $\nu = n\omega^{2/3}$ with coefficients dependent on s was proposed by *Lazutkin* [7.14]. The polynomials α_{-3}, α_{-2}, α_{-1}, α_0 and β_0 were found earlier by *Ivanov* [7.15], *Buslaev* [7.16] and *Buldyrev* [7.17] using the parabolic equation method. The possibility of approximately satisfying the impedance boundary condition by introducing the term $\beta_1(s)\omega^{-1/3}$ was first pointed out by *Molotkov* [7.18].

The imprecise nature of the definitions of Sect. 7.5 is due to the fact that there is as yet no rigorous theory of eigenfunctions of the type considered here. It can be easily proved that when $c \equiv 1$, the Green's function $G(M_0, M, k)$ that satisfies (7.5.3) is analytically continuable in k not only into the upper half-plane Im $k \geq 0$, but also into an infinitely sheeted Riemann surface $-\infty <$ arg $k < \infty$ of logarithmic type, where it can have no singularities other than poles, whose position is independent of M and M_0 [7.3]. In the three-dimensional case $G(M_0, M, k)$ is a meromorphic function simply on the k-plane rather than on a logarithmic Riemann surface.

For a convex region Ω it has been shown for the planar case that the region $0 \geq \text{Im } k \geq -c(\text{Re } k)^{1/3}$, where $c > 0$, does not contain eigenvalues [7.19]. In the three-dimensional case, for a class of regions somewhat more general than convex, *Lax* et al. [7.20] obtained results from which it follows that there is a strip $0 \geq \text{Im } k \geq -c_1$ (where $c_1 > 0$ is some constant) in which there are no eigenvalues. Later, *Babič* and *Grigor'eva* [7.21] showed that there are no eigenvalues even in the larger region $0 \geq \text{Im } k \geq -c_2(\text{Re } k)^{1/3}$ for the case of the Neumann boundary condition and positive Gaussian curvature.

The proof that the asymptotic expansion for the eigenvalues (Sect. 7.6) is indeed asymptotic was outlined by *Lazutkin* [7.14]. The method of proof is analogous to *Maslov*'s [7.22].

8. Eigenfunctions Concentrated in the Neighborhood of an Extremal Ray of a Region

In Sect. 5.4 the ray method in the small was used to study eigenfunctions concentrated in the vicinity of a ray meeting the boundary Σ of a region Ω at two points C and D, and orthogonal to Σ at these points.

In this chapter this same class of eigenfunctions will be considered by the etalon problem method. All the constructions are centered around two tasks: 1) construction of a formal solution of the equation

$$\Delta u + \frac{\omega^2}{c^2(x,y)} u = 0 \quad ,$$

which is concentrated in the vicinity of a finite segment of an arbitrary ray, and 2) satisfaction of the boundary conditions. It is of interest to note that the boundary conditions can be satisfied, even in the first approximation, only in the case when the extremal ray under consideration in this region is stable (Sect. 5.4), although this stability is not necessary in order to construct a formal solution of the Helmholtz equation in the vicinity of an arbitrary ray (without regard to boundary conditions). The ideas of this chapter will be further developed in Chap. 10, where the more general problem for a three-dimensional region is considered, the eigenfunctions being concentrated in the vicinity of a closed ray path reflected from the boundary of the region at N points. The mathematical apparatus of Chap. 10 can also be applied to solution of the two-dimensional problem considered in the present chapter.

8.1 The Etalon Problem

In accordance with the etalon problem method (Sect. 7.1), we must first of all consider the simplest etalon problem that yields an exact solution concentrated in the vicinity of some ray as $\omega \to \infty$. We arrive at such a problem by considering the equation

$$\Delta u + \frac{\omega^2}{c^2(x,y)} u = 0 \tag{8.1.1}$$

and setting the function $c(x,y)$ equal to

$$c(x,y) = c(y) = c_0 \left[1 - \left(\frac{y}{y_0} \right)^2 \right]^{-1/2} \quad ,$$

$$0 < c_0 = \text{const} \quad . \tag{8.1.2}$$

For this velocity distribution, the axis OX is a ray, and the solution of the wave equation (8.1.1) can be constructed by the method of separation of variables. We will seek solutions of (8.1.1) in the form

$$u(x, y) = X(x)Y(y) \quad . \tag{8.1.3}$$

Substituting (8.1.3) into (8.1.1), and after separating variables, we get the equations

$$X''(x) + \zeta^2 X(x) = 0 \quad ,$$
$$Y''(y) + \left[\frac{\omega^2}{c_0^2} \left(1 - \frac{y^2}{y_0^2} \right) - \zeta^2 \right] Y(y) = 0 \quad . \tag{8.1.4}$$

We write the solutions of the first equation in the form

$$X(x) = e^{\pm i\zeta x} \quad ,$$

and in the second we convert to the new variable

$$\nu = \frac{\omega^{1/2}}{\sqrt{c_0 y_0}} y \quad .$$

As a result we get the equation

$$\frac{d^2 Y}{d\nu^2} + (2q + 1 - \nu^2) Y = 0 \quad ,$$
$$2q + 1 = y_0 \left(\frac{\omega}{c_0} - \frac{c_0}{\omega} \zeta^2 \right) \quad ,$$

with solution equal to $D_q(\sqrt{2}\nu)$ ($D_q(z)$ is a parabolic cylinder function[1]). The parabolic cylinder functions $D_q(z)$ tend to zero as $z \to \pm\infty$ when and only when q is a nonnegative integer: $q = 0, 1, 2\ldots$ Thus if we wish to construct a solution of (8.1.1) that approaches zero as $y \to \pm\infty$ we must set

$$y_0 \left(\frac{\omega}{c_0} - \frac{c_0}{\omega} \zeta^2 \right) - \frac{1}{2} = q \quad .$$

Hence for the separation constant ζ we get the expression

$$\zeta = \frac{\omega}{c_0} \sqrt{1 - \frac{c_0}{y_0} \frac{2q+1}{\omega}} \quad .$$

This enables us to write the solution in the form

$$u_q(x, y) = \exp \left[\pm i \frac{\omega}{c_0} \sqrt{1 - \frac{c_0}{y_0} \frac{2q+1}{\omega}} x \right] D_q \left(\frac{2^{1/2} \omega^{1/2}}{\sqrt{c_0 y_0}} y \right) \quad . \tag{8.1.5}$$

---

[1] We have already encountered the parabolic cylinder functions $D_q(z)$ in Chap. 6. They form a complete set of eigenfunctions for the boundary value problem $y''_{zz} + (\lambda - z^2/4)y = 0$; $y \to 0$ as $z \to \pm\infty$ which obviously has a discrete spectrum, the eigenvalue corresponding to $D_q(z)$ being equal to $q+1/2$. We have $D_q(z) = H_q(z/\sqrt{2})e^{-z^2/4}$, $D_q(\sqrt{2}\nu) = H_q(\nu)e^{-\nu^2/2}$ where $H_q(\nu)$ is a Hermite polynomial [8.1].

If $\omega<(2q+1)c_0/y_0$, the solutions $u_q(x,y)$ will increase or decrease exponentially as x increases; on the other hand if $\omega>(2q+1)c_0/y_0$, we get oscillatory solutions. If $\omega\gg(2q+1)c_0/y_0$, the square root in the exponential function can be expanded in a series, and then

$$u_q(x,y) = \exp\left[\pm i\frac{\omega}{c_0}\left(x - \frac{c_0}{y_0}\frac{2q+1}{2\omega}x - \frac{c_0^2}{y_0^2}\frac{(2q+1)^2}{8\omega^2}x + \dots\right)\right]$$
$$\times D_q\left(\frac{2^{1/2}\omega^{1/2}}{\sqrt{c_0 y_0}}y\right) . \tag{8.1.6}$$

By virtue of the fact that $D_q(\sqrt{2}\nu)$ oscillates when $|\nu|\leq\sqrt{2q+1}$ and is exponentially damped when $|\nu|>\sqrt{2q+1}$, the solutions (8.1.6) are appreciably different from zero only in the narrow strip

$$|y|\leq\sqrt{c_0 y_0}\frac{\sqrt{2q+1}}{\omega^{1/2}} ,$$

i.e., they are concentrated in the vicinity of the ray $y=0$ that coincides with the OX axis.

The conclusion that solutions of the Helmholtz equation concentrated in the neighborhood of an arbitrary ray are expressed in terms of parabolic cylinder functions could have been drawn from the results of Sects. 6.4, 5 which dealt with the parabolic equation method, however, our study of the etalon problem reveals the analytical nature of not only the first approximation but all subsequent ones as well.

8.2 Construction of the Principal Terms of the Formal Series

For solutions of the Helmholtz equation, concentrated in general in the vicinity of a segment of an arbitrary ray, we now construct the formal series and find their first few terms.

First of all, on the basis of the analogy between the general problem and the etalon problem, let us find the form of solutions of the Helmholtz equation which are concentrated in the vicinity of a segment of an arbitrary ray in the general case. Let S be a segment of an arbitrary ray in a nonhomogeneous medium. On the basis of geometrical optics constructions and a study of the etalon problem it is natural to expect that there exists a formal solution of (8.1.1) that is concentrated in a strip surrounding S having a width of order $O(\omega^{-1/2})$. Let us introduce the coordinate system (s,n) (see Sect. 5.3) in the vicinity of S. By virtue of the fact that the metric coefficients for this coordinate system are given by (5.3.3), Eq. (8.1.1) for points sufficiently close to S can be written in the form

$$\frac{\partial}{\partial n}\left[\left(1+\frac{n}{\varrho(s)}\right)\frac{\partial u}{\partial n}\right]+\frac{\partial}{\partial s}\left[\left(1+\frac{n}{\varrho(s)}\right)^{-1}\frac{\partial u}{\partial s}\right]$$

$$+\frac{\omega^2}{c^2(s,n)}\left(1+\frac{n}{\varrho(s)}\right)u=0 \quad . \tag{8.2.1}$$

We will assume that the velocity $c(s,n)$ has derivatives of all orders, and therefore for any $M = 1, 2, 3, \ldots$

$$c(s,n) = c_0(s) + c_1(s)n + c_2(s)n^2 + \ldots + c_M(s)n^M + O(n^{M+1}) \quad ,$$

$$c_0(s) = c(s,0) \quad , \quad c_l = \frac{1}{l!}\frac{d^l}{dn^l}c(s,n)\bigg|_{n=0} \quad , \quad l = 1, 2, \ldots \quad . \tag{8.2.2}$$

We introduce the new variable

$$\nu = \omega^{1/2}n \quad . \tag{8.2.3}$$

By virtue of the fact that in the strip of interest around S where the desired solutions of (8.2.1) are concentrated, $n = O(\omega^{-1/2})$ and for large ω the variable ν is a finite quantity: $\nu = O(1)$. In (s, ν) coordinates (8.2.1) takes the form

$$\omega A\frac{\partial^2 u}{\partial \nu^2} + \omega^{1/2}B\frac{\partial u}{\partial \nu} + C\frac{\partial^2 u}{\partial s^2} + \omega^{-1/2}D\frac{\partial u}{\partial s} + \omega^2 Eu = 0 \quad , \tag{8.2.4}$$

where

$$A = 1 + \omega^{-1/2}\frac{\nu}{\varrho(s)} \quad , \quad B = \frac{1}{\varrho(s)} \quad , \quad C = \left(1+\omega^{-1/2}\frac{\nu}{\varrho(s)}\right)^{-1} \quad ,$$

$$D = \frac{\nu}{\varrho^2(s)}\varrho'(s)\left(1+\omega^{-1/2}\frac{\nu}{\varrho(s)}\right)^{-2} \quad , \quad E = \left(1+\omega^{-1/2}\frac{\nu}{\varrho(s)}\right)c^{-2}(s,n) \quad . \tag{8.2.5}$$

Expanding C, D and E in powers of $\omega^{-1/2}$, we get

$$C = \sum_{m=0}^{M-1}(-1)^m\left(\frac{\nu}{\varrho(s)}\right)^m \omega^{-m/2} + O\left(\nu^M\omega^{-M/2}\right) \quad , \tag{8.2.6}$$

$$D = \frac{\nu}{\varrho^2(s)}\varrho'(s)$$
$$\times \left\{\sum_{m=0}^{M-1}(-1)^m(m+1)\left(\frac{\nu}{\varrho(s)}\right)^m \omega^{-m/2} + O\left(\nu^M\omega^{-M/2}\right)\right\} \quad , \tag{8.2.7}$$

$$E = \frac{1}{c_0^2(s)} + \sum_{m=1}^{M-1}e_m(s)\left(\frac{\nu}{\varrho(s)}\right)^m \omega^{-m/2} + O\left(\nu^M\omega^{-M/2}\right) \quad , \tag{8.2.8}$$

where

$$e_1 = \frac{\varrho(s)}{c_0^2(s)}\left(\frac{1}{\varrho(s)} - 2\frac{c_1(s)}{c_0(s)}\right) \quad , \tag{8.2.9}$$

183

$$e_2 = -\frac{\varrho^2(s)}{c_0^2(s)}\left(2\frac{c_1(s)}{c_0(s)}\frac{1}{\varrho(s)} + 2\frac{c_2(s)}{c_0(s)} - 3\frac{c_1^2(s)}{c_0^2(s)}\right) \quad , \tag{8.2.10}$$

$$e_3 = -\frac{\varrho^3(s)}{c_0^2(s)}\left[\left(2\frac{c_2(s)}{c_0(s)} - 3\frac{c_1^2(s)}{c_0^2(s)}\right)\frac{1}{\varrho(s)}\right.$$
$$\left. + 2\frac{c_3(s)}{c_0(s)} - 6\frac{c_1(s)c_2(s)}{c_0^2(s)} + 4\frac{c_1^3(s)}{c_0^3(s)}\right] \quad . \tag{8.2.11}$$

The coefficients of $\omega^{-m/2}$ in expansions (8.2.6–8) are polynomials in ν, whose coefficients are functions of s. In this connection, by analogy with the solutions (8.1.6) of the etalon problem, we will look for the solution of (8.2.4) in the form

$$u(s,\nu) = \text{const} \exp\left[i\sum_{m=-2}^{M-1}\alpha_m(s,\nu)\omega^{-m/2} + O\left(\omega^{-M/2}\right)\right]$$
$$\times D_q\left[\sqrt{2}\sum_{m=0}^{M-1}\beta_m(s,\nu)\omega^{-m/2} + O\left(\omega^{-M/2}\right)\right] \quad , \tag{8.2.12}$$

where the $D_q(\sqrt{2}\Psi)$ are parabolic cylinder functions satisfying

$$\frac{d^2}{d\Psi^2}D_q(\sqrt{2}\Psi) + [(2q+1) - \Psi^2]D_q(\sqrt{2}\Psi) = 0 \quad . \tag{8.2.13}$$

As in expansions (8.2.6–8), we will assume the coefficients $\alpha_m(s,\nu)$ and $\beta_m(s,\nu)$ to be polynomials in ν:

$$\alpha_m(s,\nu) = \sum_{l=0}^{N_1}\alpha_{ml}(s)\nu^l \quad , \tag{8.2.14}$$

$$\beta_m(s,\nu) = \sum_{l=0}^{N_2}\beta_{ml}(s)\nu^l \quad . \tag{8.2.15}$$

Let us find the equations that must be satisfied by $\alpha_m(s,\nu)$ and $\beta_m(s,\nu)$ for expression (8.2.12) to formally satisfy (8.2.4). After substituting the function $u(s,\nu)$ defined by (8.2.12) into (8.2.4), it will contain the function $D_q(\sqrt{2}\Psi)$ and its derivatives $(d/d\Psi)D_q(\sqrt{2}\Psi)$ and $(d^2/d\Psi^2)D_q(\sqrt{2}\Psi)$. Using (8.2.13), we can express $(d^2/d\Psi^2)D_q(\sqrt{2}\Psi)$ in terms of $D_q(\sqrt{2}\Psi)$, and then the result of substituting $u(s,\nu)$ into (8.2.4) after cancelling the exponential function will be

$$a(s,\nu;\omega)D_q\{\sqrt{2}\Psi(s,\nu,\omega)\}$$
$$+b(s,\nu;\omega)\frac{d}{d\Psi}D_q\{\sqrt{2}\Psi(s,\nu,\omega)\} = 0 \quad , \tag{8.2.16}$$

where $a(s,\nu;\omega)$ and $b(s,\nu;\omega)$ are rather complicated expressions that depend on the coefficients of (8.2.4) and the polynomials α_m and β_m. Equation (8.2.16) will be satisfied if and only if the coefficients of the linearly independent functions D_q and $(dD_q/d\Psi)$ are equal to zero, i.e.,

$$a(s, \nu; \omega) = 0 \quad , \quad b(s, \nu; \omega) = 0 \quad . \tag{8.2.17}$$

These equalities must be satisfied identically in ω. Let us expand $a(s, \nu; \omega)$ and $b(s, \nu; \omega)$ with respect to powers of $\omega^{-1/2}$:

$$a(s, \nu; \omega) = \omega \sum_{m=-4}^{M-1} a_m(s, \nu) \omega^{-m/2} + O\left(\omega^{-(M/2)+1}\right) \quad ,$$

$$b(s, \nu; \omega) = \omega \sum_{m=-2}^{M-1} b_m(s, \nu) \omega^{-m/2} + O\left(\omega^{-(M/2)+1}\right) \quad . \tag{8.2.18}$$

The coefficients $a_m(s, \nu)$ and $b_m(s, \nu)$ here contain the partial derivatives with respect to s and ν of the desired polynomials $\alpha_n(s, \nu)$ and $\beta_n(s, \nu)$. Equations (8.2.17) will be identically satisfied in ω if and only if

$$a_m(s, \nu) = 0 \quad , \quad m = -4, \, -3, \, -2, \ldots \quad ,$$
$$b_m(s, \nu) = 0 \quad , \quad m = -2, \, -1, \, 0, \ldots \quad . \tag{8.2.19}$$

Equations (8.2.19) are a recurrent system of partial differential equations that must be satisfied by the unknown polynomials $\alpha_n(s, \nu)$ and $\beta_n(s, \nu)$. Let us find the principal terms $\alpha_{-2}(s, \nu)$, $\alpha_{-1}(s, \nu)$, $\alpha_0(s, \nu)$ and $\beta_0(s, \nu)$ of the formal series. This will permit the equations of (8.2.19) for $m \geq 1$ to be substantially simplified. After making the necessary calculations we get

$$a_{-4}(s, \nu) \equiv -\left(\frac{\partial \alpha_{-2}}{\partial \nu}\right)^2 = 0 \quad , \tag{8.2.20}$$

$$a_{-3}(s, \nu) \equiv -2\frac{\partial \alpha_{-2}}{\partial \nu}\frac{\partial \alpha_{-1}}{\partial \nu} - \left(\frac{\partial \alpha_{-2}}{\partial \nu}\right)^2 \frac{\nu}{\varrho} = 0 \quad , \tag{8.2.21}$$

$$a_{-2}(s, \nu) \equiv -\left(\frac{\partial \alpha_{-1}}{\partial \nu}\right)^2 - 2\frac{\partial \alpha_{-2}}{\partial \nu}\frac{\partial \alpha_0}{\partial \nu} - 2\frac{\partial \alpha_{-2}}{\partial \nu}\frac{\partial \alpha_{-1}}{\partial \nu}\frac{\nu}{\varrho}$$
$$+ i\frac{\partial^2 \alpha_{-2}}{\partial \nu^2} - \left(\frac{\partial \alpha_{-2}}{\partial s}\right)^2 + \frac{1}{c_0^2(s)} = 0 \quad , \tag{8.2.22}$$

$$a_{-1}(s, \nu) \equiv -2\frac{\partial \alpha_{-1}}{\partial \nu}\frac{\partial \alpha_0}{\partial \nu} - \left(\frac{\partial \alpha_{-1}}{\partial \nu}\right)^2 \frac{\nu}{\varrho} + i\frac{\partial^2 \alpha_{-1}}{\partial \nu^2}$$
$$- 2\frac{\partial \alpha_{-2}}{\partial s}\frac{\partial \alpha_{-1}}{\partial s} + \left(\frac{\partial \alpha_{-2}}{\partial s}\right)^2 \frac{\nu}{\varrho} + e_1(s)\frac{\nu}{\varrho} = 0 \quad , \tag{8.2.23}$$

$$b_{-2}(s, \nu) \equiv 2i\frac{\partial \alpha_{-2}}{\partial \nu}\frac{\partial \beta_0}{\partial \nu} = 0 \quad , \tag{8.2.24}$$

185

$$b_{-1}(s,\nu) \equiv 2i\frac{\partial \alpha_{-1}}{\partial \nu}\frac{\partial \beta_0}{\partial \nu} + 2i\frac{\partial \alpha_{-2}}{\partial \nu}\frac{\partial \beta_1}{\partial \nu}$$

$$+ 2i\frac{\partial \alpha_{-2}}{\partial \nu}\frac{\partial \beta_0}{\partial \nu}\frac{\nu}{\varrho} = 0 \quad . \tag{8.2.25}$$

Equation (8.2.20) implies that $\alpha_{-2}(s,\nu)$ does not depend on ν, and consequently is a polynomial of zero degree in ν:

$$\alpha_{-2}(s,\nu) = \alpha_{-20}(s) \quad . \tag{8.2.26}$$

By virtue of (8.2.26), Eqs. (8.2.21, 24) are identically satisfied, and (8.2.25) reduces to

$$2i\frac{\partial \alpha_{-1}}{\partial \nu}\frac{\partial \beta_0}{\partial \nu} = 0 \quad .$$

This equation will be satisfied if either $(\partial \alpha_{-1})/\partial \nu = 0$, or $\partial \beta_0/\partial \nu = 0$, or both. However, assuming $\partial \beta_0/\partial \nu = 0$ would lead to a contradiction with the special case considered in Sect. 8.1 by separation of variables, where $\partial \beta_0/\partial \nu \neq 0$. Assuming that $\partial \alpha_{-1}/\partial \nu = 0$, we arrive at

$$\alpha_{-1}(s,\nu) = \alpha_{-10}(s) \quad . \tag{8.2.27}$$

Substituting (8.2.26 and 27) into (8.2.22) we get

$$\left(\frac{d\alpha_{-20}}{ds}\right)^2 = \frac{1}{c_0^2(s)} \quad ,$$

whence

$$\alpha_{-20}(s) = \pm \int_{d_{-2}}^{s} \frac{d\tau}{c_0(\tau)} \quad . \tag{8.2.28}$$

where d_{-2} is some arbitrary constant. Subsequent calculations are carried out assuming the + sign in (8.2.28). From (8.2.9, 23) and the equalities

$$\frac{\partial \alpha_{-1}}{\partial \nu} = 0 \quad , \quad \frac{\partial \alpha_0}{\partial s} = \frac{1}{c_0(s)} \quad ,$$

$$\frac{1}{P(s)} = \frac{1}{\varrho(s)} - \frac{c_1(s)}{c_0(s)} = 0 \tag{8.2.29}$$

[the relation $1/P(s) = 0$ means that the effective radius of curvature $P(s)$ is equal to infinity, i.e. the curve S is a ray (Sects. 5.3, 4)] it follows that

$$\frac{\partial \alpha_{-1}}{\partial s} = 0 \quad ,$$

whence

$$\alpha_{-1}(s,\nu) = d_{-1} \quad , \tag{8.2.30}$$

where d_{-1} is an arbitrary constant.

By the equalities

$$\frac{\partial \alpha_{-2}}{\partial \nu} = \frac{\partial \alpha_{-1}}{\partial \nu} = \frac{\partial \alpha_{-1}}{\partial s} = 0 \quad,$$

$$\frac{\partial \alpha_{-2}}{\partial s} = \frac{1}{c_0(s)} \tag{8.2.31}$$

we have now established, the equations of (8.2.19) for $m = 0$ take the form

$$a_0(s, \nu) \equiv -\left(\frac{\partial \alpha_0}{\partial \nu}\right)^2 + i\frac{\partial^2 \alpha_0}{\partial \nu^2} - 2\frac{1}{c_0(s)}\frac{\partial \alpha_0}{\partial s} - \frac{1}{c_0^2(s)}\frac{\nu^2}{\varrho^2}$$

$$+ e_2(s)\frac{\nu^2}{\varrho^2} - \left(\frac{\partial \beta_0}{\partial \nu}\right)^2 (2q + 1 - \beta_0^2) - i\frac{c_0'(s)}{c_0^2(s)} = 0 \quad, \tag{8.2.32}$$

$$b_0(s, \nu) \equiv 2i\frac{\partial \alpha_0}{\partial \nu}\frac{\partial \beta_0}{\partial \nu} + \frac{\partial^2 \beta_0}{\partial \nu^2} + 2i\frac{1}{c_0(s)}\frac{\partial \beta_0}{\partial s} = 0 \quad. \tag{8.2.33}$$

On the basis of (8.2.10 and 29) the coefficient of ν^2 in (8.2.32) can be written as

$$-\frac{1}{c_0^2(s)}\frac{1}{\varrho^2(s)} + \frac{e_2(s)}{\varrho^2(s)} = -2\frac{c_2(s)}{c_0^3(s)} \quad.$$

Equations (8.2.32, 33) are used to determine the polynomials $\alpha_0(s, \nu)$ and $\beta_0(s, \nu)$. Equation (8.2.33) implies that $\alpha_0(s, \nu)$ is a second-degree polynomial:

$$\alpha_0(s, \nu) = \alpha_{00}(s) + \alpha_{01}(s)\nu + \alpha_{02}(s)\nu^2 \quad. \tag{8.2.34}$$

If $\alpha_0(s, \nu)$ is a polynomial of second degree, then it follows from (8.2.32) that $\beta_0(s, \nu)$ is a first-degree polynomial:

$$\beta_0(s, \nu) = \beta_{00}(s) + \beta_{01}(s)\nu \quad. \tag{8.2.35}$$

Substituting (8.2.34, 35) into (8.2.32, 33), and expanding the result in powers of ν, we get

$$\left(-4\alpha_{02}^2 - 2\frac{1}{c_0}\alpha_{02}' - 2\frac{c_2}{c_0^3} + \beta_{01}^4\right)\nu^2$$

$$+ \left(-4\alpha_{02}\alpha_{01} - 2\frac{1}{c_0}\alpha_{01}' + 2\beta_{01}^3\beta_{00}\right)\nu$$

$$+ \left(-\alpha_{01}^2 + 2i\alpha_{02} - 2\frac{1}{c_0}\alpha_{00}' - (2q + 1)\beta_{01}^2 + \beta_{01}^2\beta_{00}^2 - i\frac{c_0'}{c_0^2}\right) = 0 \tag{8.2.36}$$

and

$$\left(2\alpha_{02}\beta_{01} + \frac{1}{c_0}\beta_{01}'\right)\nu + \left(\alpha_{01}\beta_{01} + \frac{1}{c_0}\beta_{00}'\right) = 0 \quad. \tag{8.2.37}$$

These two equations will be satisfied identically in ν if and only if all terms in parentheses are equal to zero. From (8.2.37) we find

$$\alpha_{02} = -\frac{1}{2c_0}\frac{\beta'_{01}}{\beta_{01}} \quad , \quad \alpha_{01} = -\frac{1}{c_0}\frac{\beta'_{00}}{\beta_{01}} \quad . \tag{8.2.38}$$

Substituting (8.2.38) into (8.2.36) and setting the coefficients of ν^2, ν^1 and ν^0 equal to zero, we get equations for determining β_{01}, β_{00} and α_{00}:

$$\frac{1}{c_0}\left(\frac{1}{c_0}\frac{\beta'_{01}}{\beta_{01}}\right)' - \left(\frac{1}{c_0}\frac{\beta'_{01}}{\beta_{01}}\right)^2 + \beta_{01}^4 - 2\frac{c_2}{c_0^3} = 0 \quad , \tag{8.2.39}$$

$$2\frac{1}{c_0}\left(\frac{1}{c_0}\frac{\beta'_{00}}{\beta_{01}}\right)' - 2\frac{1}{c_0^2}\frac{\beta'_{01}}{\beta_{01}^2}\beta'_{00} + 2\beta_{01}^3\beta_{00} = 0 \quad , \tag{8.2.40}$$

$$2\frac{1}{c_0}\alpha'_{00} = -i\frac{c''_0}{c_0^2} - i\frac{1}{c_0}\frac{\beta'_{01}}{\beta_{01}} - (2q+1)\beta_{01}^2$$

$$-\frac{1}{c_0^2}\left(\frac{\beta'_{00}}{\beta_{01}}\right)^2 + \beta_{01}^2\beta_{00}^2 \quad . \tag{8.2.41}$$

First of all we construct the solution of (8.2.39). Introducing a new unknown function $F(s) = c_0^{-1/2}(s)\beta_{01}^{-1}(s)$ into (8.2.39), we get the equation

$$\frac{F''}{F} - \frac{1}{F^4} + \frac{1}{2}\frac{c''_0}{c_0} - \frac{3}{4}\left(\frac{c'_0}{c_0}\right)^2 + 2\frac{c_2}{c_0} = 0 \quad \text{or}$$

$$F'' + K(s)F = \frac{1}{F^3} \quad , \tag{8.2.42}$$

where

$$K(s) = \frac{1}{2}\frac{c''_0(s)}{c_0(s)} - \frac{3}{4}\left(\frac{c'_0(s)}{c_0(s)}\right)^2 + 2\frac{c_2(s)}{c_0(s)} \quad .$$

The coefficient $K(s)$ in (8.2.42) is exactly the same as the coefficient $K(s)$ in the linearized Euler equation describing rays close to S in the first approximation (Sect. 5.4):

$$f'' + K(s)f = 0 \quad . \tag{8.2.43}$$

Any positive solution of (8.2.42) can be represented in the form

$$F(s) = \sqrt{\sum_{r,t=1}^{2} a_{rt}f_r(s)f_t(s)} \quad , \tag{8.2.44}$$

where (f_1, f_2) is a fundamental system of solutions to (8.2.43), and $\|a_{rt}\|$ is a symmetric matrix that satisfies the normalization condition

$$\det\|a_{rt}\|[W(f_1, f_2)]^2 = 1 \quad ; \quad W(f_1, f_2) = f_1 f'_2 - f'_1 f_2 \tag{8.2.45}$$

(Sect. 6.5, and Appendix A.3).

We will assume that the matrix $\|a_{rt}\|$ is positive definite, and we will understand the square root in (8.2.44) to be taken always positive.

In (8.2.40) we set

$$\beta_{00}(s) = \beta_{01}(s)c_0^{1/2}(s)f(s) = \frac{f(s)}{F(s)}$$

and for $f(s)$ we get

$$f'' + \left(-\frac{F''}{F} + \frac{1}{F^4}\right)f = 0 \quad,$$

which, since $F(s)$ satisfies (8.2.42), may be rewritten in the form (8.2.43), and is thus the linearized Euler equation mentioned above.

If (8.2.42, 43) are solved, and the functions $F(s)$ and $f(s)$ (in terms of which $\beta_0(s, \nu)$ is expressed) have been found, then the coefficients $\alpha_{02}(s)$ and $\alpha_{01}(s)$ of $\alpha_0(s, \nu)$ will be determined from (8.2.38), and (8.2.41) for the coefficient $\alpha_{00}(s)$ will take the form

$$\alpha'_{00} = i\left(-\frac{1}{4}\frac{c'_0}{c_0} + \frac{1}{2}\frac{F'}{F}\right) - \left(q + \frac{1}{2}\right)\frac{1}{F^2}$$

$$-\frac{1}{2}\left(f' - \frac{F'}{F}f\right)^2 + \frac{1}{2}\frac{f^2}{F^4} \quad,$$

whence

$$\alpha_{00}(s) = -\frac{i}{2}\ln(c_0^{1/2}F^{-1}) - \left(q + \frac{1}{2}\right)\int\limits_{d_0}^{s}\frac{d\tau}{F^2(\tau)}$$

$$+\frac{1}{2}\int\limits_{d_0}^{s}\left[-\left(\frac{F'(\tau)}{F(\tau)} - \frac{f'(\tau)}{f(\tau)}\right)^2 + \frac{1}{F^4(\tau)}\right]f^2(\tau)d\tau \quad,$$

where d_0 is an arbitrary constant.

*8.3 Construction of the Polynomials α_m and β_m, $m \geq 1$

Let us begin with the determination of the polynomials $\alpha_1(s, \nu)$ and $\alpha_2(s, \nu)$. Setting $m = 1$ in (8.2.19), and taking (8.2.31) into consideration, we get the two equations

$$a_1(s, \nu) \equiv -2\frac{\partial\alpha_0}{\partial\nu}\frac{\partial\alpha_1}{\partial\nu} + i\frac{\partial^2\alpha_1}{\partial\nu^2} - 2\frac{1}{c_0(s)}\frac{\partial\alpha_1}{\partial s}$$

$$-4\frac{\partial\beta_0}{\partial\nu}\frac{\partial\beta_1}{\partial\nu}\left(q + \frac{1}{2} - \frac{1}{2}\beta_0^2\right) + 2\beta_0\beta_1\left(\frac{\partial\beta_0}{\partial\nu}\right)^2$$

$$+\left[\frac{1}{c_0^2(s)} + e_3(s)\right]\frac{\nu^3}{\varrho^3(s)} + \left[-\left(\frac{\partial\alpha_0}{\partial\nu}\right)^2 + i\frac{\partial^2\alpha_0}{\partial\nu^2}\right.$$

---

* This section can be skipped on first reading.

$$+2\frac{1}{c_0(s)}\frac{\partial\alpha_0}{\partial s} - 2\left(\frac{\partial\beta_0}{\partial\nu}\right)^2\left(q+\frac{1}{2}-\frac{1}{2}\beta_0^2\right)$$

$$+ i\frac{c_0'(s)}{c_0^2(s)} + i\frac{1}{c_0(s)}\frac{\varrho'(s)}{\varrho(s)}\Bigg]\frac{1}{\varrho(s)}\nu + i\frac{\partial\alpha_0}{\partial\nu}\frac{1}{\varrho(s)} = 0 \quad . \tag{8.3.1}$$

$$b_1(s,\nu) \equiv 2i\frac{\partial\alpha_0}{\partial\nu}\frac{\partial\beta_1}{\partial\nu} + 2i\frac{\partial\alpha_1}{\partial\nu}\cdot\frac{\partial\beta_0}{\partial\nu} + 2i\frac{1}{c_0(s)}\frac{\partial\beta_1}{\partial s}$$

$$+ \frac{\partial^2\beta_1}{\partial\nu^2} + \left[2i\frac{\partial\alpha_0}{\partial\nu}\cdot\frac{\partial\beta_0}{\partial\nu} + \frac{\partial^2\beta_0}{\partial\nu^2} - 2i\frac{1}{c_0(s)}\frac{\partial\beta_0}{\partial s}\right]\frac{1}{\varrho(s)}\nu$$

$$+ \frac{\partial\beta_0}{\partial\nu}\frac{1}{\varrho(s)} = 0 \quad .$$

$$\tag{8.3.2}$$

Leaving the unknown functions $\alpha_1(s,\nu)$ and $\alpha_2(s,\nu)$ on the left sides of the equations, and moving all other terms to the right sides, we have

$$-2\frac{\partial\alpha_0}{\partial\nu}\cdot\frac{\partial\alpha_1}{\partial\nu} + i\frac{\partial^2\alpha_1}{\partial\nu^2} - 2\frac{1}{c_0(s)}\frac{\partial\alpha_1}{\partial s}$$

$$-4\frac{\partial\beta_0}{\partial\nu}\frac{\partial\beta_1}{\partial\nu}\left(q+\frac{1}{2}-\frac{1}{2}\beta_0^2\right) + 2\beta_0\beta_1\left(\frac{\partial\beta_0}{\partial\nu}\right)^2$$

$$= \gamma_{03}(s)\nu^3 + \gamma_{02}(s)\nu^2 + \gamma_{01}(s)\nu + \gamma_{00}(s) \quad , \tag{8.3.3}$$

$$2i\frac{\partial\alpha_1}{\partial\nu}\frac{\partial\beta_0}{\partial\nu} + 2i\frac{\partial\alpha_0}{\partial\nu}\frac{\partial\beta_1}{\partial\nu} + 2i\frac{1}{c_0(s)}\frac{\partial\beta_1}{\partial s} + \frac{\partial^2\beta_1}{\partial\nu^2}$$

$$= \delta_{02}(s)\nu^2 + \delta_{01}(s)\nu + \delta_{00}(s) \quad . \tag{8.3.4}$$

The coefficients of the polynomials $\gamma_{0j}(s)$ and $\delta_{0j}(s)$ on the right sides of the equations are easily expressed in terms of already known functions of the arc length:

$$\gamma_{03}(s) = -4\frac{c_1(s)c_2(s)}{c_0^4(s)} + 2\frac{c_3(s)}{c_0^3(s)} + \frac{1}{\varrho(s)}\left[4\alpha_{02}^2(s) - 2\frac{\alpha_{02}'(s)}{c_0(s)} - \beta_{01}^4(s)\right] \quad ,$$

$$\gamma_{02}(s) = \frac{1}{\varrho(s)}\left[4\alpha_{02}(s)\alpha_{01}(s) - 2\frac{\alpha_{01}'(s)}{c_0(s)} - 2\beta_{01}^3(s)\beta_{00}(s)\right] \quad ,$$

$$\gamma_{01}(s) = \frac{1}{\varrho(s)}\left[\alpha_{01}^2(s) - 2\frac{\alpha_{00}'(s)}{c_0(s)} + 2\beta_{01}^2(s)\left(q+\frac{1}{2}-\frac{1}{2}\beta_{00}^2(s)\right)\right.$$

$$\left. - i\left(4\alpha_{02}(s) + \frac{c_0'(s)}{c_0^2(s)} + \frac{1}{c_0(s)}\frac{\varrho'(s)}{\varrho(s)}\right)\right] \quad ,$$

$$\gamma_{00}(s) = -i\frac{\alpha_{01}(s)}{\varrho(s)}$$

and

$$\delta_{02}(s) = i\frac{4}{\varrho(s)}\frac{\beta'_{01}(s)}{c_0(s)} , \qquad \delta_{01}(s) = i\frac{4}{\varrho(s)}\frac{\beta'_{00}(s)}{c_0(s)} ,$$

$$\delta_{00}(s) = -\frac{1}{\varrho(s)}\beta_{01}(s) .$$

Equation (8.3.4) implies that the degree of the polynomial $\beta_1(s,\nu)$ is one less than that of $\alpha_1(s,\nu)$. Then it follows from (8.3.3) that $\alpha_1(s,\nu)$ is a third-degree polynomial

$$\alpha_1(s,\nu) = \alpha_{13}(s)\nu^3 + \alpha_{12}(s)\nu^2 + \alpha_{11}(s)\nu + a_{10}(s) . \tag{8.3.5}$$

We write polynomial $\beta_1(s,\nu)$ in the form

$$\beta_1(s,\nu) = \beta_{12}(s)\nu^2 + \beta_{11}(s)\nu + \beta_{10}(s) . \tag{8.3.6}$$

Substituting (8.3.5,6) and also $\alpha_0(s,\nu)$ and $\beta_0(s,\nu)$ into (8.3.4) and setting the coefficients of different powers of ν equal to zero, we get

$$\alpha_{13}(s) = -\frac{1}{3}\frac{\beta'_{12}(s)}{c_0(s)\beta_{01}(s)} - \frac{4}{3}\alpha_{02}(s)\frac{\beta_{12}(s)}{\beta_{01}(s)} + \frac{1}{6i\beta_{01}(s)}\delta_{02}(s) ,$$

$$\alpha_{12}(s) = -\frac{1}{2}\frac{\beta'_{11}(s)}{c_0(s)\beta_{01}(s)} - \alpha_{02}(s)\frac{\beta_{11}(s)}{\beta_{01}(s)}$$
$$- \alpha_{01}(s)\frac{\beta_{12}(s)}{\beta_{01}(s)} + \frac{1}{4i\beta_{01}(s)}\delta_{01}(s) . \tag{8.3.7}$$

$$\alpha_{11}(s) = -\frac{\beta'_{10}(s)}{c_0(s)\beta_{01}(s)} - \alpha_{01}(s)\frac{\beta_{11}(s)}{\beta_{01}(s)} - \frac{\beta_{12}(s)}{i\beta_{01}(s)} + \frac{1}{2i\beta_{01}(s)}\delta_{00}(s) .$$

If we now substitute these expressions for $\alpha_{1j}(s)$, $j = 1, 2, 3$ into (8.3.3), then the fact that the coefficients of ν^1, ν^2 and ν^3 are equal to zero, will give us equations for determining $\beta_{1l}(s)$, $l = 0, 1, 2$. The fact that the constant term is equal to zero enables us to determine $\alpha_{10}(s)$ in the form of an integral with variable upper limit:

$$\alpha_{10}(s) = \frac{1}{4}\int_{d_1}^{s} \{ -4\alpha_{01}(\tau)\alpha_{11}(\tau) - 4(2q+1)\beta_{01}(\tau)\beta_{11}(\tau)$$
$$+ 4\beta_{00}^2(\tau)\beta_{01}(\tau)\beta_{11}(\tau) + 4\beta_{00}(\tau)\beta_{01}^2(\tau)\beta_{10}(\tau)$$
$$+ i[2\alpha_{01}(\tau)\varrho^{-1}(\tau) + 2\alpha_{12}(\tau)]\}c_0(\tau)d\tau . \tag{8.3.8}$$

If we convert to new unknown functions $f_{1l}(s)$ in the equations for $\beta_{1l}(s)$, using the formula

$$\beta_{1l}(s) = \beta_{01}^{l+1}(s)c_0^{1/2}(s)f_{1l}(s) = \frac{f_{1l}(s)}{c_0^{1/2}(s)F^{l+1}(s)} , \tag{8.3.9}$$

these equations will take the form

$$f''_{1l}(s) + \left\{ K(s) + [(l+1)^2 - 1]\frac{1}{F^4(s)} \right\}f_{1l}(s) = p_{1l}(s) . \tag{8.3.10}$$

The function $p_{12}(s)$ on the right side of the first ($l = 2$) of (8.3.10) is expressed in terms of coefficients $c_j(s)$, $j = 0, 1, 2, 3$, that appear in the expansion (8.2.2) of the velocity, the radius of curvature $\varrho(s)$ and its derivative, and also in terms of the function $F(s)$:

$$
p_{12}(s) = \frac{3}{2}c_0^{1/2}F^2\left\{-4\frac{c_1 c_2}{c_0^2} + 2\frac{c_3}{c_0} + \frac{1}{\varrho}\left[\frac{14}{3}\frac{c_2}{c_0}\right.\right.
$$
$$
\left.\left. -\frac{2}{3}\left(\frac{1}{2}\frac{c_0'}{c_0} + \frac{F'}{F}\right)^2 + \frac{4}{3}\frac{\varrho'}{\varrho}\left(\frac{1}{2}\frac{c_0'}{c_0} + \frac{F'}{F}\right) - \frac{10}{3}\frac{1}{F^4}\right]\right\} \quad .
$$

In addition to $c_0(s)$ and $\varrho(s)$, the expression for $p_{11}(s)$ includes the functions $f(s)$, $F(s)$ and $f_{12}(s)$. Once equation (8.3.10) for $l = 2$ is integrated, the function $f_{12}(s)$ is known. Note that if $f(s) = 0$, then we also have $p_{11}(s) = 0$. The function $p_{10}(s)$ that enters into the last ($l = 0$) of equations (8.3.10) depends on $c_0(s)$, $\varrho(s)$, $f(s)$, $F(s)$, $f_{12}(s)$ and $f_{11}(s)$. If the first two of (8.3.10) are integrated, the functions $f_{12}(s)$ and $f_{11}(s)$ are known.

Solutions of the nonhomogeneous equations (8.3.10) can be constructed by the method of variation of parameters. However, to do this, it is necessary first to find the general solution of the corresponding homogeneous equation

$$
\Phi''(s) + \left\{K(s) + [(l+1)^2 - 1]\frac{1}{F^4(s)}\right\}\Phi(s) = 0 \quad . \tag{8.3.11}
$$

It turns out that the general solution of (8.3.11) can be easily obtained if we know the linearly independent solutions $f_1(s)$ and $f_2(s)$ of (8.2.43), and consequently the function $F(s)$. Converting to new variables in (8.3.11):

$$
t = \int_0^s F^{-2}(\tau)d\tau \ , \quad y(t) = \Phi(s)F^{-1}(s) \ ,
$$

we get the equation

$$
y''(t) + (l+1)^2 y(t) = 0
$$

and consequently

$$
\Phi(s) = F(s)\left\{A \sin\left[(l+1)\int_0^s F^{-2}(\tau)d\tau\right]\right.
$$
$$
\left. + B \cos\left[(l+1)\int_0^s F^{-2}(\tau)d\tau\right]\right\} \quad . \tag{8.3.12}
$$

After the functions $f_{1l}(s)$ have been found, the coeffcients $\beta_{1l}(s)$ of the polynomial $\beta_1(s, \nu)$ are determined from (8.3.9). The β_{1l} in turn enable us to calculate the coefficients $\alpha_{1j}(s)$, $j = 3, 2, 1$ of $\alpha_1(s, \nu)$ from (8.3.7). The constant term $\alpha_{10}(s)$ is determined from (8.3.8), all functions in the integrand now being known.

The polynomials $\alpha_m(s, \nu)$ and $\beta_m(s, \nu)$ for $m \geq 2$ are determined by a scheme analogous to that used to construct the polynomials $\alpha_1(s, \nu)$ and $\beta_1(s, \nu)$. Consider the equations of (8.2.19) for arbitrary $m \geq 2$:

$$a_m(s, \nu) = 0 , \quad b_m(s, \nu) = 0 . \tag{8.3.13}$$

In addition to the unknown polynomials $\alpha_m(s, \nu)$ and $\beta_m(s, \nu)$, equations (8.3.13) will include the polynomials $\alpha_j(s, \nu)$, $\beta_j(s, \nu)$, $j = 0, 1, \ldots, m - 1$, already determined in preceding steps. It can be shown that $\alpha_m(s, \nu)$ and $\beta_m(s, \nu)$ are polynomials of degrees $m + 2$ and $m + 1$ respectively:

$$\alpha_m(s, \nu) = \alpha_{m,m+2}(s)\nu^{m+2} + \alpha_{m,m+1}(s)\nu^{m+1} + \ldots + \alpha_{m0}(s) ,$$
$$\beta_m(s, \nu) = \beta_{m,m+1}(s)\nu^{m+1} + \beta_{m,m}(s)\nu^m + \ldots + \beta_{m0}(s) .$$

Substituting these into (8.3.13) and setting the coefficients of different powers of ν equal to zero as before, we get $2m + 5$ equations for determining the functions $\alpha_{ml}(s)$, $l = 0, 1, \ldots, m + 2$ and $\beta_{ml}(s)$, $l = 0, 1, 2, \ldots, m + 1$. If we convert from the unknown functions $\beta_{ml}(s)$ to new unknown functions $f_{ml}(s)$ via:

$$\beta_{ml}(s) = \frac{f_{ml}(s)}{c_0^{1/2}(s)F^{l+1}(s)} ,$$

then for the functions $f_{ml}(s)$ we get equations completely analogous to (8.3.10):

$$f_{ml}''(s) + \left\{ K(s) + [(l+1)^2 - 1]\frac{1}{F^4(s)} \right\} f_{ml}(s) = p_{ml}(s) ,$$
$$l = 0, 1, \ldots, m + 1 , \tag{8.3.14}$$

where $p_{m,m+1}(s)$ is a known function, and the $p_{ml}(s)$ when $l < m + 1$ are determined from the functions $f_{mj}(s)$, $j > 1$, which were found in solving previous equations. The general solution of (8.3.14) can be constructed just as for (8.3.10) by varying the arbitrary constants in the general solution (8.3.12) of the homogeneous equation (8.3.11). After the $f_{ml}(s)$ have been found, the functions $\alpha_{ml}(s)$ are determined from formulas analogous to (8.3.7), $\alpha_{m0}(s)$, like $\alpha_{10}(s)$ in (8.3.8), being expressed by a quadrature containing an arbitrary constant. This constant, like the others that arise in the integration of differential equations, is determined from the boundary conditions, which must be formulated in some form or another when setting up the boundary value problem.

8.4 Basic Results and Some of Their Consequences

The polynomials $\alpha_m(s, \nu)$ and $\beta_m(s, \nu)$, $m \leq M - 1$, found by the scheme outlined above, enable us to construct the function (8.2.12), which, when substituted in the Helmholtz equation (8.2.1), gives a residual (uncompensated terms on the left side) of order $\omega^{-M/2+1}$.

Let us assemble the basic formulas and results found in the preceding sections of this chapter. Then we shall extract simple physical consequences from these explicit formulas.

The field in the neighborhood of S is represented in the form

$$u(s, n) = \text{const} \exp\left[i \sum_{m=-2}^{M-1} \alpha_m(s, \nu)\omega^{-m/2} + O\left(\omega^{-M/2}\right)\right]$$

$$\times D_q\left[\sqrt{2}\sum_{m=0}^{M-1} \beta_m(s, \nu)\omega^{-m/2} + O\left(\omega^{-M/2}\right)\right] . \qquad (8.4.1)$$

The coefficients $\alpha_m(s, \nu)$ are polynomials in ν of degree $m+2$, the coefficients $\beta_m(s, \nu)$ are polynomials in ν of degree $m+1$. Substituting expressions (8.4.1) into the Helmholtz equation (8.2.1) leads to a residual (uncompensated terms on the left side of the equation) of order $\omega^{-M/2+1}$. The first few coefficients are:

$$\alpha_{-2}(s, \nu) = \int_{d_{-2}}^{s} \frac{d\tau}{c_0(\tau)} , \qquad \alpha_{-1}(s, \nu) = d_{-1} ,$$

$$\alpha_0(s, \nu) = \left(\frac{F'(s)}{F(s)} + \frac{1}{2}\frac{c_o'(s)}{c_0(s)}\right)\frac{1}{2c_0(s)}\nu^2 + \left(\frac{F'(s)}{F(s)} - \frac{f'(s)}{f(s)}\right)\frac{f(s)}{c_0^{1/2}(s)}\nu$$

$$- \frac{i}{2}\ln\left[c_0^{1/2}(s)F^{-1}(s)\right] - \left(q + \frac{1}{2}\right)\int_{d_0}^{s}\frac{d\tau}{F^2(\tau)}$$

$$+ \frac{1}{2}\int_{d_0}^{s}\left[-\left(\frac{F'(\tau)}{F(\tau)} - \frac{f'(\tau)}{f(\tau)}\right)^2 + \frac{1}{F^4(\tau)}\right]f^2(\tau)d\tau ,$$

$$\beta_0(s, \nu) = \frac{1}{c_0^{1/2}(s)F(s)}\nu + \frac{f(s)}{F(s)} , \qquad \nu = \omega^{1/2}n .$$

In these formulas

1) $F(s) = \sqrt{a_{11}f_1^2(s) + 2a_{12}f_1(s)f_2(s) + a_{22}f_2^2(s)}$,

where $f_1(s)$ and $f_2(s)$ are linearly independent real solutions of the linearized Euler equation for the rays

$$f''(s) + K(s)f(s) = 0 , \qquad K(s) = \frac{1}{2}\frac{c_0''(s)}{c_0(s)} - \frac{3}{4}\left(\frac{c_0'(s)}{c_0(s)}\right)^2 + 2\frac{c_2(s)}{c_0(s)} , \qquad (8.4.2)$$

a_{rt} are the elements of a symmetric matrix that satisfy the conditions

$\det\|a_{rt}\|[f_1 f_2' - f_1' f_2]^2 = 1 , \quad a_{11} > 0 ,$

$a_{11}a_{22} - a_{12}a_{21} > 0$.

2) $f(s)$ is any solution of (8.4.2).
3) d_{-2}, d_{-1}, d_0 are arbitrary constants.

If we set $f(s) = 0$ and take the fixed limits of integration d_{-2}, d_{-1}, d_0, ...
so that $\alpha_{m0}(0) = 0$, the formulas take the form

$$\alpha_{-2}(s, \nu) = \int\limits_0^s \frac{d\tau}{c_0(\tau)} \quad , \qquad \alpha_{-1}(s, \nu) = 0 \quad ,$$

$$\begin{aligned}
\alpha_0(s, \nu) = {}& \left(\frac{F'(s)}{F(s)} + \frac{1}{2} \frac{c_0'(s)}{c_0(s)} \right) \frac{1}{2c_0(s)} \nu^2 \\
& - \frac{i}{2} \ln \left(\frac{c_0^{1/2}(s)F(0)}{c_0^{1/2}(0)F(s)} \right) - \left(q + \frac{1}{2} \right) \int\limits_0^s \frac{d\tau}{F^2(\tau)} \quad ,
\end{aligned}$$
(8.4.3)

$$\beta_0(s, \nu) = \frac{1}{c_0^{1/2}(s)F(s)} \nu \quad .$$
(8.4.4)

The solutions $u(s, n)$ thus constructed approach zero as $\nu = n\omega^{1/2} \to \pm\infty$ only
if the index q is a non-negative integer. Thus, when $q = 0, 1, 2, \ldots$ formula
(8.4.1) gives a sequence of solutions $u_q(s, n)$, $q = 0, 1, 2, \ldots$ such that

$$|u_q(s, n)| \to 0 \quad \text{when} \quad |n\omega^{1/2}| \to \pm\infty \quad .$$

Let us determine the width of the strip surrounding S in which the function
$u_q(s, n)$ oscillates, and beyond which it decays exponentially. Since the func-
tion $D_q(\sqrt{2}\Psi)$ is exponentially damped when $|\Psi| > \sqrt{2q+1}$, the strip beyond
which the function $u_q(s, n)$ is exponentially damped is determined in the first
approximation by the inequality

$$|n| \leq \sqrt{2q+1} c_0^{1/2}(s)F(s)\omega^{-1/2} \quad .$$
(8.4.5)

If q is too large, the width of the strip becomes so great that function $u_q(s, n)$ no
longer describes the wave field correctly. Therefore in (8.4.1) we must assume
that $q < N$, where the number N does not depend on ω.

8.5 Formulation of the Boundary Value Problem and Derivation of the Eigenvalue Equation

Let us now find the asymptotic expansion of the eigenvalues and eigenfunctions
in the problem involving an extremal ray of a region. The methods of this chapter
enable us to construct the asymptotic expansions of eigenvalues and eigenfunctions, generally speaking, to any arbitrary degree of accuracy in orders of powers
of $1/\omega$ as $\omega \to \infty$ (Sect. 5.4).

Let a nonhomogeneous medium fill a bounded region with boundary $\sum$ and
let the curve S, intersecting the boundary $\sum$ orthogonally at points C and D,
be a ray (Fig. 5.10). As in the preceding section, we introduce the coordinates
s and n in the vicinity of S. We will measure the arc length s along S from

C and take the length of S to be h. We will assume that in the vicinity of C and D the boundary of the region $\sum$ is described by the equations $s = \varphi(n)$ and $s = h + \psi(n)$ respectively, in which the functions $\varphi(n)$ and $\psi(n)$ can be represented by the series

$$\varphi(n) = a_2 n^2 + a_3 n^3 + \dots , \quad \psi(n) = b_2 n^2 + b_3 n^3 + \dots .$$

There is no first-order term in n in these expressions since the curves S and $\sum$ intersect orthogonally, and $a_2 = 1/2r_1$, $b_2 = -1/2r_2$, where r_1 and r_2 are the radii of curvature of $\sum$ at C and D respectively. As in the case of the function $c(s, n)$, it would have been sufficient for what follows to assume that $\phi(n)$ and $\psi(n)$ can be represented by truncated Taylor series.

Let us formulate the problem of finding solutions of the wave equation

$$\Delta u + \frac{\omega^2}{c^2(s, n)} u = 0 \quad , \tag{8.5.1}$$

that vanish on the boundary of the region

$$u|_{\Sigma} = 0 \tag{8.5.2}$$

and are concentrated around S, i.e. that satisfy the condition

$$u(s, n) \to 0 \quad \text{when} \quad |\omega^{1/2} n| \to \infty \quad . \tag{8.5.3}$$

Besides the functions $u(s, n)$ we will have to find the eigenvalues of (8.5.1–3).

The eigenfunctions concentrated around S can be constructed only with certain limitations on $c(s, n)$ and on the radii of curvature of $\sum$ at C and D. We will see later on that these limitations are precisely the conditions under which the system of rays in the vicinity of S is stable in the first approximation.

We have constructed solutions of (8.5.1) that are concentrated in the neighborhood of S in the form:

$$u(s, n) = \text{const} \exp \left[i \sum_{m=-2}^{M-1} \alpha_m(s, \nu) \omega^{-m/2} + O\left(\omega^{-M/2}\right) \right]$$

$$\times D_q \left[\sqrt{2} \sum_{m=0}^{M-1} \beta_m(s, \nu) \omega^{-m/2} + O\left(\omega^{-M/2}\right) \right] \quad , \tag{8.5.4}$$

$$\nu = n\omega^{1/2} , \quad q = 0, 1, 2, \dots .$$

For the solutions (8.5.4), boundary condition (8.5.2) reduces to the conditions[2]

$$u[\varphi(n), n] = 0 , \quad u[h + \psi(n), n] = 0 \quad . \tag{8.5.5}$$

The polynomials $\alpha_m(s, \nu)$ and $\beta_m(s, \nu)$ in (8.5.4) were determined from a recurrent system of differential equations up to arbitrary constants. We will now show

---

[2] On sections of Σ far from C and D, the solutions (8.5.4) are exponentially small, so that on the whole, when conditions (8.5.5) are satisfied, boundary condition (8.5.2) will be satisfied to within exponentially small terms.

that boundary conditions (8.5.5) enable us to determine these constants, and in addition to arrive at an equation from which the eigenvalues of the problem can be found.

We will seek the eigenfunctions $U(s, n)$ of (8.5.1–3) in the form

$$U(s, n) = Au(s, n) + Bu^*(s, n) \quad , \tag{8.5.6}$$

where A and B are constants, and the asterisk denotes the complex conjugate. Substituting (8.5.6) into (8.5.5) we get a system of two equations for determining A and B:

$$Au[\varphi(n), n] + Bu^*[\varphi(n), n] = 0 \quad ,$$
$$Au[h + \psi(n), n] + Bu^*[h + \psi(n), n] = 0 \quad . \tag{8.5.7}$$

We will determine the unknown constants A and B on the basis of the requirement that equalities (8.5.7) hold with an accuracy of $O(\omega^{-M/2})$. This means that in substituting (8.5.4) into (8.5.7) we can drop all terms in these expressions that decrease as $\omega \to \infty$ faster than $\omega^{-M/2}$. After doing this, (8.5.7) can be written as

$$A \exp\left[i \sum_{m=-2}^{M-1} \alpha_m(s, \nu)\omega^{-m/2}\right] D_q\left[\sqrt{2} \sum_{m=0}^{M-1} \beta_m(s, \nu)\omega^{-m/2}\right]$$

$$+ B \exp\left[-i \sum_{m=-2}^{M-1} \alpha_m^*(s, \nu)\omega^{-m/2}\right]$$

$$\times D_q\left[\sqrt{2} \sum_{m=0}^{M-1} \beta_m^*(s, \nu)\omega^{-m/2}\right]\Bigg|_{\substack{s=\varphi(n) \\ s=h+\psi(n)}} = 0 \quad . \tag{8.5.8}$$

To satisfy (8.5.8) it is necessary that the arguments of the two parabolic cylinder functions D_q be the same:

$$\sum_{m=0}^{M-1} \beta_m(s, \nu)\omega^{-m/2} = \sum_{m=0}^{M-1} \beta_m^*(s, \nu)\omega^{-m/2} \quad ,$$

for both $s = \phi(n)$ and $s = h + \psi(n)$, i.e., that

$$\text{Im}\left[\sum_{m=0}^{M-1} \beta_m(s, \nu)\omega^{-m/2}\right]\Bigg|_{\substack{s=\varphi(n) \\ s=h+\psi(n)}} = 0 \quad . \tag{8.5.9}$$

If these conditions are satisfied, then after cancelling

$$\exp\left[-i \sum_{m=-2}^{M-1} \alpha_m^*(s, \nu)\omega^{-m/2}\right] D_q\left[\sum_{m=0}^{M-1} \beta_m^*(s, \nu)\omega^{-m/2}\right] \quad ,$$

from (8.5.8), we get

$$A \exp\left\{i \sum_{m=-2}^{M-1} [\alpha_m(s, \nu) + \alpha_m^*(s, \nu)]\omega^{-m/2}\right\} + B\Bigg|_{\substack{s=\varphi(n) \\ s=h+\psi(n)}} = 0 \quad .$$

Here, the constants A and B are independent of ν and therefore the coefficient of A cannot depend on ν. In other words, it is necessary that the sum

$$\text{Re}\left\{\sum_{m=-2}^{M-1}\alpha_m(s,\nu)\omega^{-m/2}\right\}\Bigg|_{\substack{s=\varphi(n)\\s=h+\psi(n)}} \tag{8.5.10}$$

be independent of ν when $s=\varphi(n)$ and $s=h+\psi(n)$.

We now show that (8.5.9) and the requirement that (8.5.10) be independent of ν lead to boundary conditions that must be satisfied by the coefficients in the polynomials

$$\beta_m(s,\nu) = \sum_{j=0}^{m+1}\beta_{mj}(s)\nu^j \quad ,$$

$$\alpha_m(s,\nu) = \sum_{j=0}^{m+2}\alpha_{mj}(s)\nu^j,$$

when $s=0$ and $s=h$, i.e., to the functions $\beta_{mj}(s)$ and $\alpha_{mj}(s)$.

First consider (8.5.9). We represent the coefficients in $\beta_{mj}(s)$ by Taylor series:

$$\beta_{mj}(s) = \sum_{k=0}^{\infty}\beta_{mjk}(0)s^k$$

$$\beta_{mj}(s) = \sum_{k=0}^{\infty}\beta_{mjk}(h)(s-h)^k \quad ,$$

where in place of s and $s-h$ we substitute the expansions

$$s = \sum_{l=2}^{\infty}a_l\nu^l\omega^{-l/2} \quad , \quad s-h = \sum_{l=2}^{\infty}b_l\nu^l\omega^{-l/2} \quad . \tag{8.5.11}$$

As a result, the summation appearing in (8.5.9) becomes a double expansion in ν and $\omega^{-1/2}$:

$$\sum_{j=0}^{\infty}\nu^j\sum_{m=j-1}^{M+j-1}\chi_{mj}(s)\omega^{-m/2}\Bigg|_{s=0,h} \quad (\chi_{mj}\equiv 0 \quad\text{when } m<0) \quad .$$

The values of the first few coefficients are:

$$\chi_{m0}(s)|_{s=0,h} = \beta_{m0}(s)|_{s=0,h} \quad , \quad m = 0,1,2,\ldots,$$

$$\chi_{m1}(s)|_{s=0,h} = \beta_{m1}(s)|_{s=0,h} \quad , \quad m = 0,1,2,\ldots,$$

$$\chi_{m2}(0) = \beta_{m2}(0) + \beta'_{m-2,0}(0)a_2 \quad , \quad m = 1,2,3,\ldots,$$

$$\chi_{m2}(h) = \beta_{m2}(h) + \beta'_{m-2,0}(h)b_2 \quad , \quad m = 1,2,3,\ldots,$$

$$\chi_{m3}(0) = \beta_{m3}(0) + \beta'_{m-2,1}(0)a_2 + \beta'_{m-3,0}(0)a_3, \quad m = 2,3,\ldots,$$

$$\chi_{m3}(h) = \beta_{m3}(h) + \beta'_{m-2,1}(h)b_2 + \beta'_{m-3,0}(h)b_3, \quad m = 2,3,\ldots. \tag{8.5.12}$$

(In these formulas β_{lj} must be set equal to zero when $l<0$.) Equations (8.5.9) will be identically satisfied in ω and ν only if the coefficients $\chi_{mj}(s)$ are real when $s = 0$ and $s = h$, i.e. the imaginary part of χ_{mj} must be equal to zero:

$$\text{Im}\{\chi_{mj}(s)\}|_{s=0,h} = 0 \ , \quad j \geq 0 \ , \quad m \geq j - 1 \ . \tag{8.5.13}$$

Conditions (8.5.13) can also be written in the following way:

$$\text{Im}\{\beta_{mj}(s)\}|_{s=0,h} = \text{Im}\left\{\vartheta_{mj}^{(0,h)}\right\} \ , \quad j \geq 0 \ , \quad m \geq j - 1 \ , \tag{8.5.14}$$

where the constants $\vartheta_{mj}^{(0,h)}$ are expressed in terms of the values of the derivatives $\beta'_{nj}(s)$, $0 \leq n < m$, $0 \leq j \leq m + 1$, at the points $s = 0$ and $s = h$, and in terms of the coefficients a_l and b_l. Using formulas (8.5.12) we easily find that

$$\vartheta_{mj}^{(0,h)} = 0 \ , \quad j = 0, 1, \ m = 0, 1, 2, \ldots$$

$$\vartheta_{12}^{(0)} = 0 \ , \quad \vartheta_{m2}^{(0)} = -\beta'_{m-2,0}(0)a_2 \ , \quad m = 2, 3, \ldots,$$

$$\vartheta_{12}^{(h)} = 0 \ , \quad \vartheta_{m2}^{(h)} = -\beta'_{m-2,0}(h)b_2 \ , \quad m = 2, 3, \ldots,$$

$$\vartheta_{23}^{(0)} = -\beta'_{01}(0)a_2 \ , \quad \vartheta_{m3}^{(0)} = -\beta'_{m-2,1}(0)a_2 - \beta'_{m-3,0}a_3 \ ,$$
$$m = 3, 4, \ldots,$$

$$\vartheta_{23}^{(h)} = -\beta'_{01}(h)b_2 \ , \quad \vartheta_{m3}^{(h)} = -\beta'_{m-2,1}(h)b_2 - \beta'_{m-3,0}(h)b_3 \ ,$$
$$m = 3, 4, \ldots \ . \tag{8.5.15}$$

Thus (8.5.9) has led to conditions (8.5.14) on the imaginary parts of the coefficients in $\beta_{mj}(s)$.

Let us turn now to the requirement that (8.5.10) be independent of ν. We expand the coefficients $\alpha_{mj}(s)$, $0 \leq j \leq m+2$, of $\alpha_m(s, \nu)$ in power series in s and $s - h$, and substitute (8.5.11) in place of s and $s - h$. After this, we rearrange the terms in (8.5.10), and as a result, in place of (8.5.10) we get a double expansion in powers of ν and $\omega^{-1/2}$ of the form

$$\text{Re}\left\{ \sum_{m=-2}^{M-1} \alpha_{m0}(s)\omega^{-m/2} + \sum_{j=1}^{\infty} \nu^j \sum_{m=j-2}^{M+j-2} \kappa_{mj}(s)\omega^{-m/2} \right\}\bigg|_{s=0,h} \tag{8.5.16}$$

The first few coefficients $\kappa_{mj}(s)|_{s=0,h}$ have the values

$$\kappa_{m1}(s)|_{s=0,h} = \alpha_{m1}(s)|_{s=0,h} \ , \quad m = -1, 0, 1, \ldots,$$

$$\kappa_{m2}(0) = \alpha_{m2}(0) + \alpha'_{m-2,0}(0)a_2 \ ,$$

$$\kappa_{m2}(h) = \alpha_{m2}(h) + \alpha'_{m-2,0}(h)b_2 \ , \quad m = 0, 1, 2, \ldots,$$

$$\kappa_{m3}(0) = \alpha_{m3}(0) + \alpha'_{m-2,1}(0)a_2 + \alpha'_{m-3,0}(0)a_3 \ ,$$

$$\kappa_{m3}(h) = \alpha_{m3}(h) + \alpha'_{m-2,1}(h)b_2 + \alpha'_{m-3,0}(h)b_3, \quad m = 1, 2, 3, \ldots \ . \tag{8.5.17}$$

Formula (8.5.16) implies that (8.5.10) will not depend on ν if

$$\text{Re}\{\kappa_{mj}(s)\}|_{s=0,h} = 0 \ , \quad j \geq 1 \ , \quad m \geq j - 2 \ . \tag{8.5.18}$$

Conditions (8.5.18) can be written as follows:

$$\text{Re}\{\alpha_{mj}(s)\}|_{s=0,h} = \text{Re}\{\eta_{mj}^{(0,h)}\} \ , \ j \geq 1 \ , \ m \geq j - 2 \ , \tag{8.5.19}$$

where the constants $\eta_{mj}^{(0,h)}$ depend on the values of the derivatives of $\alpha_{nk}(s)$, $-2 \leq n < m$, $0 \leq k \leq n+2$, at 0 and h, as well as on the coefficients a_l and b_l. From (8.5.17) we see that

$$\eta_{m1}^{(0,h)} = 0 \ , \ m = -1, 0, 1, \ldots,$$

$$\eta_{m2}^{(0)} = -\alpha'_{m-2,0}(0)a_2 \ ,$$

$$\eta_{m2}^{(h)} = -\alpha'_{m-2,0}(h)b_2 \ , \ m = 0, 1, 2, \ldots,$$

$$\eta_{m3}^{(0)} = -\alpha'_{m-2,1}(0)a_2 - \alpha'_{m-3,0}(0)a_3 \ ,$$

$$\eta_{m3}^{(h)} = -\alpha'_{m-2,1}(h)b_2 - \alpha'_{m-3,0}(h)b_3 \ , \ m = 1, 2, \ldots \ . \tag{8.5.20}$$

Thus the requirement that (8.5.10) be independent of ν has led to boundary conditions (8.5.19) for the real parts of $\alpha_{mj}(s)$. When we satisfy conditions (8.5.14) and (8.5.19), in other words when we have satisfied (8.5.9) and the condition that (8.5.10) be independent ν, (8.5.8) takes the form

$$A \exp\left[i2\,\text{Re}\left\{\sum_{m=-2}^{M-1} \alpha_{m0}(s)\omega^{-m/2}\right\}\right] + B\Bigg|_{s=0,h} = 0. \tag{8.5.21}$$

As we shall now see, the requirement that (8.5.21) be solvable for the constants A and B leads to an equation from which we can find the eigenvalues $\omega_{p,q}$ of the problem.

We recall that the constant terms of $\alpha_m(s, \nu)$, i.e. the functions $\alpha_{m0}(s)$, are expressed by integrals with variable upper limit and arbitrary constant lower limit. By our choice of the constants A and B we can always arrange things so that the equalities

$$\alpha_{m0}(0) = 0 \ , \ m = -2, -1, 0, \ldots \tag{8.5.22}$$

are satisfied. In what follows we will assume that (8.5.22) holds and consequently the lower limits of integration in the above-mentioned integrals are equal to zero. With (8.5.22) satisfied, (8.5.21) takes the very simple form

$$A + B = 0 \ , \ A \exp\left[i2\,\text{Re}\left\{\sum_{m=-2}^{M-1} \alpha_{m0}(h)\omega^{-m/2}\right\}\right] + B = 0 \ .$$

In order for this homogeneous system to have a non-zero solution, its determinant

$$\Delta = 1 - \exp\left[i2\,\text{Re}\left\{\sum_{m=-2}^{M-1} \alpha_{m0}(h)\omega^{-m/2}\right\}\right]$$

must be equal to zero. The equation $\Delta = 0$ is equivalent to the equation

$$\text{Re} \left\{ \sum_{m=-2}^{M-1} \alpha_{m0}(h)\omega^{-m/2} \right\} = \pi p \quad , \tag{8.5.23}$$

where p is an integer. We must assume that $p \gg 1$ since all our constructions relate to the case of large ω.

Equation (8.5.23) serves to determine the eigenvalues $\omega_{p,q}$ of the given problem.

8.6 Formulas for Eigenvalues and Eigenfunctions in the First Approximation

We now determine the boundary conditions for the functions $F(s)$ and $f(s)$ in terms of which $\beta_{01}(s)$ and $\beta_{00}(s)$ are expressed. Refer to conditions (8.5.14, 19); when $m = -1$, condition (8.5.19) is automatically satisfied since, see (8.2.30),

$$\alpha_{-1}(s,\nu) = d_{-1} \quad \text{and} \quad d_{-1} = 0$$

by virtue of (8.5.22).

Setting $m = 0$ in (8.5.14 and 19), and using formulas (8.5.15 and 20) we get

$$\text{Im}\{\beta_{00}(s)\}|_{s=0,h} = 0 \quad \text{and} \quad \text{Im}\{\beta_{01}(s)\}|_{s=0,h} = 0 \quad , \tag{8.6.1}$$

$$\text{Re}\{\alpha_{01}(s)\}|_{s=0,h} = 0 \quad \text{and} \quad \text{Re}\{\alpha_{02}(0)\} = -\frac{a_2}{c_0(0)} \quad . \tag{8.6.2}$$

Since

$$\text{Re}\{\alpha_{02}(h)\} = -\frac{b_2}{c_0(h)} \quad ,$$

$$\alpha_{01} = -\frac{1}{c_0}\frac{\beta'_{00}}{\beta_{01}} \quad \text{and} \quad \alpha_{02} = -\frac{1}{2c_0}\frac{\beta'_{01}}{\beta_{01}} \quad ,$$

conditions (8.6.2) reduce to the requirements

$$\text{Re}\left\{\frac{\beta'_{00}(s)}{\beta_{01}(s)}\right\}\bigg|_{s=0,h} = 0 \quad \text{and} \quad \text{Re}\left\{\frac{\beta'_{01}(0)}{\beta_{01}(0)}\right\} = 2a_2 \quad ,$$

$$\text{Re}\left\{\frac{\beta'_{01}(h)}{\beta_{01}(h)}\right\} = 2b_2. \tag{8.6.3}$$

These are imposed, as are conditions (8.6.1), on the functions $\beta_{00}(s)$ and $\beta_{01}(s)$. We have seen that $F(s) = c_0^{-1/2}(s)\beta_{01}^{-1}(s)$ satisfies the nonlinear equation (8.2.42):

$$F''(s) + K(s)F(s) = \frac{1}{F^3(s)} \quad . \tag{8.6.4}$$

From (8.6.1, 3) we see that we must solve equation (8.6.4) subject to the following boundary conditions:

$$\text{Im}\{F(s)\}|_{s=0,h} = 0$$

and

$$\text{Re}\left\{\frac{F'(0)}{F(0)} + \frac{1}{2}\frac{c_0'(0)}{c_0(0)} + 2a_2\right\} = 0 \ ,$$

$$\text{Re}\left\{\frac{F'(h)}{F(h)} + \frac{1}{2}\frac{c_0'(h)}{c_0(h)} + 2b_2\right\} = 0 \ . \tag{8.6.5}$$

Since $K(s)$ is real, the function $F(s)$ can also be taken to be real, to obey (8.6.4) and to satisfy the boundary conditions

$$F'(s) + g(s)F(s)|_{s=0,h} = 0 \ , \tag{8.6.6}$$

where

$$g(0) = \frac{1}{2}\frac{c_0'(0)}{c_0(0)} + \frac{1}{r_1} \quad \text{and} \quad g(h) = \frac{1}{2}\frac{c_0'(h)}{c_0(h)} - \frac{1}{r_2} \ ,$$

$r_1 = (1/2a_2)$ and $r_2 = -(1/2b_2)$ are the radii of curvature of the boundary of Σ at the points C and D.

We have also seen that the function $f(s) = \beta_{00}(s)F(s)$ satisfies a second-order homogeneous differential equation:

$$f''(s) + K(s)f(s) = 0 \ . \tag{8.6.7}$$

Since $F(s) \neq 0$, and conditions (8.6.1, 3) together with (8.6.7) will be satisfied only when $f(s) \equiv 0$[3], it is necessary that $\beta_{00}(s) \equiv 0$.

If the function $F(s)$ is known, and the function $f(s) \equiv 0$, the polynomials $\alpha_0(s, \nu)$ and $\beta_0(s, \nu)$ are given by (8.4.3 and 4), and take the form

$$\alpha_0(s, \nu) = \left(\frac{F'(s)}{F(s)} + \frac{1}{2}\frac{c_0'(s)}{c_0(s)}\right)\frac{1}{2c_0(s)}\nu^2$$

$$- \frac{i}{2}\ln\frac{c_0^{1/2}(s)F(0)}{c_0^{1/2}(0)F(s)} - \left(q + \frac{1}{2}\right)\int_0^s F^{-2}(\tau)d\tau \ ,$$

$$\beta_0(s, \nu) = \frac{1}{c_0^{1/2}(s)F(s)}\nu \ .$$

(The boundary conditions for $\beta_m(s, \nu)$ when $m \geq 1$ will be derived in the next section.)

Let us now derive formulas for the eigenvalues and eigenfunctions in the first approximation, using the polynomials already found — α_{-2}, α_{-1}, α_0 and β_0. Equation (8.5.23) implies that

---

[3] We will not consider here the case where (8.6.7) has a nonzero solution that satisfies the homogeneous conditions (8.6.1, 3).

$$\omega_{p,q} = \frac{1}{\mathrm{Re}\{\alpha_{-20}(h)\}} \left\{ \pi p - \mathrm{Re}\{\alpha_{00}(h)\} + O\left(\frac{1}{p^{1/2}}\right) \right\} \ .$$

Below (Sect. 8.7) it will be shown that the correction term is actually of order of magnitude $O(1/p)$. Thus, substituting the values of $\alpha_{-20}(h)$ and $\alpha_{00}(h)$, we get

$$\omega_{p,q} = \left(\int\limits_0^h \frac{d\tau}{c_0(\tau)} \right)^{-1} \left[\pi p + \left(q + \frac{1}{2} \right) \int\limits_0^h \frac{d\tau}{F^2(\tau)} + O\left(\frac{1}{p}\right) \right] \ . \tag{8.6.8}$$

Hence, determination of the eigenvalues $\omega_{p,q}$ has been reduced to the evaluation of the integral $\int_0^h d\tau/F^2(\tau)$, which will be carried out at the end of the section. For now, we notice one important point. If the equality

$$\int\limits_0^h \frac{d\tau}{F^2(\tau)} = \pi \frac{P}{Q} \quad (P, Q - \text{are integers}) \ , \tag{8.6.9}$$

holds, then the eigenvalues $\omega_{p,q}$ in the first approximation will be multiply degenerate in the sense that

$$\omega_{p,q} = \omega_{p-Pj,q+Qj} \ , \quad j = \pm 1, \pm 2, \ldots \ .$$

Degeneracy of eigenvalues in the first approximation, as we know from perturbation theory, leads to difficulties in constructing higher approximations. We will see below (Sect. 8.7) that in fact, when (8.6.9) holds, the construction of higher approximations runs up against serious difficulties that can be overcome only by modifying our entire line of reasoning.

We now construct the function $F(s)$ satisfying (8.6.6), and clarify the role played in construction of $F(s)$ by the condition of stability in the first approximation of a system of multiply reflected rays.

In Sect. 8.2 we gave formula (8.2.44) for $F(s)$ (Appendix A.3):

$$F(s) = [a_{11}f_1^2(s) + 2a_{12}f_1(s)f_2(s) + a_{22}f_2^2(s)]^{1/2} \ , \tag{8.6.10}$$

in which $f_1(s)$ and $f_2(s)$ are linearly independent solutions of the equation

$$f''(s) + K(s)f(s) = 0$$

and the elements of the symmetric positive definite matrix $\|a_{rt}\|$ satisfy the normalization condition

$$\det\|a_{rt}\| W^2 = 1 \quad (W = f_1 f_2' - f_1' f_2) \ . \tag{8.6.11}$$

We shall choose a_{rt} such that the function $F(s)$ satisfies the boundary conditions (8.6.6). Substituting (8.6.10) into (8.6.6), we arrive at the two equations

$$\sum_{r,t=1}^2 [f_t'(s) + g(s)f_t(s)]a_{rt}f_r(s)|_{s=0,h} = 0 \ , \tag{8.6.12}$$

where $g(0) = \frac{1}{2}[c_0'(0)/c_0(0)] + (1/r_1)$ and $g(h) = \frac{1}{2}[c_0'(h)/c_0(h)] - (1/r_2)$. Equations

(8.6.12), with condition (8.6.11), form a system of three equations for determining a_{11}, a_{22} and a_{12}.

We will assume that there are no non-zero solutions of (8.6.7) that satisfy the conditions $f'(s) + g(s)f(s)|_{s=0,h} = 0$. This assumption enables us to choose linearly independent solutions $f_1(s)$ and $f_2(s)$ such that the conditions

$$f_1'(0) + g(0)f_1(0) = 0 \qquad (f_1(0) \neq 0) ,$$
$$f_2'(h) + g(h)f_2(h) = 0 \qquad (f_2(h) \neq 0) , \tag{8.6.13}$$

are satisfied. Equations (8.6.13) significantly simplify equations (8.6.12), which now take the form:

$$a_{12}f_1(0) + a_{22}f_2(0) = 0 ,$$
$$a_{11}f_1(h) + a_{21}f_2(h) = 0 . \tag{8.6.14}$$

Let $f_1(h) \neq 0$ and $f_2(0) \neq 0$; then

$$\frac{a_{22}}{a_{12}} = -\frac{f_1(0)}{f_2(0)} , \qquad \frac{a_{11}}{a_{12}} = -\frac{f_2(h)}{f_1(h)} ,$$

and from (8.6.11) we see that

$$a_{12}^2 \left(\frac{f_1(0)}{f_1(h)} \frac{f_2(h)}{f_2(0)} - 1 \right) W^2 = 1 . \tag{8.6.15}$$

This equation has a real solution only if the condition

$$\frac{f_1(0)}{f_1(h)} \frac{f_2(h)}{f_2(0)} > 1 \tag{8.6.16}$$

holds. When taking the square root in (8.6.15), the sign of a_{12} must be taken opposite to that of $[f_1(0)/f_2(0)]W$. Then the quadratic form in (8.6.10) will be positive definite, and $F(s)$ will be strictly positive. If inequality (8.6.16) does not hold, it will not be possible to construct a real function $F(s)$.

If $f_1(h) = 0$, $f_2(0) \neq 0$ or $f_1(h) \neq 0$, $f_2(0) = 0$ and $W[f_1, f_2] \neq 0$, then the system of equations (8.6.11, 14) has only a trivial solution, and in this case the function $F(s)$ cannot be constructed either. The impossibility of constructing $F(s)$ means that there are no eigenfunctions of the form (8.5.4) which are concentrated in the vicinity of S.

On the other hand, if

$$f_1(h) = f_2(0) = 0 , \tag{8.6.17}$$

then both of equations (8.6.14) are satisfied when $a_{12} = a_{21} = 0$, and only (8.6.11) is left to determine the two elements a_{11} and a_{22}. In this case, we can obviously construct a whole family of functions $F(s)$ that depend on a single parameter. The eigenfunctions of our problem cannot, however, be continuously dependent on the parameter. Therefore the value of this parameter must be completely determined in the process of constructing the higher approximations for the eigenfunctions.

Consider now the case when (8.6.7) has a non-zero solution satisfying the conditions $f'(s) + g(s)f(s)|_{s=0,h} = 0$, and consequently any two solutions f_1, f_2

of (8.6.7) satisfying (8.6.13) are linearly dependent. In this case, the solutions f_1 and f_2 of (8.6.7) used to construct F [see (8.6.10)] cannot be chosen in the way just described. Let us now take as f_1 the non-zero solution of (8.6.7) satisfying $f_1'(s) + g(s)f_1(s)|_{s=0,h} = 0$, and as f_2 any solution of (8.6.7) that is linearly independent of f_1. Equations (8.6.12) will take the form

$$a_{12}f_1(0) + a_{22}f_2(0) = 0 \ ,$$
$$a_{12}f_1(h) + a_{22}f_2(h) = 0 \ . \tag{8.6.18}$$

The system (8.6.18) will have a non-zero solution only if

$$f_1(0)f_2(h) - f_1(h)f_2(0) = 0 \ . \tag{8.6.19}$$

If (8.6.19) is satisfied, we can construct, as in the previous case, a whole family of functions $F(s)$ which continuously depend on a single parameter.

Note that conditions (8.6.16, 17 and 19), which when satisfied enable us to construct $F(s)$, are precisely the same as the conditions of stability in the first approximation: (5.4.15, 18 and 19), which were found for rays close to the extremal ray of the region. Thus $F(s)$ can be constructed in all cases where the system of rays is stable in the first approximation. On the other hand, if the system of rays is not stable, it will not be possible to construct $F(s)$ nor the corresponding eigenfunctions concentrated in the neighborhood of the extremal ray.

Proceeding now to the computation of the integral $\int_0^h d\tau / F^2(\tau)$, we will assume that one of the conditions (8.6.16, 17 or 19) is satisfied, and the function $F(s)$ has been constructed.

The integral

$$\int_0^h \frac{d\tau}{F^2(\tau)} = \int_0^h \frac{d\tau}{a_{11}f_1^2(\tau) + 2a_{12}f_1(\tau)f_2(\tau) + a_{22}f_2^2(\tau)} \tag{8.6.20}$$

has already been computed in Chap. 5 when inequality (8.6.16) holds. The method of evaluating (8.6.20) described in Chap. 5 could have been extended as well to the case of (8.6.17 or 19). Here, however, we will find the value of (8.6.20) by a somewhat different method.

Let $f_1(s)$ vanish N times at the points s_k, $k = 1, 2, \ldots, N$ in the open interval $(0, h)$. Let us represent integral (8.6.20) as the sum of integrals

$$\int_0^h \frac{d\tau}{F^2(\tau)} = \int_0^{s_1} \frac{d\tau}{F^2(\tau)} + \sum_{k=1}^{N-1} \int_{s_k}^{s_{k+1}} \frac{d\tau}{F^2(\tau)} + \int_{s_N}^h \frac{d\tau}{F^2(\tau)} \ . \tag{8.6.21}$$

In each integral we convert to a new variable of integration θ, setting

$$\theta(\tau) = \arctan\left[W\frac{a_{22}f_2(\tau) + a_{12}f_1(\tau)}{f_1(\tau)}\right] \ . \tag{8.6.22}$$

Formula (8.6.22) uses the principal branch of the arc tangent whose values lie in the interval $(-\pi/2, \pi/2)$. The function $\theta(\tau)$ is a continuous function on each interval. Its differential, as can be easily shown using (8.6.11), is

$$d\theta = \frac{1}{a_{11}f_1^2(\tau) + 2a_{12}f_1(\tau)f_2(\tau) + a_{22}f_2^2(\tau)}\, d\tau \quad .$$

The quadratic form in the denominator is positive definite, therefore the function $\theta(\tau)$ is monotonically increasing on each interval, and consequently the value $\theta = -\pi/2$ corresponds to the points s_k in the lower limits of integration, and $\theta = \pi/2$ corresponds to the s_k in the upper limits of integration. Since $a_{12}f_1(0) + a_{22}f_2(0) = 0$, $\theta = 0$ corresponds to the point $s = 0$, and

$$\theta_h = \arctan\left[W\left(a_{22}\frac{f_2(h)}{f_1(h)} + a_{12} \right) \right]$$

corresponds to the point $s = h$.

Replacing a_{12} and a_{22} by their values, we have

$$\theta_h = \arctan\left[\operatorname{sgn}\left(W\frac{f_1(0)}{f_2(0)} \right) \sqrt{\frac{f_1(0)f_2(h)}{f_2(0)f_1(h)} - 1} \right] \quad . \tag{8.6.23}$$

When (8.6.17) holds, the function $f_1(s)$ vanishes at the upper limit of integration, and therefore $\theta_h = \pi/2$. In the case of (8.6.19), obviously $\theta_h = 0$. Formula (8.6.23) can be rewritten as

$$\theta_h = \operatorname{sgn}\left(g(0) + \frac{f_2'(0)}{f_2(0)} \right) \arccos\sqrt{\frac{f_1(h)f_2(0)}{f_2(h)f_1(0)}} \quad . \tag{8.6.24}$$

In the latter formula the arc cosine ranges over the interval $(0, \pi/2)$.

Having found the limits of integration corresponding to each integral in (8.6.21) for the new variable of integration, we arrive at

$$\int_0^h \frac{d\tau}{F^2(\tau)} = \int_0^{\pi/2} d\theta + \sum_{k=1}^{N-1} \int_{-\pi/2}^{\pi/2} d\theta + \int_{-\pi/2}^{\theta_h} d\theta \quad ,$$

whence

$$\int_0^h \frac{d\tau}{F^2(\tau)} = \operatorname{sgn}\left(g(0) + \frac{f_2'(0)}{f_2(0)} \right) \arccos\sqrt{\frac{f_1(h)f_2(0)}{f_2(h)f_1(0)}} + \pi N \quad . \tag{8.6.25}$$

Having computed $\int_0^h d\tau/F^2(\tau)$, formula (8.6.8) enables us to determine the eigenvalues $\omega_{p,q}$ and after that, the eigenfunctions $u_{p,q}$ from (8.5.4).

Substituting (8.6.25) into (8.6.8), we get

$$\omega_{p,q} = \left[\int_0^h \frac{d\tau}{c_0(\tau)} \right]^{-1} \left\{ \pi p + \left(q + \frac{1}{2} \right) \left[\operatorname{sgn}\left(g(0) + \frac{f_2'(0)}{f_2(0)} \right) \right.\right.$$

$$\left.\left. \times \arccos\sqrt{\frac{f_1(h)f_2(0)}{f_2(h)f_1(0)}} + \pi N \right] + O\left(\frac{1}{p}\right) \right\} \quad . \tag{8.6.26}$$

We recall that $f_1(s)$ and $f_2(s)$ are chosen so that (8.6.13) is satisfied, and N is the number of zeros of $f_1(s)$ on the interval $(0, h)$.

In the case of condition (8.6.17) the first term in quadratic brackets is equal to $\pi/2$, while in the case of (8.6.19) it is equal to zero. In these two cases formula (8.6.26) is respectively simplified:

$$\omega_{p,q} = \left(\int_0^h \frac{d\tau}{c_0(\tau)} \right)^{-1} \left[\pi p + \left(q + \frac{1}{2} \right) \left(\frac{1}{2} + N \right) \pi + O\left(\frac{1}{p} \right) \right] \quad,$$

$$\omega_{p,q} = \left(\int_0^h \frac{d\tau}{c_0(\tau)} \right)^{-1} \left[\pi p + \left(q + \frac{1}{2} \right) N \pi + O\left(\frac{1}{p} \right) \right]. \qquad (8.6.27)$$

Despite the fact that in both cases the first approximation for the eigenfunctions depends continuously on a single parameter, formulas (8.6.27) for the natural frequencies contain no parameter. (Recall that the function F appearing in the solution depends on three parameters – see (8.6.10) – and is subjected to one linear relation and the normalizing condition (8.6.11).) This means that corresponding to the one eigenvalue obtained in the first approximation is a continuum of eigenfunctions. Using the terminology of quantum mechanics we can say that the eigenvalues of the first approximation are infinitely degenerate. Of course, the spectrum of our problem is discrete, and there can be no continuum of eigenvalues. This infinite degeneracy must be removed in the construction of higher approximations.

Finally, let us write out the expression for the eigenfunctions $u_{p,q}$. To do this, we substitute the eigenvalues we have found into formula (8.5.4). Taking into consideration that $A = -B$, we get

$$u_{p,q}(s, n) = \sqrt{\frac{c_0^{1/2}(s)}{F(s)}} \left\{ \sin\left[\omega_{p,q} \int_0^s \frac{d\tau}{c_0(\tau)} - \left(q + \frac{1}{2} \right) \int_0^s \frac{d\tau}{F^2(\tau)} \right. \right.$$

$$+ \left(\frac{1}{2} \frac{c_0'(s)}{c_0(s)} + \frac{F'(s)}{F(s)} \right) \frac{\nu^2}{2 c_0(s)} \right] + O\left(\frac{1}{\omega_{p,q}^{1/2}} \right) \right\}$$

$$\times \left\{ D_q \left(\frac{\sqrt{2}\nu}{c_0^{1/2}(s) F(s)} \right) + D_q' \left(\frac{\sqrt{2}\nu}{c_0^{1/2}(s) F(s)} \right) \right.$$

$$\times O\left(\frac{1}{\omega_{p,q}^{1/2}} \right) \right\} \quad , \qquad \nu = n \omega^{1/2} \quad , \qquad (8.6.28)$$

for the eigenfunctions in the first approximation.

Below, in Sect. 8.8, formula (8.6.26) will be used to calculate the natural frequencies of an open resonator filled with a nonhomogeneous medium whose propagation velocity is

$$c(s, n) = c_0 + c_2 n^2 \quad .$$

In the next section, we derive the boundary conditions for $f_{ml}(s)$ in terms of which we express the polynomials $\alpha_m(s,\nu)$ and $\beta_m(s,\nu)$ for $m \geq 1$ which enter into the higher approximations for the eigenfunctions.

8.7 Procedure for Constructing the Polynomials $\alpha_m(s,\nu)$ and $\beta_m(s,\nu)$ for $m \geq 1$

Consider (8.5.14 and 19) when $m = 1$. Using (8.5.15 and 20) we get

$$\text{Im}\{\beta_{1j}(s)\}|_{s=0,h} = 0 \ , \quad j = 0,1,2 \ , \tag{8.7.1}$$

$$\text{Re}\{\alpha_{1j}(s)\}|_{s=0,h} = 0 \ , \quad j = 1,2 \ ,$$

$$\text{Re}\{\alpha_{13}(0)\} = -\frac{1}{c_0(0)}a_3 \ , \quad \text{Re}\{\alpha_{13}(h)\} = -\frac{1}{c_0(h)}b_3 \ . \tag{8.7.2}$$

It was established in Sect. 8.3 that the functions $\alpha_{1j}(s)$, $j = 1,2,3$, are expressed in terms of $\beta_{1,j-1}(s)$, (8.3.7). Thus, conditions (8.7.2) are, in essence, imposed on the $\beta_{1,j-1}(s)$, and considering formulas (8.6.3 and 7.1), they take the form

$$\text{Re}\{\beta'_{10}(s)\}|_{s=0,h} = 0 \ ,$$

$$\text{Re}\{\beta'_{11}(0) - 2a_2\beta_{11}(0)\} = 0 \ ,$$

$$\text{Re}\{\beta'_{11}(h) - 2b_2\beta_{11}(h)\} = 0 \ ,$$

$$\text{Re}\{\beta'_{12}(0) - 4a_2\beta_{12}(0)\} = \left[3a_3 + 4\frac{a_2}{\varrho(0)}\right]\beta_{01}(0) \ ,$$

$$\text{Re}\{\beta'_{12}(h) - 4b_2\beta_{12}(h)\} = \left[3b_3 + 4\frac{b_2}{\varrho(h)}\right]\beta_{01}(h) \ .$$

It can be shown that in the general case of $m \geq 2$, conditions (8.5.19) reduce to the conditions

$$\text{Re}\{\beta'_{ml}(0) - 2la_2\beta_{ml}(0)\} = \tau^{(0)}_{ml} \ , \quad l = m+1, m, \ldots, 1, 0 \ ,$$

$$\text{Re}\{\beta'_{ml}(h) - 2lb_2\beta_{ml}(h)\} = \tau^{(h)}_{ml} \ , \quad l = m+1, m, \ldots, 1, 0 \ , \tag{8.7.3}$$

where the $\tau^{(0,h)}_{ml}$ are constants defined in terms of the coefficients $\beta_{nj}(s,\nu)$, $0 \leq n \leq m$, $j > l$, found in previous steps, and are expressed in terms of a_l, b_l, and the derivatives of $c_l(s)$ and $\varrho(s)$ at the points $s = 0$ and $s = h$.

In determining the functions $\beta_{ml}(s)$ we have introduced the new unknown functions $f_{ml}(s) = c_0^{1/2}(s)F^{l+1}(s)\beta_{ml}(s)$ and have for them equations (8.3.14). It follows from (8.7.3 and 8.6.6) that the functions $f_{ml}(s)$ at the ends of interval $[0,h]$ must satisfy the conditions

$$\text{Re}\{[f'_{ml}(s) + g(s)f_{ml}(s)]\}|_{s=0,h} = c_0^{1/2}(s)F^{l+1}(s)|_{s=0,h}\tau^{(0,h)}_{ml} \ . \tag{8.7.4}$$

To these, we must also add the conditions on the imaginary parts

$$\text{Im}\{f_{ml}(s)\}|_{s=0,h} = \text{Im}\{\vartheta_{ml}^{(0,h)}\} \quad , \tag{8.7.5}$$

stemming from (8.5.14).

Thus (8.3.14) must be solved subject to the boundary conditions (8.7.4, 5). Let us consider the case $m = 1$ in a little more detail. To determine the functions $f_{12}(s)$ we have the nonhomogeneous equation (8.3.10) with known real right side $p_{12}(s)$, the nonhomogeneous condition (8.7.4) and the homogeneous condition (8.7.5). If zero is not an eigenvalue of the resultant Sturm-Liouville problems that arise for $\text{Re}\{f_{12}(s)\}$ and $\text{Im}\{f_{12}(s)\}$, then $f_{12}(s)$ is a real function that is uniquely defined.

The second equation (8.3.10) for $f_{11}(s)$ is homogeneous since $f(s) = 0$, and consequently $p_{11}(s) = 0$. Conditions (8.7.4 and 5) for $f_{11}(s)$ are also homogeneous, and therefore $f_{11}(s) = 0$. Once more we assume that zero is not an eigenvalue of the corresponding boundary value problem. The fact that $f_{11}(s)$ vanishes implies that $\beta_{11}(s) = 0$. In conjunction with the equality $f(s) = 0$, this latter equation leads to the conclusion [see (8.3.7 and 8)] that

$$\alpha_{12}(s) = 0 \quad \text{and} \quad \alpha_{10}(s) = 0 \tag{8.7.6}$$

and consequently $\alpha_{10}(h) = 0$. We have used the latter result in the previous section when we substituted $O(1/p)$ for the correction term $O(1/p^{1/2})$ in formula (8.6.8) for the eigenvalues. Equation (8.3.10) is again nonhomogeneous when $l = 0$. By solving (8.3.10) with the homogeneous conditions (8.7.4 and 5), we find a non-zero function $f_{10}(s)$ for determining $\beta_{10}(s)$ and $\alpha_{11}(s)$. Thus the polynomial $\beta_1(s, \nu)$ contains only even powers of ν, while $\alpha_1(s, \nu)$ contains only odd powers.

It can be shown that in the general case, the polynomials $\beta_{2j-1}(s, \nu)$ and $\alpha_{2j}(s, \nu)$, $j = 1, 2, 3, \ldots$ contain only even powers, while the polynomials $\beta_{2j}(s, \nu)$ and $\alpha_{2j-1}(s, \nu)$ contains only odd powers. Therefore equation (8.5.23) for the eigenvalues will contain only even powers. Solving this equation by the method of iterations, we get an expansion for the eigenvalues $\omega_{p,q}$ in negative powers of the integer $p \gg 1$. The number of terms in this expansion is determined by the number of polynomials that have been constructed[4]. After the eigenvalues have been found, approximate expressions can be constructed for the eigenfunctions as well. The residual in the Helmholtz equation, resulting from these approximate eigenfunctions, can be made of the order of any negative power of $\omega_{p,q}$. It can be rigorously proved that the expansion for the eigenvalues that arises in this way will be asymptotic. The method of proof is analogous to that used in Sect. 7.6 to prove the asymptotic nature of the expansions eigenvalues in the whispering gallery case.

Let us now find the conditions under which the general homogeneous equation (8.3.11) has a non-zero solution that satisfies the homogeneous conditions (8.7.4 and 5). The solution of (8.3.11), as we have seen, can be written in the form

---

[4] In the next section a formula will be derived for the natural frequencies of an open resonator up to terms proportional to $1/p$.

$$f_{ml}(s) = F(s)\left\{ A \sin\left[(l+1)\int_0^s F^{-2}(\tau)d\tau\right]\right.$$

$$\left. + B \cos\left[(l+1)\int_0^s F^{-2}(\tau)d\tau\right]\right\} \ .$$

Substituting $f_{ml}(s)$ into (8.7.4,5), and also taking conditions (8.6.6) for $F(s)$ into account, we get a system of equations for determining the constants A and B

$$\mathrm{Re}\frac{l+1}{F(s)}\left\{ A \cos\left[(l+1)\int_0^s F^{-2}(\tau)d\tau\right]\right.$$

$$\left. -B \sin\left[(l+1)\int_0^s F^{-2}(\tau)d\tau\right]\right\}\Bigg|_{s=0,h} = 0 \ ,$$

$$\mathrm{Im}\left\{ A \sin\left[(l+1)\int_0^s F^{-2}(\tau)d\tau\right]\right.$$

$$\left. +B \cos\left[(l+1)\int_0^s F^{-2}(\tau)d\tau\right]\right\}\Bigg|_{s=0,h} = 0 \ .$$

Setting $s = 0$, we find $\mathrm{Re}\{A\} = \mathrm{Im}\{B\} = 0$. Then at $s = h$, we get

$$\mathrm{Re}\left\{ B \sin\left[(l+1)\int_0^h F^{-2}(\tau)d\tau\right]\right\} = 0 \ ,$$

$$\mathrm{Im}\left\{ A \sin\left[(l+1)\int_0^h F^{-2}(\tau)d\tau\right]\right\} = 0 \ .$$

If

$$\int_0^h F^{-2}(\tau)d\tau \neq \frac{P}{l+1}\pi \ , \tag{8.7.7}$$

where P is an integer, then

$$\mathrm{Re}\{B\} = \mathrm{Im}\{A\} = 0 \ .$$

When

$$\int_0^h F^{-2}(\tau)d\tau = \frac{P}{l+1}\pi \tag{8.7.8}$$

$\mathrm{Re}\{A\}$ and $\mathrm{Im}\{B\}$ remain arbitrary.

Obviously with condition (8.7.7) the corresponding nonhomogeneous problem (8.3.14) and (8.7.4,5) is solvable, while with condition (8.7.8), it does not, generally speaking, have a solution.

What we have said implies that the process of constructing the polynomials $\alpha_m(s, \nu)$ and $\beta_m(s, \nu)$ depends on the arithmetical nature of the integral $\int_0^h F^{-2}(\tau)d\tau$ and may continue indefinitely if

$$\int\limits_0^h F^{-2}(\tau)d\tau \neq \frac{P}{Q}\pi \quad , \tag{8.7.9}$$

where P/Q is a fraction in lowest terms.

Otherwise, in determining the leading coefficient $\beta_{Q-2,Q-1}(s)$ of the polynomial $\beta_{Q-2}(s, \nu)$ we have a Sturm-Liouville problem at an eigenvalue[5], and the procedure described above cannot, in general, be used to construct this polynomial. Recall that if condition (8.7.9) is not satisfied, then the eigenvalues are multiply degenerate in the first approximation.

8.8 Natural Frequencies of an Open Resonator (Inhomogeneous Filling, Higher Approximations)

As an example of the application of (8.6.26), let us determine the natural frequencies $\omega_{p,q}$ in the special case where the ray S is straight, and the velocity of wave propagation in (s, n) coordinates is described by the formula

$$c(s, n) = c_0 + c_2 n^2 \quad , \tag{8.8.1}$$

where c_0 and c_2 do not depend on s. This problem arises in determining the natural frequencies of an open resonator filled with an inhomogeneous medium. When a laser operates, its active medium loses its homogeneity, and the velocity of wave propagation becomes dependent on the transverse coordinate as in (8.8.1). For eigenfunctions concentrated close to the resonator axis, radiation losses can be neglected, and this leads allow us to use the same technique as for closed regions when constructing the eigenfunctions of an open resonator.

For the velocity given by (8.8.1), the coefficient $K(s)$ in equation (8.6.4) is constant: $K(s) = 2(c_2/c_0) \equiv k_0^2$, and the coefficients $g(0)$ and $g(h)$ in (8.6.13) reduce to $1/r_1$ and $-1/r_2$ respectively. We choose the functions $f_1(s)$ and $f_2(s)$ equal to

$$f_1(s) = -\frac{1}{r_1}\frac{\sin k_0 s}{k_0} + \cos k_0 s \quad ,$$

$$f_2(s) = \frac{1}{r_2}\frac{\sin k_0(s-h)}{k_0} + \cos k_0(s-h) \quad ,$$

assuming that

---

[5] By this we mean that a nontrivial solution of the corresponding homogeneous Sturm-Liouville problem exists.

$$W[f_1, f_2] = \left(k_0^2 - \frac{1}{r_1 r_2}\right)\frac{\sin k_0 h}{k_0} + \left(\frac{1}{r_1} + \frac{1}{r_2}\right)\cos k_0 h \neq 0 \quad .$$

Then

$$\frac{f_1(h)f_2(0)}{f_2(h)f_1(0)} = \cos^2 k_0 h - \left(\frac{1}{r_1} + \frac{1}{r_2}\right)\frac{\cos k_0 h \sin k_0 h}{k_0} + \frac{1}{r_1 r_2}\frac{\sin^2 k_0 h}{k_0^2} \quad ,$$

and the condition to be satisfied for the existence of eigenfunctions concentrated close to the axis takes the form

$$0 < \cos^2 k_0 h - \left(\frac{1}{r_1} + \frac{1}{r_2}\right)\frac{\cos k_0 h \sin k_0 h}{k_0} + \frac{1}{r_1 r_2}\frac{\sin^2 k_0 h}{k_0^2} < 1 \quad , \quad (8.8.2)$$

and the natural frequencies $\omega_{p,q}$ are described by the formula

$$\omega_{p,q} = \frac{c_0}{h}\left[\pi p + \left(q + \frac{1}{2}\right)\left\{\text{sgn}\left(\frac{1}{r_1} - \frac{\cos k_0 h + k_0 r_2 \sin k_0 h}{k_0^{-1} \sin k_0 h - r_2 \cos k_0 h}\right)\right.\right.$$

$$\times \arccos\left[\cos^2 k_0 h - \left(\frac{1}{r_1} + \frac{1}{r_2}\right)\frac{\sin k_0 h \cos k_0 h}{k_0}\right.$$

$$\left.\left.+ \frac{1}{r_1 r_2}\frac{\sin^2 k_0 h}{k_0^2}\right]^{1/2} + \pi N\right\} + O\left(\frac{1}{p}\right)\right] \quad . \quad (8.8.3)$$

In (8.8.2, 3) we may take the limit as $k_0 \to 0$, which is equivalent to the transition to a homogeneous medium. As a result of doing this, we get

$$0 < \left(1 - \frac{h}{r_1}\right)\left(1 - \frac{h}{r_2}\right) < 1 \quad (8.8.4)$$

and

$$\omega_{p,q} = \frac{c_0}{h}\left[\pi p + \left(q + \frac{1}{2}\right)\arccos\sqrt{\left(1 - \frac{h}{r_1}\right)\left(1 - \frac{h}{r_2}\right)} + O\left(\frac{1}{p}\right)\right] \quad (8.8.5)$$

if

$$1 - \frac{h}{r_1} > 0 \quad \text{and} \quad 1 - \frac{h}{r_2} > 0 \quad ,$$

$$\omega_{p,q} = \frac{c_0}{h}\left\{\pi p + \left(q + \frac{1}{2}\right)\right.$$

$$\left.\times \left[\pi - \arccos\sqrt{\left(\frac{h}{r_1} - 1\right)\left(\frac{h}{r_2} - 1\right)}\right] + O\left(\frac{1}{p}\right)\right\} \quad , \quad (8.8.6)$$

if

$$\frac{h}{r_1} - 1 > 0 \quad \text{and} \quad \frac{h}{r_2} - 1 > 0 \quad .$$

Inequality (8.8.4) and formulas (8.8.5,6) are well known in the theory of open resonators. Let us note that in (8.8.2, 3) the quantity k_0 can be replaced by ik_0, and then all trigonometric functions are replaced by hyperbolic functions of the argument $k_0 h$. Such a substitution means that we have gone over to the case $c_2 < 0$ where the inhomogeneity of the medium has a defocusing effect.

For a homogeneous medium, it is comparatively easy in (8.8.5) to calculate the next term A_{-1}/p, proportional to p^{-1}.

We will assume that the mirrors of a symmetric resonator are described by the equations

$$s = \frac{1}{2r_1} n^2 + a_4 n^4 + \dots ,$$

$$s = h - \frac{1}{2r_2} n^2 + b_4 n^4 + \dots$$

and $r_1 \neq h$, $r_2 \neq h$ (the resonator is not confocal).

Carrying out calculations by the procedure outlined above, we arrive at the following formula for the coefficient A_{-1}:

$$
\begin{aligned}
A_{-1} = \; & \frac{1}{16\pi} \left[\left(q + \frac{1}{2} \right)^2 + \frac{9}{4} \right] \left(\frac{1}{r_1} + \frac{1}{r_2} \right) \\
& - \frac{3}{16\pi} \left[\left(q + \frac{1}{2} \right)^2 + \frac{1}{4} \right] \frac{h}{r_1 + r_2 - h} \left[\left(\frac{1}{r_1^3} - 8a_4 \right) r_1^2 \frac{h - r_2}{h - r_1} \right. \\
& \left. + \left(\frac{1}{r_2^3} + 8b_4 \right) r_2^2 \frac{h - r_1}{h - r_2} \right] .
\end{aligned}
\tag{8.8.7}
$$

Note that the ray method in the small as further developed in Chap. 10 and the construction of a system of reflected rays in the quadratic approximation has enabled us to obtain only the leading terms proportional to $(q + \frac{1}{2})^2$ in the formula for A_{-1}.

8.9 Notes on the Literature

Solutions of the Helmholtz equation concentrated near the axis of a waveguide that have the form of exponentials multiplied by parabolic cylinder functions whose arguments are an infinite series were constructed by *Buldyrev* [8.2] and *Lazutkin* [8.3]. The first one to run into the insolubility of the eigenvalue problem when $\int_0^h [1/F^2(\tau)] d\tau$ is a rational multiple of π, was *Lazutkin* [8.4], in a study of eigenfunctions of the "bouncing ball" type in a homogeneous medium. Eigenfunctions of this type in an inhomogeneous medium were considered by *Buldyrev* in [8.5]. He also obtained a formula for the eigenfrequencies of an open resonator filled with an inhomogeneous medium. Corrections to the formula for the eigenfrequencies of a nonconfocal resonator were found by *Lazutkin* [8.6]. Related work has been done by *Smith* [8.7].

9. Eigenfunctions Concentrated in the Vicinity of a Closed Geodesic

Let l be a closed geodesic on an $m+1$-dimensional compact orientable Riemannian manifold[1]. Here we will construct a set of eigenfunctions of the Laplacian that are concentrated in the neighborhood of l. It is assumed that l has a stability property totally analogous to that of stability of the diameter of a region (Chap. 5). In the construction of higher approximations it becomes necessary to impose a restriction on the arithmetical nature of certain parameters α_j that are important characteristics of the geodesic l. We have already encountered a similar situation in Chap. 8.

The other problem considered in this chapter is that of constructing the eigenfunctions of a three-dimensional region bounded by a surface S. It is assumed that the eigenfunctions are concentrated in the vicinity of a geodesic l located on S.

9.1 Formulation of the Problem and Derivation of the Parabolic Equation

Let us take up the investigation of those eigenfunctions of the Laplacian operator Δ on a Riemannian manifold which are concentrated in the neighborhood of a closed geodesic l. We can construct a series u which formally satisfies the equation

$$(\Delta + k^2)u = 0 \quad , \tag{9.1.1}$$

$$\Delta u = \frac{1}{\sqrt{g}} \frac{\partial}{\partial y^i} \left(\sqrt{g} g^{ij} \frac{\partial u}{\partial y^j} \right) \quad , \quad g = \frac{1}{\det \|g^{ij}\|} \quad . \tag{9.1.2}$$

The components of the tensor $\|g^{ij}\|$ are related to the metric tensor $\|g_{ij}\|$ by the expression $\|g^{ij}\| = \|g_{ij}\|^{-1}$. Formula (9.1.2) is the classical expression for the Laplacian on a Riemannian manifold. The Laplacian thus defined is invariant to a change of the variables y^i.

---

[1] All constructions of this chapter relate only to the neighborhood of geodesic l. We require compactness so as not to complicate the exposition with a discussion of what is meant by an eigenfunction of the Laplacian (such a dicussion is necessary in the non-compact case). The requirement of orientability of the manifold can, of course, be dispensed with at the cost of making certain formulas more complicated.

To find the asymptotic expansions of the eigenfunctions in the first approximation, the parabolic equation method is very convenient. We shall now proceed to derive this parabolic equation. First, we will describe a coordinate system in which it will be convenient to carry out the subsequent constructions. We will characterize points on l by the arc length s measured along l on either side of some fixed point. It will be convenient for us to assume that $-\infty < s < +\infty$. Let $O = O(s)$ be some point on l. Through $O = O(s)$, we draw all geodesics which are orthogonal to l at $O(s)$. On the submanifold f_s which the weaving of these geodesics forms, we introduce the so-called Riemannian normal coordinates. Let $\vec{e}_1, \vec{e}_2, \ldots, \vec{e}_m$ be some orthonormal system of vectors orthogonal to l at $O(s)$[2]. To specify an arbitrary point M on f_s it is sufficient to give the geodesic passing through $O(s)$ and M (and orthogonal to l) and the arc length $|OM|$ of this geodesic. To do this it is sufficient in turn to specify the m numbers

$$y_j = |OM| \cos(\vec{\eta}_0, \vec{e}_j)$$

where $\vec{\eta}_0$ is a unit vector at $O = O(s)$ tangent to OM and directed from O toward M. From now on for the sake of brevity we will simply write "the point s" instead of "the point $O(s)$." The quantities y_i are called *Riemannian normal coordinates* [9.1,2]. By a parallel displacement of $\vec{e}_j$ along l (as understood in Riemannian space), for $j = 1, 2, \ldots, m$, one can introduce Riemannian coordinates on any submanifold $f_{s'}, s' \in l$. By virtue of the fact that the dot product of any vector pair is not altered by parallel displacement, the vectors $\vec{e}_0, \vec{e}_1, \ldots, \vec{e}_m$ (where $\vec{e}_0$ is a unit vector tangent to l) will always be an orthonormal vector system. Now, when we perform a parallel displacement of the unit vector $\vec{e}_0$ along l, recall that since $\vec{e}_0$ is tangent to l at s, we end up with a unit vector tangent to l at every point of l. This is implied by the fact that l is a geodesic.

If the y_i vary in a small neighborhood of zero (for instance within the sphere $\sqrt{y_1^2 + \ldots + y_m^2} < \varepsilon$ for sufficiently small ε), and $-\infty < s_1 < s < s_2 < \infty$, then $s, y_1, \ldots, y_m$ can be used as a coordinate system in the vicinity of the segment $s_1 < s < s_2$ of l. The limitation on ε above is necessary, as otherwise we could not assume a reciprocal one-to-one correspondence between the points of the Riemannian manifold in the vicinity of the segment $s_1 < s < s_2$ of l and the points of the cylinder

$$s_1 < s < s_2 \quad, \quad \sqrt{y_1^2 + \ldots + y_m^2} < \varepsilon \quad.$$

At points on the geodesic l, the system of coordinates is obviously orthonormal, therefore on l the conditions

$$g_{ij} = \delta_{ij} = \begin{cases} 1 & , \quad i = j \\ 0 & , \quad i \neq j \end{cases} , \quad i, j = 0, 1, \ldots, m \qquad (9.1.3)$$

---

[2] Recall that the given Riemannian space is $m + 1$-dimensional. This discussion is limited to some neighborhood of $O(s)$ on the submanifold $f(s)$. Appropriate refinements will be made later on.

hold (g_{ij} are components of the metric tensor in y^j coordinates, where $y_0 = s$, $y^j = y_j$, $j>0$).

Generally speaking, such a coordinate system will not be defined in an entire neighborhood of l since after a parallel displacement of the vectors $\vec{e}_1, \vec{e}_2, \ldots, \vec{e}_m$ along l, when we finally return to the starting point, the vectors $\vec{e}_1, \ldots, \vec{e}_m$ are transformed into the vectors $\vec{e}_1', \ldots, \vec{e}_m'$, and the two systems of vectors are the same only accidentally. The orientability of the manifold implies that the orientations of the vector systems $\vec{e}_1, \ldots, V\vec{e}_m$ and $\vec{e}_1', \ldots, \vec{e}_m'$ are identical. Let there exist an orthogonal matrix T_{ij} which transforms the basis vectors $\vec{e}_j$ to $\vec{e}_j'$:

$$\vec{e}_i' = \sum_{j=1}^{m} T_{ij}\vec{e}_j \quad , \quad \vec{e}_j = \sum_{i=1}^{m} T_{ij}\vec{e}_i' \quad .$$

The identical orientations imply that $\det(T_{ij}) = 1$. Let $\vec{\mu}$ be the arbitrary vector

$$\vec{\mu} = \sum \mu_i \vec{e}_i = \sum \mu_i' \vec{e}_i' \quad .$$

We obviously have the following relationship between the components of $\vec{\mu}$ in the coordinate systems $\{\vec{e}_i\}$ and $\{\vec{e}_i'\}$:

$$\begin{pmatrix} \mu_1' \\ \vdots \\ \mu_m' \end{pmatrix} = \begin{pmatrix} T_{11} & \cdots & T_{1m} \\ \vdots & \vdots & \vdots \\ T_{m1} & \cdots & T_{mm} \end{pmatrix} \begin{pmatrix} \mu_1 \\ \vdots \\ \mu_m \end{pmatrix} = T \begin{pmatrix} \mu_1 \\ \vdots \\ \mu_m \end{pmatrix} \quad , \quad T = (T_{ij}),$$

which will play an important part in the development.

In the two-dimensional case $m = 1$, and identical orientations of the orthonormalized systems $\{\vec{e}_i\}$ and $\{\vec{e}_i'\}$ simply means that $\vec{e}_i = \vec{e}_i'$.

Let us derive some important properties of the coordinates

$$(s, y_1, \ldots, y_m) \equiv (y^0, y^1, \ldots, y^m) \quad .$$

The set of points $\{s, y_1\gamma, \ldots, y_m\gamma; \; 0{\leq}\gamma{\leq}1\}$ is a segment of a geodesic orthogonal to l at s, and therefore the equations [9.1]

$$\frac{d^2\hat{y}^j}{d\gamma^2} = -\Gamma_{ab}^j \frac{d\hat{y}^a}{d\gamma} \frac{d\hat{y}^b}{d\gamma} \quad ,$$

$$\Gamma_{ab}^j = \frac{1}{2}g^{jc}\left(\frac{\partial g_{ac}}{\partial y^b} + \frac{\partial g_{bc}}{\partial y^a} - \frac{\partial g_{ab}}{\partial y^c}\right) \quad ,$$

$$(\hat{y}^0, \hat{y}^1, \ldots, \hat{y}^m) = (0, y_1\gamma, \ldots, y_m\gamma) \quad , \quad 0{\leq}\gamma{\leq}1 \quad ,$$

are satisfied, where Γ_{ab}^j are the Christoffel symbols of the second kind; in other words

$$\sum_{a,b=1}^{m} \Gamma_{ab}^j y_a y_b = 0 \quad .$$

Differentiating this equation with respect to y_a and y_b, we get

$$\Gamma_{ab}^j|_l = 0 \quad , \quad a, b = 1, 2, \ldots, m \quad , \quad j = 0, 1, 2, \ldots, m \tag{9.1.4}$$

216

on l. A parallel displacement of the unit vector $\vec{e}_j = (0, \ldots, 1, \ldots, 0)$ along l will always yield a vector with the same components. The classical formula for the infinitesimal changes δe_j^κ in the components of $\vec{e}_j = \{e_j^\kappa\} = (0, \ldots, 1, \ldots, 0)$ due to parallel displacement along an infinitesimal vector $\{\delta y^a\}$ gives

$$\delta e_j^\kappa = -\Gamma_{ab}^\kappa \delta y^a e_j^b \quad .$$

Considering that $e_j^b = 1$ when $j = b$, $e_j^b = 0$ when $j \neq b$, $\delta e_j^\kappa = 0$, and assuming that the infinitesimal vector δy^a is parallel to a tangent to l, we get in addition to (9.1.4)

$$\Gamma_{0j}^\kappa|_l = 0 \quad , \quad \kappa, j = 0, 1, \ldots, m \quad . \tag{9.1.5}$$

Using the known formula [9.2]

$$\frac{\partial g_{ik}}{\partial y^j} = g_{ka}\Gamma_{ij}^a + g_{ia}\Gamma_{kj}^a$$

and (9.1.3–5), we get

$$\left.\frac{\partial g_{ik}}{\partial y^j}\right|_l = 0 \quad , \quad i, j, k = 0, 1, \ldots, m \quad ; \quad \frac{\partial}{\partial y^0} = \frac{\partial}{\partial s} \quad . \tag{9.1.6}$$

The classical formula for the components of the curvature tensor

$$R_{ijkr} = \frac{1}{2}\left(\frac{\partial^2 g_{ri}}{\partial y^j \partial y^k} - \frac{\partial^2 g_{ik}}{\partial y^j \partial y^r} - \frac{\partial^2 g_{rj}}{\partial y^i \partial y^k} + \frac{\partial^2 g_{jk}}{\partial y^i \partial y^r}\right)$$
$$+ g_{ab}(\Gamma_{ir}^b\Gamma_{jk}^a - \Gamma_{jr}^b\Gamma_{ik}^a)$$

gives

$$R_{0i0j} = -\frac{1}{2}\left.\frac{\partial^2 g_{00}}{\partial y_i \partial y_j}\right|_l \quad ,$$

i.e.,

$$g_{00} = 1 - \sum_{i,j=1}^m K_{ij}(s)y_i y_j + O\left[\left(\sum_{j=1}^m y_j^2\right)^{3/2}\right] \quad ,$$
$$K_{ij}(s) = K_{ji}(s) = R_{0i0j}|_l \quad . \tag{9.1.7}$$

We now easily get the desired parabolic equation for the problem. Let us write the eigenfunction equation $(\Delta + k^2)u = 0$ in $\{s, y_1, \ldots, y_m\}$ coordinates:

$$\Delta u + k^2 u \equiv \frac{1}{\sqrt{g}}\frac{\partial}{\partial y^i}\left(\sqrt{g}g^{ij}\frac{\partial u}{\partial y^j}\right) + k^2 u = 0 \quad , \tag{9.1.8}$$

$$(s, y_1, \ldots, y_m) = (y^0, \ldots, y^m) \quad , \quad \|g^{ij}\| = \|g_{ij}\|^{-1} \quad , \quad g = \det\|g_{ij}\| \quad .$$

We seek a solution of this equation in the form

$$u = e^{iks}U(s, y_1, \ldots, y_m, k) \quad , \tag{9.1.9}$$

where U is a new unknown function – the *attenuation function* (in Fock's terminology).

In deriving the parabolic equation for U we will assume that k is a large quantity, and

$$U = O(1) \quad , \quad \frac{\partial^{r_1+r_2+r_3} U}{\partial s^{r_1} \partial y_i^{r_2} \partial y_j^{r_3}} = O\left(k^{(r_2+r_3)/2}\right) \quad ,$$

$$|y_j| = O(k^{-1/2}) \quad . \tag{9.1.10}$$

Relations (9.1.3 and 6) will play an important part in what follows. From these expressions we readily find

$$\sqrt{g} = \sqrt{1 + O(|y|^2)} = 1 + O(|y|^2) \quad ,$$

$$\frac{1}{\sqrt{g}} = 1 + O(|y|^2) \quad , \quad |y|^2 = \sum_{j=1}^{m} y_j^2 \quad ,$$

$$\|g^{ij}\| = \|g_{ij}\|^{-1} = \|I + g_{ij} - I\|^{-1}$$
$$= I - (\|g_{ij}\| - I) + (\|g_{ij}\| - I)^2 + \dots \quad ,$$

where I is the unit matrix.

From the latter expression for $\|g^{ij}\|$ and formulas (9.1.6 and 7) we get

$$\|g^{ij}\| = I + O(|y|^2) \quad , \quad g^{00} = 1 + \sum_{i,j=1}^{m} K_{ij}(s) y_i y_j + O(|y|^3) \quad . \tag{9.1.11}$$

Substituting (9.1.9) into (9.1.8), taking (9.1.10 and 11) into account, and disregarding terms of order $O(\sqrt{k})$ and lower, we get the following equation for U:

$$2ik \frac{\partial U}{\partial s} + \sum_{j=1}^{m} \frac{\partial^2 U}{\partial y_j^2} - k^2 \sum_{j,h=1}^{m} K_{jh}(s) y_j y_h U = 0 \tag{9.1.12}$$

[it can be readily seen that each term in equation (9.1.12) is of order $O(k)$]. Let us divide both sides of equation (9.1.12) by k and introduce the new variables

$$\mu_j = \sqrt{k} y_j \quad .$$

Equation (9.1.12) takes the form

$$\mathcal{L} U = 2i \frac{\partial U}{\partial s} + \sum_{j=1}^{m} \frac{\partial^2 U}{\partial \mu_j^2} - \sum_{j,h=1}^{m} K_{jh} \mu_j \mu_h U = 0 \quad . \tag{9.1.13}$$

The solutions of equation (9.1.13) in which we are interested must "tend to zero at infinity." By virtue of the fact that the coordinate system $(s, y_1, \dots, y_m)$ is regular only for small y_j, the conditions at infinity cannot be formulated directly in these coordinates. We will have to carry out some auxiliary constructions.

If we associate with each point s of the geodesic l an m-dimensional hyperplane Φ_s that is normal to l and passes through s, we get a geometric object Ξ known as a normal fibering. To specify an element of Ξ it is sufficient to give the point $s \in l$ and a vector $\vec{\zeta} \in \Phi_s$ (i.e. Ξ is the set of pairs $(s, \vec{\zeta})$, $s \in l$, $\vec{\zeta} \in \Phi_s$).

On the fibering Ξ, s varies over a segment of length L, where L is the length of l. It will prove convenient to assume that s varies from $-\infty$ to $+\infty$. The corresponding set of pairs $(s, \vec{\zeta})$ $(-\infty < s < +\infty$, $\vec{\zeta} \in \Phi_s)$ forms another fibering Ξ'. By identifying those points in this fibering for which $s' - s''$ is a multiple of L, we arrive at the fibering Ξ again. (Ξ is a bundle space with m-dimensional vector *fibers* (see *Bishop* and *Crittenden* [9.3]). The geodesic l is what we might call the *base* of the fibering, and each fiber is an m-dimensional Euclidean space Φ_s. The fibering Ξ' is a *universal covering* (see [9.4]) for Ξ.

We associate a point $M(s, y_1, \ldots, y_m)$ lying on the Riemannian manifold in the neighborhood of l with a point $N \in \Xi$ according to the following rule: the introduction of the coordinates $s, y_1, \ldots, y_m$ assumes that the vectors $\vec{e}_1, \ldots, \vec{e}_m$ have already been determined (see the beginning of this section); we take these vectors as a basis on the hyperplane Φ_s that is normal to l at s, and we associate with $M(s, y_1, \ldots, y_m)$ the point $N(s, \vec{\mu}) \in \Xi$, where $\vec{\mu} = \sum_{i=1}^m \mu_i \vec{e}_i$, $\mu_i = \sqrt{k} y_i$.

In place of equation (9.1.13) (which is satisfied on the Riemannian manifold in the vicinity of l) we will consider an equation on the fibering Ξ':

$$2i\frac{DU}{ds} + \Delta U - (K(s)\vec{\mu}, \vec{\mu})U = 0 \quad , \tag{9.1.14}$$

where D/ds is the covariant derivative of the function U, that should be understood as the limit

$$\frac{DU(s, \vec{\mu})}{ds} = \lim_{\Delta s \to 0} \frac{U(s + \Delta s, \vec{\mu}_{\Delta s}) - U(s, \vec{\mu})}{\Delta s} \quad . \tag{9.1.15}$$

Here $\vec{\mu}_{\Delta s}$ is the result of a parallel displacement of the vector $\vec{\mu}$ along l from the point $s \in l$ to the point $(s + \Delta s) \in l$, Δ is the Laplacian on the Euclidean hyperplane Φ_s, $(K(s)\vec{\mu}, \vec{\mu})$ is a quadratic form on Φ_s in the coordinate system $(s, \mu_1, \ldots, \mu_m)$ that takes the form $\sum_{j,h=1}^m K_{jh}\mu_j\mu_h$.

The problem consists in finding on Ξ' the solutions of equation (9.1.14) which satisfy the following conditions:

$$\max_{|\mu|=R} |U(s, \vec{\mu})| \xrightarrow[R \to \infty]{} 0, \quad U \neq 0 \quad , \tag{9.1.16}$$

$$U(s + L, \vec{\mu}) = e^{i\kappa}U(s, \vec{\mu}) \quad . \tag{9.1.17}$$

Here L is the length of l.

Having solved the problem (9.1.14–17), we get the desired approximation to the attenuation function U [see (9.1.9) *et seq.*] if we set $\mu_j = \sqrt{k} y_j$ in the solution of (9.1.14–17), where y_j are the coordinates introduced at the beginning of the section that characterize points on the submanifolds f_s. The coordinates y_j on f_s vary in a neighborhood of zero.

By introducing Ξ' and examining equation (9.1.14) with conditions (9.1.16, 17) thereon, we get a precisely formulated boundary value problem, whereas if we had carried out our study on the Riemannian manifold without introducing Ξ', it would have been difficult to give a clear meaning to the condition that U decays at infinity, which appears here simply as (9.1.16).

Let us note that fibering could also have been introduced in other places; Chaps. 8 and 10, for example. However, in the presence of reflecting boundaries the description of fiberings is fairly complicated and their introduction would have made the exposition more awkward.

By analogy with the theory of ordinary linear differential equations we will call the solutions of (9.1.14, 16, 17) *Floquet solutions*, and the indices κ − *Floquet indices*. As we shall see later on, by using the equations of the ray method in the small this problem can, in a certain sense, be solved, or more accurately, reduced to the solution of ordinary differential equations. To make such a solution possible, the geodesic l must satisfy the stability condition in the first approximation (Chaps. 5 and 8).

It turns out that there is a denumerable set of Floquet indices κ to which there correspond solutions of (9.1.14, 16, 17). Corresponding to each κ is a finite set of Floquet solutions.

If the solution of problem (9.1.14, 16, 17) is constructed, it is not difficult to get the asymptotic expansion of the eigenvalues by requiring that the eigenfunction [see (9.1.9)] be L-periodic with respect to s: we can accomplish this by making u single-valued (in the first approximation).

Formulas (9.1.9, 17) and the periodicity condition give the following expressions[3] for the eigenvalue k in the first approximation:

$$k \approx k_p = \frac{1}{L}(2\pi p - \kappa) \quad , \tag{9.1.18}$$

where $p \gg 1$ is an integer.

9.2 The Jacobi Equation for the Geodesic l

Geodesics play the part of rays in the problem we are considering. It will be very important for use to study the differential equations of rays (i.e. geodesics) close to l, i.e. the Euler equations for the ray method in the small.

Let $y_j = y_j(s)$ be the equation of a geodesic near l. In the fundamental variational equation for geodesics

$$\delta \int \sqrt{\sum_{i,j=0}^{m} g_{ij} \frac{dy^i}{ds} \frac{dy^j}{ds}} \, ds = 0 \quad ,$$

$$(y^0, y^1, \ldots, y^m) = (s, y_1, \ldots, y_m) \tag{9.2.1}$$

---

[3] Here and in the following we will call k an eigenvalue, although it would be more correct to call k^2 the eigenvalue.

we will assume that y_i and (dy_j/ds) are small of first order. Using relations (9.1.3, 6, 7) and dropping all terms smaller than second order, we obtain:

$$\int \sqrt{\sum_{i,j=0}^{m} g_{ij} \frac{dy^i}{ds} \frac{dy^j}{ds}} \, ds$$

$$\approx \int \sqrt{1 - \sum_{i,j=1}^{m} K_{ij}(s) y_i y_j + \sum_{j=1}^{m} \left(\frac{dy^j}{ds}\right)^2} \, ds$$

$$\approx \int \left(1 + \frac{1}{2} \sum_{j=1}^{m} \left(\frac{dy^j}{ds}\right)^2 - \frac{1}{2} \sum_{i,j=1}^{m} K_{ij}(s) y_i y_j \right) ds \quad . \tag{9.2.2}$$

The Euler equations of the functional (9.2.2) take the form

$$\frac{d^2 y_j}{ds^2} + \sum_{i=1}^{m} K_{ij}(s) y_i = 0 \quad , \quad j = 1, 2, \ldots, m \quad . \tag{9.2.3}$$

The system of equations (9.2.3) can be written in a convenient covariant form. Let $\vec{e}_1, \ldots, \vec{e}_m \in \Phi_s$, be the basis vectors introduced in Sect. 9.1. Let us set

$$\vec{y} = \sum_{j=1}^{m} y_j(s) \vec{e}_j \quad . \tag{9.2.4}$$

The vector $\vec{y}$ satisfies the equation

$$\frac{D^2 \vec{y}}{ds^2} + K(s) \vec{y} = 0 \quad , \quad -\infty < s < +\infty \quad , \tag{9.2.5}$$

where D/ds is covariant differentiation:

$$\frac{D\vec{y}(s)}{ds} = \lim_{\Delta s \to 0} \frac{\vec{y}_{s+\Delta s}(s + \Delta s) - \vec{y}(s)}{\Delta s} \quad ,$$

$$\frac{D^2 \vec{y}(s)}{ds^2} = \frac{D}{ds}\left(\frac{D\vec{y}}{ds}\right) \quad .$$

The vector

$$\vec{y}_{s+\Delta s}(s + \Delta s) = \sum_{j=1}^{m} y_j(s + \Delta s) \vec{e}_j(s)$$

is the result of parallel displacement of the vector $\vec{y}(s+\Delta s)$ from the point $s+\Delta s$ to s, $K(s)$ is a linear operator acting from Φ_s to Φ_s; in the basis $\vec{e}_1, \ldots, \vec{e}_m$

$$K(s)\vec{y} = K \sum_{j=1}^{m} y_j(s) \vec{e}_j = \sum_{j,r=1}^{m} K_{rj}(s) y_j \vec{e}_r \quad ,$$

$$K_{rj} = K_{jr} \quad . \tag{9.2.6}$$

By virtue of the fact that $\|K_{rj}\|$ is real and symmetric [cf. (9.1.7)], K is a self-adjoint operator. Equation (9.2.5) for a ray in the first approximation $y = y(s)$

is the well known Jacobi equation for a geodesic [9.5]. (The Jacobi equation is defined as the Euler equation for the second variation of a functional, and (9.2.2), for which the Euler equation is (9.2.5), differs from the second variation of the functional $\int d\sigma = \int \sqrt{g_{ij} dy^i dy^j}$ only by a constant term.)

The geodesic l will be called *stable in the first approximation* if equation (9.2.5) does not have unbounded solutions when $s \to \pm\infty$. The solution given later on (Sect. 9.3) for problem (9.1.14, 16, 17) makes sense only for a stable geodesic.

We note that the definition of stability for l agrees completely with the concept of stability of rays in the first approximation, which played a fundamental role in Chaps. 5, 7 and 8.

The system of equations (9.2.3) [or what amounts to the same thing, the vector equation (9.2.5)] can be conveniently written in Hamiltonian form:

$$\frac{d\vec{y}}{ds} = \vec{p} \quad , \quad \frac{d\vec{p}}{ds} = -K\vec{y} \quad . \tag{9.2.7}$$

Let $\{\vec{X}(s)\}$ be the set of pairs of vectors $\vec{X} = (\vec{x}_1(s), \vec{x}_2(s))$ where $\vec{x}_j(s)$ lie in the hyperplane Φ_s which is normal to l at s.

Let us introduce the operators J and H:

$$J\vec{X} = J\begin{pmatrix}\vec{x}_1 \\ \vec{x}_2\end{pmatrix} = \begin{pmatrix}\vec{x}_2 \\ -\vec{x}_1\end{pmatrix} \quad , \quad H\begin{pmatrix}\vec{x}_1 \\ \vec{x}_2\end{pmatrix} = \begin{pmatrix}K\vec{x}_1 \\ \vec{x}_2\end{pmatrix} \quad .$$

The system of equations (9.2.7) is equivalent to the equation

$$\frac{D\vec{Z}}{ds} = \begin{pmatrix}D\vec{z}_1/ds \\ D\vec{z}_2/ds\end{pmatrix} = JH\vec{Z} \quad \left(\vec{Z} = \begin{pmatrix}\vec{y} \\ \vec{p}\end{pmatrix} \in \{\vec{X}(s)\}\right) \quad . \tag{9.2.8}$$

By introducing the basis vectors $\vec{e}_1, \ldots, \vec{e}_m$, the vector equation (9.2.8) can be written in coordinate form:

$$\begin{pmatrix}dz_1/ds \\ \vdots \\ dz_{2m}/ds\end{pmatrix} = J_{2m}H_{2m}\begin{pmatrix}z_1 \\ \vdots \\ z_{2m}\end{pmatrix} \quad , \tag{9.2.9}$$

$$z_1, \ldots, z_m = y_1, \ldots, y_m \quad , \quad z_{m+1}, \ldots, z_{2m} = p_1, \ldots, p_m \quad ,$$

$$\vec{p} = \sum_{j=1}^{m} p_j \vec{e}_j \quad , \quad \vec{y} = \sum_{j=1}^{m} y_j \vec{e}_j \quad .$$

Here J_{2m} and H_{2m} are the matrices:

$$H_{2m} = \begin{Vmatrix} \|K_{ij}\| & 0 \\ 0 & I \end{Vmatrix} \quad , \quad J_{2m} = \begin{Vmatrix} 0 & I \\ -I & 0 \end{Vmatrix} \quad ,$$

where I (0) are the unit (zero) $m \times m$ matrices respectively. Equation (9.2.8), like (9.2.5) will be considered over the entire axis $-\infty < s < +\infty$.

It can be easily seen that for a stable geodesic (and only in this case) any solution of equation (9.2.8) is uniformly bounded on the entire s-axis. Let us derive necessary and sufficient criteria for the stability of l that will be convenient for the following development. We introduce in the usual way the scalar product of vectors $\vec{X}^{(1)}$ and $\vec{X}^{(2)}$ from $\{\vec{X}(s)\}$:

$$(\vec{X}^{(1)}, \vec{X}^{(2)}) = \sum_{j=1}^{2m} x_j^{(1)}\overline{x_j^{(2)}} \quad , \quad (\vec{X}^{(1)}, \vec{X}^{(2)}) = \overline{(\vec{X}^{(2)}, \vec{X}^{(1)})} \quad .$$

With respect to this scalar product, the operators J and H satisfy the relations

$$J^* = -J \quad , \quad H^* = H$$

(the asterisk denotes the Hermitian conjugate), i.e.

$$(J\vec{X}, \vec{Z}) = -(\vec{X}, J\vec{Z}) \quad , \quad (H\vec{X}, \vec{Z}) = (\vec{X}, H\vec{Z}) \quad .$$

Moreover, $J^2\vec{X} = -\vec{X}$.

Let Π be a $2m \times 2m$ matrix in which the columns are linearly independent vectors-solutions of (9.2.8):

$$\frac{d\Pi}{ds} = JH\Pi \quad . \tag{9.2.10}$$

Such matrices are called *fundamental matrices* of the system of equations (9.2.10). Let us derive two identities involving Π that we will need in what follows.

1) $\det \Pi = \text{const}$ $\qquad\qquad\qquad\qquad\qquad\qquad$ (9.2.11)

(i.e. $\det \Pi$ is not dependent on s).

Identity (9.2.11) is easily deduced from the Liouville equation [9.6]

$$\det \Pi(s) = \det \Pi(s_0)\exp \int_{s_0}^{s} \text{Tr}(J_{2m}H_{2m})ds \quad ,$$

where $\text{Tr}(J_{2m}H_{2m})$ is the trace of the matrix $J_{2m}H_{2m}$, which is obviously equal to zero. Identity (9.2.11) expresses the classical theorem of conservation of volume in phase space for Hamiltonian systems.

2) Let us prove that

$$\Pi^* J \Pi = \text{const} \quad . \tag{9.2.12}$$

To do this, we calculate $(d/ds)\Pi^* J\Pi$:

$$\frac{d}{ds}\Pi^* J\Pi = \frac{d\Pi^*}{ds}J\Pi + \Pi^* J\frac{d\Pi}{ds} = \left(\frac{d\Pi}{ds}\right)^* J\Pi + \Pi^* J\frac{d\Pi}{ds}$$

$$= (JH\Pi)^* J\Pi + \Pi^* J\frac{d\Pi}{ds} = \Pi^* H^* J^* J\Pi + \Pi^* J^2 H\Pi$$

$$= \Pi^* H^* \Pi - \Pi^* H\Pi = \Pi^*(H^* - H)\Pi = 0 \quad .$$

Formulas (9.2.11 and 12) play an important part in what follows.

The set of solutions of equation (9.2.8) forms a $2m$-dimensional linear space. By virtue of the fact that the geodesic l is closed, by substituting $s + L$ for s (L

is the length of l) we get a linear one-to-one mapping of this space onto itself. The operator E that performs this mapping we will call a *monodromy operator* (the accepted term in the theory of linear Hamiltonian systems with periodic coefficients).

We now introduce the important concept of a solution of Floquet type. A solution $\vec{Z}(s)$ of (9.2.8) satisfying

$$\vec{Z}(s+L) = \lambda \vec{Z}(s) \quad , \tag{9.2.13}$$

is called a *Floquet solution,* and the number λ is a *multiplier* of equation (9.2.8). By virtue of (9.2.13) and the fact that

$$\vec{Z}(s+L) = E\vec{Z}(s) \quad ,$$

multipliers (and only multipliers) are eigenvalues of the monodromy operator E. We have the following

Theorem. *For stability of the geodesic l it is necessary and sufficient that the absolute values of all eigenvalues of the monodromy operator be equal to 1, and that they have no associated vectors*[4].

Proof: Necessity. Let $\vec{Z}$ be a Floquet solution, then the sequence

$$E^n \vec{Z}(s) = \lambda^n \vec{Z}(s) = \vec{Z}(s+nL) \quad , \quad n = 0, \pm 1, \pm 2, \ldots \quad ,$$

is bounded as $n \to \pm\infty$ when and only when $|\lambda| = 1$. If $\vec{Z}_0$ is a Floquet solution, and $\vec{Z}_1$ is the first associated vector, then it can be readily shown that

$$\vec{Z}_1(s+nL) = \lambda^n \vec{Z}_1(s) + n\lambda^{n-1} \vec{Z}_0(s) \quad .$$

If $|\lambda| = 1$ and $n \to \pm\infty$, then $\|\vec{Z}_1(s+nL)\| \to \infty$, which contradicts stability. Sufficiency follows from the fact that if the conditions of the theorem are satisfied, then the Floquet solutions are bounded on the entire axis, and form a fundamental systems of solutions. Any solution of equation (9.2.8) is bounded as a linear combination of bounded solutions.

In each basis (specification of the basis in the solution space is the same as specifying a fundamental system of solutions) the operator E is represented by a $2m \times 2m$ matrix. The roots of the characteristic polynomial

$$\mathcal{J}(\lambda) = \det(E - \lambda I) \quad ,$$

where I is the $(2m \times 2m)$ identity matrix, will be eigenvalues of the monodromy operator E, i.e. multipliers of equation (9.2.8).

Let us prove two simple properties of the polynomial $\mathcal{J}(\lambda)$.

1) *All coefficients of $\mathcal{J}(\lambda)$ are real.*
In fact, by virtue of the fact that the coefficients of the Jacobi equation (9.2.8) are real, the monodromy operator converts real solutions into real solutions. Hence

---

[4] The sequence of vectors $\vec{Z}_0, \vec{Z}_1, \ldots, \vec{Z}_\eta$ forms a series corresponding to the eigenvalue λ if $\vec{Z}_0 \neq 0$, $E\vec{Z}_0 = \lambda\vec{Z}_0$, $E\vec{Z}_1 = \lambda\vec{Z}_1 + \vec{Z}_0$, $\ldots$, $E\vec{Z}_\eta = \lambda\vec{Z}_\eta + \vec{Z}_{\eta-1}$. The vectors $\vec{Z}_j$, $j > 0$, are called *associated* (or *root*) vectors (see [9.6]).

it follows that in a real basis (i.e. when a real fundamental system of solutions is chosen), the matrix of the monodromy operator will be real. On the other hand, the polynomial $\mathcal{J}(\lambda)$ does not depend on the choice of basis (by changing the basis we will replace the matrix of E by a similar matrix), therefore $\mathcal{J}(\lambda)$ will be a polynomial with real coefficients in any basis.

2) *The characteristic polynomial is palindromic, i.e.*

$$\det(E - \lambda I) = \mathcal{J}(\lambda) = \lambda^{2m} \mathcal{J}(1/\lambda) \quad . \tag{9.2.14}$$

Let $\Pi(s)$ (respectively $\Pi(s + L)$) be a fundamental matrix with columns corresponding to selection of the orthonormalized system $\vec{e}_j(s)$ (and respectively $\vec{e}_i' = \vec{e}_i(s + L)$) as the basis on the hyperplane Φ_s. After simple calculations we get

$$\Pi(s + L) = \begin{pmatrix} T & 0 \\ 0 & T \end{pmatrix} \Pi(s)E \quad ,$$

where 0 is a zero $(m \times m)$-matrix, T is the orthogonal $(m \times m)$-matrix that accomplishes conversion from $\vec{e}_j(s)$ to $\vec{e}_i' = \vec{e}_i(s + L)$ (Sect. 9.1), E is the matrix representation of the monodromy operator, and the columns of $\Pi(s)$ serve as the basis.

Taking determinants and using the fact that

$$(\det T)^2 = 1 \quad , \quad \det \Pi(s) = \text{const} = \det \Pi(s + L) \neq 0 \quad ,$$

we get

$$\det E = 1 \quad . \tag{9.2.15}$$

Identity (9.2.12) and the formula relating $\Pi(s + L)$ to $\Pi(s)$ yield

$$\Pi^*(s + L)J\Pi(s + L) = \left[\begin{pmatrix} T & 0 \\ 0 & T \end{pmatrix} \Pi(s)E\right]^* \cdot J \begin{pmatrix} T & 0 \\ 0 & T \end{pmatrix} \Pi(s)E$$
$$= E^* \Pi^*(s)J\Pi(s)E = \Pi^*(s)J\Pi(s) \quad ,$$

whence

$$E = D^{-1}E^{*-1}D \quad , \quad D = \Pi^*(s)J\Pi(s) \quad .$$

Consequently

$$E - \lambda I = D^{-1}(E^{*-1} - \lambda I)D = \lambda D^{-1}\left(\frac{1}{\lambda}I - E^*\right)E^{*-1}D \quad ,$$

where I is the $2m \times 2m$ identity matrix. Taking determinants and using the fact that $\mathcal{J}(\lambda)$ is real, we get (9.2.14).

The equation $\mathcal{J}(\lambda) = 0$ can be easily reduced to an equation of degree m: in fact it is well known that in the case where $\mathcal{J}(\lambda)$ is a palindromic polynomial of degree $2m$, it can be represented in the form $\mathcal{J}(\lambda) = \lambda^m \mathcal{J}_1(\lambda_1)$, where $\mathcal{J}_1$ is a polynomial of degree m, and $\lambda_1 = (\lambda + \lambda^{-1})/2$.

All the following constructions (which make sense only in the case of a geodesic that is stable in the first approximation) can be carried out conveniently if the Floquet solutions are chosen in a special form; specifically, as indicated in the following

Theorem. *A fundamental system of solutions of the vector differential equation (9.2.8) can be constructed from Floquet solutions $\vec{Z}_j$, $j = 1, 2, \ldots, 2m$ in such a way that the formula for the matrix formed from the elements $(J\vec{Z}_j, \vec{Z}_r)$:*

$$\|(J\vec{Z}_j, \vec{Z}_r)\| = i \left\| \begin{matrix} I & 0 \\ 0 & -I \end{matrix} \right\| \quad , \quad j, r = 1, \ldots, 2m \quad , \tag{9.2.16}$$

will hold, where I is the identity matrix, the matrix element $(J\vec{Z}_j, \vec{Z}_r)$ is the scalar product of the vectors $J\vec{Z}_j$ and $\vec{Z}_r$,

$$\vec{Z}_{j+m} = \vec{Z}_j^* \quad (j = 1, 2, \ldots, m) \tag{9.2.17}$$

($\vec{Z}_j^*$ is the vector whose components are the complex conjugates of those of $\vec{Z}_j$).

First let us note that no matter what the solutions $\vec{Z}_j$ and $\vec{Z}_r$ of equation (9.2.8) may be, the components of matrix

$$A = \|(J\vec{Z}_j, \vec{Z}_r)\|$$

do not depend on s. (This follows immediately from (9.2.12) if we assume that the vectors $\vec{Z}_j$ form columns of the matrix Π). Hence it follows that if $\vec{Z}_j$ and $\vec{Z}_k$ are Floquet solutions corresponding to different multipliers λ_j and λ_k, then

$$(J\vec{Z}_j, \vec{Z}_k) = 0 \quad . \tag{9.2.18}$$

In fact,

$$\begin{aligned}(J\vec{Z}_j(s), \vec{Z}_k(s)) &= (J\vec{Z}_j(s + L), \vec{Z}_k(s + L)) \\ &= \lambda_j \overline{\lambda}_k (J\vec{Z}_j, \vec{Z}_k) \quad .\end{aligned}$$

Considering that $|\lambda_k| = 1$ and $\overline{\lambda}_k = 1/\lambda_k$, we have

$$\begin{aligned}(J\vec{Z}_j(s), \vec{Z}_k(s)) &- (J\vec{Z}_j(s + L), \vec{Z}_k(s + L)) \\ &= (1 - \lambda_j \lambda_k^{-1})(J\vec{Z}_j(s), \vec{Z}_k(s)) = 0 \quad ,\end{aligned}$$

which implies (9.2.18) when $\lambda_j \neq \lambda_k$.

Replacing the Floquet solutions that correspond to the same multiplier by their linear combinations, we can arrange it so that the matrix

$$A = \|(J\vec{Z}_j, \vec{Z}_r)\| \quad , \quad j, r = 1, 2, \ldots, 2m \quad ,$$

is diagonal.

The rest of the construction of the fundamental system reduces to multiplying the Floquet solutions by normalizing factors and renumbering the solutions. We shall omit the details of these procedures.

Recalling that the vector $\vec{Z} = \left(\begin{matrix} \vec{y} \\ D\dot{y}/ds \end{matrix} \right) = \left(\begin{matrix} \vec{y} \\ \vec{p} \end{matrix} \right)$, where $\vec{y}$ is a solution of (9.2.5), equations (9.2.16, 17) can be rewritten in the form

$$\left(\frac{D\vec{y}_j}{ds}, \vec{y}_h\right) - \left(\vec{y}_j, \frac{D\vec{y}_h}{ds}\right) = (\vec{p}_j, \vec{y}_h) - (\vec{y}_j, \vec{p}_h) = i\delta_{jh} \quad,$$
$$j, h = 1, 2, \ldots, m \quad, \tag{9.2.19}$$

$$\left(\frac{D\vec{y}_j}{ds}, \vec{y}_h^*\right) - \left(\vec{y}_j, \frac{D\vec{y}_h^*}{ds}\right) = (\vec{p}_j, \vec{y}_h^*) - (\vec{y}_j, \vec{p}_h^*) = 0 \quad,$$
$$j, h = 1, 2, \ldots, m \quad. \tag{9.2.20}$$

The components of the vectors $\vec{y}_h$ and $\vec{y}_h^*$ are complex conjugates. We introduce the matrices Y and P whose columns are made up of the components of the vectors $\vec{y}_1, \ldots, \vec{y}_m$ and $D\vec{y}_1/ds, \ldots, D\vec{y}_m/ds = \vec{p}_1, \ldots, \vec{p}_m$. Equations (9.2.19 and 20) can be rewritten as two matrix relations

$$Y^* P - P^* Y = iI \quad, \tag{9.2.21}$$

$$Y^\mathsf{T} P - P^\mathsf{T} Y = 0 \tag{9.2.22}$$

(the symbol T denotes the transpose).
   Formula (9.2.19) implies that

$$\det Y(s) \neq 0 \quad, \quad s \in l \quad, \tag{9.2.23}$$

over the whole of the geodesic l.
   In fact, we need only prove the linear independence of the vectors $\vec{y}_j(s)$ at each point $s \in l$. Let

$$\sum_{j=1}^{m} a_j \vec{y}_j(s) = 0 \quad. \tag{9.2.24}$$

We multiply (9.2.19) by $a_j \bar{a}_h$ and sum the result over j and h. Considering (9.2.24), we get

$$\sum_{j=1}^{m} |a_j|^2 = 0 \quad,$$

whence

$$a_j = 0 \quad, \quad j = 1, 2, \ldots, m \quad.$$

This proves the linear independence of the $\vec{y}_j(s)$ and consequently, also proves (9.2.23).

9.3 The Zero-Order Approximation

We will seek solutions of the parabolic equation (9.1.14) in the form

$$U = S(s)\exp\left[\frac{i}{2}(\varGamma(s)\vec{\mu}, \vec{\mu})\right] \quad, \tag{9.3.1}$$

where $S(s)$ is a function of s, $(\boldsymbol{\Gamma}(s)\vec{\mu}, \vec{\mu})$ is a quadratic form on the m-dimensional hyperplane $\boldsymbol{\Phi}_s$ normal to l at s.

In the coordinate system $(s, \mu_1, \ldots, \mu_m)$ ($\mu_1, \ldots, \mu_m$ are the components of the vector $\vec{\mu}$ on $\boldsymbol{\Phi}_s$), the quadratic form $(\boldsymbol{\Gamma}\vec{\mu}, \vec{\mu})$ can be written as

$$\sum_{i,j=1}^{m} \Gamma_{ij}(s)\mu_i\mu_j \quad , \tag{9.3.2}$$

where

$$\boldsymbol{\Gamma} = \|\Gamma_{ij}(s)\| \quad , \quad i, j = 1, 2, \ldots, m \quad ,$$

is a symmetric matrix. Substituting expression (9.3.1) into (9.1.14), we get a matrix Riccati equation for $\boldsymbol{\Gamma}$:

$$\boldsymbol{\Gamma}' + \boldsymbol{\Gamma}^2 + \boldsymbol{K} = 0 \quad , \tag{9.3.3}$$

where $\boldsymbol{K} = \|K_{ij}(s)\| = \|R_{0i0j}(s)\|$ is an $m \times m$ matrix [see (9.1.7)]. For $S(s)$ we get the equation

$$2S' + S\operatorname{Tr}(\boldsymbol{\Gamma}) = 0 \tag{9.3.4}$$

[$\operatorname{Tr}(\boldsymbol{\Gamma})$, as usual, denotes the trace of $\boldsymbol{\Gamma}$]. The matrix Riccati equation (just like its scalar analog) reduces to a system of ordinary linear equations. We will look for the solution of (9.3.3) in the form

$$\boldsymbol{\Gamma} = Y'Y^{-1} = \frac{dY}{ds}Y^{-1} \quad , \tag{9.3.5}$$

where Y is a new unknown matrix. Substituting (9.3.5) into (9.3.3), and using the obvious identity

$$Y\frac{dY^{-1}}{ds} + \frac{dY}{ds}Y^{-1} = \frac{d}{ds}YY^{-1} \equiv 0 \quad ,$$

we get the equation

$$\frac{d^2Y}{ds^2} + KY = 0 \quad . \tag{9.3.6}$$

Equation (9.3.6) means that the equations

$$\frac{d^2y_{hj}}{ds^2} + \sum_{a=1}^{m} K_{ha}(s)y_{aj} = 0 \quad , \quad h, j = 1, \ldots, m \quad , \tag{9.3.7}$$

are satisfied, which for fixed j and $h = 1, 2, \ldots, m$ are identical with equations (9.2.5) written in coordinate form if we set

$$\vec{y} = (y_1, \ldots, y_m) = (y_{1j}, y_{2j}, \ldots, y_{mj}) = \vec{y}_j \quad .$$

Thus we get a solution for equation (9.3.3) if we take as our matrix $\boldsymbol{\Gamma}$ the expression (9.3.5), where Y is a matrix whose columns $\vec{y}_j$ are solutions of equation (9.2.5). Let $\vec{y}_j$ be solutions of the Floquet equation (9.2.5) [see formulas (9.2.8, 13, 16–24)].

Let us turn now to equation (9.3.4). Using (9.3.5) it is easily shown that S is equal to

$$S(s) = \exp\left[-\frac{1}{2}\int_0^s \mathrm{Tr}(Y'Y^{-1})ds\right]$$

to within a constant factor. Using the known formula for the derivative of a determinant (see [9.7])

$$\frac{d}{ds}\det Y = \mathrm{Tr}(Y'Y^{-1})\det Y \quad,$$

we get (again up to a constant factor)

$$S(s) = \frac{1}{\sqrt{\det Y}} \quad. \tag{9.3.8}$$

Recall that $\det Y \neq 0$ (9.2.23). Formulas (9.3.1, 5, and 8) now give

$$\begin{aligned}
U_0 &= \frac{1}{\sqrt{\det Y}}\exp\left[\frac{i}{2}(Y'Y^{-1}\vec{\mu},\vec{\mu})\right] \\
&= \frac{\exp\left[\frac{i}{2}(PY^{-1}\vec{\mu},\vec{\mu})\right]}{\sqrt{\det Y}} \quad, \quad P = Y' \quad.
\end{aligned} \tag{9.3.9}$$

We must show that U_0 given by this formula actually is one of the solutions of (9.1.14, 16, and 17). We will first show that formula (9.3.5) defines a symmetric matrix. Let us turn to relation (9.2.22):

$$Y^\mathsf{T}Y' - Y'^\mathsf{T}Y \equiv Y^\mathsf{T}P - P^\mathsf{T}Y = 0 \quad, \quad P = Y' \quad.$$

Multiplying this equality on the left by $(Y^\mathsf{T})^{-1}$ and on the right by Y^{-1} we get

$$PY^{-1} - (Y^\mathsf{T})^{-1}P^\mathsf{T} = 0 \quad,$$
$$\Gamma = PY^{-1} = (PY^{-1})^\mathsf{T} = \Gamma^\mathsf{T} \quad,$$

i.e. the matrix (9.3.5) is actually symmetric.

To prove property (9.1.16) it is sufficient to show that

$$\mathrm{Im}\,\Gamma = \frac{1}{2i}(\Gamma - \Gamma^*) \tag{9.3.10}$$

is a positive definite matrix.

Let us use formula (9.2.21):

$$Y^*P - P^*Y = iI \quad.$$

Multiplying this equality on the left by $(Y^*)^{-1}$ and on the right by Y^{-1} we get

$$PY^{-1} - Y^{*-1}P^* = iY^{*-1}Y^{-1} \quad,$$

or

$$\Gamma - \Gamma^* = 2i\,\mathrm{Im}\,\Gamma = i(YY^*)^{-1} \quad, \tag{9.3.11}$$

which implies that matrix (9.3.10) is positive definite.

It remains for us to show that U_0 of (9.3.9) satisfies (9.1.17). When shifted by L (L is the length of l), the vectors $\vec{y}_j$ acquire the factors $\lambda_j = \exp(i\alpha_j)$. We call the numbers α_j *Floquet indices* (they are determined up to an integer multiple of 2π). Let the basis vectors $\vec{e}_j(s)$, after a complete circuit of the geodesic l, be converted into the basis vectors $\vec{e}'_i(s) = \vec{e}_i(s + L)$, and let the relation between $\vec{e}_j$ and $\vec{e}'_i$ be given by the matrix $T = (T_{ij})$ (see Sect. 9.1). Assuming that the columns of $Y(s)$ (or $Y(s+L)$) are components of the vectors $\vec{y}_j(s)$ (or $\vec{y}_j(s+L)$, respectively) in the orthonormalized basis $\{\vec{e}_i\}$ (or $\{\vec{e}'_i\}$), we get

$$Y(s + L) = TY(s)\Upsilon(\alpha) \quad ,$$
$$P(s + L) = TSV(s)GY(\alpha) \quad , \quad P = Y' \quad , \tag{9.3.12}$$

where

$$\Upsilon(\alpha) = \left\|
\begin{matrix}
e^{i\alpha_1} & & & 0 \\
& e^{i\alpha_2} & & \\
& & \ddots & \\
0 & & & e^{i\alpha_m}
\end{matrix}
\right\| \quad . \tag{9.3.13}$$

The vector $\vec{\mu} = \sum_{i=1}^{m} \mu_i \vec{e}_i$ in the basis $\{\vec{e}'_i\}$ will have components μ'_i, where (see Sect. 9.1)

$$\begin{pmatrix} \mu'_1 \\ \vdots \\ \mu'_m \end{pmatrix} = T \begin{pmatrix} \mu_1 \\ \vdots \\ \mu_m \end{pmatrix} \quad .$$

From this and from formulas (9.3.5, 12 and 13) it follows that

$$(\Gamma(s)\vec{\mu}, \vec{\mu}) = \sum_{i,j=1}^{m} \Gamma_{ij}(s)\mu_i\mu_j = \sum_{i,j=1}^{m} \Gamma_{ij}(s + L)\mu'_i\mu'_j \quad . \tag{9.3.14}$$

The relation $\det T = 1$ [5] and formulas (9.3.8, 9, 12, and 13) lead to the relation:

$$U_0(s + L, \mu) = e^{i\kappa_0} U_0(s, \mu) \quad , \tag{9.3.15}$$

where

$$\kappa_0 = -\tfrac{1}{2}(\alpha_1 + \ldots + \alpha_m) \quad , \tag{9.3.16}$$

i.e. U_0 really is a Floquet solution of parabolic equation (9.1.14). We recall that Floquet indices have so far been defined only up to integral multiples of 2π. However, for formula (9.3.16) to be valid, they must satisfy the condition

$$\exp\left[-\frac{i}{2}(\alpha_1 + \ldots + \alpha_m) \right] = \exp(i\kappa_0) = \frac{S(L)}{S(0)} \quad , \tag{9.3.17}$$

---

[5] The relation $\det T = 1$ is a consequence of the orientability of the Riemannian manifold. Even if the manifold is non-orientable, all the constructions of Chap. 9 can still be carried out. However, when $\det T = -1$ holds, it becomes necessary to make slight changes in certain formulas [for instance, in formula (9.3.16)]. Let us take this opportunity to note that all the constructions of *Babič*'s paper [9.8] apply to the case $\det T = 1$, which unfortunately was nowhere stipulated therein.

which can be considered as a normalization condition (note that the right side of (9.3.17) is uniquely defined). We will assume that the α_j are chosen so that (9.3.17) is satisfied.

Based on the Floquet solution constructed above, we can find other Floquet solutions of parabolic equation (9.1.14). We introduce the operators $\Lambda_1^*, \ldots, \Lambda_m^*$ such that for any nonnegative integers $q_1, \ldots, q_m$ the function

$$U_{q_1}, \ldots, q_m = \Lambda_1^{*q_1}, \ldots, \Lambda_m^{*q_m} U_0$$

will be a Floquet solution of equation (9.1.14).

Let us note first of all that, as can be easily established by direct calculation, any operator of the form

$$\Lambda = \frac{1}{i} \sum_{\gamma=1}^{m} y_\gamma \frac{\partial}{\partial \mu_\gamma} - (\vec{y}', \vec{\mu}) = \frac{1}{i}(\vec{y}, \nabla_\mu) - (\vec{p}, \vec{\mu}) \quad (\vec{p} = \vec{y}') \quad , \qquad (9.3.18)$$

where $y_\gamma(s)$ are components of a vector $\vec{y}$ that satisfies equation (9.2.5), commutes with the operator $\mathcal{L}$.

We now construct the operators

$$\Lambda_j = \frac{1}{i}(\vec{y}_j, \nabla_\mu) - (\vec{p}_j, \vec{\mu}) \quad , \quad j = 1, 2, \ldots, m \quad , \qquad (9.3.19)$$

and

$$\Lambda_j^* = \frac{1}{i}(\vec{y}_j^*, \nabla_\mu) - (\vec{p}_j^*, \vec{\mu}) \quad , \quad j = 1, 2, \ldots, m \quad , \qquad (9.3.20)$$

where the $\vec{y}_i$ are Floquet solutions. Operators (9.3.19, 20) satisfy the commutation relations

$$\Lambda_j \Lambda_h - \Lambda_h \Lambda_j = 0 \quad , \quad \Lambda_j^* \Lambda_h^* - \Lambda_h^* \Lambda_j^* = 0 \quad , \qquad (9.3.21)$$

$$\Lambda_j \Lambda_h^* - \Lambda_h^* \Lambda_j = \delta_{hj} \; (\delta_{hj} \text{ is the Kronecker delta}) \quad . \qquad (9.3.22)$$

Moreover

$$\Lambda_j U_0 = S(s) \exp\left[\tfrac{i}{2}(\Gamma\vec{\mu}, \vec{\mu})\right] [(\Gamma\vec{y}_j, \vec{\mu}) - (\vec{y}_j', \vec{\mu})] \equiv 0 \quad . \qquad (9.3.23)$$

The commutativity of $\mathcal{L}$ and Λ_j^* implies that the functions

$$\begin{aligned}
U_q &= U_{q_1 q_2 \ldots q_m} = (\Lambda_1^*)^{q_1} (\Lambda_2^*)^{q_2} \ldots (\Lambda_m^*)^{q_m} U_0 \\
&= Q_{q_1 \ldots q_m}(\vec{\mu}) \exp\left[\tfrac{i}{2}(\Gamma\vec{\mu}, \vec{\mu})\right] = Q_q(\vec{\mu}) \exp\left[\tfrac{i}{2}(\Gamma\vec{\mu}, \vec{\mu})\right]
\end{aligned} \qquad (9.3.24)$$

will be solutions of the parabolic equation $\mathcal{L}U = 0$, where q_j are integers ≥ 0, $q = (q_1, q_2, \ldots, q_m)$, $Q_q(\vec{\mu})$ is a polynomial in $\mu_1, \ldots, \mu_m$ of degree $|q| = q_1 + \ldots + q_m$ with coefficients that depend on s.

Applying the operators $\Lambda_1^{q_1}, \ldots, \Lambda_m^{q_m}$ to U_0 gives the trivial zero solution of the equation

$$\mathcal{L}U = 0 \quad .$$

Let us show that the function U_q is a Floquet solution of the equation $\mathcal{L}U = 0$ with index

$$\kappa_q = -\left[\frac{1}{2}(\alpha_1 + \ldots + \alpha_m) + \sum_{j=1}^{m} \alpha_j q_j\right] \quad , \quad q = (q_1, \ldots, q_m). \qquad (9.3.25)$$

In fact, it has already been mentioned that U_q is a solution of the equation $\mathcal{L}U = 0$; the equality

$$U_q(s + L, \vec{\mu}) = e^{i\kappa_q} U_q(s, \vec{\mu})$$

follows from (9.3.15), the equalities

$$\vec{y}_j^*(s + L) = e^{-i\alpha_j} \vec{y}_j^* \quad , \quad \vec{p}_j^*(s + L) = e^{-i\alpha_j} \vec{p}_j^*(s) \quad ,$$
$$\vec{p}_j = \vec{y}_j' \quad ,$$

and the form of the operators Λ_j^*; the inequalities $U_q \not\equiv 0$ are implied by the orthogonality relations

$$\int_{\Phi_s} U_q \overline{U}_{q'} d\vec{\mu} = \int_{-\infty}^{+\infty} \ldots \int_{-\infty}^{+\infty} U_{q_1 \ldots q_m} \overline{U_{q_1' \ldots q_m'}} d\mu_1 \ldots d\mu_m$$

$$= \begin{cases} 0 \text{ when } q \neq q' \quad , \text{ i.e. when } (q_1, \ldots, q_m) \neq (q_1', \ldots, q_m') \quad ; \\ \\ q_1! \ldots q_m! \int_{-\infty}^{+\infty} \frac{|\exp\left[(i/2)(\Gamma \vec{\mu}, \vec{\mu})\right]|}{|\det Y|} d\vec{\mu} = q_1! \ldots q_m!(2\pi)^{m/2} \\ \\ \text{ when } q = q' \quad , \text{ i.e. when } (q_1, \ldots, q_m) = (q_1', \ldots, q_m') \end{cases}$$

$$\qquad (9.3.26)$$

which are deduced below. Here Φ_s is the hyperplane normal to the geodesic l at s.

The equality $q = q'$ (or the inequality $q \neq q'$, respectively) means that all equalities $q_1 = q_1', \ldots, q_m = q_m'$ are simultaneously satisfied (or, respectively, that at least one of these inequalities is *not* satisfied). The proof of (9.3.26) is easily demonstrated by using integration by parts and formulas (9.3.21, 22 and 23). Thus, the operators Λ_j in essence annihilate the solution U_0, and therefore they can be called *annihilation operators*. Conversely, application of the operators Λ_j^* enables us to find new solutions, and these operators can be called *creation operators*.

Let us present some additional data on the Floquet solutions of (9.1.14, 16, and 17). We easily see from (9.3.21–23) that U_q is an eigenfunction of the elliptic differential operator

$$\sum_{j=1}^{m} \Lambda_j^* \Lambda_j \quad , \qquad (9.3.27)$$

which acts on functions defined on Φ_s. Corresponding to an eigenfunction U_q is the eigenvalue

$$|q| = q_1 + q_2 + \ldots + q_m \quad .$$

Orthogonality relations (9.3.26) imply the linear independence of the polynomials $Q_q(\vec{\mu})$, see (9.3.24). Therefore, any polynomial of the variables $(\mu_1, \ldots, \mu_m) = \vec{\mu}$ is a linear combination of the Q_q. In fact, the set of polynomials in $\mu_1, \ldots, \mu_m$ whose degrees do not exceed $q_1 + \ldots + q_m$ is a linear space for which the monomials $\mu_1^{q_1} \mu_2^{q_2} \ldots \mu_m^{q_m}$ form a linearly independent basis. There is actually the same number of $Q_{q_1 \ldots q_m}$ polynomials as monomials $\mu_1^{q_1} \ldots \mu_m^{q_m}$. By virtue of the linear independence of the $Q_{q_1 \ldots q_m}$, these polynomials also form a basis in the set of polynomials of degree $q_1 + q_2 + \ldots + q_m$. By virtue of the fact that functions of the form

$$\hat{Q}(\mu_1, \mu_2, \ldots, \mu_m) U_0 \quad ,$$

where $\hat{Q}$ is a polynomial, are dense in $L_2(\Phi_s)$, the functions U_q form a complete system in $L_2(\Phi_s)$.

The completeness of the system of functions U_q implies that in some sense they form a complete set of Floquet solutions of the equation $\mathcal{L}U = 0$. In fact, by multiplying the equation $\mathcal{L}U = 0$ by $\overline{U}_q$ and integrating with respect to Φ_s, we get (with very general assumptions on U)

$$\frac{\partial}{\partial s} \int\limits_{\Phi_s} U \overline{U}_q d\vec{\mu} = 0 \quad . \tag{9.3.28}$$

Expanding U in the U_q (which can also be done with very broad assumptions on U), we get

$$U = \sum_q C_q U_q \quad , \quad q = (q_1, \ldots, q_m) \quad , \quad q_j \geq 0 \quad . \tag{9.3.29}$$

From the orthogonality relations (9.3.26) and formula (9.3.28) we see that $C_q = $ const. From the equality

$$U(s + L) - e^{i\kappa} U(s) = 0$$

and (9.3.29) we get the relation

$$\sum_q C_q U_q(s)(e^{i\kappa_q} - e^{i\kappa}) = 0 \quad ,$$

which because of the orthogonality of the U_q implies that the equalities

$$C_q(e^{i\kappa_q} - e^{i\kappa}) = 0$$

hold for all $q = (q_1, \ldots, q_m)$.

At least one of the numbers C_q is different from zero; therefore at some $q, e^{i\kappa_q} = e^{i\kappa}$ and that number κ must necessarily have the form (9.3.25) (within an additive multiple of 2π). If q' is such that $e^{i\kappa_{q'}} \neq e^{i\kappa}$, then C_q is equal to zero. Formula (9.3.29) now implies that U is equal to a linear combination (possibly infinite) of the U_q such that

$$U_q(s + L) = e^{i\kappa} U_q(s) \quad , \quad e^{i\kappa} = e^{i\kappa_q} \quad .$$

Now let $\pi, \alpha_1, \ldots, \alpha_m$ be linearly independent over the ring of integers, that is, let the equality

$$\pi \gamma_0 + \alpha_1 \gamma_1 + \ldots + \alpha_m \gamma_m = 0 \quad (\gamma_0, \ldots, \gamma_m \text{ are integers}) \qquad (9.3.30)$$

imply that

$$\gamma_1 = \gamma_2 = \ldots = \gamma_m = 0 \quad .$$

Then all the multipliers $e^{i\kappa_q}$ are distinct, and the equality $e^{i\kappa} = e^{i\kappa_q}$ can hold for no more than one choice of q, see (9.3.25). As implied by (9.3.29), for this q

$$U = \text{const}\, U_q \quad .$$

Substituting the Floquet solutions U_q into (9.1.9), we get the principal term of the asymptotic expansion of the eigenfunctions. Formula (9.1.18) will give the principal term of the asymptotic expansion of the corresponding eigenvalue.

If $\alpha_1, \ldots, \alpha_m$ are all distinct, and in addition the numbers $\pi, \alpha_1, \ldots, \alpha_m$ are linearly independent over the ring of integers, i.e., (9.3.30) implies that all $\gamma_j = 0$, then the set of numbers $(q_1, \ldots, q_m) = q$ uniquely determines the eigenvalue in the first approximation:

$$k_{pq} = k_{pq_1\ldots q_m} = \left[2\pi p + \frac{1}{2}(\alpha_1 + \ldots + \alpha_m) + \sum_{j=1}^{m} q_j \alpha_j \right] \frac{1}{L}$$

$$= \left[2\pi p + \sum_{j=1}^{m} \left(q_j + \frac{1}{2} \right) \alpha_j \right] \frac{1}{L} \quad .$$

Otherwise (for instance, when the polynomial $\mathcal{J}(\lambda)$ has multiple roots, Sect. 9.2), then different sets $q_1, \ldots, q_m$ may correspond to the same eigenvalue, but will have different expressions for the eigenfunctions.

In fact, if we assume that not all γ_j are simultaneously equal to zero in (9.3.30), then when

$$p' - p = \gamma_0 \zeta \quad , \quad q'_j - q_j = 2\gamma_j \zeta \quad (\zeta \text{ is an integer}) \quad ,$$

we obviously have

$$k_{pq_1\ldots q_m} = k_{p'q'_1\ldots q'_m} \quad .$$

In the terminology of quantum mechanics, we say that the eigenvalue $k_{pq_1\ldots q_m}$ is degenerate. In the case when $\pi, \alpha_1, \ldots, \alpha_m$ are linearly independent, we readily find (Sect. 9.4) higher approximations both for the eigenvalues and for the eigenfunctions. When these numbers are linearly dependent, difficulties arise just as they do in the perturbation method of quantum mechanics if the eigenvalue is degenerate.

We note further that the functions

$$U_q = U_{q_1\ldots q_m}(\vec{\mu}) = Q_{q_1}\ldots{}_{q_m}(\vec{\mu})\exp\left[\frac{i}{2}(\Gamma\vec{\mu}, \vec{\mu}) \right]$$

$$= \Lambda_1^{*q_1}\ldots\Lambda_m^{*q_m} \frac{1}{\sqrt{\det Y}} \exp\left[\frac{i}{2}(\Gamma\vec{\mu}, \vec{\mu}) \right] \quad ,$$

with any Γ can be considered entirely separate from the problem (9.1.14, 16, and 17) as a certain multidimensional generalization of the Hermite functions. In fact, in deriving orthogonality relations and proving that the eigenfunctions of the operator (9.3.27) are $U_{q_1...q_m}$, we use only the commutation relations (9.3.21 and 22) which are obtained in their turn from (9.2.21 and 22). Any symmetric matrix Γ with positive definite imaginary part, as is readily proved, can be represented in the form $\Gamma = PY^{-1}$, where P and Y are matrices that satisfy relations (9.2.21 and 22).

9.4 Construction of the Higher Approximations

Assuming the linear independence of π, α_1, ..., α_m over the ring of integers, let us describe a method of constructing eigenvalues and eigenfunctions in any approximation. We will assume the components of the metric tensor g_{ij} in s, y_1, ..., y_m to be infinitely differentiable.

We seek the solution of the Helmholtz equation $u = u_q$, $q = (q_1, ..., q_m)$ as the formal series

$$u_q = e^{ik_q s}\left(U_q^0 + \frac{1}{\sqrt{k_q}}U_q^1 + \frac{1}{k_q}U_q^2 + \frac{1}{k_q^{3/2}}U_q^3 + ... \right) , \qquad (9.4.1)$$

where

$$U_q^0 = U_q = \Lambda_1^{*q_1} ... \Lambda_m^{*q_m} U_0 , \quad U_0 = \frac{1}{\sqrt{\det Y}}\exp\left[\frac{i}{2}(PY^{-1}\vec{\mu}, \vec{\mu}) \right]$$

$$k_q = \frac{1}{L}\left[2\pi p + \sum_{j=1}^m \left(q_j + \frac{1}{2} \right)\alpha_j \right] , \quad q = (q_1, ..., q_m) , \qquad (9.4.2)$$

and the functions

$$U_q^j = U_q^j(s, \mu_1, ..., \mu_m) \qquad (9.4.3)$$

($\mu_h = \sqrt{k_q}y_h$; s, y_1, ..., y_m are coordinates on the Riemannian manifold in the neighborhood of l) are to be determined.

Let the eigenvalue k (for simplicity of notation we will omit the subscripts) corresponding to the eigenfunction (9.4.1) have the following asymptotic expansion

$$k = k_q + \frac{\delta_1}{\sqrt{k_q}} + \frac{\delta_2}{k_q} + ..., \qquad (9.4.4)$$

where the δ_j are numbers to be determined.

Replacing u and k in equation $(\Delta + k^2)u = 0$ in accordance with formulas (9.4.1 and 4), and setting the coefficients of like powers of k_q equal to zero, we arrive at the equations

$$\mathcal{L}_0 U_q^0 = 0 \quad ,$$
$$\mathcal{L}_0 U_q^1 + \mathcal{L}_1 U_q^0 = 0 \quad ,$$
$$\mathcal{L}_0 U_q^h + \mathcal{L}_1 U_q^{h-1} + \ldots + \mathcal{L}_h U_q^0 = 0 \quad . \tag{9.4.5}$$

The vanishing of the expression $\mathcal{L}_0 U_q^h + \ldots$ corresponds to the vanishing of the coefficient of $k_q^{1-h/2}$.

By using the formulas of Sect. 9.1 we readily find that the operator $\mathcal{L}_0$ coincides with the parabolic operator $\mathcal{L}$, see (9.1.3), and when $h \geq 1$

$$\mathcal{L}_h U = R_h^0 \frac{\partial U}{\partial s} + \sum_{j=1}^m R_{h+1}^j \frac{\partial U}{\partial \mu^j} + R_{h-2}^{00} \frac{\partial^2 U}{\partial s^2}$$

$$+ \sum_{j=1}^m R_{h-1}^{0j} \frac{\partial^2 U}{\partial s \partial \mu_j} + \sum_{ij=1}^m R_h^{ij} \frac{\partial^2 U}{\partial \mu^i \partial \mu^j}$$

$$+ (\psi_h + R_{h+2})U \quad (R_{-1}^{00} \equiv 0) \quad . \tag{9.4.6}$$

Here

$$\psi_h = \delta_h + \delta_1 \delta_{h-3} + \delta_2 \delta_{h-4} + \ldots + \delta_{h-3}\delta_1 + \delta_h \tag{9.4.7}$$

and R_{h+2}, R_h^0, R_{h+1}^j, R_h^{ij} ... are polynomials in $\mu_1, \ldots \mu_m$ with coefficients that are infinitely differentiable with respect to s. The degree of these polynomials does not exceed the subscript. In addition, their parity[6] coincides with that of the subscript, i.e. the polynomial R and its subscript are simultaneously even or odd.

The polynomials R have one more important property: when s is increased by the length L of the geodesic l, each term in (9.4.6) is transformed into itself if U is a single-valued function in the neighborhood of l. (If the basis vectors $\vec{e}_1, \ldots, \vec{e}_m$ introduced at the beginning of Sect. 9.1 had been transformed into themselves after parallel displacement over the length L along l, then this latter property of the polynomials R could have been formulated quite simply: the coefficients of the polynomials R have period L with respect to s).

We shall find the U_q^h by means of a recurrence procedure described later. We will now consider (s, μ_i) as coordinates on the tangent space Ξ' (Sect. 9.1). In (9.4.2, 3) the μ_i have the different meaning $\mu_i = \sqrt{k}y_i$ but this should not lead to any misunderstanding.

Suppose the functions $U_q^1, \ldots, U_q^{h_0-1}$ and numbers $\delta_1, \ldots, \delta_{h_0-1}$ have been found, and let U_q^0 have the form (9.4.2). We will look for the function $U_q^{h_0}$ as the solution of the $(h_0 + 1)$-th equation of the recurrent system (9.4.5): $\mathcal{L}_0 U_q^{h_0} + \ldots = 0$, which is considered as an equation on Ξ'.

We further require: 1) L-periodicity of function $U_q^{h_0} \exp(ik_q s)$ on Ξ'; 2) that $U_q^{h_0} \to 0$ as $\mu \to \infty$.

---

[6] A polynomial $R(\vec{\mu})$ is called *even* (or *odd*) if $R(-\vec{\mu}) \equiv R(\vec{\mu})$ (or $R(-\vec{\mu}) \equiv -R(\vec{\mu})$).

Solving this boundary value problem and setting $\mu_i = \sqrt{k}y_i$, we get the desired functions, defined on the Riemannian manifold near l. In the process of finding U_q^{ho}, the number δ_{h_0} will be determined as well.

Let us assume that when $h < h_0$

$$U_q^h = \Phi_q^h(\vec{\mu}, s)U_0 \quad , \quad U_0 = \frac{1}{\sqrt{\det Y}}\exp\left[\frac{i}{2}(PY^{-1}\vec{\mu}, \vec{\mu})\right] , \tag{9.4.8}$$

where Φ_q^h is a polynomial in $\vec{\mu}$ with coefficients that are infinitely differentiable with respect to s.

If we carry out all the necessary differentiations, we can give the equation for U_q^{ho} the form

$$\mathcal{L}U_q^{ho} = -\mathcal{L}_1 U_q^{ho-1} - \ldots - \mathcal{L}_{h_0}U_q^0 = \Psi_q^{ho}(\vec{\mu}, s)U_0 \quad . \tag{9.4.9}$$

Here U_0 has the same expression as in (9.4.8), Ψ_q^{ho} is a polynomial in $\vec{\mu}$ with coefficients infinitely differentiable with respect to s.

After multiplication by $\exp(ik_q s)$, the right side of (9.4.9) will be an L-periodic function by virtue of the fact that U_q^h has this property. It can be represented as a finite linear combination (with coefficients that are infinitely differentiable with respect to s) of the functions

$$U_r = \Lambda_1^{*r_1} \ldots \Lambda_m^{*r_m} U_0 = Q_r(\vec{\mu}, s)U_0 \quad ,$$
$$r = (r_1, \ldots, r_m) \quad , \tag{9.4.10}$$

since any polynomial can be represented in the form of a linear combination Q_r (see Sect. 9.3). So

$$\Psi_q^{ho}(\vec{\mu}, s)U_0 = \sum_{(r)} A_{qr}^{ho}(s)U_r \quad . \tag{9.4.11}$$

By virtue of the orthogonality of the U_r, see (9.3.26),

$$A_{qr}^{ho}(s) = \frac{1}{r_1!\ldots r_m!(2\pi)^{m/2}} \int_{-\infty}^{+\infty} \Psi_q^{ho}U_0\overline{U}_r d\vec{\mu} \quad . \tag{9.4.12}$$

From this formula and the properties of $\Psi_q^{ho}U_0$ and U_r we easily get the equality

$$A_{qr}^{ho}(s + L) \equiv e^{i(\kappa_q - \kappa_r)}A_{qr}^{ho}(s) \quad , \tag{9.4.13}$$

where

$$\kappa_q = -\sum_{j=1}^m (1/2 + q_j)\alpha_j \quad , \quad q = (q_1, \ldots, q_m) \quad ,$$

$$\kappa_r = -\sum_{j=1}^m (1/2 + r_j)\alpha_j \quad , \quad r = (r_1, \ldots, r_m) \quad . \tag{9.4.14}$$

We will seek the function U_q^{ho} in the form of a linear combination of the U_r:

237

$$U_q^{h_0} = \sum_{(r)} B_{qr}^{h_0}(s) U_r \quad . \tag{9.4.15}$$

In order for $U_q^{h_0} \exp(\mathrm{i}k_q s)$ to be an L-periodic function, it is sufficient that

$$B_{qr}^{h_0}(s + L) = B_{qr}^{h_0}(s) \exp \mathrm{i}(\kappa_q - \kappa_r) \quad . \tag{9.4.16}$$

We will look for $B_{qr}^{h_0}(s)$ that satisfy this condition.

Substituting (9.4.11, 15) into (9.4.9) and using the fact that $\mathcal{L}U_r = 0$, we easily arrive at the very simple differential equations

$$2\mathrm{i}\frac{d}{ds} B_{qr}^{h_0}(s) = A_{qr}^{h_0}(s), \tag{9.4.17}$$

where the functions $B_{qr}^{h_0}$ are to satisfy conditions (9.4.16). The right sides $A_{qr}^{h_0}$ satisfy the analogous conditions (9.4.13). Problem (9.4.16, 17) is a self-adjoint problem on the interval $(0, L)$. It is easily shown (by direct computation, without reference even to the general theory) that for this problem to have a single-valued solution, it is necessary and sufficient that the corresponding homogeneous problem have only the trivial solution, for which in turn it is necessary and sufficient that

$$\exp\left[\mathrm{i}(\kappa_q - \kappa_r)\right] \neq 1 \quad .$$

From expressions (9.4.14) for κ_q, κ_r and the linear independence of π, α_1, ..., α_m over the ring of integers we find immediately when $r \neq q$ ($r \neq q$ is equivalent to the inequality $\sum_{j=1}^{m} |q_j - r_j| > 0$, $r = (r_1, ..., r_m)$, $q = (q_1, ..., q_m)$) that all the boundary value problems (9.4.16, 17) are uniquely solvable.

Let us turn to the case $r = q$; then the boundary value problem reduces to finding the L-periodic function $B_{qq}^{h_0}$, that satisfies (9.4.17) with an L-periodic right side. For this equation to be solvable it is necessary and sufficient that

$$\int_0^L A_{qq}^{h_0}(s) ds = 0 \quad . \tag{9.4.18}$$

It turns out that condition (9.4.18) can always be satisfied by properly choosing the yet undetermined parameter δ_{h_0}. In fact, taking into consideration (9.4.6, 7, 9, and 12), condition (9.4.18) can be given the form

$$\int_0^L (\delta_{h_0} + D_q^{h_0}(s)) ds = 0 \quad , \tag{9.4.19}$$

where $D_q^{h_0}$ is some already known function. Equation (9.4.19) uniquely determines δ_{h_0}.

We shall prove by induction that: 1) all the δ_h with odd subscripts are equal to zero; 2) U_q^h it can always be arranged that the parity of the number $h + |q| = h + q_1 + ... + q_m$ and that of the polynomial $\Phi_q^h(\vec{\mu}, s)$ [see (9.4.8)] are the same.[7] (footnote [7] see p. 239)

Let these statements be true when $h < h_0$; then in formula (9.4.9), as implied by the properties of the operator $\mathcal{L}_j$ and the induction hypothesis, the right side will take the form

$$-\psi_{h_0} U_q + H_q^{h_0}(\vec{\mu}, s) U_0 \quad , \tag{9.4.20}$$

where $H_q^{h_0}$ is a polynomial with the same parity as

$$h_0 + |q| = h_0 + q_1 + \ldots + q_m \quad .$$

Equations (9.4.12) imply that $A_{qr}^{h_0}(s)(r \neq q)$ are different from zero only in the case where the parities of $h_0 + |q|$ and $|r|$ are the same. When $r = q$, using formula (9.4.20) we get

$$A_{qq}^{h_0}(s) = -\psi_{h_0} + \int_{-\infty}^{+\infty} \frac{H_q^{h_0}(\vec{\mu}, s) U_q(\vec{\mu}, s) d\vec{\mu}}{q_1! \ldots q_m! (2\pi)^{m/2}} \quad . \tag{9.4.21}$$

If h_0 is odd, the integral in this formula is equal to zero, being the integral of an odd function over all space. Equation (9.4.7) and the induction hypothesis for h_0 odd give $\psi_{h_0} = 2\delta_{h_0}$. Equation (9.4.21) takes the form

$$A_{qq}^{h_0}(s) = -2\delta_{h_0} \quad , \tag{9.4.22}$$

and from (9.4.18) we find that $\delta_{h_0} = 0$. Setting $B_{qq}^{h_0} = 0$, we get a polynomial $\Phi_q^{h_0}$, whose parity is equal to that of $h_0 + |q|$. This is implied by formulas (9.4.15, 17), and the fact that the $A_{qr}^{h_0}$ are different from zero (recall that h_0 is odd) only when the difference $|q| - |r|$ is odd. The fact that the parities of $h_0 + |q|$ and the polynomial $\Phi_q^{h_0}$ with h_0 even are the same is a consequence of (9.4.15, 17) and the fact that the $A_{qr}^{h_0}$ are equal to zero when difference $|q| - |r|$ is odd.

9.5 The Eigenfunction Problem in a Three-Dimensional Region

Let Ω be a region in three-dimensional Euclidean space bounded by an infinitely differentiable surface S.

Consider the following eigenfunction problem:

$$(\Delta + k^2)u = 0 \quad , \quad u \not\equiv 0 \quad , \quad (x, y, z) \in \Omega \quad , \tag{9.5.1}$$

---

[7] The function U_q^h is uniquely defined because the solution of (9.4.16, 17) for $r = q$ is determined to within an arbitrary positive constant. Ambiguity in the determination of an eigenfunction can be easily foreseen: multiplying the eigenfunction by a constant having the form

$$1 + \frac{C_1}{k_q^{1/2}} + \frac{C_2}{k_q} + \ldots, \quad C_i = \text{const.} \quad ,$$

we again get an eigenfunction with constants C_i appearing in the asymptotic expansion. (Translator's note: See the footnote on p. 41 of [9.9]).

$$u|_S = 0 \quad \text{or} \quad \left.\frac{\partial u}{\partial n}\right|_S = 0 \quad . \tag{9.5.2}$$

We will study the eigenfunctions of (9.5.1,2) which are concentrated in the neighborhood of a closed geodesic l on S.

The constructions for the case of the internal problem are easily extended to the external problem (see the end of this section).

Let us introduce the so called semi-geodesic coordinate systems (*see* [9.10]) on S in the vicinity of l which is a special case of the coordinate system introduced in Sect. 9.1 when $m = 1$. We will characterize points on l by their arc length s (measured from some fixed point). A point M on S has the coordinates (s, y) if it lies at a distance y from l on a geodesic which is orthogonal to l and passes through the point $s \in l$. On l, the coordinate $y = 0$, and we let $y > 0$ on one side of l and $y < 0$ on the other side. The conventional techniques of variational calculus show the orthogonality of (s, y) coordinate system (see Sect. 2.2). We will characterize a point M along the normal to S from a point (s, y) by the quantity n – the distance along the normal – taking $n > 0$ outside of S and $n < 0$ inside (i.e., in Ω). In this way we have defined a regular system of curvilinear coordinates s, y, n near l. It is convenient to express the relation between (s, y, n) coordinates and cartesian coordinates (x_1, x_2, x_3) by the vector equation

$$\vec{x} = \vec{r}(s, y) + n\vec{n}^0 \quad , \quad \vec{x} = (x_1, x_2, x_3) \quad . \tag{9.5.3}$$

Here $\vec{n}^0$ is the outward unit normal to S, and $\vec{r} = \vec{r}(s, y)$ is the vector notation for the parametric equation of S near l. We have the following expression for a differential arc length along an arbitrary curve in the (s, y, n) coordinate system:

$$d\sigma^2 = |d\vec{x}|^2 = \left[\left(\frac{\partial \vec{r}}{\partial s} + n\frac{\partial \vec{n}^0}{\partial s} \right) ds + \left(\frac{\partial \vec{r}}{\partial y} + n\frac{\partial \vec{n}^0}{\partial y} \right) dy + \vec{n}^0 dn \right]^2$$

$$= g_{ih} d\zeta^i d\zeta^h \quad , \quad (s = \zeta^1, \ y = \zeta^2, \ n = \zeta^3) \quad , \tag{9.5.4}$$

from which we readily get

$$g_{11} = g_{ss} = \vec{r}_s \cdot \vec{r}_s + 2n\vec{r}_s \cdot \vec{n}_s^0 + n^2 \vec{n}_s^0 \cdot \vec{n}_s^0$$
$$g_{12} = g_{21} = g_{sy} = g_{ys} = n[\vec{n}_s^0 \cdot \vec{r}_y + \vec{n}_y^0 \cdot \vec{r}_s] + n^2 \vec{n}_s^0 \cdot \vec{n}_y^0$$
$$g_{22} = g_{yy} = \vec{r}_y \cdot \vec{r}_y + 2n\vec{r}_y \cdot \vec{n}_y^0 + n^2 \vec{n}_y^0 \cdot \vec{n}_y^0$$
$$g_{23} = g_{32} = g_{yn} = g_{ny} = 0 \quad , \quad g_{33} = g_{nn} = 1 \quad , \tag{9.5.5}$$

for the components of the metric tensor.

Equation (9.5.1) in the coordinate system $(\zeta^1, \zeta^2, \zeta^3) = (s, y, n)$ is written in the form (see Appendix A.2)

$$\frac{1}{\sqrt{g}} \frac{\partial}{\partial \zeta^i} \left(g^{ij} \sqrt{g} \frac{\partial}{\partial \zeta^j} u \right) + k^2 u = 0 \quad , \quad (s, y, n) = (\zeta^1, \zeta^2, \zeta^3) \quad , \tag{9.5.6}$$

(the matrix $\|g^{ij}\|$ is the inverse of $\|g_{ij}\|$).

240

Introducing b_{ss}, b_{sy} and b_{yy} – the coefficients of the second quadratic form of Gauss for the surface S in the (s, y) coordinate system and using an ellipsis to denote quantities that do not exceed

$$\text{const}(|y|^3 + |y| \cdot |n| + n^2) \quad,$$

in order of magnitude, we get

$$g^{11} = g^{ss} = 1 + K(s)y^2 + 2nb_{ss}(s,0) + \cdots \quad,$$
$$g^{22} = g^{yy} = 1 + 2nb_{yy}(s,0) + \cdots \quad,$$
$$g^{12} = g^{sy} = -2nb_{sy}(s,0) + \cdots \quad,$$
$$g^{13} = g^{sn} = 0 \quad, \quad g^{23} = g^{yn} = 0 \quad, \quad g^{nn} = 1 \quad. \tag{9.5.7}$$

Here $K(s)$ is the Gaussian curvature of S at $(s, 0)$.

The rest of our presentation will be based on the assumption that

$$b_{ss}(s,0) < 0 \quad. \tag{9.5.8}$$

When inequality (9.5.8) is satisfied in Ω close to l a whispering gallery effect will occur (see Chap. 5). Using the classical expression for the coefficients of the second quadratic form of Gauss, inequality (9.5.8) can be rewritten as

$$\vec{r}_s \cdot \vec{n}_s^0 = -b_{ss}(s,0) > 0 \quad. \tag{9.5.9}$$

The vector $\vec{r}_s$ is a unit vector tangent to l and pointed in the direction of increasing arc length. Using formulas for the curvature of a curve on a surface, we find that inequality (9.5.9) is equivalent to the inequality

$$-b_{ss}(s,0) = \frac{1}{\varrho} > 0 \quad, \tag{9.5.10}$$

where ϱ is the radius of curvature of a plane normal cross section of the surface S along l at $(s, 0)$. The latter inequality means that the center of curvature of the given plane normal cross section of surface S lies on the side of l where $n < 0$. (We recall that points near S where $n < 0$ are points in Ω). Inequality (9.5.10) in the case of a plane curve is the condition for the onset of the whispering gallery effect (see Sect. 5.3) on the side of l where $n < 0$.

As is to be expected, the behavior of the eigenfunctions for any fixed n must be analogous to the behavior of the eigenfunctions of the problem considered in Sects. 9.1–4 with $m = 1$. On the other hand, at fixed y it is natural to expect that the eigenfunctions will behave (in $n < 0$) analogously to eigenfunctions of the whispering gallery type (see Chap. 7). The following constructions confirm these assumptions.

We will seek the function in the form

$$u = e^{ik_0 s + i\varphi_1(s)k_0^{1/3} + i\varphi_0(s)} V(s, \mu, \nu, k_0) \quad,$$
$$\mu = \sqrt{k_0}\, y \quad, \quad \nu = k_0^{2/3} n \quad, \tag{9.5.11}$$

where k_0 is a first approximation to the eigenvalue. We will calculate this approximation later on. We can seek the eigenvalue k as a formal series:

$$k = k_0 + \sum_{j=1}^{\infty} \delta_j k_0^{-j/6} \quad , \tag{9.5.12}$$

Furthermore, we assume that the function V takes the form

$$V = \sum_{j=0}^{m} V_j(s, \mu, \nu) k_0^{-j/6} \quad . \tag{9.5.13}$$

We substitute expressions (9.5.11–13) into (9.5.6). The leading term will be of order $k_0^{4/3}$. Setting this term equal to zero, we get

$$-2\frac{\partial \varphi_1}{\partial s} V_0 - 2b_{ss}\nu V_0 + \frac{\partial^2 V_0}{\partial \nu^2} = 0 \quad . \tag{9.5.14}$$

Equation (9.5.14) must be solved subject to the boundary conditions

$$V_0|_{\nu=0} = 0 \quad \text{or} \quad \frac{\partial V_0}{\partial \nu}\bigg|_{\nu=0} = 0 \quad , \quad V_0 \xrightarrow[\nu \to -\infty]{} 0 \quad . \tag{9.5.15}$$

(Recall that the internal problem (9.5.11) is being solved here, and solutions are being sought that decay with increasing distance from surface S.) Equation (9.5.14) is an ordinary linear equation in V_0 in which ν acts as the independent variable. This equation is easily reduced to the Airy equation. Considering the condition at infinity on V_0, we get

$$V_0 = A_0(s, \mu)v(-\nu\beta(s) - t) \quad , \quad \beta(s) = \sqrt[3]{-2b_{ss}(s)} > 0 \quad , \tag{9.5.16}$$

$$t = -\frac{2\varphi_1'(s)}{\beta^2(s)} \quad . \tag{9.5.17}$$

Here v is an Airy function.

Taking the boundary condition at $\nu = 0$ into account, we find that t must be a root of the function $v(-t)$ for the boundary condition $u|_S = 0$ (or of $v'(-t)$ for $\partial u/\partial n|_S = 0$). From this and (9.5.17) we get:

$$\varphi_1(s) = t\frac{1}{\sqrt[3]{2}} \int_0^s \frac{ds}{\varrho^{2/3}} \quad , \tag{9.5.18}$$

where t is a root of the Airy function $v(-t)$ or its derivative $v'(-t)$ and $\varrho(s) = -1/b_{ss}(s, 0)$, see (9.5.9, 10). We have set the constant of integration equal to zero, which does not limit the generality of our solutions.

Now to get the first approximation to the solution it remains for us to find the function $\varphi_0(s)$, see (9.5.11), and the function $A_0(s, \mu)$ in expression (9.5.16) for V_0. We will determine these functions from the solvability conditions for the boundary value problems arising from the higher approximations.

Setting the terms of order $k_0^{7/6}$ and k_0 equal to zero, we get

$$V_1 = A_1(s, \mu)v(-\nu\beta(s) - t) \quad , \tag{9.5.19}$$

242

$$2i\frac{\partial V_0}{\partial s} + \frac{\partial^2 V_0}{\partial \mu^2} - K(s)\mu^2 V_0 - 2\frac{d\varphi_0}{ds}V_0$$

$$+\frac{\partial^2 V_2}{\partial \nu^2} - \left(2\frac{d\varphi_1}{ds} + 2b_{ss}\nu\right)V_2 = 0 \quad . \tag{9.5.20}$$

We substitute our expression for V_0 into (9.5.20):

$$\left(2i\frac{\partial A_0}{\partial s} + \frac{\partial^2 A_0}{\partial \mu^2} - K(s)\mu^2 A_0\right)v(-\nu\beta(s) - t)$$

$$-2\frac{d\varphi_0}{ds}A_0 v(-\nu\beta(s) - t) + 2iA_0 \cdot \nu\beta' \cdot v'(-\nu\beta(s) - t)$$

$$+\frac{\partial^2 V_2}{\partial \nu^2} - 2\left(\frac{d\varphi_1}{ds} + b_{ss}\nu\right)V_2 = 0 \quad . \tag{9.5.21}$$

We transfer all terms not containing V_2 to the right side. In this way we get an ordinary linear differential equation for V_2 with a nonzero right side, which we will solve under the requirement that V_2 satisfy (9.5.15). Thus, to find V_2 we must solve a boundary value problem of the Sturm-Liouville type on an infinite interval with boundary conditions (9.5.15). This is a problem at an eigenvalue (i.e., the corresponding homogeneous problem has a nontrivial solution), therefore for it to be solvable it is necessary and sufficient that the right side be orthogonal to the solution of the homogeneous equation, i.e. that it be orthogonal to $v(-\beta(s)\nu - t)$:

$$\left(2i\frac{\partial A_0}{\partial s} + \frac{\partial^2 A_0}{\partial \mu^2} - K(s)\mu^2 A_0\right)\int_{-\infty}^{0} v^2(-\nu\beta(s) - t)d\nu$$

$$+\int_{-\infty}^{0}\left(2\frac{d\varphi_0}{ds}A_0 v + 2iA_0\nu\beta'v'\right)v(-\nu\beta(s) - t)d\nu = 0 \quad . \tag{9.5.22}$$

We choose φ_0 so that the second integral vanishes, i.e., we require that φ_0 satisfy the equation

$$\frac{d\varphi_0}{ds}\int_{-\infty}^{0} v^2(-\nu\beta(s) - t)d\nu$$

$$+i\int_{-\infty}^{0} \nu\beta'v(-\nu\beta(s) - t)v'(-\nu\beta(s) - t)d\nu = 0 \quad .$$

Hence, after some straightforward calculations, we find an expression for φ_0 up to a constant term:

$$\varphi_0(s) = \frac{i}{6}\ln\varrho(s) \quad \left(\varrho(s) = -\frac{1}{b_{ss}}\right) \quad . \tag{9.5.23}$$

Equation (9.5.22) now reduces to a parabolic equation for A_0:

$$2i\frac{\partial A_0}{\partial s} + \frac{\partial^2 A_0}{\partial \mu^2} - K(s)\mu^2 A_0 = 0 \quad ,$$

which is solved by the methods of Sects. 9.1–4:

$$A_0 = \Lambda^{*q} \frac{1}{\sqrt{Y}} \exp\left[\frac{i}{2} \frac{Y'}{Y} \mu^2\right] \quad ,$$

$$\Lambda^* = \frac{1}{i} Y^*(s) \frac{\partial}{\partial \mu} - Y^{*'} \mu \quad , \qquad (9.5.24)$$

where $Y(s)$ is the Floquet solution of the equation

$$Y'' + K(s)Y = 0 \quad , \quad K(s+L) \equiv K(s) \quad ; \quad Y(s+L) = e^{i\alpha} Y(s) . \qquad (9.5.25)$$

[Equation (9.5.25) is a special case of (9.2.5). Here $K(s)$ is the Gaussian curvature of S at $(s, 0)$.] As before, we will require L-periodicity of each approximation. The requirement of L-periodicity on the first approximation leads to an equation that determines k_0:

$$k_0 L - t k_0^{1/3} \int_0^L \varrho^{-2/3}(s) ds \frac{1}{\sqrt[3]{2}} - \left(q + \frac{1}{2}\right) \alpha = 2\pi p \quad , \qquad (9.5.26)$$

where α is the Floquet index of (9.5.25).

The function $A_1(s, \mu)$, see (9.5.19), is found analogously to $A_0(s, \mu)$ from the solvability requirement for the boundary value problem for $V_3(s, \mu, \nu)$. We arrive immediately at the boundary value problem for V_3 by equating to zero terms of order $k_0^{5/6}$.

Further approximations are found in the same way. Assume that we know $V_0, V_1, \ldots, V_{h_0-1}$, which have the form

$$V_h = \left(\sum_{j=0}^{N_h} D_{jh} \nu^j\right) v(-\nu\beta(s) - t)$$

$$+ \left(\sum_{j=0}^{N_h} \tilde{D}_{jh} \nu^j\right) v'(-\nu\beta(s) - t) \quad , \qquad (9.5.27)$$

where D_{jk} and $\tilde{D}_{jk}$ are polynomials in μ multiplied by $\exp[(i/2)(Y'/Y)\mu^2]$. We also assume that V_{h_0} and V_{h_0+1} have the form of (9.5.27), but that the coefficients D_{0h_0} and D_{0,h_0+1} in them are still undetermined.

Let us show how to find D_{0h_0} and V_{h_0+2} (the functions V_{h_0+3} and D_{0,h_0+1} are found analogously). Turning to (9.5.6) and setting the coefficient of $k_0^{1-h_0/6}$ equal to zero, we get

$$-2\frac{d\varphi_0}{ds} V_{h_0} + \mathcal{L}V_{h_0} + \frac{\partial^2 V_{h_0+2}}{\partial \nu^2} - \left(2\frac{d\varphi_1}{ds} + 2b_{ss}\nu\right) V_{h_0+2} + 2\delta_{h_0} V_0$$

$$= U_0 \left[\sum_{j=0}^{N_{h_0+2}-1} C_{jh_0} \nu^j v(-\nu\beta(s) - t)\right.$$

$$+ \sum_{j=0}^{N_{h_0+2}} \tilde{C}_{jh_0} \nu^j v'(-\nu\beta(s) - t) \Bigg] \quad ; \tag{9.5.28}$$

$$U_0 = \frac{1}{\sqrt{Y}} \exp\left[\frac{i}{2}\frac{Y'}{Y}\mu^2\right] \quad ,$$

where C_{jh_0} and $\tilde{C}_{jh_0}$ are polynomials in μ with coefficients that depend on s. We subject V_{h_0+2} to conditions (9.5.15). Equation (9.5.28) together with (9.5.15) is a boundary value problem at an eigenvalue for V_{h_0+2}. Let us set up the integral solvability condition for it. Substituting (9.5.27) for V_{h_0} in (9.5.28), we get a parabolic equation with a nonzero right side for the yet unknown coefficient D_{0h_0}. From its solvability as we did in Sect. 9.1–4, we find δ_{h_0}.

To determine V_{h_0+2} a solvable boundary value problem is obtained: 1) the equation takes the form

$$\frac{\partial^2 V_{h_0+2}}{\partial \nu^2} - \left(2\frac{d\varphi_1}{ds} + 2b_{ss}\nu\right)V_{h_0+2}$$

$$= \sum_{j=0}^{N_{h_0+2}-1} e_{jh_0} \nu^j v(-\nu\beta(s) - t) + \sum_{j=0}^{N_{h_0+2}} \tilde{e}_{jh_0} \nu^j v'(-\nu\beta(s) - t) \quad ,$$
$$\tag{9.5.29}$$

(e_{jh_0} and $\tilde{e}_{jh_0}$ are polynomials in μ multiplied by $\exp[(i/2)(Y'/Y)\mu^2]$); 2) the boundary conditions are given by (9.5.15), where V_0 must be replaced by V_{h_0+2}.

To find V_{h_0+2} we could have used the generalized Green's function (see [9.11]); however, it is convenient here to find V_{h_0+2} by the method of undetermined coefficients in the form (9.5.27), by which all coefficients except $D_{0\,h_0+2}$ are found. We will skip over all these calculations. The process of constructing V_h and δ_h can be continued indefinitely if the ratio of the Floquet index α to π is irrational. Just as in the preceding section, it can be shown that all coefficients δ_h with odd subscripts h are equal to zero. On this note we end our examination of the interior problem (9.5.1).

Analogously to problem (9.5.1) we could have examined the exterior problem for the region located outside of region Ω. This problem is formulated just like problem (9.5.1), except that the solution of equation $(\Delta + k^2)u = 0$ is sought at points that do not belong to Ω. Some additional refinements relating to formulation of the external problem could be made as was done in Sect. 7.5 for the two-dimensional case.

For the exterior problem, we can also construct formal series for the eigenvalues and eigenfunctions, except that the part of the Airy function v will now be played by the Airy function w_1, and the roots t of the function $v(-t)$ (or $v'(-t)$) are replaced by the roots ξ of the function $w_1(\xi)$ (or $w_1'(\xi)$).

In conclusion we present expressions for the eigenfunctions in the first approximation. For the interior problem (9.5.1),

$$u \approx e^{ik_0 s} \varrho^{-1/6}(s)\theta^{-1/2}(s)\exp\left[ik_0 s - i\left(q + \frac{1}{2}\right)\int\limits_0^s \frac{ds}{\theta^2(s)}\right.$$

$$\left. - t\frac{ik_0^{1/3}}{\sqrt[3]{2}}\int\limits_0^s \left(\frac{1}{\varrho}\right)^{2/3} ds + i\frac{\theta'}{2\theta}\mu^2\right] D_q\left(\sqrt{2}\frac{\mu}{\theta}\right) v\left(-\left(\frac{2}{\varrho}\right)^{1/3}\nu - t\right),$$

where $\hspace{8cm}$ (9.5.30)

$$\theta = |Y| \quad , \quad \varrho = -\frac{1}{b_{ss}} \quad , \quad D_q(\sqrt{2}x) = e^{-x^2/2}\,H_q(x) \quad ;$$

H_q is a Hermite polynomial, $t = t_r$ (or $= t'_r$) is the r-th root of the Airy function $v(-t)$ (or $v'(-t)$) for the boundary condition $u|_S = 0$ (or $\partial u/\partial n|_S = 0$), and k_0 is determined from (9.5.26).

To obtain the asymptotic expansion of eigenfunctions and eigenvalues in the first approximation for the exterior problem, we must replace v in (9.5.30) by w_1 and t by $-\xi$, where ξ is the root of the Airy function $w_1(\xi)$ in the case of boundary condition $u|_S = 0$, or its derivative for boundary condition $\partial u/\partial n|_S = 0$.

Formula (9.5.26) holds for the eigenvalues, where once more we must substitute $-\xi$ for t.

We omit the calculations leading to (9.5.30).

9.6 Asymptotic Solution of a System of Elliptic Equations on a Riemannian Manifold, Concentrated Near a Ray

In the constructions of the foregoing sections for the eigenfunctions of the Laplacian concentrated near a closed geodesic, the construction of the formal asymptotic series (9.4.1) (in negative half-integral powers of the wavenumber k) satisfying the Helmholtz equation (9.1.1) was of fundamental importance. The terms of this series decay exponentially outside a neighborhood of order $k^{-1/2}$ around the geodesic. In this sense we can speak of the concentration of the series (9.4.1) in the neighborhood of the geodesic. This same geodesic satisfies a Hamiltonian system corresponding to the Helmholtz equation, and consequently can be called a phase trajectory, or ray. In this section we will generalize these constructions and obtain a formal asymptotic solution to a system of second-order elliptic equations on a Riemannian manifold, concentrated in the neighborhood of some phase trajectory of this system. This allows us to find an asymptotic subset of the eigenvalues of the elliptic operators which correspond to eigenfunctions concentrated near a closed phase trajectory. This is done in the same way as for the Laplacian, and so we will limit ourselves here only to the construction of the dominant terms of the asymptotic solution. Let us note that these have other applications as well. For example, when integrated with respect to a parameter describing the phase trajectory, they allow us to describe the field near a singular manifold on which the geometrical divergence of the corresponding ray field vanishes and on which the ordinary ray method is inapplicable.

Let V_{m+1} be a smooth $(m + 1)$-dimensional manifold, and $g_{ij}(x)$ for $x \in V_{m+1}$ the Riemann metric defined on V_{m+1}. Further, let $A^{is}_{jr}(x)$ and $B_{jr}(x)$ be smooth, real tensors of the fourth and second rank respectively, defined on a region $\Omega \in V_{m+1}$. We will assume that the tensors $A^{is}_{jr}(x)$ and $B_{jr}(x)$ are symmetric and positive definite, i.e.,

$$A^{is}_{jr}(x) = A^{si}_{rj}(x) \quad , \quad B_{jr}(x) = B_{rj}(x) \tag{9.6.1}$$

and

$$A^{is}_{jr}\eta^j_i\eta^r_s \geq a \sum_{i,j=0}^{m}(\eta^j_i)^2 \quad , \quad a > 0$$

$$B_{jr}\xi^j\xi^r \geq b \sum_{i=0}^{m}(\xi^i)^2 \quad , \quad b > 0 \tag{9.6.2}$$

(here and in what follows we adopt the summation convention that repeated upper and lower indices are to be summed over). Consider the following system of equations on Ω for the vector $U(x) = \{U^0(x), U^1(x), \ldots, U^m(x)\}$:

$$\nabla_i[A^{is}_{jr}(x)\nabla_s U^r(x)] + \omega^2 B_{jr}(x)U^r(s) = 0 \quad . \tag{9.6.3}$$

In (9.6.3), ω is a parameter and ∇_i is the operator of covariant differentiation, expressed in terms of the Christoffel symbols

$$\Gamma^r_{is} = \frac{1}{2}g^{rt}\left(\frac{\partial g_{it}}{\partial x^s} + \frac{\partial g_{st}}{\partial x^i} - \frac{\partial g_{is}}{\partial x^t}\right) \quad , \quad \|g^{rt}\| = \|g_{rt}\|^{-1}$$

by the formula

$$\nabla_i U^{r_1,r_2,\ldots,r_p}_{s_1,s_2,\ldots,s_q} = \partial U^{r_1,r_2,\ldots,r_p}_{s_1,s_2,\ldots,s_q}/\partial x^i$$
$$+ \Gamma^{r_1}_{it} U^{t,r_2,\ldots,r_p}_{s_1,s_2,\ldots,s_q} + \ldots + \Gamma^{r_p}_{it} U^{r_1,r_2,\ldots,t}_{s_1,s_2,\ldots,s_q}$$
$$- \Gamma^t_{is_1} U^{r_1,r_2,\ldots,r_q}_{t,s_2,\ldots,s_q} - \ldots - \Gamma^t_{is_q} U^{r_1,r_2,\ldots,r_p}_{s_1,s_2,\ldots,t} \quad .$$

(In particular, $\nabla_i U^r = (\partial U^r/\partial x^i) + \Gamma^r_{it}U^t$). Note that when (9.6.2) is satisfied, system (9.6.3) will be elliptic. The parameter ω has the meaning of an angular frequency and appears in (9.6.3) as the result of seeking a solution for the nonstationary system

$$\nabla_i[A^{is}_{jr}(x)\nabla_s V^r(t, x)] - B_{jr}(x)\frac{\partial^2 V^r(t, x)}{\partial t^2} = 0 \quad , \quad j = 0, 1, 2, \ldots, m \tag{9.6.4}$$

in the time-harmonic form $V(t, x) = e^{-i\omega t}U(x)$. Equations of the form (9.6.3 and 4) are encountered in many branches of physics, e.g., the anisotropic theory of elasticity, magnetic electrodynamics and gravitational theory.

In this section we will construct a formal asymptotic solution to (9.6.3) which is concentrated near a phase trajectory. In order to do this, we must first define what we mean by a phase trajectory of the system (9.6.3), and then carry out some auxiliary constructions. If we seek the solution to the system as the ray expansion

$$U^r(x) = e^{i\omega\tau(x)} \sum_{j=0}^{\infty} \frac{U_j^r(x)}{\omega^{j+\gamma}} \quad ; \quad \gamma = \text{const}$$

then by the usual means we obtain the following equation for the function $\tau(x)$:

$$\det\| A_{jr}^{is}(x)p_i p_s - B_{jr}(x)\| = 0 \tag{9.6.5}$$

where $p_i = \nabla_i\tau$. Equation (9.6.5) is of $(m+1)$th order in p_i, and in the following all calculations will be performed for an arbitrary fixed root of this equation. We rewrite this equation (for some fixed root) for $\tau(x)$ in a form more convenient for the subsequent manipulations. The equation

$$\det\| A_{jr}^{is}(x)p_i p_s - H^2 B_{jr}(x)\| = 0 \tag{9.6.6}$$

has, as a function of H^2, $m+1$ positive roots (including multiplicity) by virtue of (9.6.1, 2). Let $H = H(x, p)$, $p = p_0, p_1, \ldots, p_m$ be one of these positive roots of (9.6.6). Clearly, the function $H(x, p)$ is a homogeneous function of first degree in p. When $H = 1$, Eqs. (9.6.5, 6) are identical, and therefore the function $\tau(x)$ corresponding to the root $H(x, p)$ can be described as the solution of

$$H(x, \nabla\tau) = 1 \quad ; \quad \nabla\tau = \nabla_0\tau, \nabla_1\tau, \ldots, \nabla_m\tau \quad . \tag{9.6.7}$$

The first-order partial differential equation (9.6.7) for $\tau(x)$ is usually called the Hamilton-Jacobi equation. The Hamiltonian system corresponding to (9.6.7) is

$$\frac{dx^i}{dt} = \frac{\partial H}{\partial p_i} \quad , \quad \frac{dp_i}{dt} = -\frac{\partial H}{\partial x^i} \quad , \quad i = 0, 1, 2, \ldots m \quad . \tag{9.6.8}$$

The solutions of the Hamiltonian system (9.6.8)

$$x^i = x^i(t) \quad , \quad p_i = p_i(x) \quad , \quad i = 0, 1, 2, \ldots, m \tag{9.6.9}$$

are called bicharacteristics or phase trajectories of (9.6.7). The projections $x^i = x^i(t)$, $i = 0, 1, 2, \ldots, m$ of the phase trajectories (9.6.9) in coordinate space will be called the rays of system (9.6.3).

The solution $\tau(x)$ of the Hamilton-Jacobi equation (9.6.7) and the soltuions $p_i(t)$ of the Hamiltonian system (9.6.8) are related by[8]

$$\frac{\partial\tau}{\partial x^i} = p_i \tag{9.6.10}$$

---

[8] In particular, thanks to relation (9.6.10), we can use the same notations $p_0, p_1, \ldots, p_m$ both for the partial derivatives for τ and for the solutions of (9.6.8).

and therefore along a phase trajectory we have

$$H(x, p) = 1 \quad \text{and} \quad \frac{d\tau}{dt} = p_i \frac{\partial H}{\partial p_i} \quad .$$

By Euler's theorem on homogeneous functions, $p_i(\partial H/\partial p_i) = H(x,p)$, and consequently $d\tau/dt = H(x,p) = 1$. Thus, τ can be taken as the parameter along a phase trajectory.

Let Γ be a phase trajectory and γ its projection into coordinate space, i.e., a ray. In phase space, we introduce the local coordinates

$$x^0, x^1, \ldots, x^m \quad ; \quad p_0, p_1, \ldots, p_m$$

in the neighborhood of Γ,[9] such that in these local coordinates the phase trajectory Γ is described by the equations

$$x^0 = \tau \quad , \quad x^i = 0 \quad ; \quad p_0 = 1 \quad , \quad p_i = 0 \quad , \quad i = 1, 2, \ldots, m \quad . \qquad (9.6.11)$$

(The equality $p_0 = 1$ follows from the relations $p_0 = dx^0/dt$ and $d\tau/dt = 1$).

We call the local coordinates $x^0, x^1, \ldots, x^m$ on the manifold V_{m+1} transverse coordinates with respect to the ray γ. Evidently, an atlas of V_{m+1} can always be chosen such that the local coordinate map curving through a ray γ will be transverse with respect to it. Let us prove this simple

Lemma 1. In order that the curve Γ given by (9.6.11) be a phase trajectory, it is necessary and sufficient that

$$\left.\frac{\partial H}{\partial p_0}\right|_\Gamma = 1 \quad , \quad \left.\frac{\partial H}{\partial p_i}\right|_\Gamma = 0 \quad ; \left.\frac{\partial H}{\partial x^0}\right|_\Gamma = 0 \quad ; \quad \left.\frac{\partial H}{\partial x^i}\right|_\Gamma = 0$$

$$i = 1, 2, \ldots, m \quad . \qquad (9.6.12)$$

The necessity of this assertion is obvious. To prove sufficiency, note that when (9.6.12) is satisfied, the Hamiltonian system (9.6.8) along Γ has the form

$$\frac{dx^0}{d\tau} = 1 \quad , \quad \frac{dx^i}{d\tau} = 0 \quad ; \quad \frac{dp_0}{d\tau} = 0 \quad , \quad \frac{dp_i}{d\tau} = 0 \quad , \quad i = 1, 2, \ldots, m \quad .$$

The curve Γ defined by (9.6.11) satisfies these relations and thus Γ is a phase trajectory.

Next we construct the transverse foliation of a ray. At each point of a ray γ let us consider the vector

$$p(\tau) = \{p_0(\tau), p_1(\tau), \ldots, p_m(\tau)\} \quad .$$

Vectors orthogonal to $p(\tau)$ in the g_{ij} metric form an m-dimensional linear space $\Phi_\gamma(\tau)$. We will call the set of pairs $\{\tau, \Phi_\gamma(\tau)\}$ the transverse foliation of the ray γ.

Much of the following will be carried out, not on the manifold, but on the transverse foliation of a ray. Thus, we will consider a Sturm-Liouville prob-

---

[9] We will use the previous notation for the local coordinates for ease of writing.

lem with conditions at infinity on the foliation. The correct treatment of such a problem on the manifold V_{m+1} is impossible.

Of fundamental importance in the construction of concentrated solutions of (9.6.3) will be the solution of the linearized Hamiltonian system, where the linearization is carried out in the coordinates transverse to Γ. We will denote the specialization of functions $F(x, p)$ and $f(x)$, defined on the phase and coordinate spaces respectively, to the phase trajectory Γ and the ray γ by $\overset{\circ}{F}$ and $\overset{\circ}{f}$.

Let us evaluate the matrices $H_{p_i p_s}$, $H_{p_i x^s}$ and $H_{x^i x^s}$ in x^i and p_i coordinates $(i = 0, 1, 2, \ldots, m)$ transverse to Γ. We use the notation

$$\overset{\circ}{H}_{p_i p_s} = R(\tau) \quad , \quad \overset{\circ}{H}_{p_i x^s} = L(\tau) \quad , \quad \overset{\circ}{H}_{x^i x^s} = T(\tau)$$

and consider the linear system

$$\frac{dy}{d\tau} = R(\tau)q + L(\tau)y \qquad y = (y^1, y^2, \ldots, y^m)$$

$$\frac{dq}{d\tau} = -L^\mathsf{T}(\tau)q - T(\tau)y \quad q = (q_1, q_2, \ldots, q_m) \tag{9.6.13}$$

where L^T is the transpose of L. It is evident that (9.6.13) is a linearization of the Hamiltonian system (9.6.8) in the coordinates transverse to Γ. The system (9.6.13) has the trivial solution $y = 0$, $q = 0$ corresponding to the phase trajectory Γ. The curve $\{\tau, y(\tau)\}$ on the transverse foliation of γ will be called a ray in the linear approximation.

We can now proceed directly to the construction of a formal solution to (9.6.3) concentrated in the neighborhood of the ray γ. We will assume that (9.6.3) is written in the coordinates x^i, $i = 1, 2, \ldots, m$ and $x^0 = \tau$, transverse to γ [we can do this since (9.6.3) has been written in invariant form]. Just as we did when constructing the solution for the Laplacian in the neighborhood of a geodesic, we introduce the scaled coordinates

$$\nu^i = \omega^{1/2} x^i \quad , \quad i = 1, 2, \ldots, m \quad , \quad \tau = x^0$$

in the neighborhood of γ.

A solution of (9.6.3) which is concentrated near γ will be sought in the form:

$$U(\tau, \nu) = e^{i\omega\tau} \sum_{k=0}^{\infty} \omega^{-k/2} W_k(\tau, \nu) \tag{9.6.14}$$

where

$$W_k(\tau, \nu) = (W_k^0(\tau, \nu), W_k^1(\tau, \nu), \ldots, W_k^m(\tau, \nu))$$

are as yet unknown vectors. We expand the tensors $A_{jr}^{is}(x)$, $B_{jr}(x)$ and the Christoffel symbols $\Gamma_{ij}^l(x)$ in (9.6.3) in Taylor series in powers of x^i, $i = 1, 2, \ldots, m$; we rewrite the result in ν^i coordinates and substitute the series (9.6.14) into the system. On the left side of the equation we group the terms associated with $\omega^{-k/2}$ arising from this procedure. Equating to zero the coefficients of these various powers of ω leads to a recurrent system of equations for determining the vectors W_k. Without presenting the rather lengthy calculations we give only the recurrent system itself:

$$\overset{\circ}{C}W_0 = 0 \qquad (9.6.15)$$

$$\overset{\circ}{C}W_k = \sum_{j=1}^{k} \mathcal{L}_j W_{k-j}, \quad k = 1, 2, \ldots \ , \qquad (9.6.16)$$

where $\overset{\circ}{C}$ is the specialization of the matrix $C_{jr}(x) = A_{jr}^{is}(x)p_i p_s - H^2(x,p)B_{jr}(x)$ to the ray γ and $\mathcal{L}_j$ are matrix differential operators given by

$$\mathcal{L}_1 = \mathrm{i}\{\overset{\circ}{C}_p, \nabla\} - \{\overset{\circ}{C}_x, \nu\} \qquad (9.6.17)$$

$$\mathcal{L}_2 = \frac{1}{2}\{\overset{\circ}{C}_{pp}\nabla, \nabla\} + \mathrm{i}\{\overset{\circ}{C}_{px}\nu, \nabla\} - \frac{1}{2}\{\overset{\circ}{C}_{xx}\nu, \nu\}$$
$$+ \frac{1}{2}\overset{\circ}{C}_{p_s x^s} + \overset{\circ}{B}\mathcal{L} + \mathrm{i}\left[\overset{\circ}{C}_{p0}\frac{\partial}{\partial\tau} + \frac{1}{\sqrt{g}}\frac{d}{d\tau}(\overset{\circ}{B}\sqrt{g}) + \overset{\circ}{M}(\tau)\right] \ . \qquad (9.6.18)$$

In (9.6.17 and 18), the following notations are used:

$$\nu = (\nu^1, \nu^2, \ldots, \nu^n) \ , \quad \nabla = \left(\frac{\partial}{\partial\nu^1}, \frac{\partial}{\partial\nu^2}, \ldots, \frac{\partial}{\partial\nu^n}\right) \ ,$$

$$g(\tau) = \det\|\overset{\circ}{g}_{ij}\|$$

$$\{\overset{\circ}{C}_p, \nabla\} = \sum_{s=1}^{n} \overset{\circ}{C}_{p_s}\frac{\partial}{\partial\nu^s} \ , \quad \{\overset{\circ}{C}_x, \nu\} = \sum_{s=1}^{n} \overset{\circ}{C}_{x^s}\nu^s$$

$$\{\overset{\circ}{C}_{pp}\nabla, \nabla\} = \sum_{s,l=1}^{n} \overset{\circ}{C}_{p_s p_l}\frac{\partial}{\partial\nu^s}\frac{\partial}{\partial\nu^l}$$

$$\{\overset{\circ}{C}_{px}\nu, \nabla\} = \sum_{s,l=1}^{n} \overset{\circ}{C}_{p_s x^l}\nu^l\frac{\partial}{\partial\nu^s}$$

$$\{\overset{\circ}{C}_{xx}\nu, \nu\} = \sum_{s,l=1}^{n} \overset{\circ}{C}_{x^s x^l}\nu^s\nu^l$$

$$\mathcal{L} = 2\mathrm{i}\frac{\partial}{\partial\tau} + (R\nabla, \nabla) + \mathrm{i}(L\nu, \nabla) - (T\nu, \nu) + \mathrm{i}\,\mathrm{Tr}\,L$$

$$M = \|M_{jr}\| \ ,$$

$$M_{jr} = \frac{1}{2}\frac{\partial}{\partial x^i}(A_{jr}^{i0} - A_{rj}^{i0}) + \frac{1}{2}(A_{jr}^{i0} - A_{rj}^{i0})\Gamma_{li}^{l} + A_{lj}^{i0}\Gamma_{ir}^{l} - A_{lr}^{i0}\Gamma_{ij}^{l} \ .$$

The operator $\mathcal{L}$, since it contains no second derivatives with respect to τ, can be called parabolic. The matrices R, L and T which appear in it are the same ones which appear in the linearized Hamiltonian system (9.6.13). This important fact will be used later on in a crucial way. Lastly, it is readily seen that $C = \|C_{jr}\|$ is a symmetric matrix, while $M = \|M_{ir}\|$ is antisymmetric.

The recurrent system of equations (9.6.15, 16) will be considered on a transverse foliation of the ray γ under the additional conditions

$$|W_k| \to 0 \quad, \quad k = 0, 1, 2, \ldots \text{ as } |\nu| \to \infty \quad,$$

which guarantee the concentrated nature of the solution. We denote by $I_H(x, p)$ the kernel of the operator $C(x, p)$, i.e., the set of solutions of the homogeneous equation $C(x, p)\Psi = 0$. The symbol $\operatorname{Ker} C(x, p) = I_H(x, p)$ will be employed for the kernel of an operator.

In what follows, we will assume that the dimensionality of $\overset{\circ}{I}_H(x, p) = \operatorname{Ker} \overset{\circ}{C}(x, p)$ is preserved along the ray γ. The subspace of vectors representable in the form $w(\tau, \nu)\overset{\circ}{\Psi}(\tau)$, where $w(\tau, \nu)$ is an arbitrary smooth function and $\overset{\circ}{\Psi}(\tau) \in \overset{\circ}{I}_H$ will be denoted by J_H, and $J_H^\perp$ will be its orthogonal complement in the metric g_{ij}.

Consider Eq. (9.6.15). It is a homogeneous equation because of $\det \|\overset{\circ}{C}\| = 0$, and has a nontrivial solution

$$W_0 = w_0(\tau, \nu)\overset{\circ}{\Psi}_0(\tau), \tag{9.6.19}$$

where $\overset{\circ}{\Psi}_0(\tau) \in \overset{\circ}{I}_H$ and $w_0(\tau, \nu)$ is an arbitrary smooth function. Thus $W_0 \in J_H$.

Equations (9.6.16) for $k \geq 1$ are inhomogeneous, and are solvable only if the right sides satisfy certain conditions. In order to obtain these conditions, we prove the following[10].

Lemma 2. For any vector Z, the vector $\overset{\circ}{C} Z$ lies in $J_H^\perp$.

To prove this, we expand $\overset{\circ}{C} Z$ as the sum of two vectors, $\overset{\circ}{C} Z = x + y$, where $x \in J_H$ and $y \in J_H^\perp$. We need to show that $x = 0$. Clearly, $x^2 = (x, \overset{\circ}{C} Z)$ and by the symmetry of $\overset{\circ}{C}$ we have $x^2 = (\overset{\circ}{C} x, Z) = 0$, since $x \in J_H$.

From Lemma 2 it follows that in order for the inhomogeneous equations (9.6.16) to be solvable, it is necessary and sufficient that their right sides $\sum_{j=1}^k \mathcal{L}_j W_{k-j}$ belong to $J_H^\perp$.

Lemma 3. The operator $\mathcal{L}_1$ maps the space J_H into the space $J_H^\perp$, i.e., $\mathcal{L}_1 J_H C \subset J_H^\perp$.

To show this, we need to show that for any vectors $\overset{\circ}{\Psi}$ and $w\overset{\circ}{\Psi}$ lying in J_H, the equation $(\overset{\circ}{\Psi}, \mathcal{L}_1 w\overset{\circ}{\Psi}) = 0$ is satisfied. We re-express $\mathcal{L}_1 w\overset{\circ}{\Psi}$ as:

$$\mathcal{L}_1 w\overset{\circ}{\Psi} = i\{\overset{\circ}{C}_p p, \nabla\} w\overset{\circ}{\Psi} - \{\overset{\circ}{C}_x, \nu\} w\overset{\circ}{\Psi}$$
$$= i\{\overset{\circ}{C}_p \overset{\circ}{\Psi}, \nabla w\} - w\{\overset{\circ}{C}_x \overset{\circ}{\Psi}, \nu\} \quad.$$

---

[10] This lemma is completely analogous to the theorem on linear equations that we used in Chap. 2 in the study of the ray method for Maxwell's equations.

But differentiating the identity $C(x,p)\Psi(x,p) = 0$ by x and by p, we get

$$C_x\Psi = -C\Psi_x \quad \text{and} \quad C_p\Psi = -C\Psi_p \tag{9.6.20}$$

and consequently,

$$\mathcal{L}_1 w\overset{\circ}{\Psi} = -i\overset{\circ}{C}\{\overset{\circ}{\Psi}_p, \nabla w\} + \overset{\circ}{C}\, w\{\overset{\circ}{\Psi}_x, \nu\} \quad . \tag{9.6.21}$$

Now, evaluating the inner product presents no difficulty:

$$
\begin{aligned}
(\overset{\circ}{\Psi}, \mathcal{L}_1 w\overset{\circ}{\Psi}) &= (\overset{\circ}{\Psi}, \overset{\circ}{C}[\,-i\{\overset{\circ}{\Psi}_p, \nabla w\} + \{\overset{\circ}{\Psi}_x, \nu\}]) \\
&= (\overset{\circ}{C}\overset{\circ}{\Psi}, -i\{\overset{\circ}{\Psi}, \nabla w\} + \{\overset{\circ}{\Psi}_x, \nu\}) = 0
\end{aligned}
$$

since $\overset{\circ}{C}\Psi = 0$.

From Lemma 3 it follows that the first equation of the recurrent system (9.6.16), that is,

$$\overset{\circ}{C} W_1 = \mathcal{L}_1 W_0 \tag{9.6.22}$$

is solvable for any vector W_0. In fact, since $\overset{\circ}{C}W_0 = 0$, we have $W_0 \in J_H$, and the solvability condition $\mathcal{L}_1 W_0 \in J_{\overline{H}}^{\perp}$ of (9.6.22) is a simple consequence of Lemma 3.

The vector W_1, and the subsequent vectors $W_2, W_3, \ldots$, are sought in the form

$$W_k = w_k(\tau, \nu)\overset{\circ}{\Psi}_k(\tau) + W_k^{\perp} \tag{9.6.23}$$

where $\overset{\circ}{\Psi}_k \in J_H$ and $W_k^{\perp} \in J_{\overline{H}}^{\perp}$. From the solvability of (9.6.22) it follows that

$$W_1^{\perp} = \overset{\circ}{C}^{-1} \mathcal{L}_1 W_0 \tag{9.6.24}$$

where $\overset{\circ}{C}^{-1}$ is the operator inverse to $\overset{\circ}{C}$ in the subspace $J_{\overline{H}}^{\perp}$. We turn now to the solvability condition on (9.6.16) for $k = 2$:

$$\mathcal{L}_1 W_1 + \mathcal{L}_2 W_0 \in J_{\overline{H}}^{\perp} \quad . \tag{9.6.25}$$

We can rewrite (9.6.25) in view of (9.6.23) and (9.6.24) as

$$\mathcal{L}_1(w_1\overset{\circ}{\Psi}_1) + (\mathcal{L}_1\overset{\circ}{C}^{-1}\mathcal{L}_1 + \mathcal{L}_2)W_0 \in J_{\overline{H}}^{\perp} \quad .$$

Since by Lemma 3, $\mathcal{L}_1(w_1\overset{\circ}{\Psi}_1) \in J_{\overline{H}}^{\perp}$, this condition is equivalent to

$$(\mathcal{L}_1\overset{\circ}{C}^{-1}\mathcal{L}_1 + \mathcal{L}_2)W_0 \in J_{\overline{H}}^{\perp} \quad . \tag{9.6.26}$$

To transform (9.6.26), we rewrite (9.6.21) using $w = w_0$ and $\overset{\circ}{\Psi} = \overset{\circ}{\Psi}_0$, as

$$\overset{\circ}{C}\left(\overset{\circ}{C}^{-1}\mathcal{L}_1(w_0\overset{\circ}{\Psi}_0) + \left[i\left\{\frac{\partial\overset{\circ}{\Psi}_0}{\partial p}, \nabla\right\} + \left\{\frac{\partial\overset{\circ}{\Psi}_0}{\partial x}, \nu\right\}\right]w_0\right) = 0.$$

It follows from this that the vector in the outermost parentheses belongs to the kernel of $\overset{\circ}{C}$. Applying Lemma 2, we have

$$\mathcal{L}_1 \overset{\circ}{C}{}^{-1} \mathcal{L}_1(w_0\overset{\circ}{\Psi}_0) + \mathcal{L}_1\left[i\left\{\frac{\partial\overset{\circ}{\Psi}_0}{\partial p},\nabla\right\} + \left\{\frac{\partial\overset{\circ}{\Psi}_0}{\partial x},\nu\right\}\right]w_0 \in J_{\overset{\perp}{H}} \quad . \tag{9.6.27}$$

Upon calculating the vectors on the left sides of (9.6.26 and 27), we transform the solvability condition into

$$-\mathcal{L}_1\left[i\left\{\frac{\partial\overset{\circ}{\Psi}_0}{\partial p},\nabla\right\} + \left\{\frac{\partial\overset{\circ}{\Psi}_0}{\partial x},\nu\right\}\right]w_0 + \mathcal{L}_2(w_0\overset{\circ}{\Psi}_0) \in J_{\overset{\perp}{H}} \quad .$$

Substituting at this point the expressions for the operators $\mathcal{L}_1$ and $\mathcal{L}_2$ and simplifying the result using (9.6.20), we obtain

$$\overset{\circ}{B}\overset{\circ}{\Psi}_0\mathcal{L}w_0 + 2i\overset{\circ}{B}\frac{d\overset{\circ}{\Psi}_0}{d\tau}w_0 + \frac{i}{\sqrt{g}}\frac{d}{d\tau}(\overset{\circ}{B}\sqrt{g})\overset{\circ}{\Psi}_0 w_0$$

$$+i(\overset{\circ}{N} + \overset{\circ}{M})\overset{\circ}{\Psi}_0 w_0 \in J_{\overset{\perp}{H}} \tag{9.6.28}$$

where

$$\overset{\circ}{N} = \frac{1}{2}(\overset{\circ}{C}_{x^\bullet}\overset{\circ}{C}{}^{-1}\overset{\circ}{C}_{p^\bullet} - \overset{\circ}{C}_{p^\bullet}\overset{\circ}{C}{}^{-1}\overset{\circ}{C}_{x^\bullet})$$

is an antisymmetric matrix.

Now, using relation (9.6.28), we deduce an equation for the scalar function $w_0(\tau,\nu)$. Multiplying the vector on the left side of (9.6.28) scalarly by $\overset{\circ}{\Psi}_0$ and using the selfadjointness of $\overset{\circ}{B}$, we arrive at the equation

$$(\overset{\circ}{B}\overset{\circ}{\Psi}_0,\overset{\circ}{\Psi}_0)\mathcal{L}w_0 + \frac{i}{\sqrt{g}}\frac{d}{d\tau}[(\overset{\circ}{B}\overset{\circ}{\Psi}_0,\overset{\circ}{\Psi}_0)\sqrt{g}]w_0$$

$$+i((\overset{\circ}{N} + \overset{\circ}{M})\overset{\circ}{\Psi}_0,\overset{\circ}{\Psi}_0)w_0 = 0 \quad . \tag{9.6.29}$$

Since $\overset{\circ}{N} + \overset{\circ}{M}$ is antisymmetric, the last term in (9.6.29) is equal to zero.

Evidently, the vector $\overset{\circ}{\Psi}_0 \in \overset{\circ}{I}_H$ can be normalized in an arbitrary way. It is convenient to put

$$\overset{\circ}{\Psi}_0 = \frac{1}{\sqrt[4]{g}}\overset{\circ}{\xi}$$

where $\overset{\circ}{\xi} \in \overset{\circ}{I}_H$ and $(\overset{\circ}{B}\overset{\circ}{\xi},\overset{\circ}{\xi}) = 1$. Then by (9.6.29) we obtain the parabolic equation

$$\mathcal{L}w_0 = 0 \tag{9.6.30}$$

for w_0. The vector $\overset{\circ}{\xi}$ defines the polarization of the solution U in the zeroth approximation. Let us show that (9.6.28) determines the vector $\overset{\circ}{\xi}$ up to its initial

254

value $\overset{\circ}{\xi}(\tau_0)$. Let dim $\overset{\circ}{I}_H = r$; Let $l_1, l_2, \ldots, l_r$ be a basis in $\overset{\circ}{I}_H$ satisfying

$$(\overset{\circ}{B}l_i, l_j) = \begin{cases} 0 & i \neq j \\ 1 & i = j \end{cases} \tag{9.6.31}$$

and let $\alpha^1, \alpha^2, \ldots, \alpha^r$ be the coordinates of $\overset{\circ}{\xi}$ in this basis. Clearly (9.6.28) is equivalent to the system of equations

$$\left(2\overset{\circ}{B}\frac{d\overset{\circ}{a}{}^s l_s}{d\tau} + \frac{d\overset{\circ}{B}}{d\tau}\alpha^s l_s + (\overset{\circ}{N} + \overset{\circ}{M})\alpha^s l_s, l_j \right) = 0 \quad , \; j = 1, 2, \ldots, r \tag{9.6.32}$$

which by (9.6.31) can be transformed into

$$\frac{d\alpha^j}{d\tau} = \mathcal{P}_{js}\alpha^s \tag{9.6.33}$$

where

$$\mathcal{P}_{js} = \frac{1}{2}\left[\left(\overset{\circ}{B}\frac{d l_j}{d\tau}, l_s \right) - \left(\overset{\circ}{B}\frac{d l_s}{d\tau}, l_j \right) - ((\overset{\circ}{N} + \overset{\circ}{M})l_s, l_j) \right]$$

is an antisymmetric matrix. Because $\mathcal{P}_{js} = -\mathcal{P}_{sj}$ the system (9.6.33) has the first integral

$$\sum_{s=1}^{r} (\alpha^s)^2 = \text{const}$$

which can be rewritten as $(\overset{\circ}{B}\overset{\circ}{\xi}, \overset{\circ}{\xi}) = \text{const}$. Thus, if the initial value $\overset{\circ}{\xi}(\tau_0)$ of $\overset{\circ}{\xi}$ satisfies $(\overset{\circ}{B}\overset{\circ}{\xi}, \overset{\circ}{\xi}) = 1$, then this condition will also be satisfied for any other value of τ.

If $H(x, p)$ is a simple root of (9.6.6), the space J_H is one-dimensional, and $\overset{\circ}{\xi} = \alpha^1 l_1$. Since $\mathcal{P}_{11} = 0$, the system (9.6.33) reduces to $d\alpha^1/d\tau = 0$. Thus $\alpha^1 = \text{const}$, and the condition $(\overset{\circ}{B}\overset{\circ}{\xi}, \overset{\circ}{\xi}) = 1$ will be satisfied if we put $\alpha^1 = 1$.

Thanks to the fact that the matrices R, L and T appearing in $\mathcal{L}$ are the same as those in the Hamiltonian system (9.6.13), the parabolic equation (9.6.30) can be solved by the same means as the parabolic equation of Sect. 9.3. Equation (9.6.30), just as (9.3.1), allows a sequence $w_0 = w_{0,n}$ of solutions ($n = n_1, n_2, \ldots, n_m$ is a multi-index and the n_i are nonnegative integers) satisfying

$$w_{0,n} \to 0 \text{ as } |\nu| \to \infty \quad .$$

The fundamental solution $w_{0,0} = w_{0(0,0,\ldots,0)}$ is expressed by (9.3.9) in which the matrices Y and P must be replaced by the fundamental matrices Y and Q of solutions to the Hamiltonian system (9.6.13):

$$w_{0,0} = \frac{\text{const}}{\sqrt{\det Y}} \exp\left[\frac{i}{2}(QY^{-1}\nu, \nu) \right]. \tag{9.6.34}$$

The solutions w_{0n} are constructed using the creation operator

$$\Lambda_j^* = \frac{1}{i}(Y_j^*, \nabla) - (Q_j^*, \nu) \tag{9.6.35}$$

(where Y_j and Q_j are the jth columns of the matrices Y and Q, and $*$ indicates complex conjugation) according to the formula

$$w_{0n}(\tau, \nu) = (\Lambda_1^*)^{n_1}(\Lambda_2^*)^{n_2} \ldots (\Lambda_m^*)^{n_m} w_{0,0}$$

$$= H_n(\tau, \nu)\exp\left[\frac{i}{2}(QY^{-1}\nu, \nu)\right] \ . \tag{9.6.36}$$

Here $H_n(\tau, \nu)$ is a polynomial in $\nu_1, \nu_2, \ldots, \nu_m$ of degree $|n| = n_1 + n_2 + \ldots + n_m$, whose coefficients depend on τ. Note that (9.6.35 and 36) are completely analogous to (9.3.20 and 24).

Thus, in the zeroth-order approximation, the solution of (9.6.3) concentrated near the ray γ is

$$U \sim \frac{e^{i\omega\tau}}{\sqrt[4]{g(\tau)}} w_{0n} \overset{\circ}{\xi} \ ,$$
$$n = (n_1, n_2, \ldots, n_m) \ , \quad n_i \geq 0 \ .$$

Higher-order approximations can be found by the techniques of Sect. 9.4.

9.7 Notes on the Literature

The starting point for the study of eigenfunctions concentrated in the neighborhood of a closed geodesic was the work of *Vainshtein* [9.12, 13] and *Bykov* [9.14], and heuristic considerations due to *Buslaev*, who suggested that corresponding to every[11] closed geodesic l on a closed surface S (in three-dimensional euclidean space) is a particular set of eigenvalues for the infinite region external to S (see the paper by *Babič* [9.15] containing remarks by *Buslaev*). *Buslaev* was the first to suggest the hypothesis that the principal term of the asymptotic expansion for the eigenvalues of the region exterior to S has the form

$$k_{pq} = \frac{2\pi q}{L} - \xi_p \left(\frac{\pi q}{L}\right)^{1/3} \int_0^L \frac{ds}{L\varrho^{2/3}}$$

where $q \gg 1$, $p = O(1)$ and p, q are integers. The ray method in the small enables us to find the next approximation for k_{pq} as $q \to \infty$, see [9.16].

Sections 9.1–3 present the work of *Babič* [9.8]. The case $m = 1$ was considered earlier in a paper by *Babič* and *Lazutkin* [9.17]. The properties of the

---

[11] It seems more likely to us that, as in the problem of natural oscillations of an ellipsoid, the eigenfunctions of the region external to S can be concentrated only in the neighborhood of a stable geodesic. The methods of Chap. 9 do not permit determination of the asymptotic expansions of the eigenfunctions if the geodesic is unstable, and it seems to us that this serves (to some extent) to confirm our point of view.

coordinate system $(s, y_1, ..., y_m)$ that are used in Sects. 9.1–3 are well known (see, e.g., the monograph of *Milnor* [9.5]).

The exposition of the theory of the Jacobi equation (Sect. 9.2) follows that of the classical theory of linear hamiltonian systems with periodic coefficients [9.18]. Formula (9.3.9) reflects the fact that the Schrödinger equation with a quadratic potential is satisfied exactly by the quasiclassical approximation. For the solution of the Schrödinger equation in this case, see [9.19, 20]. The reduction of the problem of finding the asymptotic expansions of the eigenfunctions and eigenvalues to one of finding Floquet solutions of (9.1.14), and the derivation of the formulas for these solutions is due to *Babič*, see Sect. 9.3). The solution of matrix Riccati equations analogous to (9.3.3) can be found in the textbook by *Gel'fand* and *Fomin* [9.21].

In quantum mechanics, operators which satisfy the commutation relations (9.3.21, 22), and the operator (9.3.27) play an important role. It is not difficult to see the analogy between the operators Λ_j and Λ_j^*, and the creation and annihilation operators used previously by *Lazutkin* in [9.22]. Functions analogous to $U_{q_1,...,q_m}$ (Sect. 9.3) have repeatedly been encountered by researchers in the development of the mathematical apparatus of quantum mechanics (see for example [9.20]).

The proof that (9.1.14) has no other Floquet solutions besides U_q (Sect. 9.3) is the work of *Lazutkin*, and is published here for the first time. *Pyshkina* [9.23] constructed the higher approximations (Sect. 9.4). The idea of finding the $U_q^{(j)}$ in the form of a linear combination of functions U_r was suggested by *Lazutkin*. He had used another method earlier to study the case $m = 1$, see [9.24].

The natural oscillations of a three-dimensional region Ω which are concentrated in the neighborhood of a closed geodesic l lying on its surface were studied in the first approximation by *Babič* and *Lazutkin* [9.17]. The higher-order approximations for this problem (Sect. 9.5) were found by *Dymchenko* using a technique developed by *Kirpičnikova* [9.25] in connection with the theory of narrow-beam propagation of elastic surface waves. Eigenfunctions concentrated near closed ray trajectories in anisotropic media were considered by *Buldyrev* [9.26]. Further generalization of these constructions was carried out by *Buldyrev* and *Nomofilov*, and is described in Sect. 9.6.

Investigations in recent years have shown that there is a close relation between the solutions considered in this chapter and ray solutions with complex eikonals [9.27–29]. The monograph of *Maslov* [9.27] contains many analogs and generalizations (particularly nonlinear ones) of the solutions studied in Chaps. 9 and 10.

10. Multiple-Mirror Resonators

This chapter will examine the natural oscillations of three-dimensional open multiple-mirror resonators filled with an inhomogeneous medium of propagation velocity $c = c(M)$ without considering diffraction losses.

Interest in problem of this type has arisen in recent years in connection with the development of lasers, in particular gas lasers. In the mathematical sense, the study of natural oscillations of resonators leads approximately to eigenvalue/eigenfunction problems for the scalar Helmholtz equation. In particular, the natural oscillations of a two-mirror resonator are eigenfunctions of the bouncing-ball type that were considered in Chaps. 5 and 8.

The principal terms of the asymptotic formulas for the eigenfunctions are derived by the parabolic equation method. To solve the boundary value problem that arises here for the parabolic equation, we use the technique developed previously for a closed geodesic (see Sects. 7.1–3). The chapter concludes with construction of higher approximations for the natural oscillations of a multiple-mirror resonator.

10.1 The Multiple-Mirror Resonator and Formulation of the Problem

Consider the multiple-mirror resonator filled with a nonhomogeneous medium shown in Fig. 10.1. Let $I = \int d\sigma/c(M)$ be the functional of geometric optics, where $d\sigma$ is an element of length in the three-dimensional Euclidean space R_3.

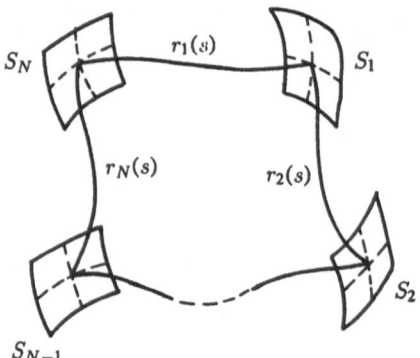

Fig. 10.1. Multiple-mirror resonator

258

Let $\vec{r} = \vec{r}_1(s)$, $\vec{r} \in R^3$ be an extremal of this functional, and let s be the arc length along this extremal measured from some point. Let us assume that where $s = s_1$ the initial extremal is incident on an ideally reflecting mirror S_1 and is reflected therefrom in accordance with the laws of geometrical optics, generating the extremal $\vec{r} = \vec{r}_2(s)$, which is reflected in turn at $s = s_2$ from the surface S_2 of the second mirror and so forth. Finally, after reflection from the $N - 1$-th mirror S_{N-1} an extremal $\vec{r} = \vec{r}_N(s)$ arises that is reflected at $s = s_N$ from the N-th mirror S_N. Let us assume that the mirrors S_k are arranged to that the extremal reflected from S_N coincides with the initial extremal $\vec{r} = \vec{r}_1(s)$. In this way we get a closed curvilinear N-gon l_N with vertices $\vec{r} = \vec{r}_k(s_k)$ located on the mirrors S_k, $k = 1, 2, \ldots, N$, and sides formed by the extremals of I. Obviously, on l_N the first variation of the functional is $\delta I = 0$, and therefore l_N can be called an *extremal closed chain*.

Using the analogy that exists between geometrical optics and the classical mechanics of a point mass, we can say that l_N is a trajectory in R^3 of the periodic motion of a point mass in a corresponding problem in mechanics. This analogy will be very useful for us in what follows.

We will call the closed piecewise-smooth extremal $\vec{r} = \vec{r}(s)$ that makes up the sides of l_N the *resonator axis*.

We will assume that along the axis the arc length s varies over a range from 0 to s_N so that s_N is the length of the resonator axis and $\vec{r}(s_N) = \vec{r}(0)$, and we will call the k-th side of the N-gon $\vec{r} = \vec{r}_k(s)$, $s_{k-1} \le s \le s_k$ the *axis of the k-th arm of the resonator*.

On each mirror S_k (Fig. 10.2) the vectors

$$\vec{t}_k = \frac{d\vec{r}_k}{ds}\bigg|_{s=s_k} \quad \text{and} \quad \vec{t}_{k+1} = \frac{d\vec{r}_{k+1}}{ds}\bigg|_{s=s_k}$$

determine the plane of incidence. The normal to the surface of the k-th mirror lies in this plane and forms equal angles δ_k with the vectors $-\vec{t}_k$ and $\vec{t}_{k+1}$.

In the vicinity of each mirror we introduce an orthogonal coordinate system ζ_{1k}, ζ_{2k}, ζ_{3k} in the following way: the ζ_{1k} axis is directed along the bisector

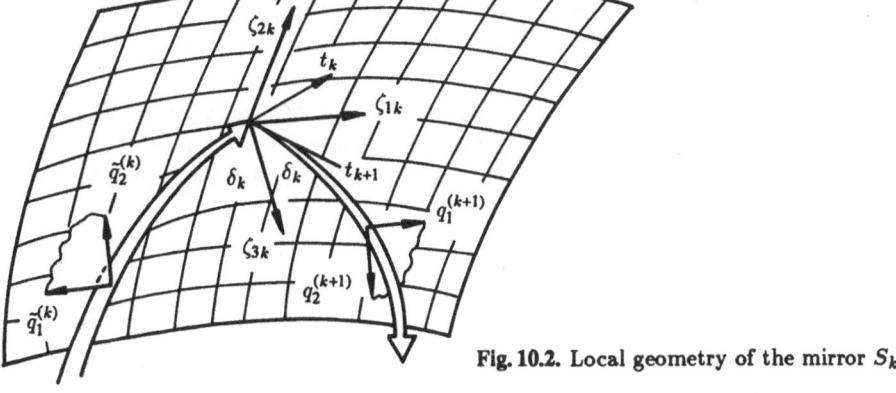

Fig. 10.2. Local geometry of the mirror S_k

of the angle formed by the vectors $\vec{t}_k$ and $\vec{t}_{k+1}$, the ζ_{3k} axis is directed along the bisector of the angle between $-\vec{t}_k$ and $\vec{t}_{k+1}$ (i.e. along the normal to S_k), and the ζ_{2k} axis is directed perpendicular to the plane of incidence and in such a way that $(\zeta_{1k}, \zeta_{2k}, \zeta_{3k})$ form a right-handed coordinate system. In the vicinity of the k-th vertex we will assume that the surface of mirror S_k is an infinitely differentiable surface (this means that the function $\zeta_{3k} = \zeta_{3k}(\zeta_{1k}, \zeta_{2k})$, that gives the explicit equation of S_k has continuous derivatives of all orders in the vicinity of $\zeta_{1k} = \zeta_{2k} = 0$). However, for most of the arguments of this chapter it will be enough if the functions $\zeta_{3k} = \zeta_{3k}(\zeta_{1k}, \zeta_{2k})$ have a finite number of derivatives. For the constructions of Sects. 10.2–7 we require only the quadratic terms in the expansion of the ζ_{3k} in the neighborhood of the origin:

$$\zeta_{3k} = D_{11}^{(k)} \zeta_{1k}^2 + 2D_{12}^{(k)} \zeta_{1k}\zeta_{2k} + D_{22}^{(k)} \zeta_{2k}^2 + \ldots = (D^{(k)} \vec{\zeta}_k, \vec{\zeta}_k) + \ldots \; ;$$

(10.1.1)

$$D^{(k)} = \begin{Vmatrix} D_{11}^{(k)} & D_{12}^{(k)} \\ D_{12}^{(k)} & D_{22}^{(k)} \end{Vmatrix} \; , \quad \vec{\zeta}_k = (\zeta_{1k}, \zeta_{2k}) \; .$$

Assuming that the field in the resonator can be described by a scalar function u, the problem for the eigenfunctions of the resonator can be formulated as follows: find functions u which oscillate in the neighborhood of the resonator axis, and that rapidly approach zero with increasing distance from the resonator axis within terms of order $O(\sqrt{\omega})$ (or higher), satisfying the equation

$$\left[\Delta + \frac{\omega^2}{c^2(M)} \right] u(M) = 0 \; ; \quad u(M) \neq 0$$

(10.1.2)

and the boundary conditions

$$u|_{S_k} = 0 \; \text{(I)} \quad \text{or} \quad \left. \frac{\partial u}{\partial n} \right|_{S_k} = 0 \; \text{(II)} \quad , \quad k = 1, 2, \ldots, N \; .$$

(10.1.3)

The real number ω is taken as a large parameter of the problem. It is clear that there will not be a subsequence of the eigenfunctions of (10.1.2, 3) concentrated in the vicinity of the axis for any arbitrary resonator (i.e., for any arbitrary extremal chain l_N). However, later on we shall see that in a resonator which is stable in the first approximation we can construct solutions of (10.1.2, 3) which are asymptotically concentrated in the neighborhood of the resonator axis as $\omega \to \infty$.

Thus the concept of resonator stability in the first approximation plays a fundamental part in studying the natural oscillations of resonators [as in other analogous problems of finding asymptotic expansions for eigenfunctions (see Chaps. 5–9)].

10.2 Conditions of Resonator Stability in the First Approximation

Near each arm of the resonator we can introduce a coordinate system (s, ν_1, ν_2):

$$\vec{r}(M) = \vec{r}(s) + \nu_1 \vec{n}(s) + \nu_2 \vec{b}(s) \quad , \tag{10.2.1}$$

where $\vec{r}(M)$ is the radius vector of the arbitrary point M, $\vec{r}(s)$ is the radius vector of some point on the axis, and $\vec{n}(s)$ and $\vec{b}(s)$ are the principal normal and binormal of the axis respectively. This coordinate system is regular in the neighborhood of the resonator axis. Generally speaking, the (s, ν_1, ν_2) coordinate system is not orthogonal, and therefore it is inconvenient to use.

Let us introduce a coordinate system (s, q_1, q_2) by means of the formula

$$\vec{r}(M) = \vec{r}(s) + q_1 \vec{e}_1(s) + q_2 \vec{e}_2(s) \quad , \tag{10.2.2}$$

where the unit vectors $\vec{e}_1, \vec{e}_2$ are rotated relative to vectors $\vec{n}$ and $\vec{b}$ through an angle $\vartheta = \vartheta(s)$:

$$\vec{e}_1 = \vec{n} \cos \vartheta - \vec{b} \sin \vartheta \quad ,$$
$$\vec{e}_2 = \vec{n} \sin \vartheta + \vec{b} \cos \vartheta \quad . \tag{10.2.3}$$

We determine the function $\vartheta = \vartheta(s)$ from the condition [compare (2.5.18)]

$$\frac{d\vartheta}{ds} = \frac{1}{T(s)} \quad , \quad \vartheta(s) = \int_0^s \frac{ds}{T(s)} + \vartheta(0) \quad , \tag{10.2.4}$$

where $1/T(s)$ is the torsion of the axis at the given point. Using the Frenet formula [10.1]

$$\frac{d^2 \vec{r}}{ds^2} \equiv \frac{d\vec{t}}{ds} = K(s)\vec{n} \quad ,$$

$$\frac{d\vec{n}}{ds} = \frac{\vec{b}}{T(s)} - K(s)\vec{t}$$

$$\frac{d\vec{b}}{ds} = -\frac{\vec{n}}{T(s)}$$

[where $K(s)$ is the curvature of the axis] and (10.2.3, 4), we get for the derivatives of the unit vectors $\vec{e}_1, \vec{e}_2$:

$$\frac{d\vec{e}_1}{ds} = -K(s) \cos \vartheta \vec{t} \quad , \quad \frac{d\vec{e}_2}{ds} = -K(s) \sin \vartheta \vec{t} \quad . \tag{10.2.5}$$

Formulas (10.2.2 and 5) imply:

$$d\sigma^2 = d\vec{r}(M) \cdot d\vec{r}(M) = [1 - K(s)(q_1 \cos \vartheta + q_2 \sin \vartheta)]^2 ds^2$$
$$+ dq_1^2 + dq_2^2 = G^2 ds^2 + dq_1^2 + dq_2^2 \quad . \tag{10.2.6}$$

Thus the coordinate system (s, q_1, q_2)[1] is orthogonal, and its metric coefficients are equal to $G, 1, 1$, where

$$G = 1 - K(s)(q_1 \cos \vartheta + q_2 \sin \vartheta) \quad . \tag{10.2.7}$$

Formulas (10.2.3 and 4) enable us to construct orthogonal coordinate systems independently in each arm of the resonator (in the k-th arm of the resonator $s_{k-1} \leq s \leq s_k$, $q_1 = q_1^{(k)}$, $q_2 = q_2^{(k)}$, $k = 1, 2, \ldots, N$) which we will use later on. Let us note that, as is implied by (10.2.4), the coordinate system $(s, q_1^{(k)}, q_2^{(k)})$ in each arm of the resonator is uniquely determined by the value of $\vartheta_k(s)$ for any s.

In the following it will be convenient to assume that s varies over the entire axis: $-\infty < s < +\infty$. Then functions that are defined and are single-valued in the neighborhood of the resonator axis can be treated as periodic functions of s with period s_N.

The equation for the axis of the k-th arm of the resonator in this coordinate system takes the form $q_1^{(k)} = 0$, $q_2^{(k)} = 0$.

Let us turn to the Fermat functional

$$I = \int \frac{1}{c(s, \vec{q})} \frac{d\sigma}{ds} ds \quad , \quad \vec{q} = (q_1, q_2) \quad ,$$

where $d\sigma$ is an element of arc length. In the following discussion we will use the expression for I in Hamiltonian form:

$$I = \int [p_1 dq_1 + p_2 dq_2 - H(s, \vec{q}, \vec{p}) ds] \quad , \tag{10.2.8}$$

where the p_i are generalized momenta conjugate to the coordinates q_i, and the Hamiltonian H is defined by the formulas

$$p_i = \frac{1}{c(s, \vec{q})} \frac{\partial}{\partial \dot{q}_i} \frac{d\sigma}{ds} \quad , \quad \dot{q}_i = \frac{dq_i}{ds} \quad , \quad i = 1, 2 \quad , \tag{10.2.9}$$

$$H(s, \vec{q}, \vec{p}) = -\frac{G}{c(s, \vec{q})} \sqrt{1 - c^2(s, \vec{q})(p_1^2 + p_2^2)} \quad . \tag{10.2.10}$$

Once again, G has the form (10.2.7). In the vicinity of the resonator axis $H(s, \vec{q}, \vec{p})$ is expanded in the power series

$$H(s, \vec{q}, \vec{p}) = -\frac{1}{c(s, 0)} + \sum_{k=2}^{\infty} H_k(s, \vec{q}, \vec{p}) \quad , \tag{10.2.11}$$

where $H_k(s, \vec{q}, \vec{p})$ is a homogeneous polynomial of degree k in the canonical variables $\vec{q}$ and $\vec{p}$, with coefficients that depend on s. Let us note that in expansion

---

[1] These coordinates are completely analogous to the coordinates introduced in Sect. 9.1. Formulas (10.2.5) imply that the unit vectors $\vec{e}_1, \vec{e}_2$ undergo parallel displacement along a ray if by this we understand parallel displacement in the Riemannian space where the metric is defined by the integral $\int d\sigma/c$.

(10.2.11) there are no linear terms in $\vec{p}$ or $\vec{q}$. The vanishing of the derivatives of $H(s, \vec{q}, \vec{p})$ with respect to p_j and q_j ($j = 1, 2$) on the resonator axis is implied by the canonical system of equations for rays

$$\frac{dq_j}{ds} = \frac{\partial H}{\partial p_j} \quad ; \quad \frac{dp_j}{ds} = -\frac{\partial H}{\partial q_j} \quad (j = 1, 2) \tag{10.2.12}$$

and by the fact that the resonator axis $q_1 = q_2 = 0$ is itself a ray. Simple calculations lead to the formula

$$H_2(s, \vec{q}, \vec{p}) = c(s, 0)\frac{p_1^2 + p_2^2}{2} + \frac{1}{2c(s, 0)}(\Psi\vec{q}, \vec{q}) \quad , \tag{10.2.13}$$

where $\vec{q} = (q_1, q_2)$ and the real symmetric matrix Ψ takes the form

$$\|\Psi_{ij}(s)\| = \frac{1}{c(s, 0)}\left\|\frac{\partial^2 c}{\partial q_i \partial q_j}\right\|_{s,0} \quad , \quad i, j = 1, 2 \quad . \tag{10.2.14}$$

We will carry out further analysis only in the first approximation, dropping all H_k, $k \geq 3$ in (10.2.11).

In the first approximation, the functional I (10.2.8) is replaced by the functional

$$I_0 = \int \left\{ p_1 dq_1 + p_2 dq_2 + \left[\frac{1}{c(s, 0)} - H_2(s, \vec{q}, \vec{p})\right] ds \right\} \quad , \tag{10.2.15}$$

for which the Euler equation takes the form

$$\frac{d\vec{q}}{ds} = c(s, 0)\vec{p} \quad , \quad \frac{d\vec{p}}{ds} = -\frac{1}{c(s, 0)}\Psi\vec{q} \quad , \tag{10.2.16}$$

where again $\vec{q} = (q_1, q_2)$, $\vec{p} = (p_1, p_2)$. Formulas (10.2.15 and 16) hold in each arm of the resonator. The Hamiltonian system (10.2.16) will be the basis for our subsequent constructions. We shall use the vector $\vec{\chi}_{(k)}(s) = [\vec{q}^{(k)}(s), \vec{p}^{(k)}(s)]$ to denote a solution of (10.2.16) defined on the k-th side of l_N, i.e. where $s_{k-1} \leq s \leq s_k$, and we will call this solution a ray in the k-th arm of the resonator ($k = 1, 2, \ldots, N$).

Let us consider the reflection of rays from the surfaces of the resonator mirrors. We assume that in the k-th arm the vector $\vec{\chi}_k(s)$ gives a ray incident on the mirror S_k at the k-th vertex $\vec{r}(s_k)$ of the polygon l_N, while $\vec{\chi}_{k+1}(s)$ describes a ray arising in the $(k+1)$-th arm after reflection from S_k. The geometrical optics law of reflection of the rays is equivalent to vanishing of the first variation of the Fermat functional $\int d\sigma/c$ in the first approximation, under the conditions that the piecewise-continuous curves over which the integration is carried out have one point on the reflecting surface, and that their ends are fixed. We will seek the reflected rays (in the first approximation) from the condition that the first variation of I_0 be zero [10.2]

$$\delta I_0 = \left[\sum_{j=1}^{2} p_i^{(k)} \delta q_j^{(k)} + \left(\frac{1}{c(s, 0)} - H_2^{(k)}\right) \delta s^{(k)}\right]_{s_k}$$

$$-\left[\sum_{j=1}^{2} p_j^{(k+1)} \delta q_j^{(k+1)} + \left(\frac{1}{c(s,0)} - H_2^{(k+1)}\right) \delta s^{(k+1)}\right]_{s_k},$$

(10.2.17)

where $H_2^{(k)}$ in accordance with (10.2.13) is expressed in terms of the canonical variables $q_j^{(k)}$, $p_j^{(k)}$, while the variations $\delta q_j^{(k)}$, $\delta q_j^{(k+1)}$, $\delta s^{(k)}$, $\delta s^{(k+1)}$ are related by the equation of the surface of the k-th mirror (10.1.1), the subscript of S_k designating that expression (10.2.17) is being considered on the k-th mirror. Let the coordinates $\tilde{q}_1^{(k)}$, $\tilde{q}_2^{(k)}$, associated with the incident ray be selected in such a way that at the point s_k (see Fig. 10.2) the direction of the $\tilde{q}_2^{(k)}$ and ζ_{2k} axes are the same, while the $\tilde{q}_1^{(k)}$ axis that lies in the plane (ζ_{1k}, ζ_{3k}), makes an angle of $\pi - \delta_k$ with the ζ_{1k} axis. In a similar way let the coordinates $q_1^{(k+1)}$, $q_2^{(k+1)}$ be such that at s_k the directions of the $q_2^{(k+1)}$ and ζ_{2k} axes are opposite each other, while the $q_1^{(k+1)}$ axis makes an angle δ_k with the ζ_{1k} axis. Up to second-order infinitesimal terms, the coordinate systems $(\tilde{q}_1^{(k)}, \tilde{q}_2^{(k)}, s^{(k)})$ and $(q_1^{(k+1)}, q_2^{(k+1)}, s^{(k+1)})$ can be considered near $\zeta_{1k} = \zeta_{2k} = \zeta_{3k} = 0$ as rectangular cartesian coordinate systems[10]. For $\tilde{q}_j^{(k)}$ $(j = 1, 2)$ and $q_j^{(k+1)}$ we have the formulas

$$\tilde{q}_1^{(k)} = -\zeta_{1k} \cos \delta_k - \zeta_{3k} \sin \delta_k \quad , \quad \tilde{q}_2^{(k)} = \zeta_{2k}$$

(10.2.18)

and

$$q_1^{(k+1)} = \zeta_{1k} \cos \delta_k - \zeta_{3k} \sin \delta_k \quad , \quad q_2^{(k+1)} = -\zeta_{2k} \quad ,$$

(10.2.19)

which are valid up to quadratic terms. Setting

$$s^{(k)} = s_k + \tau_k \quad , \quad s^{(k+1)} = s_k + \tau_{k+1} \quad ,$$

(10.2.20)

we get, to the same accuracy,

$$\tau_k = \zeta_{1k} \sin \delta_k - \zeta_{3k} \cos \delta_k \quad ,$$
$$\tau_{k+1} = \zeta_{1k} \sin \delta_k + \zeta_{3k} \cos \delta_k \quad .$$

(10.2.21)

From (10.1.1, 10.2.18 and 19) we readily see that on the surface of the mirror S_k (also in the linear approximation)

$$q_j^{(k+1)}(s_k) + \tilde{q}_j^{(k)}(s_k) = 0 \quad .$$

(10.2.22)

We now set the coefficients of the independent variations in (10.2.17) equal to zero. Using formulas (10.1.1 and 10.2.18–22) we get

---

[2] In conformity with the notation used before, arc length varies continuously along the resonator axis in the range $0 \le s \le s_N$. Here the index k means that $s^{(k)}$ varies along the k-th side of l_N.

$$p_1^{(k+1)}(s_k) + \tilde{p}_1^{(k)}(s_k) + q_1^{(k)} \tan^2 \delta_k \left[\frac{\partial}{\partial t_k} \left(\frac{1}{c} \right)_{s_k} - \frac{\partial}{\partial t_{k+1}} \left(\frac{1}{c} \right)_{s_k} \right]$$

$$+ \frac{4}{c(s_k,0)} \left[-\frac{D_{11}^{(k)}}{\cos \delta_k} \tilde{q}^{(k)}(s_k) + D_{12}^{(k)} \tilde{q}_2^{(k)}(s_k) \right]$$

$$+ 2q_1^{(k)} \tan \delta_k \left[\frac{\partial}{\partial \tilde{q}_1^{(k)}} \left(\frac{1}{c} \right) + \frac{\partial}{\partial q_1^{(k+1)}} \left(\frac{1}{c} \right) \right]$$

$$+ O(p^2 + q^2) = 0$$

$$p_2^{(k+1)}(s_k) + \tilde{p}_2^{(k)}(s_k) - \frac{4}{c(s_k,0)}$$

$$\times \left[-D_{12}^{(k)} \tilde{q}_1^{(k)}(s_k) + D_{22}^{(k)} \cos \delta_k \tilde{q}_2^{(k)}(s_k) \right] + O(p^2 + q^2) = 0 \quad ,$$

$$(10.2.23)$$

where $\partial/\partial t_k$, $\partial/\partial t_{k+1}$ are derivatives along the vectors $\vec{t}_k$, $\vec{t}_{k+1}$ (see Fig. 10.2). In the first approximation the conditions of reflection of rays from the resonator mirrors as expressed by (10.2.22 and 23) represent a linear transformation in the vector space $\vec{\chi}_k$:

$$\vec{\chi}_{k+1}(s_k) = \gamma_k \tilde{\vec{\chi}}_k(s_k) \quad , \quad k = 1, 2, \ldots, N \quad . \tag{10.2.24}$$

The reflection matrices γ_k for which $\det \gamma_k = 1$, take the form

$$\gamma_k = \left\| \begin{matrix} -I & 0 \\ \tilde{D}^{(k)} & -I \end{matrix} \right\| \quad , \tag{10.2.25}$$

where I is the identity matrix, and $\tilde{D}^{(k)}$ denotes the second-order symmetric matrix

$$\tilde{D}^{(k)} = \frac{4}{c(s_k,0)} \left\| \begin{matrix} \dfrac{D_{11}^{(k)}}{\cos \delta_k} & -D_{12}^{(k)} \\ -D_{12}^{(k)} & D_{22}^{(k)} \cos \delta_k \end{matrix} \right\|$$

$$- \left\{ \tan^2 \delta_k \left[\frac{\partial}{\partial t_k} \left(\frac{1}{c} \right)_{s_k} - \frac{\partial}{\partial t_{k+1}} \left(\frac{1}{c} \right)_{s_k} \right] \right.$$

$$\left. + 2 \tan \delta_k \left[\frac{\partial}{\partial \tilde{q}_1^{(k)}} \left(\frac{1}{c} \right) + \frac{\partial}{\partial q_1^{(k+1)}} \left(\frac{1}{c} \right) \right] \right\} \left\| \begin{matrix} 1 & 0 \\ 0 & 0 \end{matrix} \right\| \quad . \tag{10.2.26}$$

Let us note that the explicit expression for reflection matrices γ_k given by formulas (10.2.25 and 26) is determined by the special choice of coordinate systems in which the incident and reflected rays are given.

Furthermore, let the coordinate system used be fixed by (10.2.19) for the $k+1$-th arm of the resonator at $s = s_k$. Then when $s = s_{k+1}$ the directions of the axes of the selected coordinate system will not be the same as with the direction of the axes of the special coordinate system $\tilde{q}_1^{(k+1)}$, $\tilde{q}_2^{(k+1)}$, associated with the coordinates $(\zeta_{1,k+1}, \zeta_{2,k+1}, \zeta_{3,k+1})$ by (10.2.18). Thus

$$\begin{pmatrix} \tilde{q}_1^{(k+1)} \\ \tilde{q}_2^{(k+1)} \end{pmatrix} = w_{k+1} \begin{pmatrix} q_1^{k+1} \\ q_2^{k+1} \end{pmatrix} \quad , \tag{10.2.27}$$

where w_{k+1} is an orthogonal matrix whose determinant is equal to -1 since w describes transformation from a left-handed to a right-handed coordinate system. Since the components of w do not depend on s,

$$\begin{pmatrix} \tilde{p}_1^{(k+1)} \\ \tilde{p}_2^{(k+1)} \end{pmatrix} = w_{k+1} \begin{pmatrix} p_1^{(k+1)} \\ p_2^{(k+1)} \end{pmatrix} \quad .$$

Combining this last formula with (10.2.27) we get

$$\vec{\chi}_{k+1} = \begin{Vmatrix} w_{k+1} & 0 \\ 0 & w_{k+1} \end{Vmatrix} \vec{\chi}_{k+1} = W_{k+1}\vec{\chi}_{k+1}, \tag{10.2.28}$$

where $\det W_{k+1} = 1$. The matrix W_{k+1} is determined by the properties of the resonator. In view of the fact that in the first approximation the reflection of rays from resonator mirrors is described by the linear transformation (10.2.24), the propagation of a ray in a resonator can be described in the form of a linear one-to-one mapping of some vector space R^4 onto itself.

Let us now go on to the construction of the so called *monodromy operator E* that performs this mapping. (The monodromy operator introduced in this chapter is completely analogous to the monodromy operator considered in Sect. 9.2).

Let the fourth-order matrix $\Pi_k(s)$ be a fundamental matrix of the canonical system (10.2.16) considered on the k-th side of l_N where $s_{k-1} \leq s \leq s_k$, $k = 1, 2, \ldots, N$, i.e. its columns are linearly independent solutions of (10.2.16).

By means of the $\Pi_k(s)$, a ray in each arm of the resonator is defined by specifying a four-dimensional real vector $\vec{A}_k$ which is indepedent of s, by the formula

$$\vec{\chi}_k(s) = \Pi_k(s)\vec{A}_k \quad . \tag{10.2.29}$$

The reflection of rays from the resonator mirrors is, by virtue of formulas (10.2.24, 28, and 29), described by a linear transformation performed by non-singular matrices F_k,

$$\vec{A}_{k+1} = \Pi_{k+1}^{-1}(s_k)\gamma_k W_k \Pi_k(s_k)\vec{A}_k = F_k\vec{A}_k \quad ,$$
$$F_k = \Pi_{k+1}^{-1}(s_k)\gamma_k W_k \Pi_k(s_k). \tag{10.2.30}$$

If the vector $\vec{A}_k(0)$ defines some initial ray in the k-th arm of the resonator, then after a single circuit of this ray around the resonator, we get a new ray in the k-th arm which is described by $\vec{A}_k(1)$, where

$$\vec{A}_k(1) = F_{k-1}\ldots F_2F_1F_N\ldots F_{k+1}F_k\vec{A}_k(0) \equiv E_k\vec{A}_k(0). \tag{10.2.31}$$

After n circuits of the ray around the resonator we have

$$A_k(n) = E_k^n\vec{A}_k(0) \quad . \tag{10.2.32}$$

The fourth-order, non-singular, real matrix E_k performs a mapping of the space of solutions of the canonical systems (10.2.16) considered on the k-th side of l_N, onto itself, i.e., it is a monodromy operator. Setting $k = 1, 2, \ldots, N$, we get N matrices E_k; by virtue of (10.2.31) $E_{k+1} = F_k E_k F_k^{-1}$, and therefore the eigenvalues λ_j of E_k do not depend on k.

Let us next prove that det $E_k = 1$. Let us note first of all that the known Liouville identity [10.3] and the Hamiltonian nature of (10.2.16) imply that det $\Pi_\eta(s)$, $\eta = 1, 2, \ldots, N$ does not depend on s (Sect. 9.2). Furthermore, det E_k does not depend on the choice of the fundamental matrices $\Pi_\eta(s)$. In fact, replacing one fundamental system of solutions that form the columns of matrices $\Pi_\eta(s)$ by another leads to the replacement of $\Pi_\eta(s)$ by the matrix $\tilde{\Pi}_\eta(s) = \Pi_\eta(s)C_\eta$, where $C_\eta = $ const is a nonsingular matrix. Replacing Π_η by $\Pi_\eta C_\eta$ leads to the replacement of E_η by $C_\eta^{-1} E_\eta C_\eta$.

Let us now select the fundamental system of solutions so that det $\Pi_\eta(s) \equiv 1$, $\eta = 1, 2, \ldots, N$ (which can obviously be done). From (10.2.30, 31) and the relations det $W_k = 1$, det $\gamma_k = 1$, we readily get the desired equality det $E_k = 1$.

The relations det $\Pi_\eta(s) = $ const and det $\gamma_k = 1$ are equivalent to the conservation of the volume $\int \int \int \int_\Omega dq_1\, dq_2\, dp_1\, dp_2$, if each point moves in accordance with (10.2.16) and the "boundary conditions" (10.2.24–28) at s_k.

The expression $\int \int \int \int_\Omega dq_1\, dq_2\, dp_1\, dp_2$ is called an *absolute integral invariant of fourth order* [10.4] of the canonical system of (10.2.16).

The resonator is called *stable in the first approximation* if for every fixed $\vec{A}_k(0)$, $\|\vec{A}_k(n)\| < M < +\infty$ as $n \to \infty$ (which agrees with the definition of stability in the first approximation of a closed geodesic given in Chap. 9).

In order for a resonator to be stable in the first approximation, it is necessary and sufficient that:

a) *the eigenvalues λ_j of the matrix E_k are equal to one in magnitude, i.e., $|\lambda_j| = 1$, $j = 1, 2, 3, 4$;*

b) *the elementary divisors of E_k are simple.*

The preceding chapter contains the proof of a similar assertion for the case of a closed geodesic, which is easily extended to the case of a resonator. Therefore we will restrict ourselves to put a few clarifications. If there are multiple divisors among the elementary divisors of E_k, then we reduce E_k by a similarity transformation to Jordan form, and the condition $\|\vec{A}_k(n)\| < M < \infty$ will not be satisfied even if all eigenvalues $|\lambda_j| = 1$.

The characteristic equation det $(E_k - \lambda I) = 0$ will be palindromic, which can be demonstrated by repeating the arguments of Sect. 9.2 applied to the given case.

Let us denote the coefficients of the characteristic equation of E_k by a_1, a_2; then

$$\det(E_k - \lambda I) = \lambda^4 - a_1 \lambda^3 + a_2 \lambda^2 - a_1 \lambda + 1 = 0 \quad . \tag{10.2.33}$$

Setting $\nu = (\lambda + \lambda^{-1})/2$, (10.2.33) becomes

$$\nu^2 - \frac{a_1}{2}\nu + \frac{a_2 - 2}{4} = 0 \quad . \tag{10.2.34}$$

A sufficient condition for the resonator to be stable in the first approximation is that the roots of (10.2.34) be distinct and lie in the interval $(-1,1)$. This requirement leads to the system of inequalities

$$a_1^2 - 4(a_2 - 2) > 0 \quad ,$$

$$|a_1 \pm \sqrt{a_1^2 - 4(a_2 - 2)}| < 4 \quad .$$

From now on we will consider only resonators that are stable in the first approximation. In this case E_k has four linearly independent eigenvectors $\vec{h}_k^{(j)}$, $j = 1, 2, 3, 4$, satisfying the condition

$$E_k \vec{h}_k^{(\alpha)} = e^{i\varphi_\alpha} \vec{h}_k^{(\alpha)} \quad , \quad E_k \vec{h}_k^{(2+\alpha)} = e^{-i\varphi_\alpha} \vec{h}_k^{(2+\alpha)} \quad ,$$

$$\vec{h}_k^{(2+\alpha)} = \vec{h}_k^{(\alpha)*} \quad (\alpha = 1, 2) \quad . \tag{10.2.35}$$

The components of $\vec{h}_k^*$ are the complex conjugates of those of $\vec{h}_k$.

The numbers $e^{\pm i\varphi_\alpha}$ are called the *multipliers* of (10.2.16), and the φ_α are called *Floquet indices*. The Floquet indices are defined by (10.2.35) up to a multiple of 2π. Later on, in Sect. 10.7, the determination of the φ_α will be made more precise.

For each eigenvector $\vec{h}_k$, (10.2.29) can be used to construct a solution $\vec{\chi}_k^{(j)}(s)$ of the canonical system of (10.2.16) considered on the chain l_N. This solution has the following property by virtue of (10.2.31 and 35):

$$\vec{\chi}_k^{(\alpha)}(s + s_N) = e^{i\varphi_\alpha} \vec{\chi}_k^{(\alpha)}(s) \quad , \quad \alpha = 1, 2 \quad ,$$

$$\vec{\chi}_k^{(2+\alpha)}(s + s_N) = e^{-i\varphi_\alpha} \vec{\chi}_k^{(2+\alpha)}(s) \quad ,$$

$$\vec{\chi}_k^{(2+\alpha)} = \vec{\chi}_k^{(\alpha)*} \quad . \tag{10.2.36}$$

From now on in the symbols for these solutions, which are called *Floquet solutions* of system of (10.2.16), we shall omit the subscript for the resonator arm k, understanding that in the k-th arm of the resonator, i.e. where $s_{k-1} \leq s \leq s_k$,

$$\vec{\chi}^{(j)}(s) \equiv \vec{\chi}_k^{(j)}(s) \quad .$$

In the next section we will deal with some properties of these Floquet solutions, which are required for what will come after.

10.3 Some Properties of the Solutions of (10.2.16) on l_N

In Chap. 9, in connection with the problem of a closed geodesic on a Riemannian manifold, we examined a canonical linear system of $2m$ equations. The properties of its solutions established there are general properties of solutions to any linear

Hamiltonian system of equations whose coefficients are periodic. In the problem of a multiple-mirror resonator we come to the examination of the canonical linear system of Eqs. (10.2.16) on the closed polygon l_N: a special feature in this case is that solutions of (10.2.16) on two adjacent sides of l_N must be related by (10.2.24). However, the reflection matrices are such that the general properties of solutions of linear Hamiltonian equations with periodic coefficients hold in the present case as well. In the following development we will need certain properties of Floquet solutions. We will derive them here.

Let $\vec{\chi} = (\chi_1, \chi_2, \chi_3, \chi_4) = (q_1, q_2, p_1, p_2)$ be an arbitrary vector in the four-dimensional space $(\vec{q}, \vec{p})$: $\vec{q} = (q_1, q_2)$, $\vec{p} = (p_1, p_2)$. We introduce the operator J

$$J\vec{\chi} = J(\vec{q}, \vec{p}) = (\vec{p}, -\vec{q})$$

and the dot product

$$\vec{\chi}' \cdot \vec{\chi}'' = \sum_{\eta=1}^{4} \chi_\eta' \overline{\chi}_\eta'' = \vec{q}' \cdot \vec{q}'' + \vec{p}' \cdot \vec{p}'' \quad .$$

If $\vec{\chi}(s)$ and $\vec{\chi}'(s)$ are vectors that satisfy (10.2.16) and the associated conditions at s_k (Sect. 10.2), then, as can be readily demonstrated,

$$[J\vec{\chi}(s)] \cdot \vec{\chi}'(s) = \text{const} \quad , \tag{10.3.1}$$

i.e., the J-inner product is constant over the entire chain l_N.

Let $\vec{\chi}(s)$ and $\vec{\chi}'(s)$ be two Floquet solutions with different corresponding multipliers λ and λ' ($|\lambda| = |\lambda'| = 1$); then

$$[J\vec{\chi}(s)] \cdot \vec{\chi}'(s) \equiv 0 \quad , \tag{10.3.2}$$

by virtue of the fact that after a complete "lap" around l_N, (10.3.1) becomes multiplied by the factor $\lambda\overline{\lambda}' = \lambda/\lambda'$ which is different from unity.

In the case where the four multipliers $e^{\pm i\varphi_\alpha}$ ($\alpha = 1, 2$) include some which are the same, we can satisfy (10.3.2) with any pair $\vec{\chi}, \vec{\chi}'$ ($\vec{\chi} \not\equiv \vec{\chi}'$) of the four Floquet solutions that form the fundamental system of solutions (replacing the Floquet solutions corresponding to identical multipliers by linear combinations of them where necessary). In the case $\vec{\chi} = \vec{\chi}'$ we will then have

$$[J\vec{\chi}] \cdot \vec{\chi} \neq 0 \quad . \tag{10.3.3}$$

By normalizing (and renormalizing where necessary) the Floquet solutions, we can arrange that they satisfy

$$[J\vec{\chi}^{(\alpha)}] \cdot \vec{\chi}^{(\beta)} = i\delta_{\alpha\beta} \quad , \quad \alpha, \beta = 1, 2 \quad , \tag{10.3.4}$$

and

$$[J\vec{\chi}^{(\alpha)}] \cdot \vec{\chi}^{(\beta)*} = [J\vec{\chi}^{(\alpha)}] \cdot \vec{\chi}^{(\beta+2)} = 0 \quad . \tag{10.3.5}$$

We will omit the detailed derivation of the properties of these Floquet solutions to avoid reproducing Sect. 9.2.

10.4 Formulation of the Parabolic Equation for the Problem

Based on problem (10.1.2 and 3), assuming that $\omega \to \infty$, let us formulate a parabolic equation for this problem; the exact solution of this equation will then be found.

The Helmholtz operator in (s, q_1, q_2) coordinates given by (10.2.2) can, by considering (10.2.6), be written in the form

$$\Delta + \frac{\omega^2}{c^2(M)} = \frac{1}{G} \left\{ \frac{\partial}{\partial s} \left(\frac{1}{G} \frac{\partial}{\partial s} \right) + \sum_{i=1}^{2} \frac{\partial}{\partial q_i} \left(G \frac{\partial}{\partial q_i} \right) + \frac{\omega^2 G}{c^2(s, q)} \right\} \ .$$

In accordance with the parabolic equation method (Chap. 6) let us convert from q_i coordinates to x_i coordinates using

$$x_i = \sqrt{\omega} q_i \ , \quad i = 1, 2 \ , \tag{10.4.1}$$

and introduce in place of u the function U:

$$u = \sqrt{\frac{c(s, 0)}{c(0, 0)}} U(s, x_1, x_2; \omega) \exp \left\{ i\omega \int_0^s \frac{ds}{c(s, 0)} \right\} \ . \tag{10.4.2}$$

We will assume that the coordinates x_1, x_2 and the function U along with all of their partial derivatives with respect to s and x_i are $O(1)$ as $\omega \to \infty$.

Substituting (10.4.1) and (10.4.2) into the Helmholtz equation, expanding the left side in powers of ω and dividing both sides by ω, we get after some simple calculations,

$$\mathcal{L}U + O(\omega^{-1/2}) = 0 \ ,$$

where we use the following notation:

$$\mathcal{L} = \frac{2i}{c(s, 0)} \frac{\partial}{\partial s} + \nabla^2 - \frac{1}{c^2(s, 0)} (\Psi \vec{x} \cdot \vec{x}) \tag{10.4.3}$$

$$\nabla = \left(\frac{\partial}{\partial x_1} , \frac{\partial}{\partial x_2} \right) \ , \quad \vec{x} = (x_1, x_2)$$

where $\Psi(s)$ is the real symmetric matrix defined by (10.2.14).

In this and the next two sections we will be interested only in the principal terms of the eigenfunction u. To construct this function it is convenient to introduce another function in place of U, which unlike U will satisfy

$$\mathcal{L}V = 0 \tag{10.4.4}$$

exactly. By virtue of the fact that

$$\mathcal{L}(U - V) = O(\omega^{-1/2}) \ ,$$

it is natural to expect that $U - V = O(\omega^{-1/2})$. This will be confirmed in the calculations of Sect. 10.7.

270

For the sake of definiteness, we will consider only the boundary condition $u|_S = 0$. To satisfy this condition, we will construct the waves reflected from the mirrors. We will carry out all constructions in the first approximation.

Let a wave given by the principal terms of its expansion, concentrated in the neighborhood of the k-th arm of the resonator and incident onto mirror S_k, have the form

$$v^{(k)} = \sqrt{\frac{c(s,0)}{c(0,0)}} V^{(k)}(s, \tilde{x}_1^{(k)}, \tilde{x}_2^{(k)}; \omega) \exp\left[i\omega \int_0^s \frac{ds}{c(s,0)}\right] \quad ,$$

$$(s_{k-1} \le s \le s_k) \quad , \tag{10.4.5}$$

where

$$\tilde{x}_i^{(k)} = \sqrt{\omega} \tilde{q}_i^{(k)} \quad (i = 1, 2) \quad .$$

The wave v^{k+1}, reflected from the surface of mirror S_k and propagating in the $(k+1)$-th arm of the resonator, will be sought in the same form as (10.4.5), but will be taken as a function of the coordinates $x_i^{(k+1)} = \sqrt{\omega} q_i^{(k+1)}$. In the case of the boundary condition $u|_S = 0$, we get the following condition relating the incident to the reflected wave:

$$u^{(k)}|_{S_k} + u^{(k+1)}|_{S_k} = 0 \quad , \tag{10.4.6}$$

where $u^{(k)}$ (or $u^{(k+1)}$) is a wave concentrated in the neighborhood of the k-th [or $(k+1)$-th] arm of the resonator. Expressing $u^{(k)}$ and $u^{(k+1)}$ in terms of $V^{(k)}$ and $V^{(k+1)}$ to leading order [see (10.4.5)], we get from (10.4.6) after a few calculations boundary conditions relating $V^{(k)}$ to $V^{(k+1)}$. Let us now derive these boundary conditions.

Using the notation $z_{jk} = \sqrt{\omega} \zeta_{jk}$, $j = 1, 2, 3$, and the coordinate transformation formulas (10.2.18, 19), we write $u^{(k)}$ and $u^{(k+1)}$ in z_{jk} coordinates. Considering (10.2.20, 21) and the fact that the ζ_{jk} coordinates on the surface of S_k are related by (10.1.1), $z_{1k}, z_{2k} = O(1)$, $z_{3k} = O(\omega^{-1/2})$ as $\omega \to \infty$, we expand the left side of (10.4.6) in powers of ω. For instance, if the point (s, z_{1k}, z_{2k}) lies on the surface of the mirror S_k, we have

$$\omega \int_0^s \frac{ds}{c(s,0)}\bigg|_{S_k} = \omega \int_0^{s_k} \frac{ds}{c(s,0)} + \sqrt{\omega} z_{1k} \frac{\sin \delta_k}{c(s_k, 0)} - (D^{(k)} \vec{z}_k, \vec{z}_k) \frac{\cos \delta_k}{c(s_k, 0)}$$

$$+ \frac{1}{2}\left[\frac{d}{ds}c^{-1}(s,0)\right]_{s=s_k} z_{1k}^2 \sin^2 \delta_k + O(\omega^{-1/2}) \quad ,$$

where $\vec{z}_k = (z_{1k}, z_{2k})$, and the symmetric matrix $D^{(k)}$ is determined by (10.1.1) of the mirror surface in ζ_{jk} coordinates. Moreover

$$U^{(k)}(s, \tilde{x}_1^{(k)}, \tilde{x}_2^{(k)}, \omega)|_{S_k} = U^{(k)}(s_k, -z_{1k} \cos \delta_k, z_{2k}) + O(\omega^{-1/2}) \quad .$$

Here $U^{(k)}$ is the function U, see (10.4.2) in the neighborhood of the k-th arm of the resonator. Substituting all these equations into (10.4.6) and expressing $u^{(k)}$, $u^{(k+1)}$ in terms of $V^{(k)}$ and $V^{(k+1)}$ to leading order, we get the desired boundary conditions for $V^{(k)}$ and $V^{(k+1)}$. Thus, to find V we get the following boundary value problem:

$$\mathcal{L}V = 0 \quad , \tag{10.4.7}$$

$$(V = V^{(k)}, \ \mathcal{L} = \mathcal{L}_k \text{ when } s_{k-1} \leq s \leq s_k, \ k = 1, 2, \ldots, N, N+1 \quad ,$$
$$\mathcal{L}_{N+1} \equiv \mathcal{L}_1) \quad ,$$

$$V^{(k)}(s_k, -z_{1k} \cos \delta_k, z_{2k}) \exp\left[-i\frac{\cos \delta_k}{c(s_k, 0)}(\boldsymbol{D}^k \vec{z}_k, \vec{z}_k)\right]$$
$$+V^{(k+1)}(s_k, z_{1k} \cos \delta_k, -z_{2k}) \exp\left[i\frac{\cos \delta_k}{c(s_k, 0)}(\boldsymbol{D}^{(k)} \vec{z}_k, \vec{z}_k)\right] = 0. \tag{10.4.8}$$

We will be concerned with solutions of (10.4.7) which are concentrated in the vicinity of the resonator axis, i.e. such that

$$|V| \to 0 \quad \text{when} \quad \sqrt{x_1^2 + x_2^2} \to \infty. \tag{10.4.9}$$

We impose one more important restriction on the solution to be found. The natural oscillations must be single-valued in the neighborhood of the resonator axis. To satisfy this requirement, it is sufficient to require that after reflection from the N-th mirror the corresponding wave must be identical with the wave in the vicinity of the first arm of the resonator, i.e.

$$v^{(N+1)} \equiv v^{(1)} \quad .$$

This condition can be satisfied in the following way: assume that we have constructed a function V satisfying equation (10.4.7) and conditions (10.4.8 and 9), and that this function, after a complete lap around the resonator, acquires a multiplier whose magnitude is one, i.e.

$$V(s + s_N, x) = e^{i\kappa}V(s, x) \ (V \not\equiv 0, \ \text{Im} \ \kappa = 0) \quad . \tag{10.4.10}$$

Such a function is an analog of the Floquet solution for the partial differential equation (10.4.7).

If V is the Floquet solution of (10.4.7–9) that satisfies condition (10.4.10), then for V to be single-valued in the neighborhood of the resonator axis it is necessary and sufficient that the frequency ω satisfy the equation

$$\omega \int_0^{s_N} \frac{ds}{c(s, 0)} = 2\pi m_0 - \kappa \quad (m_0 \text{ an integer}) \quad . \tag{10.4.11}$$

Since ω is a large number, $m_0 \gg 1$ in (10.4.11).

We will solve (10.4.7–10) in a certain sense in explicit form, or more precisely we will express both the function V and the Floquet index κ in terms of the Floquet solution and the corresponding Floquet indices φ_α of the Hamiltonian system considered in Sects. 10.2 and 3.

10.5 Integration of the Equation $\mathcal{L}V = 0$

The parabolic equation $\mathcal{L}V = 0$, where the operator $\mathcal{L}$ is defined by (10.4.3), differs but little from (9.1.13). This enables us to use the techniques of Chap. 9 to solve prolems (10.4.7–10), which is what we will do in this section. All the following constructions are equally applicable to each arm of the resonator, and therefore we will omit the subscript k denoting the side of l_N.

Let us take the two Floquet solutions $\chi^{(\alpha)}(s) = (\vec{q}^{(\alpha)}(s), \vec{p}^{(\alpha)}(s))$, $\alpha = 1, 2$, that were constructed in Sect. 10.2. From their components we form the following two second-order matrices whose columns are made up of the components of the vectors $\vec{q}^{(\alpha)}(s)$ and $\vec{p}^{(\alpha)}(s)$:

$$\hat{Q}(s) = \|\vec{q}^{(1)}(s), \ \vec{q}^{(2)}(s)\| \quad , \quad \hat{P}(s) = \|\vec{p}^{(1)}(s), \ \vec{p}^{(2)}(S)\| \quad . \tag{10.5.1}$$

Since the $\chi^{(\alpha)}(s)$ are solutions of the canonical system (10.2.16), the matrices $\hat{Q}$ and $\hat{P}$ satisfy the equations

$$\frac{d}{ds}\hat{Q} = c(s, 0)\hat{P} \quad , \quad \frac{d}{ds}\hat{P} = -\frac{1}{c(s, 0)}\Psi\hat{Q} \quad . \tag{10.5.2}$$

Equations (10.3.4 and 5) satisfied by the Floquet solutions imply that for all s, $\det \hat{Q}(s) \neq 0$ (for the proof, see Sect. 9.2), and the equality

$$\hat{Q}^*\hat{P} - \hat{P}^*\hat{Q} = iI \quad , \tag{10.5.3}$$

is satisfied identically in s, where $\hat{P}^*$, $\hat{Q}^*$ are the hermitian conjugates of $\hat{P}$, $\hat{Q}$, and I denotes the 2×2 identity matrix. Moreover, (10.3.5) implies

$$\hat{Q}^T\hat{P} - \hat{P}^T\hat{Q} = 0, \tag{10.5.4}$$

where $\hat{Q}^T$, $\hat{P}^T$ are the transpose matrices of $\hat{Q}$ and $\hat{P}$. Then let us set

$$\Gamma(s) = \hat{P}(s)\hat{Q}^{-1}(s) \quad . \tag{10.5.5}$$

By means of simple calculations, using (10.5.2), just as in Sect. 9.1, we can readily demonstrate that the function

$$V(s, x_1, x_2) = \frac{\text{const}}{\sqrt{\det \hat{Q}(s)}}\exp\left[\frac{i}{2}(\Gamma\vec{x}, \vec{x})\right] \tag{10.5.6}$$

satisfies the parabolic equation $\mathcal{L}V = 0$. The matrix $\Gamma(s)$, by virtue of (10.5.4 and 5), is symmetric, $\Gamma^T = \Gamma$, and from (10.5.3) we see that

$$\text{Im}\,\Gamma(s) = \frac{\Gamma - \Gamma^*}{2i} = \frac{1}{2}(\hat{Q}\hat{Q}^*)^{-1} \quad . \tag{10.5.7}$$

Since the matrix on the right side in (10.5.7) is positive definite, the function $V(s, x_1, x_2)$, defined by formula (10.5.6), is concentrated in the neighborhood of the resonator axis: condition (10.4.9).

Having the solution (10.5.6) of $\mathcal{L}V = 0$ at hand, we can now construct other solutions of this equation that satisfy condition (10.4.9). Using the Floquet solutions $\vec{\chi}^{(j)}(s) = (\vec{q}^{(j)}(s), \vec{p}^{(j)}(s))$, constructed in Sect. 10.1, we form four operators Λ_α and Λ_α^* in accordance with the following formulas:

$$\Lambda_\alpha = \frac{1}{i}(\vec{q}^{(\alpha)}(s) \cdot \nabla) - (\vec{p}^{(\alpha)}(s) \cdot \vec{x}) \quad ,$$

$$\Lambda_\alpha^* = \frac{1}{i}(\vec{q}^{(2+\alpha)}(s) \cdot \nabla) - (\vec{p}^{(2+\alpha)}(s) \cdot \vec{x}) \quad (\alpha = 1, 2) \quad , \tag{10.5.8}$$

where $\nabla = (\partial/\partial x_1, \partial/\partial x_2)$ and $\vec{x} = (x_1, x_2)$. It is easily established by direct calculation that the operators Λ_α and Λ_α^* commute with $\mathcal{L}$ (10.4.3), and therefore for any integer $n > 0$ the functions $\Lambda_\alpha^n V$ and $\Lambda_\alpha^{*n} V$ satisfy equation (10.4.7) if the function V satisfies it.

By virtue of properties (10.3.4,5) of the Floquet solutions, the operators Λ_α, Λ_α^* defined by formula (10.5.8) satisfy the commutation relations

$$\Lambda_\alpha \Lambda_\beta - \Lambda_\beta \Lambda_\alpha = 0 \quad ,$$
$$\Lambda_\alpha^* \Lambda_\beta^* - \Lambda_\beta^* \Lambda_\alpha^* = 0, \tag{10.5.9}$$

$$\Lambda_\alpha \Lambda_\beta^* - \Lambda_\beta^* \Lambda_\alpha = \delta_{\alpha\beta} \quad , \tag{10.5.10}$$

where $\alpha, \beta = 1, 2$ and $\delta_{\alpha\beta}$ is the Kronecker delta. It can be verified by direct calculations that

$$\Lambda_\alpha V_{00}(s, x_1, x_2) = 0 \quad , \quad \Lambda_\alpha^* V_{00}(s, x_1, x_2) \neq 0 \quad , \tag{10.5.11}$$

where the function $V_{00} \equiv V$ is defined by (10.6.6), and consequently Λ_1 and Λ_2 can be called *annihilation operators*, while Λ_1^* and Λ_2^* are *creation operators*. Thus, by using Λ_1^* and Λ_2^* we can construct solutions of $\mathcal{L}V = 0$, that satisfy condition (10.4.9), and take the form

$$V_{m_1 m_2}(s, x_1, x_2) = \Lambda_1^{*m_1} \Lambda_2^{*m_2} V_{00}(s, x_1, x_2)$$
$$= V_{00}(s, x_1, x_2) \Phi_{m_1 m_2}(s, x_1, x_2) \quad , \tag{10.5.12}$$

where $m_1, m_2 = 0, 1, 2, 3, \ldots$, and $\Phi_{m_1 m_2}(s, x_1, x_2)$ is a polynomial in x_1, x_2 of degree $m_1 + m_2$ with coefficients dependent on s.

10.6 Eigenfunctions and Natural Frequencies of a Multiple-Mirror Resonator in the First Approximation

In Sect. 10.5, solutions of (10.4.7) were constructed in each arm of the resonator which satisfy the condition (10.4.9) of concentratedness in the vicinity of the resonator axis. To construct (in the first approximation) the eigenfunctions and natural frequencies of a multiple-mirror resonator, we must complete the solution of (10.4.7–10), which is what we will now do. Just as in Chap. 9, we shall first construct some special solution of (10.4.7–10), which we will call the *fundamental harmonic*. We will get other solutions – *higher harmonics* – of this problem by applying the operators Λ_j^*, $j = 1, 2$ (Sect. 9.3) to the fundamental. Let us take the solution written in $(q_1^{(k)}, q_2^{(k)})$ coordinates in the k-th arm of the resonator in the form

$$V_{00}^{(k)}(s, x_1^{(k)}, x_2^{(k)}) = M_k a_k^{-1/2}(s)\exp\left[\frac{i}{2}(\Gamma_k \vec{x}^{(k)}, \vec{x}^{(k)})\right], \tag{10.6.1}$$

where $M_k = \text{const}$,

$$a_k(s) = C_k \det \hat{Q}_k(s) \quad, \tag{10.6.2}$$

C_k is some constant, and $\hat{Q}_k(s)$ is the value of matrix $\hat{Q}(s)$ in the k-th arm of the resonator. Let us substitute functions of the form (10.6.1) into condition (10.4.8), first converting in function $V_{00}^{(k)}$ from $(q_1^{(k)}, q_2^{(k)})$ to $(\tilde{q}_1^{(k)}, \tilde{q}_2^{(k)})$ coordinates (Fig. 10.2). These coordinate systems are related by the linear transformation (10.2.27) where the matrix w_k is orthogonal, therefore the quadratic form in (10.6.1) is invariant to this coordinate transformation, while the function $\det \hat{Q}_k(s)$ changes sign since it is multiplied by $\det w_k$. As a result of the substitution we get

$$\tilde{\Gamma}_k(s_k) - \Gamma_{k+1}(s_k) = \tilde{D}^{(k)}, \quad k = 1, 2, \ldots, N \quad, \tag{10.6.3}$$

where $\tilde{D}^{(k)}$ is defined by (10.2.26), and the second condition

$$M_{k+1} a_{k+1}^{-1/2}(s_k) = -M_k \tilde{a}_k^{-1/2}(s_k) \quad, \tag{10.6.4}$$

where $\tilde{\Gamma}_k$ and $\tilde{a}_k$ denote Γ_k and a_k in $(\tilde{q}_1^{(k)}, \tilde{q}_2^{(k)})$ coordinates. It can be demonstrated that relation (10.6.3) for the matrices Γ_k is automatically satisfied by virtue of reflection conditions (10.2.24), which are satisfied by the Floquet solutions constructed in Sect. 10.2. This is implied by the explicit form of the reflection matrices γ_k and by (10.5.5), which defines $\Gamma(s)$ in terms of the matrices $\hat{P}(s)$ and $\hat{Q}(s)$ that are made up of the Floquet solutions.

Let us turn now to (10.6.4). We select the constants C_k in (10.6.2) so that the function $a(s) = a_k(s)$ for $s_{k-1} \leq s \leq s_k$, $k = 1, 2, \ldots, N, N + 1$, is continuous on the resonator axis. For this it is obviously sufficient that

$$C_{k+1} = C_k (\det w_k)^{-1} \quad, \quad k = 1, 2, \ldots, N, \quad C_1 = 1 \quad. \tag{10.6.5}$$

We take the branch of the square root $\sqrt{a(s)}$ so that this function will also be continuous on the resonator axis. Then (10.6.4) is considerably simplified and takes the form

$$M_{k+1} = -M_k = e^{-\pi i} M_k \quad,$$

where

$$M_{N+1} = e^{-i\pi N} M_1 \quad; \tag{10.6.6}$$

the constant M_1 remaining arbitrary.

As was noted in Sect. 10.5, the function $\det \hat{Q}(s)$ does not vanish for any s, and therefore neither does the complex-valued function $a(s)$, and consequently $\arg a(s)$ is a continuous function of s on the resonator axis, and the increment $\Delta \arg a = \arg a(s + s_N) - \arg a(s)$ is uniquely defined. Equation (10.6.5) implies the equality

$$C_{N+1} = \prod_{k=1}^{N} (\det w_k)^{-1} = (-1)^N \quad . \tag{10.6.7}$$

Taking into consideration the properties of Floquet solutions and formulas (10.6.2 and 7), we find that the magnitude of $a(s)$ is unchanged, and

$$\Delta \arg a = \varphi_1 + \varphi_2 - \pi N \quad . \tag{10.6.8}$$

This formula uniquely determines the sum of the Floquet indices $\varphi_1 + \varphi_2$. Let us note that in the case of an even number of mirrors the Floquet indices can be predefined so that $\varphi_1 + \varphi_2 = \Delta \arg a$.

From (10.6.1, 6, 8), and the equality $\Gamma_{N+1}(s) = \Gamma_1(s)$ we see that the function $V_{00} = V_{00}^{(k)}$ for $s_{k-1} \leq s \leq s_k$ actually solves the problem (10.4.7–10), the Floquet index κ, (10.4.10), being equal to

$$\kappa = -\tfrac{1}{2}(\pi N + \varphi_1 + \varphi_2) \quad .$$

Equation (10.4.11) now gives

$$\omega_{m_0 00} \int_0^{s_N} c^{-1}(s,0)ds = 2\pi \left(m_0 + \frac{N}{4} \right) + \frac{1}{2}(\varphi_1 + \varphi_2) \quad , \tag{10.6.9}$$

where the integer $m_0 \gg 1$.

Let us now pass on to the construction of the higher harmonics (transverse modes of the resonator). In this case, in each arm of the resonator we take the solution of (10.4.7) in the form

$$V_{m_1 m_2}^{(k)} = \Lambda_1^{*m_1} \Lambda_2^{*m_2} V_{00}^{(k)} = V_{00}^{(k)} \Phi_{m_1 m_2}^{(k)}(s, \vec{x}^{(k)}) \quad ,$$
$$\vec{x}^{(k)} = (x_1^{(k)}, x_2^{(k)}) \quad . \tag{10.6.10}$$

Here $V_{00}^{(k)}$ takes the form (10.6.1); however, the frequency that appears in the expression for $\vec{x}^{(k)}$, see (10.4.1), will be determined by (10.6.15) which will be derived below, rather than by (10.6.9). The functions $\Phi_{m_1 m_2}^{(k)}$, that enter into the right side of (10.6.10) are polynomials in $x_1^{(k)}$ and $x_2^{(k)}$ of degree $m_1 + m_2$ with coefficients dependent on s. Let us note that upon transformation from $(q_1^{(k)}, q_2^{(k)})$ coordinates to $(\tilde{q}_1^{(k)}, \tilde{q}_2^{(k)})$ coordinates the values of the polynomials $\Phi_{m_1 m_2}^{(k)}(s, \vec{x}^{(k)})$ remain unchanged. This is implied by the fact that when they undergo the transformation (10.2.27) (with an orthogonal matrix w_k), the operators Λ and the quadratic form $(\Gamma_k \vec{x}^{(k)}, \vec{x}^{(k)})$ do not change. In this connection it can be assumed that polynomials $\Phi_{m_1 m_2}^{(k)}$ in (10.6.10) are given in $(\tilde{q}_1^{(k)}, \tilde{q}_2^{(k)})$

coordinates, and the sign "~" that denotes this will be omitted in the following calculations.

We recall that at the mirrors the functions (10.6.10) must satisfy the reflection conditions (10.4.8). Since the functions $V_{00}^{(k)}$ satisfy these conditions, it remains for us to verify that

$$\Phi_{nm}^{(k+1)}(s, z_{1k} \cos \delta_k, -z_{2k}) = \Phi_{nm}^{(k)}(s_k, -z_{1k} \cos \delta_k, z_{2k}) \qquad (10.6.11)$$

is satisfied identically in z_{1k} and z_{2k}.

Like equality (10.6.3) for the case of the fundamental harmonic, (10.6.11) is identically satisfied automatically for every $k = 1, 2, \ldots, N$ by virtue of (10.2.24), which are satisfied by the Floquet solutions $\vec{\chi}^{(3)}(s)$ and $\vec{\chi}^{(4)}(s)$, that enter into the expressions for the operators Λ^* (10.5.8). This is verified by induction. Let $\Lambda^{(k)} = (1/i)(\vec{q}^{(k)} \cdot \nabla) - (\vec{p}^{(k)} \cdot \vec{x}^{(k)})$ be any of the operators Λ_α^*, $\alpha = 1, 2$ [see (10.5.8)] in the k-th arm of the resonator (consequently $\vec{q}^{(k)}, \vec{p}^{(k)}$ are components of one of the four Floquet solutions[3]). It is obvious that

$$\Lambda^{(k)} V_{00}^{(k)} = V_{00}^{(k)} ([\Gamma_k \vec{q}^{(k)} - \vec{p}^{(k)}], \vec{x}^{(k)}) = V_{00}^{(k)} \Phi_1^{(k)}$$

and analogously for the $(k + 1)$-th arm of the resonator. Identity (10.6.11) is equivalent to the relation

$$\vec{p}^{(k)}(s_k) + \vec{p}^{(k+1)}(s_k) = \Gamma_k \vec{q}^{(k)}|_{s_k} + \Gamma_{k+1} \vec{q}^{(k+1)}|_{s_k} \quad , \qquad (10.6.12)$$

the validity of which is readily established by using the explicit form of the reflection matrices γ_k (10.2.24) and (10.5.5). Now, in accordance with the method of induction let us assume that identity (10.6.11) is satisfied identically by the polynomials $\Phi_{m_1 m_2}^{(k+1)}$ and $\Phi_{m_1 m_2}^{(k)}$. We have

$$\begin{aligned} \Lambda^{(k)} V_{00}^{(k)} \Phi_{m_1 m_2}^{(k)} &= V_{00}^{(k)} \left[\Phi_{m_1 m_2}^{(k)} (\Gamma_k \vec{q}^{(k)} - \vec{p}^{(k)}, \vec{x}^{(k)}) \right. \\ &\quad \left. + \frac{1}{i} (\vec{q}^{(k)} \cdot \nabla) \Phi_{m_1 m_2}^{(k)} \right] \\ &= V_{00}^{(k)} \Phi_{m_1 + m_2 + 1}^{(k)} \end{aligned} \qquad (10.6.13)$$

$$(\Phi_{m_1 + m_2 + 1}^{(k)} = \Phi_{m_1 \, m_2 + 1}^{(k)} \text{ or } \Phi_{m_1 + 1 \, m_2}^{(k)})$$

and analogously in the $(k + 1)$-th arm of the resonator. From (10.6.12 and 13) we see that the polynomials $\Phi_{m_1 + m_2 + 1}^{(k)}$ and $\Phi_{m_1 + m_2 + 1}^{(k+1)}$ satisfy (10.6.11) for all $k = 1, 2, \ldots, N$.

This proves that identity (10.6.11) is automatically satisfied for any combination of operators $\Lambda^{(k)}$ in (10.6.10) for $V_{m_1 m_2}^{(k)}$.

Considering of the explicit form (10.5.8) of the operators Λ_1^* and Λ_2^* and properties (10.2.36) of the Floquet solutions of the Hamiltonian system (10.2.16),

---

[3] Let us emphasize that the index on the vectors $\vec{q}^{(k)}, \vec{p}^{(k)}$ here denotes the number of the resonator arm, rather than the number of the Floquet solution as in (10.5.8).

we find that the functions $V_{m_1 m_2}$ defined in the k-th arm by (10.6.10) will be Floquet solutions of (10.4.7–10), where the Floquet index

$$\kappa = - \sum_{a=1}^{2} \left(m_a + \frac{1}{2} \right) \varphi_a - \frac{\pi N}{2} \quad ,$$
$$m_a = 0, 1, 2, \ldots \ (m_a = O(1)) \quad . \tag{10.6.14}$$

For the principal term ω_m of the asymptotic expansion of the eigenvalues, (10.4.11) gives

$$\omega_m \int_0^{sN} \frac{ds}{c(s,0)} = 2\pi \left(m_0 + \frac{N}{4} \right) + \left(m_1 + \frac{1}{2} \right) \varphi_1 + \left(m_2 + \frac{1}{2} \right) \varphi_2 \quad ,$$

$$\tag{10.6.15}$$

where $m = (m_0, m_1, m_2)$, the integer $m_0 \gg 1$, and $m_1, m_2 = 0, 1, 2, \ldots$ ($m_1, m_2 = O(1)$). Now let us write out the expression for the function $v_m = v_{m_0 m_1 m_2}$, the principal term of the asymptotic expansion of the eigenfunction of (10.1.2 and 3). In the k-th arm of the resonator, v_m is equal to

$$v_m^{(k)} = M_k \sqrt{\frac{c(s,0)}{c(0,0)}} \Lambda_1^{*m_1} \Lambda_2^{*m_2} \frac{1}{\sqrt{a_k(s)}}$$

$$\times \exp\left[i\omega_m \int_0^s \frac{ds}{c(s,0)} + \frac{i}{2} (\Gamma_k \vec{x}^{(k)}, \vec{x}^{(k)}) \right]$$

$$= \sqrt{\frac{c(s,0)}{c(0,0)}} V_{m_1 m_2}^{(k)} \exp\left[i\omega_m \int_0^s \frac{ds}{c(s,0)} \right] = M_k \sqrt{\frac{c(s,0)}{c(0,0)}} \frac{1}{\sqrt{a_k(s)}}$$

$$\times \exp\left[i\omega_m \int_0^s \frac{ds}{c(s,0)} + \frac{i}{2} (\Gamma_k \vec{x}^{(k)}, \vec{x}^{(k)}) \right] \Phi_{m_1 m_2}(s, x_1^{(k)}, x_2^{(k)}) \quad .$$

$$\tag{10.6.16}$$

It can be easily demonstrated that function (10.6.16) has the properties presented at the end of Sect. 9.3.

10.7 Construction of the Higher Approximations

This section will describe how to construct higher approximations both for the natural frequencies and for the eigenfunctions of a multiple-mirror resonator, assuming that the Floquet indices and the number π are linearly independent over the ring of integers. In other words, we will assume that

$$\pi g_0 + \varphi_1 g_1 + \varphi_2 g_2 = 0 \quad , \tag{10.7.1}$$

where the g_i are integers (not necessarily positive), implies that

$$g_0 = g_1 = g_2 = 0 \quad .$$

In this case the eigenfunctions and eigenvalues can be constructed as formal series in the natural frequency of the first approximation, i.e. with respect to powers of the frequency ω_m determined from (10.6.15).

Let u be the desired eigenfunction. We seek an expansion of u in the form

$$u = \sum_{r=0}^{\infty} u_r \omega_m^{-r/2} = \sum_{r=0}^{\infty} \sqrt{\frac{c(s,0)}{c(0,0)}} \exp\left(i\omega_m \int_0^s \frac{ds}{c(s,0)}\right) U_r \omega_m^{-r/2}, \quad (10.7.2)$$

where $U_r^{(k)} = U_r^{(k)}(s, x_1^{(k)}, x_2^{(k)})$; $x_i^{(k)} = \sqrt{\omega_m} q_i^{(k)}$. (As before, we give an index k to the expressions for u, u_r and U_r in the vicinity of the k-th arm of the resonator.) Furthermore, we assume that

$$u_0^{(k)} = \sqrt{\frac{c(s,0)}{c(0,0)}} \exp\left(i\omega_m \int_0^{s_r} \frac{ds}{c(s,0)}\right) U_0^{(k)} = v_m^{(k)} \quad , \qquad (10.7.3)$$

where $m = (m_0, m_1, m_2)$ and $v_m^{(k)}$ is the value of v_m (the eigenfunction in the first approximation) in the k-th arm of the resonator. From (10.6.16) we find that $U_0^{(k)} = V_{m_1 m_2}^{(k)}$ is the value for the Floquet solution $V_{m_1 m_2}$ in the k-th arm of the resonator. The corresponding Floquet index is equal to

$$\kappa_m = -\sum_{a=1}^{2}\left(m_a + \frac{1}{2}\right)\varphi_a - \frac{1}{2}\pi N \qquad (10.7.4)$$

[see (10.6.14)].

Let us now substitute (10.7.2) into (10.1.2), assuming that the Helmholtz operator is written in (s, q_1, q_2) coordinates (see Sect. 10.4), and that

$$\omega = \omega_m + \frac{\delta_1}{\sqrt{\omega_m}} + \frac{\delta_2}{\omega_m} + \ldots \qquad (10.7.5)$$

$(\delta_1, \delta_2, \ldots$ are unknown).

Setting the coefficients $\omega_m^{-r/2+1}$ equal to zero, after a little manipulation we arrive at the equations

$$\sum_{j=0}^{r} \mathcal{L}^j U_{r-j} = 0 \quad , \qquad (10.7.6)$$

where the $\mathcal{L}^j$ are second-order differential operators applied to functions defined in the vicinity of the resonator axis.

The operator $\mathcal{L}^0$ is the same as the parabolic operator $\mathcal{L}$, see (10.4.3). The operators $\mathcal{L}^j$ $(j \geq 1)$ in the k-th arm of the resonator take the following form:

$$\mathcal{L}^j U = R_j^0 \frac{\partial U}{\partial s} + R_{j-2}^{00} \frac{\partial^2 U}{\partial s^2} + \sum_{a=1}^{2} R_{j+1}^a \frac{\partial U}{\partial x_a}$$

$$+ \left(R_{j+2} + \frac{\psi_j}{c^2(s,0)} \right) U \quad ; \quad j \geq 1, \quad R_{-1} = 0 \quad . \tag{10.7.7}$$

Here the indices k are left out for the sake of simplicity; R_j^0, R_{j-2}^{00}, R_{j+1}^a, and R_{j+2} are polynomials in x_1 and x_2 (with coefficients dependent on s) whose degrees do not exceed their subscripts. The parity of the polynomials[4] is the same as that of their subscripts. For the numbers ψ_j we have the formula

$$\psi_j = \delta_j + \delta_{j-1}\delta_3 + \delta_{j-2}\delta_4 + \ldots + \delta_4\delta_{j-2} + \delta_3\delta_{j-1} + \delta_j \quad . \tag{10.7.8}$$

We will assume that s ranges from $-\infty$ to $+\infty$ (see Sect. 10.1). Functions that are single-valued in the neighborhood of the resonator axis can also be considered as functions having period s_N (s_N is the length of the resonator axis). The operators $\mathcal{L}^j$ transform such functions into functions that are also s_N-periodic.

In fact, the following more general statement is true. Let the function $T(s, \vec{x})$, $-\infty < s < +\infty$, defined in the vicinity of the resonator axis satisfy a condition analogous to the Floquet condition (10.4.10),

$$T(s + s_N, \vec{x}) = e^{i\kappa} T(s, \vec{x}) \quad ; \tag{10.7.9}$$

then

$$\mathcal{L}^j T(s + s_N, \vec{x}) = e^{i\kappa} \mathcal{L}^j T(s, \vec{x}). \tag{10.7.10}$$

We will omit the proofs of these properties of the operators $\mathcal{L}^j$ to avoid undue complication of our development with awkward (and completely elementary) constructions.

Returning to the construction of the higher approximations U_r, we require that

1) The boundary condition

$$u^{(k)} + u^{(k+1)}|_{S_k} = 0 \tag{10.7.11}$$

be satisfied on the mirror S_k;

2) a wave be transformed into itself after traversing all arms of the resonator, i.e.

$$u^{(N+1)} \equiv u^{(1)} \quad ; \tag{10.7.12}$$

3) each approximation be concentrated in the vicinity of the resonator axis.

Satisfaction of conditions 1), 2) and 3) to any order in ω_m enables us to find sequentially all the functions U_r satisfying (10.7.6), and the numbers $\delta_1, \delta_2, \ldots, \delta_r$ [see (10.7.8)].

Let the functions $U_{r'}$ where $0 \leq r' < r$ be found in the form of a finite linear combination of the functions $V_{n_1 n_2}$ with coefficients dependent on s:

$$U_{r'} = \sum_{n_1, n_2} C_{n_1 n_2}^{r'}(s) V_{n_1 n_2}(s, \vec{x}) \quad , \tag{10.7.13}$$

---

[4] A polynomial $R(x_1, x_2)$ is called *even* (or *odd*) if $R(-x_1, -x_2) \equiv R(x_1, x_2)$ [or $R(-x_1, -x_2) \equiv -R(x_1, x_2)$].

where $V_{n_1 n_2} = \Lambda_1^{*n_1} \Lambda_2^{*n_2} V_{00}$ are the Floquet solutions of problem (10.4.7–10) with Floquet index κ_n:

$$\kappa_n = - \sum_{a=1}^{2} \left(n_a + \frac{1}{2} \right) \varphi_a - \frac{\pi N}{2} \quad , \quad n = (n_1, n_2) \quad . \tag{10.7.14}$$

[It will be shown later over which $n = (n_1, n_2)$ the summation is to be carried out in (10.7.13)]. Let us assume that we have found the functions $U_{r'}$, $r' < r$, such that

1) the boundary condition (10.7.11) is satisfied to order $\omega_m^{-r'/2} (r' < r)$, i.e. the conditions

$$\sum_{h=0}^{r'} \sqrt{\frac{c(s,0)}{c(0,0)}} \exp \left(i\omega_m \int_0^s \frac{ds}{c(s,0)} \right) U_h^{(k)} \omega_m^{-h/2}$$

$$+ \sum_{h=0}^{r'} \sqrt{\frac{c(s,0)}{c(0,0)}} \exp \left(i\omega_m \int_0^s \frac{ds}{c(s,0)} \right) U_h^{(k+1)} \omega_m^{-h/2} = O \left(\left(\omega_m^{-r'+1/2} \right) \right) \quad ,$$

$$r' = 0, 1, 2, \ldots, r - 1 \tag{10.7.15}$$

are satisfied on every mirror S_k;

2) $U_{r'}(s + s_N, \vec{x}) = e^{i\kappa_m} U_{r'}(s, \vec{x}) \quad , \quad r' = 0, 1, 2, \ldots, r - 1 \quad . \tag{10.7.16}$

Condition (10.7.16) guaratees the self-consistency condition (10.7.12) up to the corresponding approximation, since conditions (10.7.16) imply the s_N-periodicity of the functions

$$U_{r'} \exp \left(i\omega_m \int_0^s \frac{ds}{c(s,0)} \right), \quad r' < r \quad , \quad \text{where}$$

$$\omega_m \int_0^{s_N} \frac{ds}{c(s,0)} = 2\pi m_0 - \kappa_m = 2\pi m_0 + \sum_{a=1}^{2} \left(m_a + \frac{1}{2} \right) \varphi_a + \frac{1}{2} \pi N \quad ;$$

3) the equations

$$\sum_{j=0}^{r'} \mathcal{L}^j U_{r'-j} = 0 \quad , \quad r' < r \tag{10.7.17}$$

are satisfied.

We will assume that the numbers $\delta_{r'}$, $r' < r$ are known. Let us construct the functions U_r and the value of δ_r so that the conditions just listed hold for $r' = r$ as well.

Equation (10.7.6) can be written in the form

$$\mathcal{L} U_r = - \sum_{j=1}^{r} \mathcal{L}^j U_{r-j} = K_r - \frac{\psi_r}{c^2(s,0,0)} V_{m_1 m_2} \quad . \tag{10.7.18}$$

The function K_r is uniquely determined by the functions $U_{r'}$, $r' < r$, and the quantity ψ_r given in (10.7.8).

By virtue of (10.7.9, 10, and 16), the function K_r will satisfy condition (10.7.19) with $\kappa = \kappa_m$:

$$K_r(s + s_N, \vec{x}) = K_r(s, \vec{x}) e^{i\kappa_m} \quad ,$$

$$\kappa_m = -\sum_{a=1}^{2} \left(m_a + \frac{1}{2} \right) \varphi_a - \frac{\pi N}{2} \quad . \tag{10.7.19}$$

We see from expression (10.7.7) for the operators $\mathcal{L}^j$ that in the k-th arm of the resonator the function K_r has the form

$$K_r = \Psi_r^{(k)}(s, \vec{x}) V_{00}(s, \vec{x}) \quad ,$$

where $\Psi_r^{(k)}$ is a polynomial with s-dependent coefficients.

Arguing in the same way as in Sect. 9.3, it can be demonstrated that any polynomial is a linear combination of the polynomials $\Phi_{n_1 n_2}$ (Sect. 10.6), and therefore, see (10.6.10),

$$
\begin{aligned}
K_r(s, \vec{x}) &= \sum_{n_1, n_2} d_{n_1 n_2}^r(s) \Phi_{n_1 n_2}(s, \vec{x}) V_{00} \\
&= \sum_{n_1, n_2} d_{n_1 n_2}^r(s) V_{n_1 n_2}(s, \vec{x}) \quad .
\end{aligned} \tag{10.7.20}
$$

From condition (10.7.19) and the fact that $V_{n_1 n_2}$ is a Floquet solution of (10.4.7–10) with indices κ_n, we have

$$
\begin{aligned}
K_r(s + s_N, \vec{x}) &= e^{i\kappa_m} K_r(s, \vec{x}) \\
&= \sum_{n_1, n_2} d_{n_1 n_2}^r(s + s_N) e^{i\kappa_n} V_{n_1 n_2}(s, \vec{x}) \\
&= e^{i\kappa_m} \sum_{n_1, n_2} d_{n_1 n_2}^r(s) V_{n_1 n_2}(s, \vec{x}) \quad .
\end{aligned} \tag{10.7.21}
$$

Equation (10.7.21) and the linear independence of the functions $V_{n_1 n_2}(s, \vec{x})$[5] imply

$$d_{n_1 n_2}^r(s + s_N) = e^{i(\kappa_m - \kappa_n)} d_{n_1 n_2}^r(s) \quad . \tag{10.7.22}$$

We will now seek U_r in the form (10.7.13) (when $r' = r$). We require that U_r satisfy condition (10.7.16). In exactly the same way as conditions (10.7.22) were derived from (10.7.20 and 21), we now get the conditions

$$C_{n_1 n_2}^r(s + s_N) = e^{i(\kappa_m - \kappa_n)} C_{n_1 n_2}^r(s) \quad , \tag{10.7.23}$$

which are necessary and sufficient in order that U_r satisfy condition (10.7.16) ($r' = r$). Replacing the $U_{r'}(r' \le r)$ in (10.7.15) ($r' = r$), by their expressions

---

[5] The linear independence of functions $V_{n_1 n_2}$ is a consequence of the fact that they are non-zero Floquet solutions corresponding to different multipliers $e^{i\kappa_n}$. The linear independence of the $V_{n_1 n_2}$ could also be deduced from their orthogonality (Sect. 9.3).

(10.7.13), after a few calculations we find that when $r' = r$, (10.7.15) is equivalent to the conditions

$$C^r_{n_1 n_2}(s_k + 0) - C^r_{n_1 n_2}(s_k - 0) = e^r_{n_1 n_2}(s_k) \quad , \tag{10.7.24}$$

where $e^r_{n_1 n_2}(s_k)$ are known.

Finally, substituting (10.7.13) into (10.7.18), using (10.7.20) and the fact that the $V_{n_1 n_2}$ are solutions of the parabolic equation, we arrive at

$$\frac{2i}{c(s,0)} \frac{d}{ds} C^r_{n_1 n_2} = d^r_{n_1 n_2}(s) \quad , \quad (n_1, n_2) \neq (m_1, m_2) \quad ,$$

$$\frac{2i}{c(s,0)} \frac{d}{ds} C^r_{m_1 m_2} = d^r_{m_1 m_2}(s) + \psi_r \frac{1}{c^2(s,0)} \quad . \tag{10.7.25}$$

We will indicate over which pairs of numbers (n_1, n_2) the summation is to be carried out in (10.7.13). These are the pairs of numbers that correspond to inhomogeneous problems in (10.7.23–25), i.e. at these (n_1, n_2):

$$\sum_{k=1}^{N} |C^r_{n_1 n_2}(s_k)| + \max_{0 \leq s \leq s_N} |d^r_{n_1 n_2}(s)| > 0 \quad .$$

Let us now show that (10.7.25) and the auxiliary conditions (10.7.23 and 24) uniquely determine $C^r_{n_1 n_2}$, $(n_1, n_2) \neq (m_1, m_2)$, and the number δ_r. Problems (10.7.23–25) are analogous to classical inhomogeneous Sturm-Liouville problems. These are Fredholm problems in the sense that absence of nontrivial solutions for the homogeneous problem is necessary and sufficient for them to be solvable for any $d^r_{n_1 n_2}(s)$ and $e^r_{n_1 n_2}$.

Let us show that all of the problems (10.7.23, 24) are uniquely solvable when $(n_1, n_2) \neq (m_1, m_2)$. In fact, let $\hat{C}^r_{n_1 n_2}(s)$ be a solution of the homogeneous problem corresponding to (10.7.23, 24). Then

$$\hat{C}^{(r)}_{n_1 n_2} = \text{const} \quad . \tag{10.7.26}$$

By a virtue of the linear independence of π, φ_1 and φ_2 over the ring of integers[6] (see the beginning of this section)

$$e^{i(\kappa_m - \kappa_n)} \neq 1 \quad . \tag{10.7.27}$$

Equations (10.7.23) for $C^r_{n_1 n_2} = \hat{C}^r_{n_1 n_2}$, (10.7.26) and (10.7.27) are compatible only when $\hat{C}^r_{n_1 n_2} \equiv 0$.

The problem (10.7.23–25) for finding $C^r_{m_1 m_2}$ is posed at an eigenvalue of its operator: the (unique) solution of the homogeneous problem is a constant. Integrating both sides of (10.7.25) for $C^r_{m_1 m_2}$ over the entire resonator axis and using the periodicity of $C^r_{m_1 m_2}$, we get the necessary condition for solvability for problem (10.7.23–25), which is also sufficient, as can be confirmed by direct calculation:

---

[6] This is the only place where this linear independence will be needed.

$$2i \sum_{k=1}^{N} e^r_{m_1 m_2}(s_k) = \int_{l_N} d^r_{m_1 m_2}(s)c(s,0)ds + \psi_r \int_{l_N} \frac{ds}{c(s,0)} \quad . \qquad (10.7.28)$$

From (10.7.28) we uniquely determine the number δ_r that defines the correction to the natural frequency. Just as in Sect. 9.4, it can be proved that all the δ_r with odd indices are equal to zero. Using the fact that the eigenvalues of self-adjoint problems are real, and that (10.7.5) are asymptotic formulas for the eigenvalues of some self-adjoint problem (Sect. 7.6), we can prove that all the δ_r are real.

10.8 Notes on the Literature

This has been a detailed exposition of the work of *Popov* [10.5–7]. A two-mirror resonator with unfolded mirrors has been considered by *Boitsov* [10.8]. *Semenov* [10.9] is responsible for the results of the last section of the chapter.

11. The Field of a Point Source Located Near a Convex Curve

This chapter deals with the construction of the shortwave asymptotic expansion in the shadow zone for the two-dimensional problem of a point source located on the side of a curve S where whispering gallery waveforms cannot arise (see Sects. 5.1 and 4).[1]

11.1 Introduction

Let S be a sufficiently smooth convex curve (closed or extending to infinity) with positive curvature $\varrho^{-1}(s)$. Located at the point $M_0 = M_0(x_0, y_0)$ is a source, while the point $M = M(x, y)$ is the observation point. For the sake of simplicity, we will assume that the wave propagation velocity is a constant. Mathematically, the problem consists in constructing a solution of the Helmholtz equation with a Dirac δ-function on the right side

$$(\Delta + k^2)G(M_0, M; k) = -\delta(M - M_0) = -\delta(x - x_0)\delta(y - y_0), \qquad (11.1.1)$$

where the solution satisfies the condition

$$G(M_0, M; k)|_S = 0 \quad . \tag{11.1.2}$$

We will seek the solution of (11.1.1 and 2) in the class of functions that permit analytic continuation into the region $\mathrm{Im}\, k > 0$ and that decay exponentially as $\sqrt{x^2 + y^2} \to \infty$, if $\mathrm{Im}\, k > 0$. In this sense, the desired solutions will satisfy the principle of limiting absorption (see [11.1]). We will call the solution of (11.1.1 and 2) a *Green's function* in accordance with accepted terminology. If the point M lies in the region illuminated by M_0, then the asymptotic expansion of the Green's function as $k \to \infty$ is found by the ray method (see Sect. 2.6). Therefore, in this chapter we will pay particular attention to the construction of the asymptotic expansion of the Green's function in the shadow zone.

The methods used in this chapter enable us to get the principal term of the asymptotic expansion of the Green's function, and also to calculate the correction

---

[1] Another method of constructing the asymptotic expansion of the Green's function in the penumbra and shadow zone will be presented in Chap. 13. The methods of this chapter and Chap. 13 complement each other; we have already indicated the relation between them in Chap. 1.

terms of any order in $1/k$. The asymptotic expansion of the Green's function subject to an impedance condition

$$\frac{\partial G}{\partial n} + ikg(s)G\big|_S = 0 \quad , \tag{11.1.3}$$

frequently encountered in physics problems, is constructed in precisely the same way as in the case of the Dirichlet condition (11.1.2). To avoid undue complication of the exposition by similar formulas and arguments, we will not consider impedance boundary conditions in detail. In Sect. 11.5 we will present a brief derivation of asymptotic formulas relating to this case.

The methods of this chapter also enable us to construct the asymptotic expansion of the Green's function in the case of the Neumann boundary condition

$$\frac{\partial G}{\partial n}\bigg|_S = 0 \quad . \tag{11.1.4}$$

Let us note that passage to the limit as $g \rightarrow \infty$ in the formulas of Sect. 11.5 yields the asymptotic expansion of solutions of the Dirichlet problem (11.1.1 and 2). However, passage to the limit as $g \rightarrow 0$ in these formulas to obtain the formulas for the Neumann problem (11.1.1, 4) is not legitimate, as we make essential use of the assumption $g^{-1}(s) = O(1)$ in the derivation of these formulas in Sect. 11.5. This circumstance shows that the Neumann problem (11.1.1, 4) is in some ways set apart from the other types of problem.

In concluding this introductory section, we emphasize that many constructions of this chapter will be based on intuitive considerations and will be formal in nature. The exact proof of the asymptotic formulae was made by *Filippov* and *Zayaev* [11.2–4]. The proof is very complicated. They reduce the problem to consideration of an integral equation, use analytical continuation of the integral equation kernel in complex space etc.

11.2 The Green's Function for the Exterior of a Circle

As in the preceding chapters, we turn first of all to the investigation of an etalon problem. It seems natural to take as our etalon problem that of the Green's function for the exterior of a circle.

Consider in polar coordinates (r, ϕ) the infinitely-sheeted Riemann surface $r \geq \varrho$, $-\infty < \phi < +\infty$. We locate the source at the point $M_0(r_0, \phi_0)$. We denote the Green's function on the infinitely-sheeted surface by $\Gamma(r_0, \phi_0; r, \phi; k)$. This function must satisfy the equation

$$\left(\frac{\partial^2}{\partial r^2} + \frac{1}{r}\frac{\partial}{\partial r} + \frac{1}{r^2}\frac{\partial^2}{\partial \phi^2} + k^2\right)\Gamma(r_0, \phi_0; r, \phi, k)$$

$$= -\frac{1}{r_0}\delta(r - r_0)\delta(\phi - \phi_0) \tag{11.2.1}$$

and the conditions

$$\Gamma(r_0, \phi_0; \varrho, \phi; k) = 0 \quad , \tag{11.2.2}$$

$$\lim_{r + |\phi| \to \infty} \{\Gamma(r_0, \phi_0; r, \phi; k)|_{\operatorname{Im} k > 0}\} = 0 \quad . \tag{11.2.3}$$

The function

$$G(r_0, \phi_0; r, \phi; k) = \sum_{j=-\infty}^{+\infty} \Gamma(r_0, \phi_0; r, \phi + 2\pi j; k) \quad , \tag{11.2.4}$$

that has period 2π in ϕ is a Green's function for the exterior of the circle $r = \varrho$ on the ordinary plane rather than our infinitely-sheeted Riemann surface of logarithmic type (see [11.5]).

The problem (11.2.1–3) permits the exact solution

$$\Gamma(r_0, \phi_0; r, \phi; k)$$
$$= \frac{i}{8} \int_C e^{i\nu|\phi - \phi_0|} H_\nu^{(1)}(kr_>) \left[H_\nu^{(2)}(kr_<) - \frac{H_\nu^{(2)}(k\varrho)}{H_\nu^{(1)}(k\varrho)} H_\nu^{(1)}(kr_<) \right] d\nu \tag{11.2.5}$$

or when $\phi \ne \phi_0$

$$\Gamma(r_0, \phi_0; r, \phi; k) = \sum_{p=0}^{\infty} \Gamma_p(r_0, \phi_0; r, \phi; k)$$
$$= \frac{\pi}{4} \sum_{p=0}^{\infty} \frac{H_{\nu_p}^{(1)}(kr) H_{\nu_p}^{(1)}(kr_0) H_{\nu_p}^{(2)}(k\varrho)}{(\partial/\partial\nu)[H_\nu^{(1)}(k\varrho)]_{\nu_p}} e^{i\nu_p|\phi - \phi_0|} \tag{11.2.6}$$

where $r_>$ and $r_<$ are the greater and lesser of the coordinates $r_0, r; \nu_0, \nu_1, \ldots$ are the roots of the equation

$$H_\nu^{(1)}(k\varrho) = 0 \quad , \tag{11.2.7}$$

lying in the upper half of the ν-plane, and C is a contour that encircles these roots in the positive direction (Fig. 11.1).

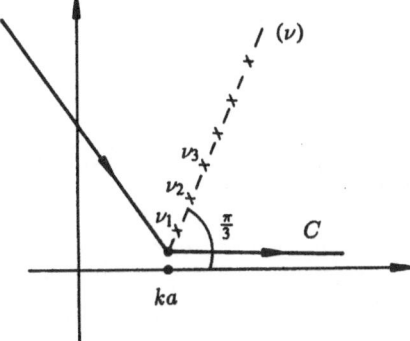

Fig. 11.1. Contour of integration for (11.2.5)

Formula (11.2.6) is obtained by applying the residue theorem to the integral (11.2.5). Substituting (11.2.6) into (11.2.4), we get the following representation for the "physical Green's function" $G(r_0, \phi_0; r, \phi; k)^2$:

$$G(r_0, \phi_0; r, \phi; k) = \sum_{p=0}^{\infty} G_p(r_0, \phi_0; r, \phi; k) \quad , \tag{11.2.8}$$

where

$$G_p(r_0, \phi_0; r, \phi, k) = \sum_{j=-\infty}^{+\infty} \Gamma_p(r_0, \phi_0; r, \phi + 2\pi j; k) \quad . \tag{11.2.9}$$

In their turn the functions $\Gamma_p(r_0, \phi_0; r, \phi; k)$, as will be seen from the following presentation, are conveniently represented when $\phi > \phi_0$ in the form

$$\Gamma_p(r_0, \phi_0; r, \phi; k)$$
$$= \frac{1}{2i(k/2)^{1/3}} \frac{1}{[w_1'(\xi_p)]^2} u_p^-(r_0, \phi_0; k) u_p^+(r, \phi; k) \quad , \tag{11.2.10}$$

where

$$u_p^{\pm}(\zeta, \eta; k) = \frac{\sqrt{\pi}}{2} e^{i(\pi/4)} \left(\frac{k}{2}\right)^{1/6} w_1'(\xi_p)$$
$$\times \sqrt{\frac{H_{\nu_p}^{(2)}(k\varrho)}{(\partial/\partial\nu)H_{\nu}^{(1)}(k\varrho)|_{\nu_p}}} H_{\nu_p}^{(1)}(k\zeta) e^{\pm i\nu_p \eta} \quad , \tag{11.2.11}$$

where $w_1(\xi)$ is an Airy function, and ξ_p, $p = 0, 1, 2, \ldots$ are its roots. If $\phi_0 > \phi$, the arguments of u_p^+ and u_p^- should be interchanged in (11.2.10).

Let us find the asymptotic formulas for $\Gamma_p(r_0, \phi_0; r, \phi; k)$ when $k \to \infty$ and $p = O(1)$. The form of these expansions depends on the relative location of the points (r_0, ϕ_0), (r, ϕ) and the circle $r = \varrho$. First we write out the asymptotic expansion for the roots of $H_{\nu}^{(1)}(k\varrho)$. The asymptotic form of the roots of $H_{\nu}^{(1)}(k\varrho)$ as $k\varrho \to \infty$ may be found in exactly the same way as that for the roots of the Bessel function $J_{\nu}(k\varrho)$. For $H_{\nu}^{(1)}(k\varrho)$ we have the asymptotic formula (7.2.4) in which the Airy function v must be replaced by the complex-valued Airy function w_1. Therefore the asymptotic formula for the roots of $H_{\nu}^{(1)}(k\varrho)$ is obtained from (7.2.5) by substituting ξ_p for $-t_p$, and takes the form

$$\nu_p = k\varrho \left[1 + \frac{\xi_p}{2} \left(\frac{2}{k\varrho}\right)^{2/3}\right.$$
$$\left. + \frac{1}{30} \left(\frac{\xi_p}{2}\right)^2 \left(\frac{2}{k\varrho}\right)^{4/3} + O\left[\left(\frac{\xi_p^{3/2}}{k\varrho}\right)^2\right]\right] \quad . \tag{11.2.12}$$

---

[2] It can be shown that the double series $\sum_{p,j} \Gamma_p(r_0, \phi_0; r, \phi + 2\pi j; k)$ converges absolutely; it is therefore legitimate to interchange the order of summation in obtaining formulas (11.2.8 and 9).

Let us turn now to the function $u_p^+(r, \phi; k)$ and get the asymptotic representation of the factor

$$H_{\nu_p}^{(1)}(kr)e^{i\nu_p\phi} \quad , \tag{11.2.13}$$

which depends on the coordinates of the observation point and appears in $u_p^+(r, \phi; k)$.

Let $r - \varrho \geq \text{const} > 0$. We use Debye's asymptotic formula

$$H_\nu^{(1)}(kr) = \sqrt{\frac{2}{\pi}} \frac{e^{-i(\pi/4)}}{\sqrt[4]{(kr)^2 - \nu^2}}$$
$$\times \exp\left[i\left(\sqrt{k^2r^2 - \nu^2} - \nu \arccos\frac{\nu}{kr}\right)\right]\left[1 + O\left(\frac{1}{k}\right)\right] \quad , \tag{11.2.14}$$

which is true (because $r - \varrho \geq \text{const} > 0$) at $\nu = \nu_p$, where ν_p is a root of (11.2.7) and $p = O(1)$. In (11.2.14) it is assumed that the square roots and the arc cosine have their principal values on the interval $-kr \leq \nu \leq kr$, and the branch cuts go from the points $\nu = \pm kr$ to infinity. According to the asymptotic formulas (11.2.12 and 14), the functions (11.2.13) are characterized by the phase factor

$$\exp\left[ik\left(-\varrho \arccos\frac{\varrho}{r} + \sqrt{r^2 - \varrho^2} + \varrho\phi\right) + i\xi_p\left(\frac{k\varrho}{2}\right)^{1/3}\left(\phi - \arccos\frac{\varrho}{r}\right)\right] \quad .$$

It is easily shown that the corresponding rays consist of arcs of the circle $r = \varrho$ and the semi-tangents to them directed along increasing ϕ (Fig. 11.2).

Thus the functions $u_\varrho^+(r, \phi; k)$ describe waves that move along the circle $r = \varrho$ with an exponential damping proportional to $k^{1/3}$ (we note that Re $i\xi_p < 0$, $(\phi - \arccos \varrho/r)\varrho = |AB|$), and shoot out along the half-tangents. It is natural to call these *creeping waves* (see Sect. 7.5). The circle $r = \varrho$ is the caustic for the corresponding rays. However, the theory of Chap. 3 cannot be applied here because boundary condition (11.2.2) must be satisfied on the circle.

A formula analogous to (11.2.14) can be derived for the factor $H_{\nu_p}^{(1)}(kr_0)$ $\times e^{-i\nu_p\phi_0}$ that enters into the function $u_p^-(r_0, \phi_0; k)$. If the coordinates r_0 and ϕ_0 are considered as variable, then the functions $u_p^-(r_0, \phi_0; k)$ will be creeping waves propagating in the opposite direction (decreasing ϕ_0).

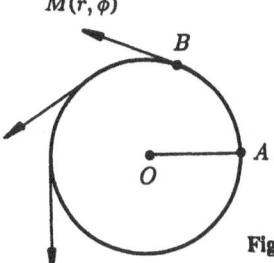

$M(r, \phi)$

Fig. 11.2. Creeping rays shed from a circular cylinder

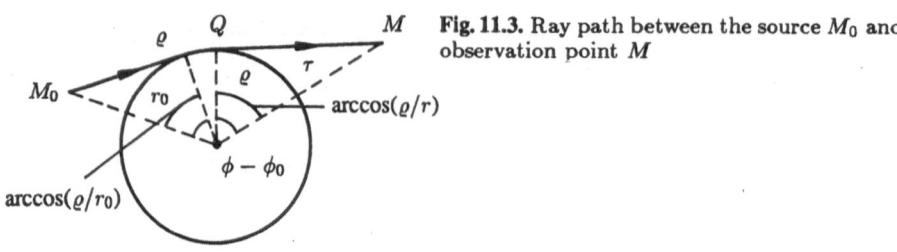

Fig. 11.3. Ray path between the source M_0 and observation point M

Using (11.2.12 and 14) we can easily find the asymptotic expansion of the first few terms of the series (11.2.6) for $r - \varrho \geq \text{const} > 0$, and $r_0 - \varrho \geq \text{const} > 0$:

$$\Gamma_p(r_0, \phi_0; r, \phi; k)$$

$$= \frac{1}{k} \left(\frac{k\varrho}{2} \right)^{1/3} \frac{1}{\sqrt{|M_0 P| \cdot |QM|}} \exp\{ik[|M_0 P| + |PQ| + |QM|]\}$$

$$\times \exp\left\{ i\xi_p \left(\frac{k}{2} \right)^{1/3} \frac{|PQ|}{\varrho^{2/3}} \right\} \cdot \left[1 + O\left(\frac{1}{(k\varrho)^{1/3}} \right) \right] \quad ; \tag{11.2.15}$$

$$p = 0, 1, \ldots, \quad p = O(1) \quad ,$$

where $|M_0 P| = \sqrt{r_0^2 - \varrho^2}$, $|QM| = \sqrt{r^2 - \varrho^2}$ are the lengths of segments $M_0 P$ and QM

$$|PQ| = \varrho \left(|\phi - \phi_0| - \arccos\frac{\varrho}{r_0} - \arccos\frac{\varrho}{r} \right)$$

is the arc length of $|PQ|$ (Fig. 11.3) along which the ray creeps before leaving the circle at Q. If the points M and M_0 are in each other's shadow, then $\Gamma_p(r_0, \phi_0; r, \phi; k)$ decreases exponentially with increasing k due to the factor

$$\exp\left[i\xi_p \left(\frac{k}{2} \right)^{1/3} \frac{|PQ|}{\varrho^{2/3}} \right] \quad . \tag{11.2.16}$$

Since $\text{Im}\,\xi_{p+1} > \text{Im}\,\xi_p$, this damping becomes more rapid the higher the p. The latter statement can be made only with reference to the function $\Gamma_p(r_0, \phi_0; r, \phi; k)$ with small parameters p for which (11.2.15) is valid. Investigation of the terms of the series for large p is cumbersome (see [11.6]) and will be omitted here. This investigation shows that for $|PQ| \geq \text{const} > 0$ the very first term of series (11.2.6), i.e. the function $\Gamma_0(r_0, \phi_0; r, \phi; k)$, gives all the algebraic terms of the asymptotic expansion of the Green's function. Subsequent terms of the series ($p = 1, 2 \ldots$) will make exponentially small contributions to the expansion, smaller the greater p is. The phase factors of the functions $\Gamma_p(r_0, \phi_0; r, \phi; k)$ are equal to

$$\exp[ik(|M_0 P| + |PQ| + |QM|)]$$

and correspond to propagation along the incident ray $M_0 P$, the arc PQ of the circle, and the shedding ray QM. The factor

$$(r_0^2 - \varrho^2)^{-1/4}(r^2 - \varrho^2)^{-1/4} = \frac{1}{\sqrt{|M_0 P| \cdot |QM|}}$$

describes the geometric divergence of the corresponding wave (see Chap. 2). We have examined the case where M_0 and M are located at a distance bounded from below away from circle $r = \varrho$.

Now let points M_0 and M be located fairly close to the circle $r = \varrho$. Using for the Hankel functions of (11.2.11) an asymptotic formula analogous to (7.2.4) (or Fock's asymptotic formulas, see [11.7]), we get

$$u_p^+(r, \phi; k) = \varrho^{-1/6} \exp\left[ik\varrho\phi + i\xi_p\left(\frac{k\varrho}{2}\right)^{1/3}\phi + \frac{i\xi_p^2}{60}\left(\frac{k\varrho}{2}\right)^{-1/3}\phi\right.$$

$$\left. - \frac{2y + \xi_p}{20(k\varrho/2)^{2/3}} + O\left(\frac{1}{k\varrho}\right)\right]$$

$$\times w_1\left[(\xi_p - y) + \frac{9y^2 - 8y\xi_p}{60(k\varrho/2)^{2/3}} + O\left(\frac{1}{(k\varrho)^{4/3}}\right)\right] \quad , \quad (11.2.17)$$

where

$$y = \frac{k(r - \varrho)}{(k\varrho/2)^{1/3}}$$

is the reduced normal distance to the circle $r = \varrho$. The formula for $u_p^-(r_0, \phi_0; k)$ is obtained from (11.2.17) by substituting $-i$ for i, ϕ_0 for ϕ and $y_0 = k(r_0 - \varrho)/(k\varrho/2)^{1/3}$ for y. Let us note that we introduced the factor $(\sqrt{\pi}/2)e^{i(\pi/4)}$ $w_1'(\xi_p)$ into (11.2.11) when defining the functions u_p so that the factor in (11.2.17) which is independent of the radius of the circle would be equal to one. Equation (11.2.17) shows that the creeping wave $u_p^+(r, \phi; k)$ close to the circle $r = \varrho$ is a wave with exponentially damped amplitude that traveling along increasing ϕ. The asymptotic formula (11.2.17) is a special case of (7.5.8).

Using (11.2.17) and the analogous formula for $u_p(r_0, \phi_0; k)$, we get the following asymptotic representation for the terms of (11.2.6):

$$\Gamma_p(r_0, \phi_0; r, \phi; k) = \frac{e^{ik\varrho|\phi - \phi_0|}}{2i(k\varrho/2)^{1/3}} \frac{1}{[w_1'(\xi_p)]^2} w_1\left[(\xi_p - y)\right.$$

$$\left. + \frac{9y^2 - 8y\xi_p}{60(k\varrho/2)^{2/3}} + O((k\varrho)^{-4/3})\right] w_1\left[(\xi_p - y_0)\right.$$

$$\left. + \frac{9y_0^2 - 8y_0\xi_p}{60(k\varrho/2)^{2/3}} + O((k\varrho)^{-4/3})\right] \exp\left[i\xi_p\left(\frac{k\varrho}{2}\right)^{1/3}|\phi - \phi_0|\right.$$

$$\left. + \frac{i\xi_p^2|\phi - \phi_0|}{60(k\varrho/2)^{1/3}} - \frac{(y + y_0) + \xi_p}{10(k\varrho/2)^{2/3}} + O((k\varrho)^{-1})\right] \quad . \quad (11.2.18)$$

Equation (11.2.18) holds only for small p. This equation implies that, as in the preceding case, the first few terms of the series damp out rapidly if points $M(r, \phi)$ and $M_0(r_0, \phi_0)$ are in each other's shadow ($|\phi - \phi_0| > \text{const} > 0$). Investigation of the functions $\Gamma_p(r_0, \phi_0; r, \phi; k)$ for $p \sim k\varrho$ and $p \gg k\varrho$ shows that, as in the preceding case, (11.2.6) converges and gives the asymptotic form of the Green's function. More precisely, the sum of the first few terms $\sum_0^{p_0} \Gamma_p$ is an asymptotic expansion of Γ, where each subsequent term is exponentially small compared with the preceding one.

In the case where $M_0(r_0, \phi_0)$ is fixed and $M(r, \phi)$ approaches the boundary between light and shadow, the terms of (11.2.6) become comparable in magnitude since the arc length $|PQ|$ approaches zero [see (11.2.12–16) and Fig. 11.3], and the first terms of series (11.2.6) no longer give an asymptotic representation of Γ. In this case (11.2.5) is more convenient for studying the function Γ than is the series (11.2.6). In the illuminated region the representation for $\Gamma(r_0, \phi_0; r, \phi; k)$ is obtained if the method of steepest descents is applied to the integral (11.2.5), substituting the asymptotic expansions (Debye formulas) for the Hankel functions in the vicinity of saddle points. As a result we get

$$\Gamma(r_0, \phi_0; r, \phi; k) = \frac{\exp[ikR + i(\pi/4)]}{2\sqrt{2\pi k R}} \left[1 + O\left(\frac{1}{k}\right)\right]$$
$$- \frac{\exp[ik(l_0 + l) + i(\pi/4)]}{2\sqrt{2\pi k \left(l_0 + l + \dfrac{2l_0 l}{\varrho \cos \theta}\right)}} \left[1 + O\left(\frac{1}{k}\right)\right] \quad (11.2.19)$$

Here the $O(1/k)$ are functions expanded in asymptotic series with respect to powers of $1/k$ and having order of magnitude $1/k$, R is the distance between points M_0 and M, l_0 and l are the lenghts of segments of the incident and reflected rays and θ is the angle of reflection. Equation (11.2.9) agrees with (2.6.3 and 6).

Consider now the poles of the Green's function $G(r_0, \phi_0; r, \phi; k)$, defined by (11.2.8 and 9) on the complex k plane. Recall (see Sect. 7.5) that the poles of the Green's function in the k-plane can be considered aś eigenvalues of the Laplacian for the exterior of a convex region; in the present case the exterior of the circle $r = \varrho$.

The Green's function G is an infinite sum of functions G_p, $p = 1, 2, \ldots$. Therefore it is natural to begin our study of the poles of G with the investigation of the poles of functions G_p at fixed p. The function $G_p(r_0, \phi_0; r, \phi; k)$ is defined by (11.2.9). By virtue of (11.2.10 and 11), the formula

$$\Gamma_p(r_0, \phi_0; r, \phi + 2\pi j; k) = u_p^-(r_0, \phi_0; k) u_p^+(r, \phi; k) e^{i2\pi\nu_p j}$$

$$(11.2.20)$$

holds for $\Gamma_p(r_0, \phi_0; r, \phi + 2\pi j; k)$. Using (11.2.20), we get

$$G_p(r_0, \phi_0; r, \phi; k) = \frac{1}{2i}\left(\frac{2}{k}\right)^{1/3} \frac{1}{[w_1'(\xi_p)]^2} \left\{ u_p^-(r_0, \phi_0; k) u_p^+(r, \phi; k) \right.$$

$$\left. + \frac{u_p^-(r_0, \phi_0; k) u_p^+(r, \phi; k) + u_p^-(r, \phi; k) u_p^+(r_0, \phi_0; k)}{1 - e^{i2\pi\nu_p}} e^{i2\pi\nu_p} \right\} \quad .(11.2.21)$$

Equation (11.2.21) is written on the assumption that $\phi > \phi_0$. If $\phi < \phi_0$ then ϕ and ϕ_0 should be interchanged in the first term of the formula. Formula (11.2.21) implies that the poles of function $G_p(r_0, \phi_0; r, \phi; k)$ in the k-plane are determined from the equation

$$\nu_p = \nu_p(k) = q \quad , \tag{11.2.22}$$

where q is an integer. For each fixed $p = 0, 1, 2, \ldots$, (11.2.22) defines a series of poles k_{pq}, $q = 0, 1, 2, \ldots$ corresponding to the term of series (11.2.8) with index p.

We will assume that the poles of G_p for any p are poles of the Green's function G, and that Green's function G has no other poles[3]. In other words, we will assume that the poles of the Green's function G are entirely determined by the roots k_{pq} of (11.2.22).

Let us note some properties of the poles k_{pq}. The smaller p is, the closer to the real axis will be the corresponding series of poles k_{pq}, $q = 0, 1, 2, \ldots$.

Using the asymptotic formula (11.2.12) for ν_p, we can get asymptotic formulas as k becomes large for series of poles of the Green's function that are located in the lower half-plane, in a sufficiently small neighborhood ($p = O(1)$) of the real axis:

$$k_{pq} = \frac{q}{\varrho}\left\{1 - \frac{\xi_p}{2}\left(\frac{2}{q}\right)^{2/3} + 3\frac{\xi_p^2}{40}\left(\frac{2}{q}\right)^{4/3} + O\left[\xi_p^3\left(\frac{2}{q}\right)^2\right]\right\}. \tag{11.2.23}$$

Equation (11.2.23) is in complete agreement with (7.5.9). The investigations of the asymptotic behavior of the Green's function of the etalon-problem as carried out in this section are the heuristic basis for constructing and studying the Green's function in the case of an arbitrary convex contour S.

11.3 Creeping Waves Near a Curve with Positive Curvature and Their Extension to Arbitrary Distances

We will use the term *creeping waves* to denote, for a sufficiently smooth curve S that has positive curvature, solutions $u_p^{\pm}(M, k)$ of the homogeneous Helmholtz equation $(\Delta + k^2)u_p^{\pm} = 0$, satisfying the homogeneous boundary condition

$$u_p^{\pm}|_S = 0 \quad \left(\text{or } \left.\frac{\partial u_p^{\pm}}{\partial n}\right|_S = 0 \quad , \quad \text{or } \left.\frac{\partial u_p^{\pm}}{\partial n} + ikgu_p^{\pm}\right|_S = 0\right) \quad ,$$

---

[3] Proof of this assumption requires a detailed investigation of the convergence of series (11.2.8) in the case of complex k, which no one has yet published, and perhaps no one has yet carried out. Nonetheless, the coincidence of the set of poles of the Green's function G and the set of sets of all poles of functions G_p is generally accepted as a known fact.

on S, and which have the properties

$$u_p^+ \to 0 \quad \text{when } k > 0, \ n > 0 \text{ and } s \to +\infty \quad ,$$
$$u_p^- \to 0 \quad \text{when } k > 0, \ n > 0 \text{ and } s \to -\infty$$

and that decay exponentially with increasing distance from S when Im $k > 0$.

The existence of creeping waves in the general case has not been proved. However, in situations where separation of variables can be used, creeping waves exist and can be constructed in explicit form. (The creeping waves for a circle were constructed in the preceding section; they can be constructed similarly for an ellipse and a parabola.)

We will assume that creeping waves exist in the general case, and that the formal solutions constructed for the Helmholtz equation in Sect. 7.5, give the asymptotic expansions of these waves.

This asymptotic expansion takes the form

$$u_p^+(M, k) = \exp[iks + iE(\xi_p, M)]w_1[T(\xi_p, M)] \quad , \tag{11.3.1}$$

$$E(\xi_p, M) = \sum_{m=-1}^{N-1} \alpha_m(s, \nu)k^{-m/3} + O\left(k^{-N/3}\right) \quad ,$$

$$T(\xi_p, M) = \sum_{m=0}^{N-1} \beta_m(s, \nu)k^{-m/3} + O\left(k^{-N/3}\right) \quad , \tag{11.3.2}$$

$$p = 0, 1, 2, \ldots, \ N = 1, 2, 3, \ldots \quad .$$

In these formulas $\xi_0, \xi_1, \ldots$, as in Sect. 11.2, are zeros of the Airy function $w_1(\xi)$, where arg $\xi_p = \pi/3$; the functions $\alpha_m(s, \nu)$, $\beta_m(s, \nu)$ are polynomials in $\nu = nk^{2/3}$ with coefficients that depend on s and ξ_p; the origin for the arc length s is chosen arbitrarily. The constant factors in the expressions for u_p^+ [see (7.5.8)] are now chosen so that in the case $\varrho = \text{const}$ the result agrees with (11.2.17). We present the formulas for the first few polynomials $\alpha_m(s, \nu)$ and $\beta_m(s, \nu)$:

$$\alpha_{-1}(s, \nu) = \frac{\xi_p}{2^{1/3}} \int_0^s \frac{ds}{\varrho^{2/3}(s)} \quad ,$$

$$\alpha_0(s, \nu) = \frac{i}{6} \ln \varrho(s) \quad ,$$

$$\alpha_1(s, \nu) = \frac{\varrho'(s)}{6\varrho(s)}\nu^2 + \alpha_{10}(s), \tag{11.3.3}$$

$$\alpha_{10}(s) = 2^{1/3}\xi_p^2 \int_0^s \varrho^{-4/3}(s)\left[\frac{1}{60} + \frac{4}{135}\varrho'^2(s) - \frac{2}{45}\varrho(s)\varrho''(s)\right] ds \quad ,$$

$$\alpha_2(s, \nu) = \alpha_{21}(s)\nu + \alpha_{20}(s) \quad ,$$

$$\alpha_{21}(s) = \frac{i}{10\varrho(s)}\left[2 - \frac{1}{3}\varrho(s)\varrho''(s) + \frac{2}{9}\varrho'^2(s)\right] \quad ,$$

$$\alpha_{20}(s) = \frac{i\xi_p}{10}\left(\frac{2}{\varrho(s)}\right)^{2/3}\left[\frac{1}{2} + \frac{1}{3}\varrho(s)\varrho''(s) - \frac{2}{9}\varrho'^2(s)\right], \tag{11.3.4}$$

$$\beta_0(s,\nu) = \xi_p - \left(\frac{2}{\varrho(s)}\right)^{1/3}\nu \quad,$$

$$\beta_1(s,\nu) = 0 \quad,$$

$$\beta_2(s,\nu) = \beta_{22}(s)\nu^2 + \beta_{21}(s)\nu \quad, \tag{11.3.5}$$

$$\beta_{22}(s,\nu) = \frac{1}{20}\left(\frac{2}{\varrho(s)}\right)^{1/3}\left[3 + \frac{1}{3}\varrho(s)\varrho''(s) - \frac{2}{9}\varrho'^2(s)\right] \quad,$$

$$\beta_{21}(s) = \frac{2\xi_p}{15\varrho(s)}\left[-2 + \frac{1}{3}\varrho(s)\varrho''(s) - \frac{2}{9}\varrho'^2(s)\right] \quad. \tag{11.3.6}$$

When $\varrho(s) = \text{const}$, (11.3.1–6) reduce to (11.2.17). Equation (11.3.1) corresponds to choosing the plus sign in the solution of the eikonal equation (7.3.16). When the minus sign is chosen in the eikonal equation, we get the asymptotic expansion of the creeping waves $u_p^-(M,k)$. The asymptotic formulas for $u_p^-(M,k)$ differ from those for $u_p^+(M,k)$ by the substitution of $-i$ for i in the exponential function in (11.3.1) and in (11.3.3, 4).

Let us go on to the derivation of expressions for $u_p^+(M,k)$ suitable for large distance from S.

The polynomials $\alpha_m(s,\nu)$ and $\beta_m(s,\nu)$ appearing in (11.3.2) have degrees $m/2$ and $m/2 + 1$ for even $m \geq 0$, and degrees $(m+3)/2$ and $(m-1)/2$ for odd $m \geq 1$ (see Sect. 7.3). With increasing ν the behavior of these polynomials begins to be determined by their leading terms. On the other hand, the truncated series (11.3.2) are suitable for calculating $u_p^+(M,k)$ so long as the terms of these series continue to decrease with increasing m. Hence we arrive at the conclusion (see Sect. 7.3) that (11.3.2) can be used not only for finite ν, but for large ν if only

$$\nu = O(k^\varepsilon) \quad, \quad \varepsilon < 2/3 \quad. \tag{11.3.7}$$

Let us rewrite the formulas for $u_p^+(M,k)$ in ray coordinates (close to S). If ν is a large quantity that satisfies (11.3.7), then the Airy function in (11.3.1) can be replaced by its asymptotic expansion. The formulas for u_p^+ that result from these transformations make sense not only in the boundary layer near S, but outside of it as well. We will assume that these formulas continue u_p^+ to arbitrary distances from S. Let us carry out the appropriate calculations. We replace the Airy function w_1 in (11.3.1) by its asymptotic expansion:

$$w_1(t) \sim \frac{e^{(2/3)t^{3/2}}}{\sqrt{\pi}t^{1/4}}\sum_{m=0}^{\infty}\frac{\Gamma(3m+\frac{1}{2})}{(2m)!}(9t^{3/2})^{-m} \quad,$$

$$\frac{\pi}{3} - \delta \geq \arg t \geq -\frac{5\pi}{3} + \delta \quad, \quad \delta > 0 \quad, \tag{11.3.8}$$

(see Appendix A.1).

Terms with various powers of ν appearing in the polynomials $\alpha_m(s, \nu)$ or $\beta_m(s, \nu)$ now have different orders of magnitude as $k \to \infty$. Using (11.3.8), we get

$$u_p^+(M, k) = \frac{e^{i(\pi/4)}}{\nu^{1/4}} \left(\frac{\varrho(s)}{2} \right)^{1/12} \exp\left[iks + iE(\xi_p, M) \right.$$
$$\left. + \frac{2}{3} T^{3/2}(\xi_p, M) \right] \chi_p(M, k) \quad , \tag{11.3.9}$$

where

$$\chi_p(M, k) = Q^{-1/4} \frac{1}{\sqrt{\pi}} \sum_{m=0}^{\infty} \frac{\Gamma(3m + \frac{1}{2})}{(2m)!} \left[9iQ^{3/2} \nu^{3/2} \left(\frac{2}{\varrho(s)} \right)^{1/2} \right]^{-m} \quad ,$$

$$Q = 1 - \frac{\xi_p}{\nu} \left(\frac{\varrho(s)}{2} \right)^{1/3} - \frac{\beta_2(s, \nu)}{k^{2/3}\nu} \left(\frac{\varrho(s)}{2} \right)^{1/3} - \frac{\beta_3(s, \nu)}{k\nu} \left(\frac{\varrho(s)}{2} \right)^{1/3} - \dots \quad .$$

The factor $\chi_p(M, k)$ as $k \to \infty$ approaches unity and may be represented in the form of a double expansion in powers of $k^{-1/3}$ and $\nu^{-1/2} = O(k^{-\epsilon/2})$. From the observation point $M(s, n)$ let us draw the tangent and normal to S (Fig. 11.4). Let σ be the arc length at the point of tangency, and let r be the length of the tangent. The position of a point can also be characterized by the ray coordinates $\tau = \sigma + r$ and σ. These coordinates are orthogonal. As $n \to 0$, $s - \sigma$ and r also approach zero. Using the natural equation of S (see Sect. 5.1), we can find the following relation between s and n and the ray coordinates σ and $\tau = \sigma + r$ as $n \to 0$ $(r \to 0)$:

$$s = \sigma + r - \frac{1}{3\varrho^2} r^3 + \frac{7}{24} \frac{\varrho'}{\varrho^3} r^4 + \dots,$$

$$n = \frac{1}{2\varrho} r^2 - \frac{\varrho'}{6\varrho^2} r^3 + \frac{2\varrho'^2 - \varrho\varrho'' - 3}{24\varrho^3} r^4 + \dots$$

$$(\varrho \equiv \varrho(\sigma) \quad , \quad \varrho' \equiv \varrho'(\sigma), \dots) \quad . \tag{11.3.10}$$

We convert in (11.3.9) to ray coordinates using (11.3.10). The resultant expression (after some cumbersome calculations) can be given the form

$$u_p^+(M, k) = \frac{q_p^+(\sigma, r, k) e^{ik\tau}}{\sqrt{r}} \quad , \tag{11.3.11}$$

where

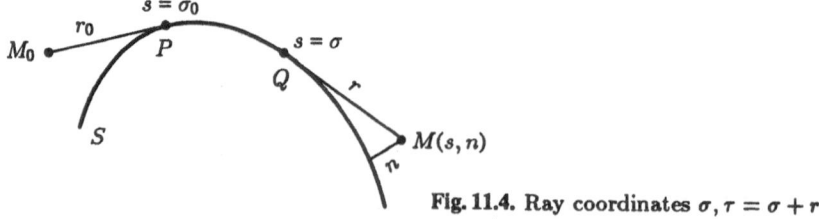

Fig. 11.4. Ray coordinates $\sigma, \tau = \sigma + r$

$$q_p^+(\sigma, r, k) = e^{i(\pi/4)} \left(\frac{2\varrho}{k}\right)^{1/6} \exp\left[i\xi_p \left(\frac{k}{2}\right)^{1/3} \int_0^\sigma \frac{ds}{\varrho^{2/3}} + \frac{i\alpha_{10}(\sigma)}{k^{1/3}}\right.$$

$$+ \frac{i\xi_p^2}{4k^{1/3}} \left(\frac{2}{\varrho}\right)^{1/3} \left(\frac{\varrho}{r} + \frac{\varrho'}{3}\right) + \left.\frac{i\alpha_{20}(\sigma)}{k^{2/3}}\right] \left\{1 + \frac{\xi_p}{k^{2/3}} \left(\frac{\varrho}{2}\right)^{1/3}\right.$$

$$\times \left.\left[\frac{\varrho}{2r^2} + \frac{\varrho'}{3r} + \frac{1}{30}\left(\frac{1}{\varrho} + \frac{\varrho'^2}{9\varrho} + \frac{7}{3}\varrho''\right)\right] + O\left(\frac{1}{k}\right)\right\} \quad .$$

$$(11.3.12)$$

It can be readily demonstrated that the length of the tangent r has the sense of a geometric divergence. Equations (11.3.11, 12) define the principal term of an asymptotic expansion for $u_p^+(M, k)$ which is analogous to the ray expansion (see Chap. 2 and the paper by *Friedlander* and *Keller* [11.8]). Equation (11.3.11) accomplishes the desired continuation of the creeping waves outside of the boundary layer. In the narrow strip near S where $\nu = O(k^\varepsilon)$, $1/3 < \varepsilon < 2/3$, (11.3.1 and 11) give the same asymptotic representation for u_p^+.

11.4 An Expression for the Green's Function in Terms of Creeping Waves

The construction of Sect. 11.2 suggest that the Green's function might be constructed by a superposition of creeping waves in the general case of a convex contour S extending to infinity. The main purpose of this section is to get an asymptotic expansion for the Green's function $\Gamma(M_0, M, k)$ by using the asymptotic formulas for $u_p^\pm$ from the preceding section.

Let us assume that in the general case the Green's function $\Gamma(M_0, M, k)$ can be represented on one side of the source M_0 as a superposition of the creeping waves u_p^+, and on the other side as a superposition of the creeping waves u_p^-:

$$\Gamma(M_0, M, k) = \begin{cases} \displaystyle\sum_{p=0}^\infty \mu_p^+ u_p^+(M, k) \quad , \\ \displaystyle\sum_{p=0}^\infty \mu_p^- u_p^-(M, k) \quad , \end{cases} \qquad (11.4.1)$$

$$\mu_p^+ = \mu_p^+(M_0, k) \quad , \quad \mu_p^- = \mu_p^-(M_0, k) \quad .$$

In the case of a circle, the line separating the regions where the first and second of representations (11.4.1) hold is the half-line going out from the center of the circle and passing through the source. In the following, we will assume that the representation $\sum \mu_p^+ u_p^+$ (or $\sum \mu_p^- u_p^-$) holds for the Green's function Γ when $s > s_0$ (or $s < s_0$) if the observation point $M(s, n)$ is not illuminated by the source point $M_0(s_0, n_0)$, i.e. if it is in the shadow zone.

In looking for the asymptotic expansion of the Green's function, we will assume that the reciprocity principle

$$\Gamma(M, M_0, k) = \Gamma(M_0, M, k)$$

holds. To satisfy this condition it is sufficient to set

$$\mu_p^+(M_0, k)u_p^+(M, k) = \mu_p^-(M, k)u_p^-(M_0, k) \quad , \quad p = 0, 1, 2, \ldots,$$

whence at points where u_p^+, $u_p^- \neq 0$,

$$\frac{\mu_p^+(M_0, k)}{u_p^-(M_0, k)} = \frac{\mu_p^-(M, k)}{u_p^+(M, k)} = B_p \quad , \quad p = 0, 1, \ldots \quad . \tag{11.4.2}$$

According to (11.4.2), B_p does not depend on M_0 or on M, i.e. it is only a function of k. In the following presentation we will use for u_p^+, u_p^- asymptotic expansions that have dimensions of length to the $-1/6$ power. By virtue of the fact that the Green's function is dimensionless (as implied by the equation $\Delta\Gamma + k^2\Gamma = -\delta(M - M_0)$, which it satisfies), the coefficient B_p must have the dimensions of length to the 1/3 power. Let us isolate a dimensionless factor in B_p, setting

$$B_p(k) = \left(\frac{k}{2}\right)^{-1/3} A_p \quad . \tag{11.4.3}$$

Combining (11.4.1, 2, and 3), we get

$$\Gamma(M_0, M, k) = \sum_{p=0}^{\infty} \Gamma_p(M_0, M, k)$$

$$= \left(\frac{k}{2}\right)^{-1/3} \sum_{p=0}^{\infty} A_p \begin{cases} u_p^-(M_0, k)u_p^+(M, k), & s > s_0 \quad , \\ u_p^+(M_0, k)u_p^-(M, k), & s < s_0 \quad , \end{cases} \tag{11.4.4}$$

(M is in the shadow zone).

It remains for us to determine the dimensionless factors A_p that do not depend on either M_0 or M. Assuming that the asymptotic formulas of Sect. 11.3 hold for u_p^+ and u_p^-, let us find an expression for the coefficients A_p for the case of the boundary condition $\Gamma|_S = 0$. To do this, we use a so-called localization principle.

Equation (11.3.11) implies that the eikonal τ corresponding to Γ_p in (11.4.4) ($p = 0, 1, \ldots, p = O(1)$) is equal to the length of the ray M_0PQM (see Fig. 11.4). It is natural to assume (the essence of the localization principle) that the asymptotic expansions of Γ_p are completely determined by the path of the ray M_0PQM (in particular the asymptotic expansion of Γ_p is dependent only on a segment PQ of the contour S, rather than on S as a whole). However, A_p cannot depend on segment PQ either, since arc PQ changes with the position of points M and M_0, and the factors A_p do not depend on M or M_0. Consequently, factors A_p are in general independent of the properties of contour S. But then, they cannot depend on k either, because if they are independent of S, no dimensionless parameter can be formed that includes k.

So we will assume that A_p is an absolute constant. It can be determined by studying the exact solution of some special problem. Let us return to (11.2.6, 10). The asymptotic expansions of the function $u_p^+(M, k)$ when $\varrho = $ const reduces to the asymptotic expansions of the functions $u_p^+(r, \phi; k)$, while the asymptotic expansions of $u_p^-(M, k)$ reduces to that of $u_p^-(r, \phi; k)$. Comparing (11.4.4, 11.3.1 and 2) with (11.2.6, 10 and 17), we find

$$A_p = \frac{1}{2i[w_1'(\xi_p)]^2}, \quad p = 0, 1, 2, \ldots . \tag{11.4.5}$$

Of course, we get the same expression for A_p when we compare wave fields far from S and those far from the circle $r = \varrho$.

Equation (11.4.5) leads us to the following final formulas for functions Γ_p [see (11.4.4)] in the case of a contour that extends to infinity:

$$\Gamma_p(M_0, M, k)$$
$$= \frac{1}{2i(\frac{k}{2})^{1/3}} \frac{1}{[w_1'(\xi_p)]^2} \begin{cases} u_p^-(M_0, k)u_p^+(M, k), & s \geq s_0 , \\ u_p^+(M_0, k)u_p^-(M, k), & s \leq s_0 . \end{cases} \tag{11.4.6}$$

If M_0 and M lie in the boundary layer [ν_0 and ν obey (11.3.7)], then when $k \to \infty$ and $p = 0, 1, \ldots, p = O(1)$ we have the formula

$$\Gamma_p(M_0, M, k) = \frac{1}{2i}\left(\frac{2}{k}\right)^{1/3} [\varrho(s_0)\varrho(s)]^{-1/6}[w_1'(\xi_p)]^{-2}$$

$$\times \exp\left[ik(s - s_0) + i\xi_p\left(\frac{k}{2}\right)^{1/3}\int_{s_0}^{s}\frac{ds}{\varrho^{2/3}(s)}\right.$$

$$+ \frac{i}{6k^{1/3}}\left[\frac{\varrho'(s)}{\varrho(s)}\nu^2 - \frac{\varrho'(s_0)}{\varrho(s_0)}\nu_0^2\right] + i\frac{\alpha_{10}(s) - \alpha_{10}(s_0)}{k^{1/3}}$$

$$\left. + i\frac{\alpha_{21}(s)\nu + \alpha_{21}(s_0)\nu_0}{k^{2/3}} + i\frac{\alpha_{20}(s) + \alpha_{20}(s_0)}{k^{2/3}} + O\left(\frac{1}{k}\right)\right]$$

$$\times w_1[T(\xi_p, M_0)]w_1[T(\xi_p, M)] . \tag{11.4.7}$$

The notation for the quantities appearing in (11.4.7) has already been introduced by (11.3.3). Equation (11.4.7) relates to the case $s \geq s_0$, otherwise s and s_0 must be interchanged.

Let M_0 lie in the boundary layer, and M be located outside of the boundary layer − in the shadow zone − at an arbitrarily great distance from S. We will characterize the position of point M by the coordinates r and σ [the length of the tangent from M to S going toward M_0, and the arc length at the point of tangency (see Fig. 11.4)]. Substituting (11.3.11 and 12) into (11.4.6), we get

$$\Gamma_p(M_0, M, k) = \frac{e^{-i\pi/4}}{\sqrt{2kr}}\left[\frac{\varrho(\sigma)}{\varrho(s_0)}\right]^{1/6}$$

$$\times \exp\left[ik(\sigma - s_0 + r) + i\xi_p \left(\frac{k}{2}\right)^{1/3} \int_{s_0}^{\sigma} \frac{ds}{\varrho^{2/3}(s)}\right.$$

$$-\frac{i\varrho'(s_0)v_0^2}{6\varrho(s_0)k^{1/3}} + i\frac{\alpha_{10}(\sigma) - \alpha_{10}(s_0)}{k^{1/3}}$$

$$\left.+\frac{i\xi_p^2}{4k^{1/3}}\left(\frac{2}{\varrho(\sigma)}\right)^{1/3}\left(\frac{\varrho(\sigma)}{r} + \frac{\varrho'(\sigma)}{3}\right) + O\left(\frac{1}{k^{2/3}}\right)\right]$$

$$\times [w_1'(\xi_p)]^{-2} w_1\left[\xi_p - \frac{2^{1/3}v_0}{\varrho^{1/3}(s_0)} + O\left(\frac{1}{k^{2/3}}\right)\right] \quad,$$

$$p = 0, 1, \ldots \quad ; \quad p = O(1) \quad . \tag{11.4.8}$$

Now let both points M_0 and M be located outside of the boundary layer. We will characterize their positions by the lengths of the tangents r_0 and r drawn from these points to S, and by the points of tangency $s = \sigma_0$ and $s = \sigma$ (Fig. 11.4). Using (11.3.11, 12 and 11.4.8), we get

$$\Gamma_p(M_0, M, k) = \frac{\sqrt[6]{\varrho(\sigma_0)\varrho(\sigma)}}{2^{1/3}k^{2/3}\sqrt{r_0 r}}[w_1'(\xi_p)]^{-2}$$

$$\times \exp\left[ik(r_0 + r + \sigma - \sigma_0) + i\xi_p(k/2)^{1/3} \int_{\sigma_0}^{\sigma} \frac{ds}{\varrho^{2/3}(s)}\right.$$

$$+ i\frac{\alpha_{10}(\sigma) - \alpha_{10}(\sigma_0)}{k^{1/3}} + \frac{i\xi_p^2}{4k^{1/3}}\left(\frac{2}{\varrho(\sigma)}\right)^{1/3}\left(\frac{\varrho(\sigma)}{r} + \frac{\varrho'(\sigma)}{3}\right)$$

$$\left.-\frac{i\xi_p^2}{4k^{1/3}}\left(\frac{2}{\varrho(\sigma_0)}\right)^{1/3}\left(\frac{\varrho(\sigma_0)}{r_0} + \frac{\varrho'(\sigma_0)}{3}\right) + O\left(\frac{1}{k^{2/3}}\right)\right] \quad,$$

$$p = 0, 1, 2, \ldots \quad ; \quad p = O(1) \quad . \tag{11.4.9}$$

The physical meaning of the various factors in (11.4.9) is discussed at the end of Sect. 13.8.

Equations (11.4.5–9) obviously do not hold when $p \to \infty$. In the case where the source and observation points are located in each other's deep shadow, the asymptotic expansion of Γ is given by the sum of the first few terms of (11.4.4):

$$\Gamma \approx \sum_{p=0}^{p_0} \Gamma_p \quad , \quad p_0 = O(1) \quad .$$

Each subsequent term in this sum is exponentially small compared with the preceding one. The residual term $\sum_{p_0+1}^{\infty} \Gamma_p$, as was to be expected from an examination of the etalon problem, is small even compared with Γ_{p_0}.

Let us turn to the case of a closed contour S. We introduce an infinitely-sheeted Riemann surface analogous to that considered in Sect. 11.2. In (s, n) coordinates, the boundary of the Riemann surface will then be described as follows: $n = 0$, $-\infty < s < +\infty$. Let $\Gamma(M_0, M, k) = \Gamma(s_0, n_0; s, n; k)$ be the Green's function of this surface. The formula

$$G(M_0, M, k) = \sum_{j=-\infty}^{\infty} \Gamma(s_0, n_0; s + jL, n; k), \qquad (11.4.10)$$

where L is the length of S, will define a physical Green's function for the exterior of the convex contour S that is L-periodic in s. Substituting the sum of the Γ_p for Γ [see (11.4.4)], we get

$$
\begin{aligned}
G &= \sum_{p=0}^{\infty} G_p(M_0, M, k) \\
&= \sum_{p=0}^{\infty} \sum_{j=-\infty}^{\infty} \Gamma_p(s_0, n_0; s + jL, n; k), \qquad (11.4.11)
\end{aligned}
$$

where

$$G_p = \sum_{j=-\infty}^{\infty} \Gamma_p(s_0, n_0; s + jL, n; k). \qquad (11.4.12)$$

The asymptotic expansions for u_p^+ and u_p^- (see Sect. 11.3) imply at least for large k that

$$
\begin{aligned}
u_p^+(s + jL, n; k) &= \exp[ij\Phi_p(k)]u_p^+(s, n; k) \quad, \\
u_p^-(s + jL, n; k) &= \exp[-ij\Phi_p(k)]u_p^-(s, n; k) \quad, \qquad (11.4.13)
\end{aligned}
$$

where

$$\Phi_p(k) = kL + \sum_{m=-1}^{N-1} (\alpha_m(L, 0) - \alpha_m(0, 0))k^{-m/3} + O(k^{-N/3}) \quad.$$

$$(11.4.14)$$

From (11.4.6 and 12–14) we get

$$
\begin{aligned}
G_p(M_0, M, k) = \frac{1}{2i(k/2)^{1/3}} \frac{1}{[w_1'(\xi_p)]^2} \Big\{ &u_p^-(M_0, k)u_p^+(M, k) \\
&+ \frac{u_p^-(M_0, k)u_p^+(M, k) + u_p^+(M_0, k)u_p^-(M, k)}{1 - e^{i\Phi_p(k)}} e^{i\Phi_p(k)} \Big\}, \\
&p = O(1) \quad.
\end{aligned}
$$

Setting the denominator equal to zero, we get the equation for the poles of G_p:

$$\Phi_p(k) = 2\pi q \quad, \quad p = 0, 1, \ldots, \quad q \gg 1 \quad \text{is an integer} \quad, \qquad (11.4.15)$$

which agrees with the equation of Sect. 7.5 defining the eigenvalues of the exterior of contour S. With increasing p, the distance of the poles from the real axis increases. The position of the poles at values of p which are $O(k)$ can no longer be determined from (11.4.15).

We remark in closing that eq. (11.4.5) can be verified by the following formal constructions: each term of the series $\sum_{p=0}^{\infty} \Gamma_p$, representing the Green's function Γ can be replaced by (11.4.7), the series summed using the residue theorem, and the resultant integral can be asymptotically calculated in the illuminated region by the method of steepest descents. The results of these calculations near the boundary between light and shadow agree with formulas given by the ray method at the corresponding points.

More confirmation is given by a formal calculation of the expression

$$(\Delta + k^2) \sum_{p=0}^{\infty} \Gamma_p \quad ,$$

near S, where the Γ_p are again replaced by (11.4.7). The results of these calculations (done in the zeroth, first and second approximations) give an expression for $-\delta(M - M_0)$ in s, n coordinates, which was to be expected.

11.5 The Green's Function for the Diffraction Problem at a Cylinder with Variable Impedance

The problem of constructing the asymptotic expansion of the Green's function in the case where the boundary condition

$$\frac{\partial \Gamma}{\partial n} + ikg(s)\Gamma \bigg|_S = 0 \quad , \tag{11.5.1}$$

is given on S is solved in exactly the same way as the problem with $\Gamma|_S = 0$[4].

As the etalon problem we use the problem for the Green's function satisfying (11.5.1) when $g = \text{const}$ on the infinite surface $r \geq \varrho$, $-\infty < \phi < \infty$ (see Sect. 11.2). The solution of this problem is known [11.9]. The Green's function $\Gamma(M_0, M; k)$, that satisfies (11.5.1) will be constructed as before in the form of a series (11.4.4) of creeping waves u_p^+ and u_p^-. The asymptotic expansion of the functions $u_p^+(M, k)$ for the exterior of a contour S with an impedance boundary condition on S was constructed in Sect. 7.5:

$$u_p^+(M, k) = \varrho^{-1/6}(s) \exp\left[iks + i\xi_p \left(\frac{k}{2}\right)^{1/3} \int_0^s \frac{ds}{\varrho^{2/3}(s)}\right.$$

$$\left. + \int_0^s \frac{ds}{\varrho(s)g(s)} + \frac{i\varrho'(s)\nu^2}{6k^{1/3}\varrho(s)} + \frac{i\alpha_{10}(s)}{k^{1/3}} + O\left(\frac{1}{k^{2/3}}\right)\right]$$

$$\times w_1[T(\xi_p, M)] \quad , \quad p = 0, 1, \ldots \quad , \quad p = O(1) \quad ;$$

$$\tag{11.5.2}$$

---

[4] In the case of boundary condition (11.5.1) the pattern of oscillations may be complicated due to a surface wave. In the case $\text{Im } g > 0$, which is of the greatest importance in physics, there are no surface waves. This is the case that we will examine here.

$$T(\xi, M) = \xi - \nu \left(\frac{2}{\varrho(s)}\right)^{1/3} - \frac{i}{(k/2)^{1/3} \varrho^{1/3}(s) g(s)} + O\left(\frac{1}{k^{2/3}}\right) \ .$$

(11.5.3)

It was assumed in deriving (11.5.2 and 3) that $1/g(s) = O(1)$, and therefore they permit passage in the limit as $g \to \infty$ to the Dirichlet condition $u_p^+|_S = 0$ but do not permit passage to the Neumann problem as $g \to 0$.

Substituting $-i$ for i and g for $-g$ here, we get

$$u_p^-(M, k) = \varrho^{-1/6}(s) \exp\left[-iks - i\xi_p \left(\frac{k}{2}\right)^{1/3} \int_0^s \frac{ds}{\varrho^{2/3}(s)}\right.$$

$$\left. - \int_0^s \frac{ds}{\varrho(s)g(s)} - \frac{i\varrho'(s)\nu^2}{6k^{1/3}\varrho(s)} - \frac{i\alpha_{10}(s)}{k^{1/3}} + O\left(\frac{1}{k^{2/3}}\right)\right]$$

$$\times w_1[T(\xi_p, M)] \quad , \quad p = 0, 1, \ldots \quad , \quad p = O(1) \ . \quad (11.5.4)$$

Equations (11.5.2–4) relate to the case where the observation point M is not too far from S, namely: the coordinate $\nu = nk^{2/3}$ of this point obeys (11.3.7). The functions (11.5.2) (cf. Sect. 11.3) can be continued to any distance from S. This results in formulas analogous to (11.3.11–12),

$$u_p^+(M, k) = \frac{q_p^+(\sigma, r, k)e^{ik\tau}}{\sqrt{r}} \quad , \quad p = 0, 1, \ldots \quad , \quad p = O(1) \ , \quad (11.5.5)$$

$$q_p^+(\sigma, r, k) = e^{i(\pi/4)} \left(\frac{2\varrho(\sigma)}{k}\right)^{1/6} \exp\left[i\xi_p \left(\frac{k}{2}\right)^{1/3} \int_0^\sigma \frac{ds}{\varrho^{2/3}(s)}\right.$$

$$\left. + \int_0^\sigma \frac{ds}{\varrho(s)g(s)} + \frac{i\alpha_{10}(\sigma)}{k^{1/3}} + \frac{i\xi_p^2}{4k^{1/3}} \left(\frac{2}{\varrho(\sigma)}\right)^{1/3}\right.$$

$$\left. \times \left(\frac{\varrho(\sigma)}{r} + \frac{\varrho'(\sigma)}{3} + \frac{1}{g(\sigma)}\right) + O\left(\frac{1}{k^{2/3}}\right)\right] \ . \quad (11.5.6)$$

The ray coordinates τ, σ and $r = \tau - \sigma$ appearing here are the same as those in Sect. 11.3.

Determination of the universal constants A_p in the expansion of the Green's function $\Gamma(M_0, M, k)$ in creeping waves $u_p^\pm$ is based, as in the preceding section, on a comparison of the expansion (11.4.4) with the analogous expansion in the etalon problem. This comparison shows that (11.4.5) for the universal constants A_p holds for the case of boundary condition (11.5.1) as well. This enables us to write out immediately the asymptotic formulas for Γ_p:

$$\Gamma_p(M_0, M, k) = \frac{1}{2i} \left(\frac{2}{k}\right)^{1/3} [\varrho(s)\varrho(s_0)]^{-1/6} [w_1'(\xi_p)]^{-2}$$

$$\times \exp\left[ik(s - s_0) + i\xi_p \left(\frac{k}{2}\right)^{1/3} \int_{s_0}^s \frac{ds}{\varrho^{2/3}(s)}\right.$$

$$+ \int_{s_0}^{s} \frac{ds}{\varrho(s)g(s)} + \frac{i}{6k^{1/3}} \left(\frac{\varrho'(s)}{\varrho(s)} \nu^2 - \frac{\varrho'(s_0)}{\varrho(s_0)} \nu_0^2 \right)$$

$$+ i \frac{\alpha_{10}(s) - \alpha_{10}(s_0)}{k^{1/3}} + O(k^{-2/3}) \bigg]$$

$$\times w_1[T(\xi_p, M_0)] w_1[T(\xi_p, M)] \quad,$$

$$p = 0, 1, \ldots \quad ; \quad p = O(1) \quad. \tag{11.5.7}$$

If both points M_0 and M are located outside of the boundary layer, each in the shadow of the other, then

$$\Gamma_p(M_0, M, k) = \frac{[\varrho(\sigma_0)\varrho(\sigma)]^{1/6}}{2^{1/3}k^{2/3}\sqrt{r_0 r}} \frac{1}{[w_1'(\xi_p)]^2} \exp\bigg[ik(r_0 + r + \sigma - \sigma_0)$$

$$+ i\xi_p \left(\frac{k}{2} \right)^{1/3} \int_{\sigma_0}^{\sigma} \frac{ds}{\varrho^{2/3}(s)} + \int_{\sigma_0}^{\sigma} \frac{ds}{\varrho(s)g(s)} + i \frac{\alpha_{10}(\sigma) - \alpha_{10}(\sigma_0)}{k^{1/3}}$$

$$+ \frac{i\xi_p^2}{4k^{1/3}} \left(\frac{2}{\varrho(\sigma)} \right)^{1/3} \left(\frac{\varrho(\sigma)}{r} + \frac{\varrho'(\sigma)}{3} + \frac{1}{g(\sigma)} \right)$$

$$- \frac{i\xi_p^2}{4k^{1/3}} \left(\frac{2}{\varrho(\sigma_0)} \right)^{1/3} \left(\frac{\varrho(\sigma_0)}{r_0} + \frac{\varrho'(\sigma_0)}{3} - \frac{1}{g(\sigma_0)} \right)$$

$$+ O\left(\frac{1}{k^{2/3}} \right) \bigg] \quad, \tag{11.5.8}$$

where $p = 0, 1, 2, \ldots, p = O(1)$, r_0, r are the lengths of the tangents drawn from M_0 and M to S towards each other (Fig. 11.4), and $s = \sigma_0$, $s = \sigma$ are the coordinates of the corresponding points of tangency.

Finally, if one of the points (for instance M_0) is in the boundary layer, and the other (M) is outside the boundary layer and also in the shadow zone, then we get

$$\Gamma_p(M_0, M, k) = \frac{e^{-i\pi/4}}{\sqrt{2kr}} \left(\frac{\varrho(\sigma)}{\varrho(s_0)} \right)^{1/6} \exp\bigg[ik(r + \sigma - s_0)$$

$$+ i\xi_p \left(\frac{k}{2} \right)^{1/3} \int_{s_0}^{\sigma} \frac{ds}{\varrho^{2/3}(s)} + \int_{s_0}^{\sigma} \frac{ds}{\varrho(s)g(s)} + i \frac{\alpha_{10}(\sigma) - \alpha_{10}(s_0)}{k^{1/3}}$$

$$+ \frac{i\xi_p^2}{4k^{1/3}} \left(\frac{2}{\varrho(\sigma)} \right)^{1/3} \left(\frac{\varrho(\sigma)}{r} + \frac{\varrho'(\sigma)}{3} + \frac{1}{g(\sigma)} \right) + O\left(\frac{1}{k^{2/3}} \right) \bigg]$$

$$\times [w_1'(\xi_p)]^{-2} w_1[T(\xi_p, M_0)] \quad,$$

$$p = 0, 1, \ldots \quad ; \quad p = O(1) \quad. \tag{11.5.9}$$

Equations (11.5.7, 8, and 9) trace the influence that different sections of the path of the ray have on the amplitude and phase of the Green's function.

In conclusion we note that all the constructions discussed at the end of Sect. 11.4 carry over to the impedance case as well.

In a number of problems in acoustics and in electromagnetic theory with two media, the impedance boundary condition (11.5.1) with an appropriate choice of $g(s)$ describes the boundary condition at the interface between the media in the first approximation. The Green's functions of such problems can be written by means of (11.4.4, 5.7–9).

11.6 Notes on the Literature

Of considerable importance for the construction of asymptotic expansions for Green's functions are the papers by *Fock* and *Leontovich*, who first conceived the parabolic equation method [11.10, 11]. Using this method they obtained a number of important formulas for the field of an electromagnetic point source. The idea of the parabolic equation method, supplemented by the method of matched asymptotic expansions, allows the shortwave asymptotics of the point source problem to be constructed either near the source or far from it [11.12–14]. Another approach, related to uniform asymptotic expansions, was developed by *Buslaev* [11.15, 16].

The asymptotics of the Green's function for the case when the contour is analytic, and the source and observation points are located on S (the planar problem) were substantiated by *Filippov* [11.2].

The Green's functions for the case of an impedance boundary condition with ϱ = const, g = const [11.9], with $\varrho^{1/3}(s)g(s)$ = const [11.17], or with $\varrho(s)$ and $g(s)$ arbitrary, smooth, positive functions [11.18] have all been found using the parabolic equation method. On the basis of the localization principle, *Babič* [11.19] had suggested the form of the leading term of the asymptotic expansion for this case. The asymptotic formulas obtained by different methods in the papers of *Babič* [11.19] and *Molotkov* [11.18] are identical. *Buldyrev* [11.20] suggested the idea of using the reciprocity theorem to construct the asymptotic expansions of Green's functions. The solutions of diffraction problems from acoustics and elasticity theory which reduce to finding asymptotic expansions of Green's functions with impedance boundary conditions are given in papers by *Molotkov* [11.21, 22].

Some other papers dealing with asymptotic expansions of the Green's functions for exterior diffraction problems are noted at the end of Chap. 13.

The present chapter is based on papers of *Molotkov* [11.18, 21]; however, the sequence of presentation was slightly altered, and a number of refinements were made.

12. Asymptotic Expansion of the Green's Function for a Surface Source (the Internal Problem)

In this chapter we investigate the high-frequency scalar field of a line source inside an inhomogeneous infinite body close to its surface.

12.1 Formulation of the Problem and Physical Assumptions

Let the boundary of the body be a cylindircal surface that extends to infinity with generatrices parallel to the z-axis, the inhomogeneity depending on the two coordinates x and y, and let the line source lie in the surface. We will assume that the field vanishes on the surface of the body except at the source. More precisely: this problem consists in finding the solution of the Helmholtz equation

$$\left(\frac{\partial^2}{\partial x^2} + \frac{\partial^2}{\partial y^2} + \frac{\omega^2}{c^2(x,y)} \right) u(x,y) = 0, \quad (x,y) \in \Omega \quad, \tag{12.1.1}$$

with boundary condition

$$u|_S = \delta(s) \quad, \tag{12.1.2}$$

where S – the directrix of the cylindrical surface – is a sufficiently smooth plane curve that forms the boundary of Ω, s is the arc length along S measured from the location of the source, and $\delta(s)$ is the Dirac delta function.

We will assume that the principle of limiting absorption is satisfied, i.e., the analytic continuation of $u(x,y;\omega)$ with respect to ω from the real axis into the upper half-plane $\operatorname{Im}\omega > 0$ must approach zero as $\sqrt{x^2 + y^2} \to \infty$ (recall that an infinite region is being considered). Moreover, let the effective radius of curvature $P(s)$ of curve S (see Sect. 5.3) satisfy the condition

$$0 < \frac{1}{P(s)} \equiv \frac{1}{\varrho(s)} - \frac{c_1(s)}{c_0(s)} < \infty \quad, \tag{12.1.3}$$

under which constructive interference of multiply reflected waves takes place close to the boundary of region Ω, i.e. a whispering gallery effect occurs. (The opposite case $P^{-1}(s) < 0$, where a shadow zone arises near the boundary of the body, was considered in the preceding chapter.)

In the following presentation the symbol u_m will denote a wave which has been reflected m times by the boundary after leaving the source. When (12.1.3) is satisfied, waves u_m with arbitrarily large m can arrive at an observation point located on S. If the observation point on S is fixed, the difference in the optical path length for the waves u_m approaches zero with increasing number of reflections m. Beginning with some number M, the waves with $m \geq M$ will interfere with each other. The higher the frequency ω, the smaller the optical path difference at which interference occurs; therefore the number M increases with increasing frequency ω. As the distance along S between the observation and source points increases, there is an increase in the optical path difference of waves with adjacent numbers, and this leads in turn to an increase in the number M as well. Obviously the interference of waves with a large number of reflections takes place not only on the boundary S itself, but also near it at interior points of Ω. As a result, a surface wave with fairly complicated structure arises that propagates near the S. Beginning at a certain depth, the optical path difference for all reflected waves will be fairly great. At this depth, the surface wave resolves completely into separate waves with well-defined geometric characteristics. Each such wave can be calculated by the formulas of the ray method of Chap. 2. At the same time, such formulas are obviously not applicable to calculation of the field of the surface wave.

The purpose of this chapter is to get a formula that describes the field of the surface wave. The formula must be such that with an increase in the frequency ω or the distance between the source and the observation point, simpler expressions must be automatically derivable from this formula that have the form of rays and describe waves u_m with a certain number of reflections m.

Some of the constructions of this chapter will be heuristic and have no pretense of mathematical rigor.

Note that neither the physical nor the mathematical literature has as yet examined any complicated cases of glancing wave incidence. The only exception is the Weyl-Van der Pol formula that describes the field of a wave propagating along an impedance plane.

12.2 The Ray Formula for Multiply Reflected Waves

In this section the ray method will be used to get a formula that describes the derivative $\partial u_m / \partial n$ on the surface of the body. By studying the phase of $\partial u_m / \partial n$, we will determine the number M where interference begins to happen for waves with a number of reflections $m > M$.

For the sake of clarity, we will do all calculations for the case $c(x, y) = c_0$ ($\omega/c_0 = k$) and only at the end will we present (without detailed calculations) the formula for a variable velocity of wave propagation.

Let the wave $u_m^{(2)}$, arriving at the observation point N be reflected from S at the points $N_1, N_2, \ldots, N_m$. If we denote the wave leaving N after reflection by $u_{m+1}^{(1)}$ then by virtue of boundary condition (12.1.2) the combined field of both these waves

$$u_m = u_m^{(2)} + u_{m+1}^{(1)}$$

must be equal to zero at the point $N(s, 0)$, and therefore we calculate at point N the value of the normal derivative

$$\frac{\partial u_m}{\partial n} = \frac{\partial u_m^{(2)}}{\partial n} + \frac{\partial u_{m+1}^{(1)}}{\partial n} \quad .$$

Let us calculate $(\partial u_m^{(2)}/\partial n)|_{n=0}$ and $(\partial u_{m+1}^{(1)}/\partial n)|_{n=0}$ by the formulas of the ray method. In the zeroth-order approximation,

$$\frac{\partial u_m^{(2)}}{\partial n}\bigg|_{n=0} = \frac{\partial u_{m+1}^{(1)}}{\partial n}\bigg|_{n=0} = A\frac{(-1)^m}{\sqrt{J_m}}ik\frac{\partial L_m}{\partial n}\bigg|_{n=0} \exp\left[i\left(kL_m - \frac{\pi}{2}m\right)\right].$$

$$(12.2.1)$$

Here A is a quantity that is constant along each ray; the source of oscillations must be given to uniquely determine this quantity. In addition, J_m is the total geometric divergence of the wave that takes into consideration the changes in the geometric divergence at points of reflection; L_m is the length of the broken curve $N_0 N_1, \ldots, N_m N$. The multiplier $(-1)^m$ is the product of the reflection coefficient ($R = -1$) at $N_1, N_2, \ldots, N_m$; the term $-\pi m/2$ in the exponential function accounts for the change in phase as the wave passes through the m caustics that are formed by the m-tuple reflection of the wave from S.

To find the quantity A, we note that in the case where Ω is the half-space $y \geq 0$, $c(x, y) = \text{const.}$, $s = x$, the problem (12.1.1–2) is easily solved explicitly. This solution is equal to

$$\tilde{u} = -\frac{i}{2}\frac{\partial}{\partial y}H_0^{(1)}(kr) \quad , \quad k = \frac{\omega}{c} \quad , \quad r = \sqrt{x^2 + y^2}$$

(here $H_0^{(1)}$ is the Hankel function). At large k and $r > 0$

$$\tilde{u} \sim \sqrt{\frac{k}{2\pi}}\exp[-(\pi i/4)] \cdot \frac{y}{r^{3/2}}\exp(ikr)\left(1 + O\left(\frac{1}{kr}\right)\right) \quad .$$

From this expression, by using the usual localization arguments (see Sect. 2.7) we find

$$A = \sqrt{\frac{k}{2\pi}}\exp[-(\pi i/4)]\sin\varepsilon_0 \quad ,$$

where ε_0 is the glancing angle of ray $N_0 N_1$. Assuming that this angle is small, we get (with an accuracy of ε_0^3)

$$A = \sqrt{\frac{k}{2\pi}}\exp[-(\pi i/4)]\varepsilon_0 \quad .$$

Let us derive approximate formulas for L_m and J_m that are valid for sufficiently large m. These formulas will express the quantities L_m and J_m in terms of the radius of curvature $\varrho(s)$ of S and the coordinate $\tilde{s}$ of the observation point N.

Let ϱ_j (see Sect. 5.1) be the radius of curvature of S at N_j, let ε_j be the glancing angle of the ray $N_j N_{j+1}$ at N_j, let Δs_j be the arc length of $N_j N_{j+1}$ along S, and let l_j be the length of the ray $N_j N_{j+1}$.

In Sect. 5.1 it was shown that

$$\varepsilon_j = \varepsilon_0 \sqrt[3]{\frac{\varrho_0}{\varrho_j}} \left[1 + O\left(\varepsilon_0^2\right) \right] \quad . \tag{12.2.2}$$

If we substitute this expression into (5.1.5), which expresses Δs_j in terms of ε_j, we get

$$\Delta s_j = 2\varrho_0^{1/3} \varrho^{2/3} \left(s_j + \frac{\Delta s_j}{2} \right) \varepsilon_0 + O\left(\varepsilon_0^3\right) \quad , \tag{12.2.3}$$

which implies that

$$\frac{(\Delta s_j)^3}{\varrho^2 \left(s_j + (\Delta s_j/2) \right)} = 2^3 \varrho_0 \varepsilon_0^3 + O\left(\varepsilon_0^5\right) \quad , \tag{12.2.4}$$

i.e., the ratio $(\Delta s_j)^3 / \varrho^2(s_j + \Delta s_j/2)$ remains unaltered to $O(\varepsilon_0^5)$ for all reflection points. Putting $j = 0, 1, \ldots, m$ $(N_{m+1} = N)$ in formula (12.2.4) and combining all the resultant equations, we get

$$\sum_{j=0}^{m} \frac{(\Delta s_j)^3}{\varrho^2 \left(s_j + (\Delta s_j/2) \right)} = (m+1)2^3 \varrho_0 \varepsilon_0^3 + (m+1)O\left(\varepsilon_0^5\right) \quad . \tag{12.2.5}$$

The sum on the left side of the equality can be approximately replaced by an integral

$$\sum_{j=0}^{m} \frac{(\Delta s_j)^3}{\varrho^2 \left(s_j + (\Delta s_j/2) \right)}$$

$$= \sum_{j=0}^{m} \frac{1}{\varrho^2 \left(s_j + (\Delta s_j/2) \right)} \left[4\varrho_0^{2/3} \varrho^{4/3} \left(s_j + \frac{\Delta s_j}{2} \right) \varepsilon_0^2 + O\left(\varepsilon_0^4\right) \right] \Delta s_j$$

$$= 4\varrho_0^{2/3} \varepsilon_0^2 \int_0^{\tilde{s}} \frac{ds}{\varrho^{2/3}(s)} + (m+1)O\left(\varepsilon_0^5\right) \quad ,$$

where $\tilde{s}$ is the arc length corresponding to the point N. Here (as in Sect. 5.1) we have used a Riemann sum formula for the approximate evaluation of the definite integral.

Now (12.2.5) can be written as

$$\varepsilon_0 = \frac{1}{2} \frac{1}{\varrho_0^{1/3}} \frac{1}{m+1} \int_0^{\tilde{s}} \frac{ds}{\varrho^{2/3}(s)} + O\left(\varepsilon_0^3\right) \quad , \tag{12.2.6}$$

whence

$$\varepsilon_0 = \frac{1}{2\varrho^{1/3}(0)} \frac{1}{m+1} \int_0^{\tilde{s}} \frac{ds}{\varrho^{2/3}(s)} + O\left[\left(\frac{1}{m+1} \frac{1}{\varrho^{1/3}(0)} \int_0^{\tilde{s}} \frac{ds}{\varrho^{2/3}(s)} \right)^3 \right] \quad . \tag{12.2.7}$$

Formula (12.2.7) gives the value of the angle ε_0 at which a ray must depart from the source to arrive at observation point N after m reflections.

We now calculate

$$L_m = \sum_{j=0}^{m} l_j \quad ,$$

where l_j is the length of the straight line segment joining the points s_j and s_{j+1}. Using formulas (5.1.2) and (5.1.3), we find

$$l_j = \sqrt{[x(s_{j+1}) - x(s_j)]^2 + [y(s_{j+1}) - y(s_j)]^2}$$

$$= (s_{j+1} - s_j) - \frac{1}{24} \frac{1}{\varrho_j^2}(s_{j+1} - s_j)^3 + \frac{1}{24} \frac{\varrho_j'}{\varrho_j^3}(s_{j+1} - s_j)^4$$

$$+ O[(s_{j+1} - s_j)^5] \quad , \quad \text{or}$$

$$l_j = \Delta s_j - \frac{1}{24} \frac{1}{\varrho^2 (s_j + (\Delta s_j/2))}(\Delta s_j)^3 + O[(\Delta s_j)^5] \quad . \tag{12.2.8}$$

Summing (12.2.8) over $j = 0, 1, 2, \ldots, m$, we get

$$L_m = \tilde{s} - \frac{1}{24} \sum_{j=0}^{m} \frac{(\Delta s_j)^3}{\varrho^2 (s_j + (\Delta s_j/2))} + (m + 1)O[(\Delta s_j)^5] \quad . \tag{12.2.9}$$

Considering (12.2.5), we rewrite (12.2.9) as

$$L_m = \tilde{s} - \tfrac{1}{24}(m + 1)2^3 \varrho_0 \varepsilon_0^3 + (m + 1)O[(\Delta s_j)^5]$$

and finally, using expression (12.2.7) for ε_0, we arrive at the final formula

$$L_m = \tilde{s} - \frac{1}{24} \frac{1}{(m + 1)^2} \left[\int_0^{\tilde{s}} \frac{ds}{\varrho^{2/3}(s)} \right]^3 + O\left[(m + 1)\varepsilon_0^5\right] \quad . \tag{12.2.10}$$

Let us calculate the derivative $(\partial L_m/\partial n)|_{n=0}$. We orient a unit vector $\vec{l}_m$ along ray $N_m N$, and write the derivative $(\partial L_m/\partial n)|_{n=0}$ in the form

$$\frac{\partial L_m}{\partial n}\bigg|_{n=0} = \frac{\partial L_m}{\partial l_m}(\vec{l}_m \cdot \vec{n}) = \sin \varepsilon_{m+1} \quad ,$$

where ε_{m+1} is the glancing angle of the ray $N_m N$ at N. Using (12.2.2) and (12.2.7), we get

$$\frac{\partial L_m}{\partial n}\bigg|_{n=0} = \varepsilon_0 \sqrt[3]{\frac{\varrho_0}{\varrho(\tilde{s})}} \left[1 + O\left(\varepsilon_0^2\right)\right]$$

$$= \frac{1}{2\varrho^{1/3}(\tilde{s})} \frac{1}{m + 1} \int_0^{\tilde{s}} \frac{ds}{\varrho^{2/3}(s)} + O\left(\varepsilon_0^3\right) \quad . \tag{12.2.11}$$

It remains for us to derive a formula for the complete geometric divergence J_m,

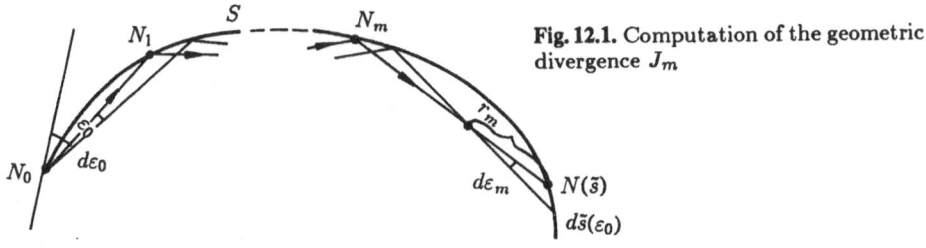

Fig. 12.1. Computation of the geometric divergence J_m

which by definition (Fig. 12.1) is equal to

$$J_m = \lim_{d\varepsilon_0 \to 0} \frac{r_m d\varepsilon_m}{d\varepsilon_0} \quad . \tag{12.2.12}$$

For a fixed number of reflections m, the position of N depends on ε_0, and consequently $\tilde{s} = \tilde{s}(\varepsilon_0)$. By virtue of the law of sines, (12.2.12) is equivalent to

$$J_m = \lim_{d\varepsilon_0 \to 0} \frac{d\tilde{s}(\varepsilon_0)}{d\varepsilon_0} \sin \varepsilon_{m+1} \quad .$$

Differentiating (12.2.7) and taking (12.2.2) into consideration, we get

$$J_m = \varrho_0^{1/3} \varrho^{1/3}(\tilde{s}) \int_0^{\tilde{s}} \frac{ds}{\varrho^{2/3}(s)} + O(\varepsilon_0) \quad . \tag{12.2.13}$$

Using (12.2.10, 11 and 13), we write the desired quantity $(\partial u_m / \partial n)|_{n=0}$ in the form

$$\left. \frac{\partial u_m}{\partial n} \right|_{n=0} \approx \frac{1}{2\sqrt{2\pi}} \exp[i(\pi/4)] \frac{k^{3/2}}{\sqrt{\varrho(0)\varrho(\tilde{s})}} \frac{1}{(m+1)^2}$$

$$\times \left[\int_0^{\tilde{s}} \frac{ds}{\varrho^{2/3}(s)} \right]^{3/2} \exp\left\{ ik\left[\tilde{s} - \frac{1}{24} \frac{1}{(m+1)^2} \right. \right.$$

$$\left. \left. \times \left(\int_0^{\tilde{s}} \frac{ds}{\varrho^{2/3}(s)} \right)^3 \right] + i\frac{\pi}{2}m \right\} \quad . \tag{12.2.14}$$

Introducing the reduced arc length

$$\gamma = (k/2)^{1/3} \int_0^{\tilde{s}} \frac{ds}{\varrho^{2/3}(s)} \quad ,$$

in place of (12.2.14) we get the following formula

$$\left. \frac{\partial u_m}{\partial n} \right|_{n=0} \approx \frac{1}{2\sqrt{\pi}} \exp[i(\pi/4)] \frac{k}{\sqrt{\varrho(0)\varrho(\tilde{s})}} \frac{\gamma^{3/2}}{(m+1)^2}$$

$$\times \exp\left\{ i\left[k\tilde{s} - \frac{1}{12} \frac{\gamma^3}{(m+1)^2} + \frac{\pi}{2}m \right] \right\} \quad . \tag{12.2.15}$$

For an inhomogeneous medium, the calculations are done completely analogously. Leaving out all intermediate steps, we give the final formula

$$
\left.\frac{\partial u_m}{\partial n}\right|_{n=0} \approx \frac{1}{2\sqrt{\pi}} \exp[i(\pi/4)] \frac{\omega}{\sqrt{c_0(0)P(0)c_0(\tilde{s})P(\tilde{s})}} \frac{\gamma^{3/2}}{(m+1)^2}
$$

$$
\times \exp\left\{i\left[\omega \int_0^{\tilde{s}} \frac{ds}{c_0(s)} - \frac{1}{12}\frac{\gamma^3}{(m+1)^2} + \frac{\pi}{2}m\right]\right\}, \quad (12.2.16)
$$

where $P(s)$ is the effective radius of curvature of S, $c_0(s)$ is the wave propagation velocity at S and

$$
\gamma = \left(\frac{\omega}{2}\right)^{1/3} \int_0^{\tilde{s}} \frac{ds}{c_0^{1/3}(s)P^{2/3}(s)}
$$

is the reduced arc length for an inhomogeneous medium. We call the reader's attention to the fact that the structure of formulas (12.2.15) and (12.2.16) (for a homogeneous and inhomogeneous medium respectively) is identical. To convert from formula (12.2.15) to formula (12.2.16) the radius of curvature $\varrho(s)$ must be replaced (as in Sect. 5.3) by the effective radius of curvature $P(s)$, the multiplying factor k in (12.2.15) must be replaced by the expression $\omega/\sqrt{c_0(0)c_0(\tilde{s})}$, and $k\tilde{s}$ in the exponent of (12.2.15) must be replaced by $\omega \int_0^{\tilde{s}} ds/c_0(s)$.

Let us note that formula (12.2.16) is symmetric with respect to the points 0 and $\tilde{s}$. This was to be expected by virtue of the fact that the principle of reciprocity must hold for the solution of (12.1.1–2).

Let us examine the phase of the quantity $(\partial u_m/\partial n)|_{n=0}$. Based on physical considerations, we can assume that two waves that have arrived at N after $m-1$ and m reflections from boundary S retain their individuality, and can be observed separately, if the difference $T_m - T_{m-1}$ between their arrival times at N exceeds the period of oscillations $T = (2\pi/\omega)$. We will assume in the following discussion that two waves may be observed indepedently of one another if the somewhat more severe condition

$$
T_m - T_{m-1} > q\left(\frac{\omega P(\tilde{s})}{2c_0(\tilde{s})}\right)^{3\delta} T, \quad (12.2.17)
$$

is satisfied, where $q \sim 1$ and δ is a small fixed number. Inequality (12.2.17) means that the delay of one wave relative to the other must exceed the period by a larger amount as ω increases. (The specific value of the quantity δ will be established in Sect. 12.3 in the error analysis of a more precise formula describing the reflected waves.)

Formula (12.2.16) implies that

$$
T_m - T_{m-1} \approx \frac{1}{\omega}\frac{\gamma^3}{12}\left[\frac{1}{m^2} - \frac{1}{(m+1)^2}\right] \approx \frac{1}{6}\frac{\gamma^3}{m^3}\frac{1}{\omega}.
$$

Substituting this expression for $T_m - T_{m-1}$ into (12.2.17), we get

$$m < \frac{\gamma}{2\sqrt{\Delta}} \left(\frac{\omega P}{2c_0} \right)^{-\delta} \quad , \tag{12.2.18}$$

where $\Delta = (\frac{3}{2}\pi q)^{2/3}$. Thus, waves that have been reflected a different number of times will be observed separately at N if the number of reflections $m < M$, where $M = [(\gamma/2\sqrt{\Delta})(\omega P/2c_0)^{-\delta}]$ is the largest integer that does not exceed $(\gamma/2\sqrt{\Delta})(\omega P/2c_0)^{-\delta}$. Waves with a number of reflections $m \geq M$ interfere, which results in a complex surface wave field.

We now find the width of the strip in which the surface wave propagates. A ray arriving at the observation point N after M reflections leaves the source at a glancing angle equal to

$$E_0 = \frac{1}{2} \frac{c_0^{1/3}(0)}{P_0^{1/3}(0)} \frac{1}{M+1} \int_0^{\tilde{s}} \frac{ds}{c_0^{1/3}(s)P^{2/3}(s)}$$

$$\approx \sqrt{\Delta} \left[\frac{\omega P(0)}{2c_0(0)} \right]^{-1/3} \left[\frac{\omega P(\tilde{s})}{2c_0(\tilde{s})} \right]^{\delta} \quad .$$

A surface wave propagating close to boundary S generates rays for which $\varepsilon_0 \leq E_0$. Let us estimate for an inhomogeneous medium the maximum distance d_j from S to an element $N_j N_{j+1}$ of a curvilinear broken ray that has left the source at an angle E_0. In estimating d_j to the first approximation we can assume that the angle of a rectilinear ray $N_j N_{j+1}$ is subtended by a circular arc of effective radius $P(s_j)$, therefore

$$d_j \approx P(s_j) - P(s_j) \cos \varepsilon_j \approx \tfrac{1}{2} P(s_j)\varepsilon_j^2 \quad ,$$

where the angle ε_j is related to E_0 by (12.2.2). Using this relation, we get

$$d_j \approx \frac{\Delta}{2} P(s_j) \left[\frac{\omega P(s_j)}{2c(s_j)} \right]^{-2/3+2\delta} \quad . \tag{12.2.19}$$

From (12.2.19) we see that the surface wave propagates in a boundary layer with thickness proportional to $\omega^{-2/3+2\delta}$. It was in just such a boundary layer that solutions (7.3.4) were constructed for the homogeneous Helmholtz equation, where these solutions had the form of series in powers of $\omega^{-1/3}$.

We obtained formulas (12.2.15) and (12.2.16) for the field by using only the zeroth approximation of the ray method. However, this approximation of the ray method is inadequate for describing reflected waves near a concave surface. The fact is that for fixed source and observation points the caustics of multiply reflected waves will be located closer to the boundary with an increasing number of reflections, and the usual formulas of the ray method are not applicable in the vicinity of caustics. The inadequacy of the zero approximation of the ray method shows up formally when estimating the higher-order approximations to (12.2.15) and (12.2.16).

Refinement of formulas (12.2.15) and (12.2.16) (see below Sect. 12.3) leads to an additional phase factor

$$\exp\left[-\mathrm{i}\frac{5}{3}\frac{(m+1)^4}{\gamma^3}\right] \qquad\qquad (12.2.20)$$

in these formulas.

12.3 Refinement of the Ray Formula[1]

Formulas (12.2.15) and (12.2.16) can be refined by using either further approximations of the ray method, or by the technique we developed in Chap. 3 specifically for describing the field in the neighborhood of caustics. We will take the second route, considering only the case of constant velocity (taken as unity).

First let us derive equations for the caustics of the multiply reflected rays. As before we will assume that the source of the wave field and the observation point are located on S at points N_0 and N (the arc length along S is measured from N_0; the arc length $\tilde{s}$ corresponds to N).

Let us inscribe within S the broken line $N_0, N_1, \ldots, N_m, N$ such that the law of reflection is satisfied at its vertices $N_j(s_j)$ $j = 1, 2, \ldots, m$; i.e., the glancing angle of an element $N_{j-1}N_j$ is equal to the glancing angle of the element N_jN_{j+1}. In the following we will call this broken curve an *extremal polygon*.

A ray emanating from N_0 along the element N_0N_1 of the extremal polygon will arrive at the observation point N after m reflections at the points $N_1, N_2, \ldots, N_m$. The angle of exit of the initial ray N_0N_1 for a predetermined coordinate $\tilde{s}$ of observation point N and a given number of reflections m is determined from formula (12.2.7):

$$\varepsilon_0 = \frac{1}{2\varrho^{1/3}(0)}\frac{1}{m+1}\int_0^{\tilde{s}}\frac{ds}{\varrho^{2/3}(s)} + O\left[\left(\frac{1}{m+1}\frac{1}{\varrho^{1/3}(0)}\int_0^{\tilde{s}}\frac{ds}{\varrho^{2/3}(s)}\right)^3\right].$$

$$(12.3.1)$$

As the angle ε_0 varies in the neighborhood of the value (12.3.1), each element of the extremal polygon $N_0, N_1, \ldots, N_m, N$ forms a family of straight lines. The families of curves generated by the elements N_jN_{j+1}, $j = 1, 2, \ldots, m$ will have envelopes – the caustics K_j of the multiply reflected rays.

Let us construct the envelope of element N_jN_{j+1}, in other words let us construct the caustic K_j of rays which have been reflected j times. To do this, we first obtain an equation of element N_jN_{j+1} in (s, n) coordinates. We use (12.2.6), which we write as

---

[1] This section can be skipped on a first reading. The only part that is essential in what follows is formula (12.3.33), which differs from formula (12.2.16) by the term (12.2.20).

$$\frac{1}{2\varrho^{1/3}(0)} \frac{1}{m+1} \int_0^{\tilde{s}} \frac{ds}{\varrho^{2/3}(s)} = \varepsilon_0 + O\left(\varepsilon_0^3\right) \; . \qquad (12.3.2)$$

Locating the observation point N at the reflection points $N_j(s_j)$, $j = 1, 2, \ldots,$ m, we will have along with (12.3.2),

$$\frac{1}{2\varrho^{1/3}(0)} \frac{1}{j} \int_0^{s_j} \frac{ds}{\varrho^{2/3}(s)} = \varepsilon_0 + O\left(\varepsilon_0^3\right) \; . \qquad (12.3.3)$$

For the glancing angle ε_j of $N_j N_{j+1}$ we get on the basis of formula (12.2.2),

$$\varepsilon_j = \frac{1}{2\varrho^{1/3}(s_j)} \frac{1}{j} \int_0^{s_j} \frac{ds}{\varrho^{2/3}(s)} + O\left[\left(\frac{1}{j} \int_0^{s_j} \frac{ds}{\varrho^{2/3}(s)}\right)^3\right] \; .$$

In (s, n) coordinates, $N_j N_{j+1}$ is described by the equation

$$n = -\frac{1}{2}\varrho^{-1/3}(s_j)\frac{1}{j} \int_0^{s_j} \frac{ds}{\varrho^{2/3}(s)}(s - s_j) + \frac{1}{2\varrho(s_j)}(s - s_j)^2$$

$$-\frac{\varrho'(s_j)}{6\varrho^2(s_j)}(s - s_j)^3 + \ldots \; , \quad j = 1, 2, \ldots \qquad (12.3.4)$$

[the derivation of (12.3.4) is analogous to the derivation of (5.1.12)]. In (12.3.4), as ε_0 varies there will be a change in the coordinate s_j of the initial point of the ray. Thus equation (12.3.4) can be considered as the equation of a family of rays generated by variation of ε_0 and dependent only on the parameter s_j. We are interested in the caustic K_j of this family of rays. Eliminating s_j from (12.3.4) and setting $(\partial n/\partial s_j) = 0$, we find the equation of the caustic that we are looking for:

$$n = -\frac{1}{2}\varrho^{1/3}(s)\frac{1}{j(j+1)} \left(\int_0^s \frac{d\tau}{\varrho^{2/3}(\tau)}\right)^2 + O\left[\left(\frac{1}{j} \int_0^s \frac{d\tau}{\varrho^{2/3}(\tau)}\right)^4\right] \; .$$

Let us introduce the function of arc length s:

$$\kappa_j(s) = \frac{1}{2j} \int_0^s \frac{d\tau}{\varrho^{2/3}(\tau)} \; . \qquad (12.3.5)$$

Then the equation of the envelope is written as

$$n = -2\varrho^{1/3}(s)\frac{j}{j+1}\kappa_j^2(s) + O\left[\kappa_j^4(s)\right] \; . \qquad (12.3.6)$$

Obviously the correction term in the equation of the envelope, (12.3.6), on the interval $0 \le s \le s_{j+l}$ (where l is an integer on the order of unity) does not exceed $O(\varepsilon_0^4)$. In the following, equation (12.3.6) for the caustic of j-tuply reflected rays will be considered on the interval $s_j \le s \le s_{j+1}$. Setting $j = 1, 2, \ldots, m$ in

(12.3.6), we get a system of caustics K_j of rays that have been reflected once, twice, and so on, and which are on adjacent segments $[s_1, s_2]$, $[s_2, s_3]$, and so on.

In Sect. 12.1 we adopted the convention that a surface wave arises as a result of interference of multiply reflected waves for which $\varepsilon_0 \leq E_0 = O(\omega^{-1/3+\delta})$. On the other hand, waves for which $\varepsilon_0 > E_0$ retain their individuality, and can be observed separately. We will now derive formulas that describe these waves for ε_0 near E_0, i.e., at the moment when they become distinguishable from the surface wave. Therefore we shall henceforth consider that the angle ε_0 and the function $\kappa_j(s)$ on $s_j \leq s \leq s_{j+1}$ are quantities of the order of $\omega^{-1/3+\delta}$.

Let us get a more precise formula of the ray type describing the field of a j-tuply reflected wave on S. (In constructing such formulas we will use the results of Chap. 3 and equations (12.3.6) of the caustics.)

The field $u_j(M)$ of a wave that has been reflected j times is described in the vicinity of its corresponding caustic K_j by the formula [cf. (3.5.12)].

$$
\begin{aligned}
u_j(M) = \frac{1}{2} \Bigg\{ & \left[\mu_j^{1/4}(M) \left(\frac{\chi_0^{(j)}(\alpha_1)}{\sqrt{J_1^{(j)}(M)}} + \frac{\chi_0^{(j)}(\alpha_2)}{\sqrt{J_2^{(j)}(M)}} \right) + O\left(\frac{1}{\omega}\right) \right] \\
& \times v\left(-\omega^{2/3} \mu_j(M) \right) \\
& + \frac{i}{\omega^{1/3}} \left[\mu_j^{-1/4}(M) \left(\frac{\chi_0^{(j)}(\alpha_2)}{\sqrt{J_2^{(j)}(M)}} - \frac{\chi_0^{(j)}(\alpha_1)}{\sqrt{J_1^{(j)}(M)}} \right) + O\left(\frac{1}{\omega}\right) \right] \\
& \times v'\left(-\omega^{2/3} \mu_j(M) \right) \Bigg\} \exp[i\omega\xi_j(M)] \cdot \omega^{-r} \quad, \qquad (12.3.7)
\end{aligned}
$$

where

$$
\mu_j = \left[\frac{3}{4} \left(\tau_2^{(j)} - \tau_1^{(j)} \right) \right]^{2/3} \quad, \quad \xi_j = \frac{\tau_1^{(j)} + \tau_2^{(j)}}{2} \quad,
$$

and $\tau_1^{(j)}$ is the eikonal of a ray approaching the caustic K_j, $\tau_2^{(j)}$ is the eikonal of a ray that has passed through the caustic, $J_1^{(j)}$, $J_2^{(j)}$ are the geometric divergences of the corresponding rays, $\chi_0^{(j)}(\alpha_1)$ is an arbitrary function which is constant on the ray approaching the caustic, which is characterized by a parameter α_1, $\chi_0^{(j)}(\alpha_2)$ is the same function calculated for the value of a parameter α_2 characterizing a ray that has passed through the caustic, and hence is constant on this ray; v and v', as always, denote the Airy function and its derivative.

Considering $\kappa_j(s)$ as a small quantity of the order of $\omega^{-1/3+\delta}$, we calculate the function $\mu_j(M)$ and the coefficients of $v(-\omega^{2/3}\mu_j)$ and $v'(-\omega^{2/3}\mu_j)$ in (12.3.7) at the point $M(s')$, $s_j \leq s' \leq s_{j+1}$ located on S, i.e. where $n = 0$. The calculations are carried out to order $O(\kappa_j^3)$.

316

Obviously when the velocity of wave propagation is taken as constant and equal to unity, $\tau_1^{(j)}$ is the length $L_{j-1}(s')$ of the extremal polygon N_0, $N_1', \ldots, N_{j-1}', M(s')$, made up of j elements, and $\tau_2^{(j)}$ is the length $L_j(s')$ of the extremal polygon $N_0, N_1'', \ldots, N_j'', M(s')$, made up of $j+1$ elements. By virtue of (12.2.10),

$$L_{j-1}(s') = s' - \frac{1}{24}\frac{1}{j^2}\left[\int_0^{s'} \frac{d\tau}{\varrho^{2/3}(\tau)}\right]^3 + O\left(\kappa_j^4\right)$$

and

$$L_j(s') = s' - \frac{1}{24}\frac{1}{(j+1)^2}\left[\int_0^{s'} \frac{d\tau}{\varrho^{2/3}(\tau)}\right]^3 + O\left(\kappa_j^4\right) \quad,$$

and consequently

$$\mu_j(s') = 2^{-2/3}\left[\frac{j\left(j+\frac{1}{2}\right)}{(j+1)^2}\right]^{2/3}\kappa_j^2(s') + O\left(\kappa_j^3\right) \quad, \tag{12.3.8}$$

where $\kappa_j(s)$ is defined by (12.3.5).

For the geometric divergences $J_j^{(1)}$ and $J_j^{(2)}$ we have the formulas

$$J_1^{(j)} = \frac{l_1}{\varrho_1^{(j)}} \quad \text{and} \quad J_2^{(j)} = \frac{l_2}{\varrho_2^{(j)}} \quad,$$

where $\varrho_1^{(j)}$ and $\varrho_2^{(j)}$ are the radii of curvature of the caustic K_j at the points of contact of the ray approaching the caustic and the ray departing from the caustic respectively; l_1 and l_2 are the lengths of these rays between the points of contact and $M(s')$. In the following discussion the quantities $\varrho_1^{(j)}$ and $\varrho_2^{(j)}$, which are constant on the corresponding rays, will be combined with the factors $\chi_0^{(j)}(\alpha_1)$ and $\chi_0^{(j)}(\alpha_2)$ and we will introduce the notation

$$\chi_0^{(j)}(\alpha_r)\sqrt{\varrho_r^{(j)}} = \chi_j(\alpha_r) \quad, \quad r = 1, 2 \quad.$$

Let $s^{(1)}$ be the coordinate of the point of contact with the caustic K_j by a ray approaching the caustic, and let $s^{(2)}$ be the coordinate of the point of contact with caustic K_j by a ray that has passed through the caustic (Fig. 12.2). Obviously

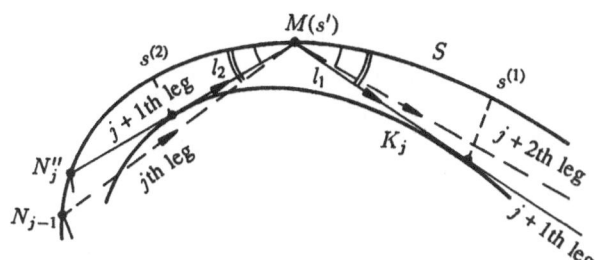

Fig. 12.2. Refinement of the ray formula for the field

the lengths l_1 and l_2 are expressed in terms of the coordinates s', $s^{(1)}$ and $s^{(2)}$ by the formulas

$$l_1 = s^{(1)} - s' + O\left[(s^{(1)} - s')^3\right] \quad,$$

$$l_2 = s' - s^{(2)} + O\left[(s' - s^{(2)})^3\right] \quad. \tag{12.3.9}$$

Let us determine $s^{(1)}$ and $s^{(2)}$. Let $n = n_1(s)$ and $n = n_2(s)$ be two rays that belong to the family (12.3.4) and that pass through $M(s')$. We write (12.3.6) of the caustic K_j in the form $n = n(s)$, where

$$n(s) = -2\varrho^{1/3}(s)\frac{j}{j+1}\kappa_j^2(s) + O\left[\kappa_j^4(s)\right] \quad.$$

The quantities $s^{(1)}$ and $s^{(2)}$ must obviously satisfy the equations $n_1(s^{(1)}) = n(s^{(1)})$ and $n_2(s^{(2)}) = n(s^{(2)})$, and solving these, we find

$$s^{(1)} - s' = \kappa_{j+1/2}(s')\varrho^{2/3}(s') + \tfrac{1}{3}\kappa_{j+1/2}^2(s')\varrho^{1/3}(s')\varrho'(s') + O(\kappa_{j+1/2}^3) \quad,$$

$$s' - s^{(2)} = \kappa_{j+1/2}(s')\varrho^{2/3}(s') - \tfrac{1}{3}\kappa_{j+1/2}^2(s')\varrho^{1/3}(s')\varrho'(s') + O(\kappa_{j+1/2}^3). \tag{12.3.10}$$

To single out the dominant terms in the desired coefficients of v and v' in (12.3.7), let us evaluate the difference $\chi_j(\alpha_2) - \chi_j(\alpha_1)$. As the parameters α_1 and α_2 we take the coordinates $s^{(1)}$ and $s^{(2)}$ of the points where the corresponding rays contact the caustic. Then

$$\chi_j(\alpha_2) - \chi_j(\alpha_1) = O(\alpha_2 - \alpha_1) = O\left(s^{(2)} - s^{(1)}\right) \quad,$$

and by virtue of (12.3.10)

$$\chi_j(\alpha_2) - \chi_j(\alpha_1) = O\left(\kappa_{j+1/2}(s')\right) \quad. \tag{12.3.11}$$

With the aid of (12.3.9–11) we get

$$\frac{\chi_0^{(j)}(\alpha_1)}{\sqrt{J_1^{(j)}(M)}} + \frac{\chi_0^{(j)}(\alpha_2)}{\sqrt{J_2^{(j)}(M)}} = \frac{1}{\kappa_{j+1/2}^{1/2}(s')\varrho^{1/2}(s')}$$

$$\times \left\{ \left[\chi_j(\alpha_1) + \chi_j(\alpha_2)\right] + O\left(\kappa_{j+1/2}^2\right) \right\} \quad,$$

$$\frac{\chi_0^{(j)}(\alpha_2)}{\sqrt{J_2^{(j)}(M)}} - \frac{\chi_0^{(j)}(\alpha_1)}{\sqrt{J_1^{(j)}(M)}} = \frac{1}{\kappa_{j+1/2}^{1/2}(s')\varrho^{1/2}(s')}$$

$$\times \left\{ \left[\chi_j(\alpha_2) - \chi_j(\alpha_1)\right] + \tfrac{1}{6}\varrho^{-1/3}(s')\varrho'(s') \left[\chi_j(\alpha_2) + \chi_j(\alpha_1)\right] \right.$$

$$\left. \times \kappa_{j+1/2}(s') + O\left(\kappa_{j+1/2}^2\right) \right\} \quad.$$

The quantities $\chi_j(\alpha_1)$ and $\chi_j(\alpha_2)$, that appear in these formulas will be determined later on.

We turn now to the Airy function $v(-\omega^{2/3}\mu)$ and its derivative $v'(-\omega^{2/3}\mu)$. Since we are assuming $\kappa_j(s) \sim \omega^{-1/3+\delta}$, the argument of the Airy function has order $\omega^{2\delta}$, and therefore the Airy function itself and its derivative can be represented by the asymptotic series (see Appendix A.1):

$$v(-\omega^{2/3}\mu) \sim \frac{\omega^{-1/6}\mu^{-1/4}}{2}\left\{\sum_{n=0}^{\infty} a_n \frac{\exp[-i(\pi/2)n]}{(\omega\mu^{3/2})^n}\exp\left(i\frac{2}{3}\omega\mu^{3/2}-i\frac{\pi}{4}\right)\right.$$

$$\left. + \sum_{n=0}^{\infty} a_n \frac{\exp[i(\pi/2)n]}{(\omega\mu^{3/2})^n}\exp\left(-i\frac{2}{3}\omega\mu^{3/2}+i\frac{\pi}{4}\right)\right\} \quad ,$$

$$v'(-\omega^{2/3}\mu) \sim \frac{\omega^{1/6}\mu^{1/4}}{2i}\left\{\sum_{n=0}^{\infty} b_n \frac{\exp[-i(\pi/2)n]}{(\omega\mu^{3/2})^n}\exp\left(i\frac{2}{3}\omega\mu^{3/2}-i\frac{\pi}{4}\right)\right.$$

$$\left. - \sum_{n=0}^{\infty} b_n \frac{\exp[i(\pi/2)n]}{(\omega\mu^{3/2})^n}\exp\left(-i\frac{2}{3}\omega\mu^{3/2}+i\frac{\pi}{4}\right)\right\} \quad ,$$

where

$$a_0 = 1, \quad a_1 = \frac{5}{48}, \quad a_n = \frac{(6n-1)!!}{144^n(2n-1)!!n!} ,$$

$$b_0 = 1, \quad b_n = a_n - \left(\frac{3}{2}n - \frac{5}{4}\right)a_{n-1} \quad .$$

Expanding the Airy function and its derivative in (12.3.7) into asymptotic series, we write out in explicit form only those terms with order lower than $\omega^{-9\delta}$, $\omega^{-2/3+2\delta}$ and $\omega^{-1/3-2\delta}$:

$$u_j(M) = u_j^{(1)}(M) + u_j^{(2)}(M) \quad , \tag{12.3.12}$$

where

$$u_j^{(1)}(M) = \frac{1}{2}\exp[i(\pi/4)]\frac{\chi_j(\alpha_1)}{\kappa_{j+1/2}^{1/2}(s')\varrho^{1/3}(s')}\left\{1 + ia_1\frac{1}{\omega\mu_j^{3/2}(s')}\right.$$

$$- a_2\frac{1}{\omega^2\mu_j^3(s')} - \frac{1}{6}\varrho^{-1/3}(s')\varrho'(s')\kappa_{j+1/2}(s') + O(\omega^{-9\delta})$$

$$\left. + O(\omega^{-2/3+2\delta}) + O(\omega^{-1/3-2\delta})\right\}\exp\left(i\omega\tau_1^{(j)}\right)\omega^{-r} \tag{12.3.13}$$

and

$$u_j^{(2)}(M) = \frac{1}{2}\exp[-i(\pi/4)]\frac{\chi_j(\alpha_2)}{\kappa_{j+1/2}^{1/2}(s')\varrho^{1/3}(s')}\left\{1 - ia_1\frac{1}{\omega\mu_j^{3/2}(s')}\right.$$

$$\left. - a_2\frac{1}{\omega^2\mu_j^3(s')} + \frac{1}{6}\varrho^{-1/3}(s')\varrho'(s')\kappa_{j+1/2}(s') + O(\omega^{-9\delta})\right.$$

$$+ O(\omega^{-2/3+2\delta}) + O(\omega^{-1/3-2\delta}) \Bigg\} \exp\left(i\omega\tau_2^{(j)}\right)\omega^{-r} .$$

$$(12.3.14)$$

The first term $u_j^{(1)}(M)$ in (12.3.12) describes a wave going through the point M toward the caustic K_j; the second term $u_j^{(2)}(M)$ describes a wave going through M away from K_j.

Now let us take two points $N_j(s = s_j)$ and $N_{j+1}(s = s_{j+1})$ on S (Fig. 12.3) such that the ray passing through these points meets the caustic K_j at some point with coordinate $s = s^{(0)}$. At each of these points the wave field can be represented in the form of a sum of two terms $u_j^{(1)}(N_j) + u_{j-1}^{(2)}(N_j)$ and $u_{j+1}^{(1)}(N_{j+1}) + u_j^{(2)}(N_{j+1})$, calculated from (12.3.13) and (12.3.14). We will assume that the wave

$$u_j^{(1)}(N_j) = \frac{1}{2}\exp[i(\pi/4)]\frac{\chi_j}{\kappa_{j+1/2}^{1/2}(s_j)\varrho^{1/3}(s_j)}\Bigg\{1 + ia_1\frac{1}{\omega\mu_j^{3/2}(s_j)}$$

$$- a_2\frac{1}{\omega^2\mu_j^3(s_j)} - \frac{1}{6}\varrho^{-1/3}(s_j)\varrho'(s_j)\kappa_{j+1/2}(s_j)$$

$$+ O(\omega^{-9\delta}) + O(\omega^{-2/3+2\delta}) + O(\omega^{-1/3-2\delta})\Bigg\}$$

$$\times \exp\left[i\omega\tau_1^{(j)}(s_j)\right]\omega^{-r} \qquad (12.3.15)$$

arises as a result of the reflection at N_j of a wave that has undergone $j - 1$ reflections on the path from the source to N_j. In propagating along the ray N_jN_{j+1}, this wave gives a field at point N_{j+1} that is described by the function

$$u_j^{(2)}(N_{j+1}) = \frac{1}{2}\exp[-i(\pi/4)]\frac{\chi_j}{\kappa_{j+1/2}^{1/2}(s_{j+1})\varrho^{1/3}(s_{j+1})}$$

$$\times \Bigg\{1 - ia_1\frac{1}{\omega\mu_j^{3/2}(s_{j+1})} - a_2\frac{1}{\omega^2\mu_j^3(s_{j+1})}$$

$$+ \frac{1}{6}\varrho^{-1/3}(s_{j+1})\varrho'(s_{j+1})\kappa_{j+1/2}(s_{j+1})$$

$$+ O(\omega^{-9\delta}) + O(\omega^{-2/3+2\delta}) + O(\omega^{-1/3-2\delta})\Bigg\}$$

$$\times \exp\left[i\omega\tau_2^{(j)}(s_{j+1})\right]\omega^{-r} . \qquad (12.3.16)$$

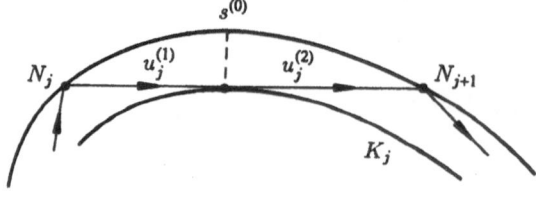

Fig. 12.3. Detail of whispering-gallery rays

We call the reader's attention to the fact that the factor χ_j in (12.3.15) and (12.3.16) takes on the same value because the function $\chi_j(\alpha)$ in both formulas must be calculated at the same value of the argument $\alpha = s^{(0)}$, corresponding to the ray $N_j N_{j+1}$. The reflection of $u_j^{(2)}$ at N_{j+1} gives rise to the wave $u_{j+1}^{(1)}$. This wave will propagate from S toward the caustic K_{j+1} and consequently will be described at N_{j+1} by (12.3.15) with j replaced by $j+1$. Subsequent reflections lead to waves for which the caustics are K_{j+2}, K_{j+3} and so on. Let us determine the relation between the eikonals $\tau_2^{(j)}$ and $\tau_1^{(j)}$ and the factors χ_j and χ_{j+1} appearing in the expression for the field in the neighborhood of the adjacent caustics K_j and K_{j+1}.

Consider the propagation of a wave along the extremal polygon N_1, N_2, N_3, ..., N_m joining the source N_0 and the observation point $N(\bar{s})$. By virtue of the boundary condition $U|_s = 0$, the equations

$$u_0 + u_1^{(1)}|_{N_1} = 0 \quad \text{and} \tag{12.3.17}$$

$$u_j^{(2)} + u_{j+1}^{(1)}|_{N_{j+1}} = 0 \quad , \quad j = 1, 2, ..., m-1 \quad , \tag{12.3.18}$$

must be satisfied at the vertices of the polygon N_0, N_1, ..., N_m, N, where u_0 is the wave emanating directly from the source. After substituting (12.3.15) and (12.3.16) into (12.3.18), since these equations must be satisfied identically in ω, we get

$$\tau_1^{(j+1)}(s_{j+1}) = \tau_2^{(j)}(s_{j+1}) \quad , \quad j = 1, 2, ..., m-1 \tag{12.3.19}$$

and

$$\exp[i(\pi/4)] \frac{\chi_{j+1}}{\kappa_{j+3/2}^{1/2}(s_{j+1})} \left\{ 1 + \frac{ia_1}{\omega \mu_{j+1}^{3/2}(s_{j+1})} - \frac{a_2}{\omega^2 \mu_{j+1}^3(s_{j+1})} \right.$$

$$-\frac{1}{6} \varrho^{-1/3}(s_{j+1}) \varrho'(s_{j+1}) \kappa_{j+3/2}(s_{j+1})$$

$$\left. +O(\omega^{-1/3-2\delta}) + O(\omega^{-2/3+2\delta}) + O(\omega^{-9\delta}) \right\}$$

$$= -\exp[-i(\pi/4)] \frac{\chi_j}{\kappa_{j+1/2}^{1/2}(s_{j+1})} \left\{ 1 - \frac{ia_1}{\omega \mu_j^{3/2}(s_{j+1})} - \frac{a_2}{\omega^2 \mu_j^3(s_{j+1})} \right.$$

$$+\frac{1}{6} \varrho^{-1/3}(s_{j+1}) \varrho'(s_{j+1}) \kappa_{j+1/2}(s_{j+1}) + O(\omega^{-1/3-2\delta})$$

$$\left. + O(\omega^{-2/3+2\delta}) + O(\omega^{-9\delta}) \right\} \quad . \tag{12.3.20}$$

Equation (12.3.20) enables us to determine the as yet unknown factors χ_j. From this, and consideration of the order relations $\omega \mu_j^{3/2} \sim \omega \mu_{j+1}^{3/2} \sim \omega^{-3\delta}$, $\kappa_{j+1/2} \sim \omega^{-1/3+\delta}$ we find that

$$\chi_{j+1} = i\frac{\kappa_{j+3/2}^{1/2}}{\kappa_{j+1/2}^{1/2}}\left\{1 - ia_1\left(\mu_j^{-3/2} + \mu_{j+1}^{-3/2}\right)\omega^{-1}\right.$$
$$- \left[a_2\left(\mu_j^{-3} - \mu_{j+1}^{-3}\right) + a_1^2\left(\mu_{j+1}^{-3} + \mu_j^{-3/2}\mu_{j+1}^{-3/2}\right)\right]\omega^{-2}$$
$$- \frac{d}{ds}\left(\ln \varrho^{-1/6}\right)\cdot\varrho^{2/3}\left(\kappa_{j+1/2} + \kappa_{j+3/2}\right) + O\left(\omega^{-1/3-2\delta}\right)$$
$$\left.+ O\left(\omega^{-2/3+2\delta}\right) + O\left(\omega^{-9\delta}\right)\right\}_{s=s_{j+1}}\chi_j \quad . \tag{12.3.21}$$

Formula (12.3.21) enables us to express χ_m in terms of χ_1. We take the logarithm of (12.3.21) and sum the result with respect to all j from 1 to $m-1$. Using the expansion $\ln(1+\alpha) = \alpha - \alpha^2/2 + O(\alpha^3)$ and considering (12.3.5) and (12.3.8), we get

$$\ln \chi_m = i\frac{\pi}{2}(m-1) + \frac{1}{2}\ln\frac{3}{2m+1}$$
$$- \sum_{j=1}^{m-1}\left[\alpha(j)\kappa_{j+1}^{-3}(s_{j+1})\omega^{-1} + \beta(j)\kappa_{j+1}^{-6}(s_{j+1})\omega^{-2}\right.$$
$$+ \gamma(j)\frac{d}{ds}\ln\varrho^{-1/6}(s)|_{s=s_{j+1}}2\varrho^{2/3}(s_{j+1})\kappa_{j+1}(s_{j+1})$$
$$\left.+ O\left(\omega^{-1/3-2\delta}\right) + O\left(\omega^{-2/3+2\delta}\right) + O\left(\omega^{-9\delta}\right)\right] + \ln\chi_1 \quad , \tag{12.3.22}$$

where

$$\alpha(j) = 4a_1 i\left[\frac{j^2}{(j+1)(2j+1)} + \frac{(j+2)^2}{(j+1)(2j+3)}\right]$$
$$= 4a_1 i + \frac{\alpha_1(j)}{j^2} \quad , \quad |\alpha_1(j)| < \text{const}$$

$$\beta(j) = \frac{\beta_1(j)}{j} \quad , \quad |\beta_1(j)| < \text{const} \quad ,$$

$$\gamma(j) = \frac{1}{2}\left(\frac{2j+2}{2j+1} + \frac{2j+2}{2j+3}\right) = 1 + \frac{\gamma_1(j)}{j^2} \quad , \quad |\gamma_1(j)| < \text{const} \quad .$$

From (12.3.3), we have

$$\kappa_{j+1}(s_{j+1}) = \varrho_0^{1/3}\varepsilon_0 + O\left(\varepsilon_0^3\right) \quad ,$$

i.e. all the $\kappa_{j+1}(s_{j+1})$ take on the same values to $O(\varepsilon_0^3) \sim O(\omega^{-1+3\delta})$. Therefore

$$\sum_{j=1}^{m-1}\alpha(j)\kappa_{j+1}^{-3}(s_{j+1})\omega^{-1} = i4a_1\frac{m-1}{\varrho_0\varepsilon_0^3\omega} + O(\omega^{-3\delta})\sum_{j=1}^{m-1}\frac{1}{j^2}$$
$$+ (m-1)O\left(\omega^{-2/3-\delta}\right)$$

and

$$\sum_{j=1}^{m-1} \beta(j)\kappa_{j+1}^{-6}(s_{j+1})\omega^{-2} = O(\omega^{-6\delta})\sum_{j=1}^{m-1}\frac{1}{j} \quad .$$

To sum the expression containing $\gamma(j)$ in (12.3.22), we recall that [see (12.2.3)]

$$2\varrho^{2/3}(s_{j+1})\varrho_0^{1/3}\varepsilon_0 = s_{j+1} - s_j + O\left(\varepsilon_0^2\right)$$

$$= \Delta s_j + O\left(\varepsilon_0^2\right) \quad . \tag{12.3.23}$$

Equality (12.3.23) enables us to replace the sum by the corresponding integral:

$$\sum_{j=1}^{m-1}\gamma(j)\frac{d}{ds}\ln\varrho^{-1/6}\bigg|_{s=s_{j+1}} \cdot 2\varrho^{2/3}(s_{j+1})\kappa_{j+1}(s_{j+1})$$

$$= \int_{s_1}^{s_m}\frac{d}{ds}\ln\varrho^{-1/6}(s)ds + O(\omega^{-1/3+\delta}) + O(\omega^{-1/3+\delta})\sum_{j=1}^{m-1}\frac{1}{j^2}$$

$$+(m-1)O(\omega^{-2/3+2\delta}) \quad .$$

Thus, since

$$\sum_{j=1}^{m-1}\frac{1}{j^2} < \sum_{j=1}^{\infty}\frac{1}{j^2} = \frac{\pi^2}{6} \quad , \quad \sum_{j=1}^{m-1}\frac{1}{j} < \log m + C$$

(where C is Euler's constant), and in view of (12.3.2),

$$m = O\left(\varepsilon_0^{-1}\right) = O(\omega^{1/3-\delta}) \quad ,$$

(12.3.22) can be written as

$$\ln\chi_m = i\frac{\pi}{2}(m-1) + \frac{1}{2}\ln\frac{3}{2m+1} - i4a_1\frac{m-1}{\varrho_0\varepsilon_0^2\omega} + \ln\frac{\varrho^{1/6}(s_m)}{\varrho^{1/6}(s_1)} + \ln\chi_1$$

$$+ O(\omega^{-1/3+\delta}) + O(\omega^{-3\delta}) + O(\omega^{1/3-10\delta}) \quad ,$$

whence

$$\chi_m = \exp[i(\pi/2)(m-1)]\sqrt{\frac{3}{2m+1}}\exp\left(-i\frac{5}{12}\frac{m-1}{\varrho_0\varepsilon_0^3\omega}\right)$$

$$\times \frac{\varrho^{1/6}(s_m)}{\varrho^{1/6}(s_1)}\chi_1[1 + O(\omega^{-1/3+\delta}) + O(\omega^{-3\delta}) + O(\omega^{1/3-10\delta})] \quad .$$

To determine χ_1 we turn to (12.3.17). The field u_0 emanating from the source along the ray N_0N_1 is described at N_1 by its ray expansion:

$$u_0 = -\frac{i\exp[i(\pi/4)]}{\sqrt{2\pi}}\sqrt{k\varepsilon_0}\frac{\exp(ikN_0N_1)}{\sqrt{N_0N_1}}\left[1 + O(\omega^{-3\delta})\right] \quad , \tag{12.3.24}$$

where $k = \omega/c_0 = \omega$, since $c_0 = 1$. [The error in (12.3.24) will be determined in the next section in the course of analyzing the exact solution for a circle.] Since

$$N_0 N_1 = 2\varrho(0)\varepsilon_0 + O\left(\varepsilon_0^3\right) = 2\varrho(0)\varepsilon_0 \left[1 + O(\omega^{-2/3+2\delta})\right] \quad ,$$

formula (12.3.24) can be rewritten as

$$u_0 = \frac{\exp[-i(\pi/4)]}{2\sqrt{\pi}} \sqrt{\frac{\omega\varepsilon_0}{\varrho(0)}} \exp(i\omega\tau_0) \left[1 + O(\omega^{-3\delta})\right] \quad , \tag{12.3.25}$$

where $\tau_0 = N_0 N_1$ is the eikonal of the wave at N_1. Substituting (12.3.15) and (12.3.25) into (12.3.17), we get

$$\tau_1^{(1)} = \tau_0 \tag{12.3.26}$$

and

$$\frac{1}{2} \exp[i(\pi/4)] \sqrt{\frac{3}{2}} \frac{\chi_1}{\kappa_1^{1/2}(s_1)\varrho^{1/3}(s_1)} \omega^{-r-1/6} \left[1 + O(\omega^{-3\delta})\right]$$

$$= -\frac{1}{2} \exp[-i(\pi/4)] \sqrt{\frac{\omega\varepsilon_0}{\pi\varrho(0)}} \left[1 + O(\omega^{-3\delta})\right] \quad . \tag{12.3.27}$$

Since

$$\kappa_1^{1/2}(s_1) = \varrho^{1/6}(0)\varepsilon_0^{1/2} \left[1 + O(\omega^{-2/3+2\delta})\right] \quad \text{and}$$

$$\varrho^{1/6}(s_1) = \varrho^{1/6}(0) \left[1 + O(\omega^{-1/3+\delta})\right] \quad ,$$

equation (12.3.27) implies that $r = -2/3$ and

$$\chi_1 = i\sqrt{\frac{2}{3\pi}} \frac{\varrho^{1/6}(s_1)}{\varrho^{1/6}(0)} \varepsilon_0 \left[1 + O(\omega^{-3\delta})\right] \quad .$$

The value found for χ_1 leads to the following formula for χ_m:

$$\chi_m = \frac{1}{\sqrt{\pi}} \exp[i(\pi/2)m] \frac{1}{\varrho^{1/6}(0)} \frac{\varepsilon_0}{\sqrt{m+1/2}} \exp\left(-i\frac{5}{12} \frac{m-1}{\varrho(0)\varepsilon_0^3\omega}\right)$$

$$\times \left[1 + O(\omega^{-1/3+\delta}) + O(\omega^{-3\delta}) + O(\omega^{1/3-10\delta})\right] \quad . \tag{12.3.28}$$

Having determined χ_m, we can calculate the wave $u_m^{(2)}$ which arrives at the observation point N along the ray $N_m N$. To do this, we must set $j = m$ in (12.3.16), and substitute $\bar{s}$ for s_{m+1}, and N for N_{m+1}. The wave $u_{m+1}^{(1)}$, that arises at N as a result of the reflection of $u_m^{(2)}$, may be calculated from (12.3.15) in which we set $j = m + 1$. The value of the factor χ_{m+1} will be given by formula (12.3.28) with $m + 1$ substituted for m.

Let us calculate the normal derivative at the point N on S of a wave that has been reflected m times:

$$\left.\frac{\partial u_m}{\partial n}\right|_N = \left.\frac{\partial u_m^{(2)}}{\partial n}\right|_N + \left.\frac{\partial u_{m+1}^{(1)}}{\partial n}\right|_N .$$

To get the quantities $(\partial u_m^{(2)}/\partial n)|_N$ and $(\partial u_{m+1}^{(1)}/\partial n)|_N$ with accuracy of $O(\omega^{-3\delta})$, it is sufficient to differentiate with respect to n only the exponential factors that appear in $u_m^{(2)}$ and $u_{m+1}^{(1)}$, and to set $\chi_{m+1} = i\chi_m$. In this way, using (12.3.15) and (12.3.16), we get

$$\left.\frac{\partial u_m}{\partial n}\right|_N = \frac{1}{\sqrt{\pi}} \exp[i(\pi/2)m] \frac{1}{\varrho^{1/6}(0)} \frac{\varepsilon_0}{\sqrt{m+1/2}}$$

$$\times \exp\left(-i\frac{5}{12}\frac{m-1}{\varrho(0)\varepsilon_0^3\omega}\right) \sqrt{\frac{2m+1}{2m+2}} \frac{i\omega^{3/2}}{2\varrho^{1/6}(\tilde{s})\varrho^{1/6}(0)\varepsilon_0^{1/2}}$$

$$\times \left[\exp[-i(\pi/4)]\left.\frac{\partial \tau_2^{(m)}}{\partial n}\right|_{n=0} \exp\left[i\omega\tau_2^{(m)}\right]\right.$$

$$\left.+ i\exp[i(\pi/4)]\left.\frac{\partial \tau_1^{(m+1)}}{\partial n}\right|_{n=0} \exp[i\omega\tau_1^{(m+1)}]\right]$$

$$\times \{1 + O(\omega^{-1/3+\delta}) + O(\omega^{-3\delta}) + O(\omega^{1/3-10\delta})\} . \qquad (12.3.29)$$

By virtue of (12.3.19) and (12.3.26), the quantity $\tau_2^{(m)}(\tilde{s}) = \tau_1^{(m+1)}(\tilde{s})$ is equal to the length of the extremal polygon $N_0, N_1, \ldots, N_m, N$ and is determined by formula (12.2.10). The quantity $[\partial\tau_2^{(m)}/\partial n]|_{n=0} = -[\partial\tau_1^{(m+1)}/\partial n]|_{n=0}$ is determined from formula (12.2.11). Since $m + 1 \sim \omega^{1/3-\delta}$ and $1/\varepsilon_0^3\omega \sim \omega^{-3\delta}$, we rewrite (12.3.29) in the form

$$\left.\frac{\partial u_m}{\partial n}\right|_N = \frac{1}{\sqrt{\pi}} \exp[i(\pi/4) + i(\pi/2)m]\frac{\omega^{3/2}\varepsilon_0^{3/2}}{\varrho^{1/2}(\tilde{s})} \frac{1}{\sqrt{m+1}}$$

$$\times \exp\left(-i\frac{5}{12}\frac{m+1}{\varrho(0)\varepsilon_0^3\omega}\right)$$

$$\times \exp\left\{i\omega\left[\tilde{s} - \frac{1}{24}\frac{1}{(m+1)^2}\left(\int_0^{\tilde{s}}\frac{d\tau}{\varrho^{2/3}(\tau)}\right)^3\right]\right\}$$

$$\{1 + O(\omega^{-1/3+4\delta}) + O(\omega^{-3\delta}) + O(\omega^{1/3-10\delta})\} . \qquad (12.3.30)$$

The correction terms show that (12.3.30) makes sense for values δ that satisfy $1/30 < \delta < 1/12$. The minimum error of order $\omega^{-1/7}$ is obtained with $\delta = 1/21$. We can give expression (12.3.30) a form that is symmetric in the positions of the source and the observer. Using (12.3.1) and taking $\delta = 1/21$, we get

$$\left.\frac{\partial u_m}{\partial n}\right|_N = \frac{1}{2\sqrt{2\pi}} \exp[i(\pi/4)]\frac{\omega^{3/2}}{\sqrt{\varrho(0)\varrho(\tilde{s})}} \frac{1}{(m+1)^2}\left[\int_0^{\tilde{s}}\frac{d\tau}{\varrho^{2/3}(\tau)}\right]^{3/2}$$

$$\times \exp\left\{ i\omega\left[\tilde{s} - \frac{1}{24}\frac{1}{(m+1)^2}\left(\int_0^{\tilde{s}} \frac{d\tau}{\varrho^{2/3}(\tau)} \right)^3 \right] \right.$$

$$\left. - i\frac{10}{3}\frac{(m+1)^4}{\omega}\left(\int_0^{\tilde{s}} \frac{d\tau}{\varrho^{2/3}(\tau)} \right)^{-3} + i\frac{\pi}{2}m \right\}\left[1 + O(\omega^{-1/7}) \right] \ .$$

$$(12.3.31)$$

Formula (12.3.31) is derived on the assumption that $c_0 = 1$. If c_0 is constant but different from unity, the frequency ω in (12.3.31) must be replaced by the wave number k ($k = \omega/c_0$). Formula (12.3.31) will then differ from (12.2.14) only in the additional factor

$$\exp\left\{ -i\frac{10}{3}\frac{(m+1)^4}{k}\left(\int_0^{\tilde{s}} \frac{d\tau}{\varrho^{2/3}(\tau)} \right)^{-3} \right\} \ , \qquad (12.3.32)$$

whose magnitude is unity. This additional factor has arisen because the small quantity $\alpha(j)(1/\kappa_{j+1}^3\omega) \sim \omega^{-3\delta}$ is accounted for in the large number of reflections $m \sim \omega^{1/3-\delta}$. It is no surprise that it was lost in the rather rough derivation of formula (12.2.14).

For the case of variable wave propagation velocity, an analogous factor can be obtained from (12.3.32) by the substitutions indicated in Sect. 12.2. By including this factor in (12.2.16) and dropping the tilde ˜ over the coordinate $\tilde{s}$ of the observation point N, we get the following more precise formula for the case of variable velocity,

$$\left. \frac{\partial u_m}{\partial n} \right|_N = \frac{1}{2\sqrt{\pi}}\exp[i(\pi/4)]\frac{\omega}{\sqrt{c_0(0)P(0)c_0(s)P(s)}}\frac{\gamma^{3/2}}{(m+1)^2}$$

$$\times \exp\left\{ i\left[\omega \int_0^s \frac{d\tau}{c_0(\tau)} - \frac{1}{12}\frac{\gamma^3}{(m+1)^2} - \frac{5}{3}\frac{(m+1)^4}{\gamma^3} + \frac{\pi}{2}m \right] \right\}$$

$$\times \left[1 + O(\omega^{-1/7}) \right] \ , \qquad (12.3.33)$$

where

$$\gamma = \left(\frac{\omega}{2} \right)^{1/3} \int_0^s \frac{d\tau}{c_0^{1/3}(\tau)P^{2/3}(\tau)} \qquad (12.3.34)$$

is the reduced arc length corresponding to the observation point $N(s,0)$.

12.4 Field of a Source Located on the Boundary of a Circle

In the preceding sections we have obtained formulas that describe multiply reflected waves at the point when they separate from the surface waves. Waves with a larger number of reflections propagating inside a boundary layer cannot be described by these formulas. As such waves interfere with each other they generate a surface wave that propagates in a boundary layer near S. To describe the surface wave we must use the solutions of the homogeneous Helmholtz equation near S obtained earlier (see Chap. 7). We will have to be able to construct a superposition of these solutions that describes a surface wave propagating from the point source. At first we will turn our attention to the surface wave near a circular boundary, where the variables are separable, and a representation can be constructed for the surface wave in the form of a contour integral. This integral will show us the pattern for the desired superposition which describes the surface wave in the general case. And so, as in the preceding chapters, we turn first of all to the simplest etalon problem, and then we extend the solution of this problem to the general case.

Consider the solution of the problem of the field of a point source lying on the boundary of a circle of radius ϱ. In other words, we will construct and study the Green's function for a circle in the special case where the source is on the boundary of the circle. The velocity of wave propagation inside the circle will be taken as constant. So that the formulated problem will have a solution at any frequency ω, we will assume that ω is a complex number with small positive imaginary part

$$\omega = \omega' + i\omega'' \quad , \quad 0 < \omega'' \ll \omega' \quad (0 < \arg \omega \ll 1) \quad .$$

We introduce polar coordinates: let (ϱ, ϕ_0) be the coordinates of the source, and (r, ϕ) be the coordinates of the observation point. We denote the desired field of the point source (Green's function) by $g(r, \phi; \varrho, \phi_0)$. The Green's function must satisfy the Helmholtz equation

$$(\Delta + k^2)g = \frac{1}{r} \frac{\partial}{\partial r} \left(r \frac{\partial g}{\partial r} \right) + \frac{1}{r^2} \frac{\partial^2 g}{\partial \phi^2} + k^2 g = 0 \quad ,$$

$$(|g(0, \phi; \varrho, \phi_0)| < \infty) \tag{12.4.1}$$

when $r < \varrho$, and the boundary condition

$$g|_{r=\varrho} = \delta[\varrho(\phi - \phi_0)] = \frac{1}{\varrho} \delta(\phi - \phi_0) \quad , \tag{12.4.2}$$

when $r = \varrho$, where $k = \omega/c$, $c = \text{const}$, and δ is the Dirac delta function.

The solution of (12.4.1, 2) is most easily constructed by the method of separation of variables in the form of a Fourier series in $\cos n(\phi - \phi_0)$, $n = 0, 1, 2, \ldots$ To study the solution as $kr \to \infty$, the Fourier series should be transformed into a contour integral by using the residue theorem (Watson's transformation).[2]

First let us construct the Fourier series for the Green's function $g(r, \phi; \varrho, \phi_0)$. The expressions

---

[2] The contour integral for the Green's function $g(r, \phi; \varrho, \phi_0)$ can also be constructed directly. This method is based on the conventional formula that represents the Green's function (in the case of separable variables) of a multidimensional problem as a contour integral of the Green's functions of corresponding one-dimensional problems (see [12.1]).

$$J_n(kr)\, \sin n\phi \quad \text{and} \quad J_n(kr)\, \cos n\phi \tag{12.4.3}$$

are special solutions of (12.4.1) which are bounded at the coordinate origin and are 2π-periodic in the angle ϕ. Here $J_n(kr)$ is a Bessel function, and $n \geq 0$ is an integer. We will seek $g(r,\, \phi;\, \varrho,\, \phi_0)$ in the form of a linear superposition of the solutions (12.4.3):

$$g(r,\, \phi;\, \varrho,\, \phi_0) = \sum_{n=0}^{\infty} (a_n \sin n\phi + b_n \cos n\phi) J_n(kr) \quad . \tag{12.4.4}$$

The coefficients a_n and b_n should be chosen so that (12.4.2) is satisfied. Comparing the right side of (12.4.4) at $r = \varrho$ with the known expansion of the delta function (see [12.2],

$$\delta(\phi - \phi_0) = \frac{1}{\pi}\left(\frac{1}{2} + \sum_{n=1}^{\infty} \cos n(\phi - \phi_0) \right) \quad ,$$

we get

$$a_n = 0 \quad , \quad b_0 = \frac{1}{2\pi\varrho}\, \frac{1}{J_0(k\varrho)} \quad , \quad b_n = \frac{1}{\pi\varrho}\, \frac{1}{J_n(k\varrho)} \quad , \quad n \geq 1 \quad .$$

Thus

$$g(r,\, \phi;\, \varrho,\, \phi_0) = \frac{1}{\pi\varrho}\left\{ \frac{1}{2}\frac{J_0(kr)}{J_0(k\varrho)} + \sum_{n=1}^{\infty} \frac{J_n(kr)}{J_n(k\varrho)} \cos n(\phi - \phi_0) \right\} \quad . \tag{12.4.5}$$

When $r < \varrho$, the series (12.4.5) converges absolutely; when $r = \varrho$ the convergence of series (12.4.5) must be understood in the sense of generalized functions.

In accordance with the residue theorem, each term of the series can be represented by a contour integral

$$\frac{J_n(kr)}{J_n(k\varrho)} \cos n(\phi - \phi_0) = \frac{1}{2\mathrm{i}} \oint_{C_n} \frac{J_\zeta(kr)}{J_\zeta(k\varrho)}\, \frac{\cos(|\phi - \phi_0| - \pi)\zeta}{\sin \pi\zeta}\, d\zeta \quad ,$$

where C_n is a circle whose center is at $\zeta = n$ and whose radius is so small that only one pole of the integrand is located inside C_n. The sum of the integrals over C_n, $n = 1, 2, \ldots,$ can be represented by a single integral over a contour $\mathcal{L}_1$ enclosing the positive real axis (Fig. 12.4).

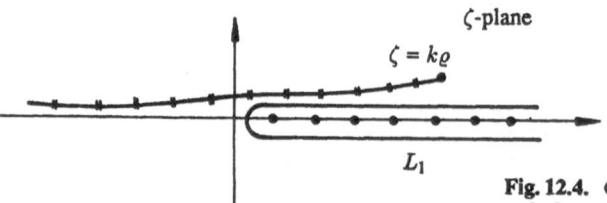

ζ-plane

$\zeta = k\varrho$

L_1

Fig. 12.4. Contour of integration and pole locations for integrand in (12.4.6)

As a result, we get the desired formula

$$g(r, \phi; \varrho, \phi_0) = \frac{1}{4\pi i \varrho} \oint_{C_0} \frac{J_\zeta(kr)}{J_\zeta(k\varrho)} \frac{\cos(|\phi - \phi_0| - \pi)\zeta}{\sin \pi \zeta} d\zeta$$

$$+ \frac{1}{2\pi i \varrho} \int_{\mathcal{L}_1} \frac{J_\zeta(kr)}{J_\zeta(k\varrho)} \frac{\cos(|\phi - \phi_0| - \pi)\zeta}{\sin \pi \zeta} d\zeta \quad . \tag{12.4.6}$$

It is assumed that there are no other poles of the integrand inside $\mathcal{L}_1$ except those at the points $\zeta = n$. Note that when k is complex, the zeros of the function $J_\zeta(k\varrho)$ (which are also poles of the integrand) in the ζ-plane are located on a curve that begins at the point $\zeta = k\varrho$ and goes out to $-\infty$. (These zeros are indicated by crosses in Fig. 12.4).

Our next task is to convert (12.4.6) into a sum of terms each of which describes high-frequency geometrical waves (as $\omega' \to \infty$) that have left the source, traveled around the coordinate origin a certain number of times, and in so doing have undergone a given number of reflections.

In the following we will assume that $\phi_0 = 0$, i.e. that the angle ϕ is measured from the source.

Let us construct two auxiliary integration contours l_1 and l_2 (Fig. 12.5). Contour l_1 begins on the circle C_0 at the point $\varepsilon \exp[-i(3\pi/4)]$ (ε is the radius of C_0), follows the right half of C_0, and then without intersecting the line of zeros of $J_\zeta(k\varrho)$ it connects the points $\varepsilon \exp[i(\pi/4)]$ and $k\varrho$ and goes out to $i\infty$, asymptotically approaching the imaginary axis. Contour l_2 is a rotation of l_1 by $180°$ about the coordinate origin, and extends from $-i\infty$ to $\varepsilon \exp[i(\pi/4)]$.

Let us integrate the following odd function of ζ

$$\frac{1}{2\pi i \varrho} \frac{1}{2} \left[\frac{J_\zeta(kr)}{J_\zeta(k\varrho)} + \frac{J_{-\zeta}(kr)}{J_{-\zeta}(k\varrho)} \right] \frac{\cos(|\phi| - \pi)\zeta}{\sin \pi \zeta} \quad , \tag{12.4.7}$$

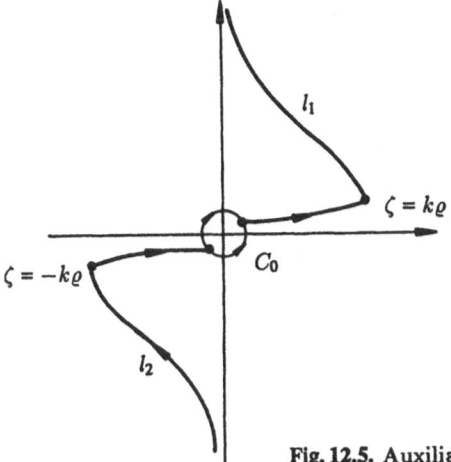

Fig. 12.5. Auxiliary contours of integration

over l_1 and l_2. Obviously the sum of these two integrals will be equal to zero:

$$\frac{1}{2\pi i \varrho} \left(\int_{l_1} + \int_{l_2} \right) \frac{1}{2} \left[\frac{J_\zeta(kr)}{J_\zeta(k\varrho)} + \frac{J_{-\zeta}(kr)}{J_{-\zeta}(k\varrho)} \right] \frac{\cos(|\phi| - \pi)\zeta}{\sin \pi \zeta} d\zeta = 0 \quad . \quad (12.4.8)$$

Let us combine (12.4.6) and (12.4.8) after having first deformed the lower branch of $\mathcal{L}_1$ so that it coincides with l_2. Since

$$\frac{1}{2\pi i \varrho} \frac{1}{2} \int_{\varepsilon \exp[i(5\pi/4)]}^{\varepsilon \exp[i(\pi/4)]} \left[\frac{J_\zeta(kr)}{J_\zeta(k\varrho)} + \frac{J_{-\zeta}(kr)}{J_{-\zeta}(k\varrho)} \right] \frac{\cos(|\phi| - \pi)\zeta}{\sin \pi \zeta} d\zeta$$

$$= \frac{1}{2\pi i \varrho} \frac{1}{2} \oint_{C_0} \frac{J_\zeta(kr)}{J_\zeta(k\varrho)} \frac{\cos(|\phi| - \pi)\zeta}{\sin \pi \zeta} d\zeta \quad , \quad (12.4.9)$$

this combination gives

$$g(r, \phi; \varrho, 0) \equiv g(r, \phi)$$

$$= -\frac{1}{2\pi i \varrho} \frac{1}{2} \int_l \left[\frac{J_\zeta(kr)}{J_\zeta(k\varrho)} + \frac{J_{-\zeta}(kr)}{J_{-\zeta}(k\varrho)} \right] \frac{\cos(|\phi| - \pi)\zeta}{\sin \pi \zeta} d\zeta$$

$$+ \frac{1}{2\pi i \varrho} \int_{l_0} \frac{J_\zeta(kr)}{J_\zeta(k\varrho)} \frac{\cos(|\phi| - \pi)\zeta}{\sin \pi \zeta} d\zeta \quad . \quad (12.4.10)$$

Contours l and l_0 are shown in Fig. 12.6. The integrand of the second integral is regular in the first quadrant above l_0 and approaches zero exponentially as $|\zeta| \to \infty$, $(r \neq \varrho, \phi \neq 0)$. Consequently, the second integral in (12.4.10) is equal to zero. We now consider the integrand of the first integral.

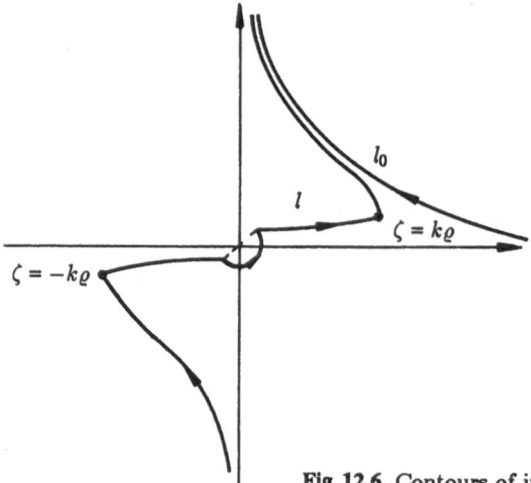

Fig. 12.6. Contours of integration for (12.4.10)

We introduce the three functions:

$$q(\zeta) = \frac{H_\zeta^{(1)}(k\varrho)}{H_\zeta^{(2)}(k\varrho)} \quad , \quad F_0(\zeta) = \frac{H_\zeta^{(2)}(kr)}{H_\zeta^{(2)}(k\varrho)} \quad ,$$

$$F_1(\zeta) = \frac{H_\zeta^{(1)}(kr)H_\zeta^{(2)}(k\varrho) - H_\zeta^{(2)}(kr)H_\zeta^{(1)}(k\varrho)}{\left[H_\zeta^{(2)}(k\varrho)\right]^2} \quad . \tag{12.4.11}$$

By virtue of the equality $H_{-\zeta}^{(1),(2)}(z) = \exp(\pm i\pi\zeta)H_\zeta^{(1),(2)}(z)$ the relations

$$q(-\zeta) = e^{2\pi i\zeta}q(\zeta) \quad , \quad F_0(-\zeta) = F_0(\zeta) \quad ,$$
$$F_1(-\zeta) = e^{2\pi i\zeta}F_1(\zeta) \tag{12.4.12}$$

are true. It can be readily demonstrated by direct calculation that the following identities are true:

$$\frac{J_\zeta(kr)}{J_\zeta(k\varrho)}\frac{\cos(|\phi| - \pi)\zeta}{\sin \pi\zeta} = \left[\frac{F_1(\zeta)}{1 + q(\zeta)} + F_0(\zeta)\right]\frac{\cos(|\phi| - \pi)\zeta}{\sin \pi\zeta}$$

$$= F_1(\zeta)\frac{e^{i\pi\zeta}[\cos|\phi|\zeta - q(\zeta)e^{i\pi\zeta}\cos(|\phi| - \pi)\zeta]}{(1 - q^2(\zeta)e^{2\pi i\zeta})\sin \pi\zeta}$$

$$+ iF_1(\zeta)\frac{-e^{i|\phi|\zeta} + q(\zeta)e^{2\pi i\zeta - i|\phi|\zeta}}{(1 - q^2(\zeta)e^{2\pi i\zeta})(1 + q(\zeta))}$$

$$+ F_0(\zeta)\frac{\cos(|\phi| - \pi)\zeta}{\sin \pi\zeta} \quad . \tag{12.4.13}$$

The first and third terms on the right side of (12.4.13) are odd functions by virtue of (12.4.12), and the second term has no singularities on the real axis.

The integrand of the first integral in (12.4.10) does not have a singularity at $\zeta = 0$; therefore the integration contour l can be drawn symmetrically through the point $\zeta = 0$ (see the dashed line on Fig. 12.6). Using (12.4.13) and the symmetry of the integration contour, we convert formula (12.4.10) for the Green's function $g(r, \phi)$ to the form

$$g(r, \phi) = \frac{1}{2\pi\varrho} \int_l F_1(\zeta)\frac{e^{i|\phi|\zeta} - q(\zeta)e^{i(2\pi - |\phi|)\zeta}}{(1 - q^2(\zeta)e^{2\pi i\zeta})(1 + q(\zeta))}d\zeta \quad . \tag{12.4.14}$$

Finally, we represent the Green's function $g(r, \phi)$ by the sum

$$g(r, \phi) = g_0(r, \phi) + g^+(r, \phi) + g^-(r, \phi) \quad ,$$

where

$$g_0(r, \phi) = \frac{1}{2\pi\varrho} \int_l F_1(\zeta)\frac{e^{i|\phi|\zeta}}{1 + q(\zeta)}d\zeta \quad ,$$

$$g^+(r, \phi) = -\frac{1}{2\pi\varrho} \int_l F_1(\zeta) \frac{q(\zeta)\, e^{i(2\pi-|\phi|)\zeta}\, d\zeta}{(1 - q^2(\zeta)\, e^{2\pi i\zeta})(1 + q(\zeta))} \quad ,$$

$$g^-(r, \phi) = \frac{1}{2\pi\varrho} \int_l F_1(\zeta) \frac{q^2(\zeta)\, e^{i(2\pi+|\phi|)\zeta}\, d\zeta}{(1 - q^2(\zeta)\, e^{2\pi i\zeta})(1 + q(\zeta))} \quad . \tag{12.4.15}$$

In the following presentation we will study only the integral $g_0(r, \phi)$, and we will show that as $\omega' \to \infty$ this integral describes waves that have traversed an angular distance $|\phi|(-\pi < \phi \leq \pi)$ and have been reflected 0, 1, 2, ... times from the boundary of the circle along a path from the source to the observation point. The same methods can be used to demonstrate that the integral $g^+(r, \phi)$ describes waves that have traversed the angular distances $2\pi m - |\phi|$, $m = 1, 2, \ldots$. It is easily shown that a wave that has covered an angular distance $2\pi m - |\phi|$ must be reflected at least $2m - 1$ times from the boundary of the circle. The integral $g^-(r, \phi)$ describes waves that have covered an angular distance $2\pi m + |\phi|$, $m = 1, 2, \ldots$, and have been reflected $2m$, $2m + 1$, ... times from the boundary of the circle.

We turn now to the investigation of the integral $g_0(r, \phi)$. The integrand of this integral in the ζ-plane has simple poles that coincide with the zeros of the functions $J_\zeta(k\varrho)$ and $H_\zeta^{(2)}(k\varrho)$. (In Fig. 12.7 the zeros of the $J_\zeta(k\varrho)$ are denoted by crosses, and the zeros of $H_\zeta^{(2)}(k\varrho)$ are denoted by circles.) Let us deform the contour l into the contour C. In the upper half of the ζ-plane the contour C is taken as the line $\arg\left[\kappa\varrho \int_1^{\zeta/k\varrho} \arccos x\, dx\right] = 0$ on which the zeros of the function $H_\zeta^{(1)}(k\varrho)$ are located and in the lower half-plane it coincides with the line $\arg\left[\kappa\varrho \int_1^{\zeta/k\varrho} \arccos x\, dx\right] = 3\pi/2$ that goes out to infinity in the third quadrant. In the integrand of (12.4.15) we replace $F_1(\zeta)$ and $q(\zeta)$ with their values from (12.4.11). Then

$$g_0(r, \phi) = \frac{1}{4\pi\varrho} \int_C \frac{H_\zeta^{(1)}(kr)H_\zeta^{(2)}(k\varrho) - H_\zeta^{(2)}(kr)H_\zeta^{(1)}(k\varrho)}{H_\zeta^{(2)}(k\varrho)J_\zeta(k\varrho)} e^{i|\phi|\zeta}\, d\zeta \quad . \tag{12.4.16}$$

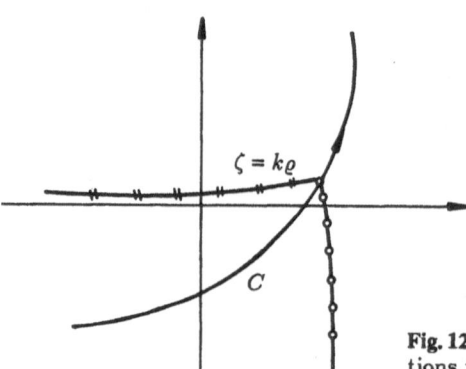

Fig. 12.7. Contour of integration and pole locations for integrand in (12.4.16)

Let us denote the integrand of (12.4.16) by $\pi(r, \phi; \zeta)$. To study the behavior of $\pi(r, \phi; \zeta)$ when $|k\varrho| \gg 1$ and $|kr| \gg 1$, we replace the cylinder functions in the complex ζ-plane by their uniform asymptotic expansions.

The asymptotic formula for the Bessel function takes the form (see Sect. 7.2)

$$J_\zeta(kr) = \sqrt{\frac{2}{\pi}} \left(\frac{T_r(\zeta)}{\zeta^2 - (kr)^2} \right)^{1/4} v\,(T_r(\zeta) + R_r(\zeta)) \left\{ 1 + O\left[\left(\frac{1}{kr}\right)^2 \right] \right\}$$

$$\left| \arg\frac{kr}{\zeta} \right| < \frac{\pi}{2} \quad . \tag{12.4.17}$$

Here

$$T_r(\zeta) = \left[\frac{3}{2}ikr \int\limits_1^{\zeta/kr} \arccos x\, dx \right]^{2/3} \quad ,$$

$$R_r(\zeta) = O\left[\left(\frac{1}{kr}\right)^{4/3} \right] \tag{12.4.18}$$

and v is an Airy function. The two branch cuts $(-\infty, -kr]$ and $[kr, \infty)$ are drawn in the ζ-plane, where $\arg T = \pi$ for real k and $-kr < \zeta < kr$. Formulas for functions $H_\zeta^{(1)}(kr)$ and $H_\zeta^{(2)}(kr)$ are obtained from formula (12.4.17) by simply replacing the function $v(T + R)$ by the Airy functions $-iw_1(T + R)$ and $iw_2(T + R)$ respectively. Formulas for the derivatives of the cylinder functions can be obtained by differentiating formula (12.4.17) and those analogous to it. If the argument of the Airy functions is sufficiently large $T_r(\zeta) \gg 1$, the Airy functions can be replaced in turn by their asymptotic formulas. This substitution leads to the usual Debye formulas for the cylinder functions.

Using these asymptotic formulas, we get

$$\pi(r, \phi; \zeta) = -i\left[\frac{T_r}{T_\varrho} \frac{\zeta^2 - (k\varrho)^2}{\zeta^2 - (kr)^2} \right]^{1/4}$$

$$\times \frac{w_1(T_r + R_r)w_2(T_\varrho + R_\varrho) - w_2(T_r + R_r)w_1(T_\varrho + R_\varrho)}{w_2(T_\varrho + R_\varrho)v(T_\varrho + R_\varrho)}$$

$$\times e^{i|\phi|\zeta}\left\{ 1 + O\left[\left(\frac{1}{k\varrho}\right)^2 \right] \right\} \quad . \tag{12.4.19}$$

Investigation of (12.4.19) when $|\zeta| \gg |kr|$, where it becomes possible to replace the Airy functions by their asymptotic expansions, shows that the function $\pi(r, \phi; \zeta)$ is exponentially damped on C when $0 < |\phi| < \pi$, and consequently when $0 < \varepsilon < |\phi| < \pi - \varepsilon$ the integral (12.4.16) converges uniformly. Let us study the rate at which $\pi(r, \phi; \zeta)$ decays in the neighborhood of point $\zeta = k\varrho$. This will enable us to determine the portion of C that is essential to the integration.

This investigation will be done in the plane of the complex variable

$$t = T_\varrho(\zeta) + R_\varrho(\zeta) \tag{12.4.20}$$

[i.e., in the plane of the arguments of the Airy functions appearing in (12.4.19)].

In the t-plane the integration contour C_t is close to the straight line $(\infty\, e^{i4\pi/3},$ $\infty\, e^{i\pi/3})$ that passes through the origin. The distance between points of C_t and the given straight line is of the order of $(1/k\varrho)^{4/3}$. By a slight deformation, C_t can be made to coincide with the straight line $(\infty\, e^{i4\pi/3}, \infty\, e^{i\pi/3})$, and in what follows we will assume that it is along such a line that the integration in the t-plane is carried out.

The function $\zeta(t)$ inverse to the function (12.4.20), is representable for sufficiently small values of the ratio $t(k\varrho)^{-2/3}$ in the form

$$\zeta = k\varrho\left\{1 + \frac{t}{2}\left(\frac{2}{k\varrho}\right)^{2/3} + \frac{t^2}{120}\left(\frac{2}{k\varrho}\right)^{4/3} + O\left[(|t|^3 + 1)\left(\frac{2}{k\varrho}\right)^2\right]\right\} .$$

(12.4.21)

We will assume that the observation point (r, ϕ) is located so close to the boundary of the circle that the reduced normal

$$\tilde{\nu} = 2\left(\frac{k\varrho}{2}\right)^{2/3}\frac{r - \varrho}{\varrho}$$

is a quantity of the order of unity. This assumption enables us to represent $T_r(\zeta)$ in a form analogous to (12.4.21)

$$T_r(\zeta) = t - \tilde{\nu} + \left(\frac{3}{20}\tilde{\nu}^2 - \frac{2}{15}\tilde{\nu}t\right)\left(\frac{2}{k\varrho}\right)^{2/3} + O\left[(|t|^2 + 1)\left(\frac{2}{k\varrho}\right)^{4/3}\right] .$$

(12.4.22)

Substituting (12.4.21) and (12.4.22) into (12.4.19), we get

$$\pi(r, \phi; \zeta)$$

$$= \frac{-i}{v(t)}\left[w_1\left\{t - \tilde{\nu} + \left(\frac{3}{20}\tilde{\nu}^2 - \frac{2}{15}\tilde{\nu}t\right)\left(\frac{2}{k\varrho}\right)^{2/3}\right.\right.$$

$$+ O\left[(|t|^2 + 1)\left(\frac{2}{k\varrho}\right)^{4/3}\right]\right\} - \frac{w_1(t)}{w_2(t)}$$

$$\times w_2\left\{t - \tilde{\nu} + \left(\frac{3}{20}\tilde{\nu}^2 - \frac{2}{15}\tilde{\nu}t\right)\left(\frac{2}{k\varrho}\right)^{2/3}\right.$$

$$\left.\left. + O\left[(|t|^2 + 1)\left(\frac{2}{k\varrho}\right)^{4/3}\right]\right\}\right]$$

$$\times \exp\left\{iks + it\gamma + i\frac{t^2}{60}\frac{s}{\varrho}\left(\frac{2}{k\varrho}\right)^{1/3}\right.$$

$$\left. - \frac{1}{20}\tilde{\nu}\left(\frac{2}{k\varrho}\right)^{2/3} + O\left[(|t|^3 + 1)\frac{2}{k\varrho}\right]\right\} ,$$

(12.4.23)

where

$$\gamma = \left(\frac{k\varrho}{2}\right)^{1/3}\frac{s}{\varrho} \qquad (12.4.24)$$

is the reduced arc length on the circle. The brackets in formula (12.4.23) can be represented as

$$\frac{2i\tilde{\nu}}{w_2(t)}\left\{1 + O(t\tilde{\nu}^2) + O\left[(|t|+1)\left(\frac{2}{k\varrho}\right)^{2/3}\right]\right\} \quad .$$

Thus for sufficiently small $\tilde{\nu}^2$ the behavior of $\pi(r, \phi; \zeta)$ in the t-plane is determined by the factor

$$\frac{1}{v(t)w_2(t)}e^{i\gamma t} \quad . \qquad (12.4.25)$$

In evaluating (12.4.25) we can use the asymptotic formulas from Appendix A.1 for the Airy functions $v(t)$ and $w_2(t)$ as soon as $t \geq 3$. The resultant estimate takes the form

$$\frac{1}{v(t)w_2(t)}e^{i\gamma t} \sim \begin{cases} 2t^{1/2}\exp(i\gamma t), & \arg t = \pi/3, \\ -2it^{1/2}\exp\left(-\frac{4}{3}t^{3/2} + i\gamma t\right), & \arg t = 4\pi/3 \quad . \end{cases}$$

The second line implies that on the lower part of the integration contour the factor (12.4.25) becomes less than unity only when

$$|t| > \left(\frac{3\sqrt{3}}{8}\gamma\right)^2 \quad ,$$

and consequently with increasing γ the rate of convergence of integral (12.4.16) becomes poorer. On the other hand, for small values of γ the integral converges slowly on the upper part of the integration contour. For values of γ that satisfy the inequality

$$\frac{2}{\sqrt{3}\Delta} \leq \gamma < 2\sqrt{\Delta}\left(\frac{k\varrho}{2}\right)^{\delta} \quad , \qquad (12.4.26)$$

where $\Delta = O(1)$ and δ is a small positive quantity[3], the main section of the integration contour C_t is the neighborhood of the coordinate origin for which

$$|t| < t_0 \quad , \quad t_0 \sim 2\Delta\left(\frac{k\varrho}{2}\right)^{2\delta} \quad . \qquad (12.4.27)$$

---

[3] For values of $k\varrho/2$ that are not too great ($k\varrho/2 \sim 10^3 - 10^6$) and small δ ($\delta = 1/21$), the factor $(k\varrho/2)^{\delta}$ is close to unity and has no essential significance. The specific value of the coefficient Δ at which the error of the calculations becomes minimal, is determined by the parameters of the problem. For the indicated values of $k\varrho/2$ and δ the coefficient Δ is close to unity, as shown by calculations done in the work of *Lanin* [12.3]. When $k\varrho \to \infty$, the factor $(k\varrho/2)^{\delta}$ gives estimates that contain negative powers of $k\varrho/2$.

At these values of t, (12.4.21) and (12.4.22) still converge fast enough so that we could really restrict ourselves to the first terms of the given expansions in evaluating the function $\pi(r, \phi; \zeta)$.

Thus, when (12.4.26) is satisfied, the integral (12.4.16) can be written as an integral in the t-plane

$$
g_0(r, \phi) = -\frac{i}{4\pi\varrho}\left(\frac{k\varrho}{2}\right)^{1/3} \int_{t_0 \exp[(4\pi/3)i]}^{t_0 \exp[(\pi/3)i]} \frac{1}{v(t)}
$$
$$
\times \left[w_1\left\{t - \tilde{\nu} + \left(\frac{3}{20}\tilde{\nu}^2 - \frac{2}{15}\tilde{\nu}t\right)\left(\frac{2}{k\varrho}\right)^{2/3}\right.\right.
$$
$$
\left. + O\left[(|t|^2 + 1)\left(\frac{2}{k\varrho}\right)^{4/3}\right]\right\} - \frac{w_1(t)}{w_2(t)}w_2\left\{t - \tilde{\nu} + \left(\frac{3}{20}\tilde{\nu}^2 - \frac{2}{15}\tilde{\nu}t\right)\right.
$$
$$
\left.\left. \times \left(\frac{2}{k\varrho}\right)^{2/3} + O\left[(|t|^2 + 1)\left(\frac{2}{k\varrho}\right)^{4/3}\right]\right\}\right]\exp\left\{iks + it\gamma\right.
$$
$$
\left. + i\frac{t^2}{60}\frac{s}{\varrho}\left(\frac{2}{k\varrho}\right)^{1/3} - \frac{1}{20}\tilde{\nu}\left(\frac{2}{k\varrho}\right)^{2/3} + O\left[(|t|^3 + 1)\left(\frac{2}{k\varrho}\right)\right]\right\}
$$
$$
\times \left\{1 + \frac{1}{30}t\left(\frac{2}{k\varrho}\right)^{2/3} + O\left[(|t|^2 + 1)\left(\frac{2}{k\varrho}\right)^{4/3}\right]\right\} dt \quad . (12.4.28)
$$

The integral (12.4.28) is a linear superposition of the solutions (7.2.6), the only difference being that the integrand contains the functions $w_1(Z)$ and $w_2(Z)$ instead of the Airy function $v(Z)$, and in addition the roots $-t_p$ of the Airy function are replaced by the variable of integration t. The weighting functions are

$$
-\frac{i}{4\pi\varrho}\left(\frac{k\varrho}{2}\right)^{1/3}\frac{1 + (1/30)t(2/k\varrho)^{2/3} + O\left[(|t|^2 + 1)(2/k\varrho)^{4/3}\right]}{v(t)} .
$$

for the first term of the integrand, and

$$
+\frac{i}{4\pi\varrho}\left(\frac{k\varrho}{2}\right)^{1/3}\frac{w_1(t)}{w_2(t)}\frac{1 + (1/30)t(2/k\varrho)^{2/3} + O\left[(|t|^2 + 1)(2/k\varrho)^{4/3}\right]}{v(t)}
$$

for the second term.

We intend to calculate the integral (12.4.16) when

$$
\gamma \geq 2\sqrt{\Delta}\left(\frac{k\varrho}{2}\right)^{\delta} .
$$

In this case the function $1/v(t)$ must be expanded in the geometric progression

$$
\frac{1}{v} = \frac{2i}{w_1 - w_2} = -2i\sum_{m=0}^{M-1}\left[\frac{w_1}{w_2}\right]^m\frac{1}{w_2} + \left[\frac{w_1}{w_2}\right]^M\frac{1}{v} .
$$

Then for the integral $g_0(r, \phi)$ we get

$$g_0(r, \phi) = \sum_{m=0}^{M-1} g_{0m}(r, \phi) + \tilde{G}_{0M}(r, \phi) \quad , \quad \text{where}$$

$$
g_{0m}(r, \phi) = -\frac{1}{2\pi\varrho} \int_C \left[\frac{T_r}{T_\varrho} \frac{\zeta^2 - (k\varrho)^2}{\zeta^2 - (kr)^2} \right]^{1/4} \left[\frac{w_1(T_\varrho + R_\varrho)}{w_2(T_\varrho + R_\varrho)} \right]^m
$$

$$
\times \left[\frac{w_1(T_r + R_r)}{w_2(T_\varrho + R_\varrho)} - \frac{w_1(T_\varrho + R_\varrho)}{w_2^2(T_\varrho + R_\varrho)} w_2(T_r + R_r) \right]
$$

$$
\times e^{i|\phi|\zeta} \left\{ 1 + O\left[\left(\frac{1}{k\varrho} \right)^2 \right] \right\} d\zeta \tag{12.4.29}
$$

and

$$
\tilde{G}_{0M}(r, \phi) = -\frac{i}{4\pi\varrho} \int_C \left[\frac{T_r}{T_\varrho} \frac{\zeta^2 - (k\varrho)^2}{\zeta^2 - (kr)^2} \right]^{1/4} \left[\frac{w_1(T_\varrho + R_\varrho)}{w_2(T_\varrho + R_\varrho)} \right]^M
$$

$$
\times \left[w_1(T_r + R_r) - \frac{w_1(T_\varrho + R_\varrho)}{w_2(T_\varrho + R_\varrho)} w_2(T_r + R_r) \right]
$$

$$
\times \frac{1}{v(T_\varrho + R_\varrho)} e^{i|\phi|\zeta} \left\{ 1 + O\left[\left(\frac{1}{k\varrho} \right)^2 \right] \right\} d\zeta \tag{12.4.30}
$$

The additional factor $[w_1/w_2]^M$ in the integrand of (12.4.30) improves the convergence of the integral on the lower part of contour C without changing the convergence on the upper part of the contour.

With an appropriate choice of the number M which depends on the reduced values of the arc length γ and the normal $\tilde{\nu}$, the main portion of the integral in (12.4.30) will as before be the section of C for which $|t| = |T_\varrho + R_\varrho| < t_0$. In this case integrals g_{0m} can be calculated by the *method of steepest descents*.

Calculation of the integrals g_{0m} by the method of steepest descents shows that these integrals can be used to describe waves that are reflected m times from S. This calculation will be carried out in detail below for the integral g_{00}. The integrals g_{0m}, $m \geq 1$ are calculated by an analogous scheme.

We will not study the integral $\tilde{G}_{0M}$ since an analogous integral will be studied in Sect. 12.5 in the general case of an inhomogeneous medium and a noncircular boundary. But let us note that the integral $\tilde{G}_{0M}$ describes a surface wave, i.e. the overall effect of waves for which the number of reflections is greater than or equal to M. The integral $\tilde{G}_{0M}$ is expressed in terms of the special function $G_M(\gamma)$ for which tables are given in Appendix A.4.

Let us turn now to calculation of g_{00} by the method of steepest descents[4]. We will get formula (12.4.36) which was used in Sect. 12.3.

---

[4] The calculation of the integral g_{00} by the method of steepest descents is not necessary in order to understand Sect. 12.5.

We introduce the functions

$$A_j(T) = \exp[i(\pi/2)(j-2)]\, T^{1/4} \exp\left[(-1)^{j-1}\tfrac{2}{3}T^{3/2}\right] w_j(T) \ ,$$
$$j = 1, 2 \ .$$

Using the asymptotic expansions of the Airy functions (Appendix A.1) we can easily get the following formulas for functions $A_j(T)$ when $|T| \gg 1$ and $\pi/3 + \varepsilon < \arg T < 5\pi/3 - \varepsilon$

$$A_j(T) = 1 + O(T^{-3/2}) \ , \quad j = 1, 2 \ . \tag{12.4.31}$$

The functions $A_j(T)$ are convenient in that

$$\lim_{|T| \to \infty} A_j(T) = 1 \ .$$

Formulas (12.4.31) imply that the functions $A_j(T)$ change fairly slowly when the T are large enough. By using the $A_j(T)$, the integral g_{00} is written as

$$g_{00}(r, \phi) = -\frac{i}{2\pi\varrho} \int_C \left[\frac{\zeta^2 - (k\varrho)^2}{\zeta^2 - (kr)^2}\right]^{1/4} \left\{ \frac{A_1(T_r)}{A_2(T_\varrho)} \exp\left[-\tfrac{2}{3}T_r^{3/2} - \tfrac{2}{3}T_\varrho^{3/2}\right] \right.$$

$$\left. - \frac{A_1(T_\varrho)}{A_2^2(T_\varrho)} A_2(T_r) \exp\left[\tfrac{2}{3}T_r^{3/2} - 2T_\varrho^{3/2}\right] \right\}$$

$$\times e^{i|\phi|\zeta} \left\{1 + O\left[\frac{1}{k\varrho}\right]\right\} d\zeta \ . \tag{12.4.32}$$

The phase functions of the first and second terms of the integrand in (12.4.32) are respectively equal to

$$k\varrho\psi_{0,0}^{(1)} = -\tfrac{2}{3}T_r^{3/2} - \tfrac{2}{3}T_\varrho^{3/2} + i|\phi|\zeta$$

$$= ik\varrho\left[|\phi|z - \frac{1}{a}\int_1^{az}\arccos x\, dx - \int_1^z \arccos x\, dx\right]$$

and

$$k\varrho\psi_{0,0}^{(2)} = \tfrac{2}{3}T_r^{3/2} - 2T_\varrho^{3/2} + i|\phi|\zeta$$

$$= ik\varrho\left[|\phi|z + \frac{1}{a}\int_1^{az}\arccos x\, dx - 3\int_1^z \arccos x\, dx\right] \ ,$$

where

$$z = \frac{\zeta}{k\varrho} \quad \text{and} \quad a = \frac{\varrho}{r} > 1 \ .$$

In the z-plane the saddle points of the phase functions $\psi_{0,0}^{(1)}$ and $\psi_{0,0}^{(2)}$ are determined from the equations $\psi_{0,0}^{(1)'}(z) = 0$ and $\psi_{0,0}^{(2)'}(z) = 0$. Let us write these equations as

338

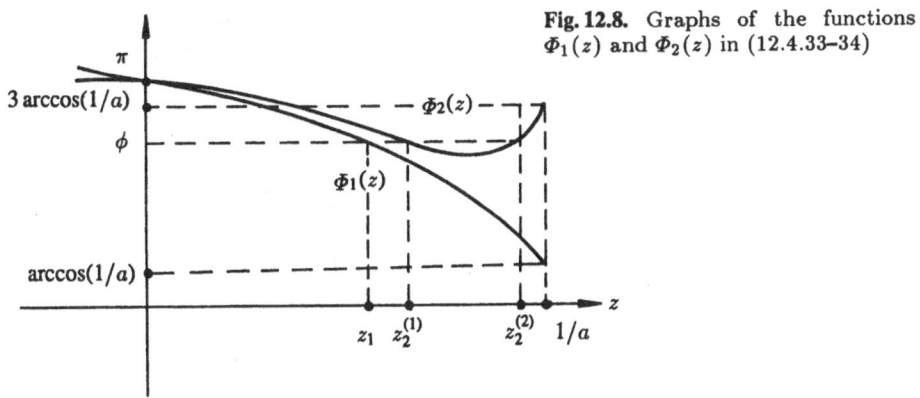

$$|\phi| = \arccos az + \arccos z = \Phi_1(z) \quad , \tag{12.4.33}$$

$$|\phi| = -\arccos az + 3\arccos z = \Phi_2(z) \quad . \tag{12.4.34}$$

The functions $\Phi_1(z)$ and $\Phi_2(z)$ have real values only on the interval $-1/a \leq z \leq 1/a$. The graphs of these functions when $a < 2$ are shown in Fig. 12.8. Obviously the phase function $\psi_{0,0}^{(1)}$ when $\phi > 0$[5] lies in $\arccos(1/a) \leq \phi \leq \pi$, and has one saddle point z_1. The phase function $\psi_{0,0}^{(2)}$ when $\phi > 0$ and

$$\min \Phi_2(z) \leq \phi \leq 3\arccos(1/a)$$

has two saddle points $z_2^{(1)}$ and $z_2^{(2)}$, of which only the first is retained for $3\arccos(1/a) < \phi \leq \pi$. It can be shown that when $\phi = 3\arccos(1/a)$ the other saddle point is transformed to a saddle point of the first term of the integrand in the next integral g_{01}. Equations (12.4.33) and (12.4.34) and the location of saddle points z_1 and $z_2^{(1)}$, $z_2^{(2)}$ are easily interpreted by ray constructions.

Let us first consider equation (12.4.33) and the saddle point z_1. We refer to Fig. 12.9. Located on the dashed circle are observation points with fixed r. In the following discussion we will call this circle the *observation circle* C_r. Ray

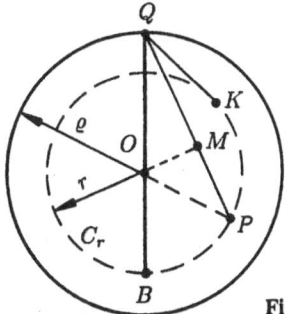

Fig. 12.9. Geometrical interpretation of the saddle point z_1

---

[5] Consideration of negative angles $\dot{\phi}$ is completely analogous.

QP arrives from the source Q at P where angular coordinate is ϕ. We draw the perpendicular QM from the center O of the circle to the ray QP. If we denote

$$\sin(\angle OQP) = z \quad , \quad \text{then}$$

$$\angle QOM = \arccos z \quad \text{and}$$

$$\cos(\angle MOP) = \frac{OM}{OP} = \frac{OQ \sin(\angle OQP)}{OP} = \frac{\varrho}{r} z = az \quad ,$$

whence $\angle MOP = \arccos az$. Consequently z satisfies (12.4.33). Thus saddle point z_1 is the sine of the angle formed by the radius OQ and the ray QP arriving at the observation point $P(r, \phi)$.

We have seen that (12.4.33) has solution z_1 when $\arccos(1/a) \leq \phi \leq \pi$. Corresponding to the angle $\phi = \arccos(1/a)$ is the saddle point $z_1 = 1/a$ and a ray tangent to the circle of radius r, while corresponding to the angle $\phi = \pi$ is the saddle point $z_1 = 0$ and a ray passing through the center of the circle. Hence we see that corresponding to different values of the saddle point z_1 are rays arriving at the observation circle of radius r between tangent ray QK and vertical ray QB (see Fig. 12.9). Calculations of the first term of g_{00} from (12.4.29) by the usual scheme of the method of steepest descents at saddle point z_1 leads to the expression

$$e^{-i(\pi/4)} \sqrt{\frac{k}{2\pi}} \, \sin \varepsilon_0 \frac{e^{ikQP}}{\sqrt{QP}} \left\{ 1 + O\left(\frac{1}{k\varrho(\phi - \varepsilon_0)^3} \right) \right\} \quad , \qquad (12.4.35)$$

where ε_0 is the glancing angle of ray QP, i.e. the angle between QP and the tangent to circle $r = \varrho$ at Q. Expression (12.4.35) obviously describes a wave u_0 that arrives at the point of observation directly from the source. The factor $\exp[-i(\pi/4)]\sqrt{(k/2\pi)} \sin \varepsilon_0$ characterizes the intensity and directional pattern of the source.

As we can see from the correction term, for small values of $\phi - \varepsilon_0$ the usual version of the method of steepest descents is inapplicable. This is because at small $\phi - \varepsilon_0$ the saddle point z_1 is located near $z = 1/a$, which is a branch point of the phase function $\psi_{00}^{(1)}(z)$. Note also that the direct wave arriving at the observation circle at points where $\phi < \arccos(1/a)$, ($\phi - \varepsilon_0 < 0$) can also be obtained from the integral $g_{00}(r, \phi)$ if it is transformed in a suitable fashion. When the observation point P is located on the boundary ($a = 1$), formula (12.4.35) can be written in the form

$$\frac{1}{2} \sqrt{\frac{k \sin \varepsilon_0}{\pi \varrho}} \exp\left[i(k\varrho 2 \sin \varepsilon_0 - \pi/4) \right]$$

$$\times \left\{ 1 + \frac{1 + O(\varepsilon_0^2)}{12 k \varrho \varepsilon_0^3} + O\left[\left(\frac{1}{k\varrho\varepsilon_0^3} \right)^2 \right] \right\} \quad . \qquad (12.4.36)$$

The correction terms show that at small glancing angles ε_0, (12.4.35) becomes inapplicable. For $\varepsilon_0 = \sqrt{\Delta}(k\varrho/2)^{-1/3+\delta}$ $\left(\gamma = 2\sqrt{\Delta}(k\varrho/2)^\delta \right)$, if the first term g_{00} is separated from the integral $g_0(r, \phi)$, the correction terms in (12.4.36) become

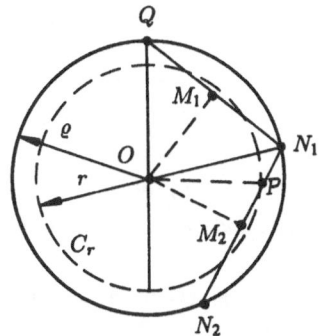

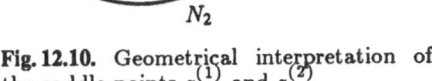

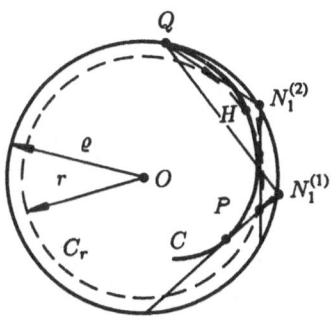

Fig. 12.10. Geometrical interpretation of the saddle points $z_2^{(1)}$ and $z_2^{(2)}$

Fig. 12.11. Geometry of reflected rays and caustic

$$\frac{1}{6\Delta^{3/2}}\left(\frac{k\varrho}{2}\right)^{-3\delta} + O\left[\frac{1}{\Delta^3}\left(\frac{k\varrho}{2}\right)^{-6\delta}\right] \quad ,$$

and consequently (12.4.36) gives a reasonably accurate description of the wave which arrives directly from the source at the boundary of the circle. Formula (12.4.36) was used in Sect. 12.3 to determine an arbitrary function in the multiply reflected wave.

We shall now give a geometric interpretation of the saddle points $z_2^{(1)}$ and $z_2^{(2)}$. To do this, we consider a ray in the vicinity of the boundary arriving at the observation circle C_r after one reflection (Fig. 12.10). If as before we denote

$$\sin(\angle OQN_1) = z \quad ,$$

then we can readily see that

$$\angle QOM_1 = \angle M_1ON_1 = \angle N_1OM_2 = \arccos z \quad \text{and}$$
$$\angle POM_2 = \arccos az \quad .$$

Thus the quantity z must satisfy the equation

$$\phi = 3\arccos z - \arccos az \quad ,$$

i.e. equation (12.4.34), and consequently the saddle points $z_2^{(1)}$ and $z_2^{(2)}$ are the sines of the angles of departure of rays from the source that arrive (after one reflection) at a point P whose coordinates are (r, ϕ). As can be seen from Fig. 12.11, there are two such rays: one ray $QN_1^{(1)}P$ arrives at the observation point before touching the caustic QHC, and the second ray $QN_1^{(2)}P$ arrives at P after touching the caustic. At the point H where the observation circle C_r intersects the caustic QHC, both rays merge. Thus, corresponding to the point H where only one ray arrives is the angle

$$\phi = \phi_{\min} \equiv \min(3\arccos z - \arccos az) \quad .$$

341

For angles $\phi < \phi_{\min}$ the observation circle falls in the shadow zone of the caustic, and obviously for such angles no singly reflected ray reaches the observation point. A ray arriving at the observation point whose coordinate angle $\phi = 3 \arccos(1/a)$, after touching the caustic, is tangent to C_r; corresponding to this ray is the maximum value $z_2^{(2)} = 1/a$. At angles $\phi > 3 \arccos(1/a)$ we can readily show that a saddle point of the first term of the next integral g_{01} corresponds to a ray that has touched the caustic. On the other hand, $z_2^{(1)}$ corresponds to a ray that has not touched the caustic, as before.

Thus in calculating the second term of the integral g_{00} (12.4.29) by the method of steepest descents, the saddle point $z_2^{(1)}$ gives a singly reflected wave that arrives at the observation point before touching the caustic, and the saddle point $z_2^{(2)}$ gives a wave that arrives at the observation point after touching the caustic. We will not give the results of the calculations as they coincide completely with the formulas of the first approximation of the ray method.

For angles ϕ near the angle $\phi = \phi_{\min}$, the usual version of the method of steepest descents is inapplicable. In this case one should use the well known modification of the method for the case of two nearby saddle points. After the appropriate calculations we would get a representation for the fields that contains the Airy function $v(Z)$, which is identical with the results of Chap. 3.

In exactly the same way, by calculating the integrals g_{0m} by the method of steepest descents we could conclude that the first term in g_{0m} describes a wave that has been reflected m times, has passed through its corresponding caustic, and is incident on the boundary of the circle, while the second term is the result of the reflection of this wave, i.e. the initial portion of a wave that has been reflected $m + 1$ times from the boundary of the circle.

Thus it falls to the integral $\tilde{G}_{0M}$ to describe the overall effect of waves for which the number of reflections from the boundary of the circle is equal to or greater than the number M. If the location of the observation point is such that rays whose number of reflections is equal to or greater than the number M do not pass through it, then in this case the integral $\tilde{G}_{0M}$ can be reduced to a sum of residues of the integrand at the poles of $1/w_2(T_\varrho + R_\varrho)$. In this case its value will be of order $\exp(-\Delta_1 k \varrho)$, $\Delta_1 > 0$.

12.5 Field of a Surface Source Close to the Concave Boundary of an Inhomogeneous Body

This is the main section of the chapter. Here we will derive formulas that describe the propagation of waves near the boundary S in the problem (12.1.1–2) formulated in Sect. 12.1. We recall that the region Ω in which the solution is sought and of which S is the boundary, extends to infinity, and it is assumed that the principle of limiting absorption holds. In the final formulas, which will be local in nature, the frequency of oscillations ω is a large real parameter. By virtue of their local nature, the resultant formulas remain valid for closed regions as well. In this case, the existence of a solution of (12.1.1–2) for arbitrary frequencies requires that we assume damping of waves in the medium.

In Sects. 12.2–4 the ray method and a refinement of it were used to get formulas describing waves that have been reflected a large but finite number of times from S, and the etalon problem for a circle was investigated. Here we will examine the general case of (12.1.1–2). We will construct the asymptotic expansion (as $\omega \to \infty$) of the solution of this problem near S in the form of a superposition of the formal solutions of the Helmholtz equation that were constructed in Chap. 7. This superposition is set up in analogy with (12.4.28), which was derived for a wave propagating near the boundary of a circle. An investigation of this superposition will show that as the frequency ω increases or the observation point becomes more distant from the source, the functions u_m become distinguished; these functions were found in Sects. 12.2 and 3, and describe waves that have experienced a certain number of reflections. This fact can be considered as confirmation that the superposition set up by analogy with the problem for a circle provides a correct description of the interference field of a point source near a boundary in the general case.

However, it should be noted that there is still no rigorous mathematical proof for the formulas derived in this section.

Let us write out solutions of the Helmholtz equation from which we will construct a linear superposition that describes the asymptotic expansion of the solution of (12.1.1–2) in the general case.

In Chap. 7 we were primarily interested in solutions of the Helmholtz equation that were concentrated in a boundary layer, and therefore the function $v(Z)$ was taken as the solution of the Airy equation. Now, however, a dominant role will be played by solutions of the Helmholtz equation that contain the complex Airy functions $w_1(Z)$ and $w_2(Z)$. Obviously the calculations and results of Sect. 7.3 are not altered by substituting $w_1(Z)$ or $w_2(Z)$ for the function $v(Z)$ in the *ansatz* (7.3.4). Therefore the solutions we want now take the form

$$
u^{(j)}(s, \nu) = \text{const} \times \exp \left\{ i \sum_{m=-3}^{M-1} \alpha_m(s, \nu) \omega^{-m/3} + O\left(\omega^{-M/3}\right) \right\}
$$
$$
\times w_j \left\{ \sum_{m=0}^{M-1} \beta_m(s, \nu) \omega^{-m/3} + O\left(\omega^{-M/3}\right) \right\} , \quad j = 1, 2,
$$

$$(12.5.1)$$

where the polynomials $\alpha_m(s, \nu)$ and $\beta_m(s, \nu)$ are defined by the formulas of Sect. 7.3.

To satisfy the boundary condition $u|_S = 0$, we set (Sect. 7.4):

$$
\beta_{00}(s) = -t_p \quad , \quad p = 0, 1, 2, \ldots \quad ; \quad \beta_{m0}(s) = 0 \quad , \quad m = 1, 2, 3 \ldots \quad .
$$

Now we will assume

$$
\beta_{00}(s) = t \quad ,
$$

$$(12.5.2)$$

where t is an arbitrary constant, and as before,

$$\beta_{m0}(s) = 0 \quad , \quad m = 1, 2, 3, \ldots \quad . \tag{12.5.3}$$

Since some coefficients of the polynomials $\alpha_m(s, \nu)$ and $\beta_m(s, \nu)$ depended on $\beta_{00}(s)$ [see for instance (7.3.24), (7.3.38) and (7.3.40)], they will now depend on the arbitrary constant t, and therefore the constant t will be introduced as a third argument in the notation for the polynomials:

$$\alpha_m = \alpha_m(s, \nu; t) \quad , \quad \beta_m = \beta_m(s, \nu; t).$$

We recall that the formulas for the constant terms of the polynomials $\alpha_m(s, \nu; t)$, i.e. for $\alpha_m(s, 0; t) = \alpha_{m0}(s; t)$, contained arbitrary constants d_m. In Chap. 7 we adopted the convention of taking these constants as zero, and accordingly were arbitrary in our assignment of the constant factor in (12.5.1). Thus the constant in (12.5.1) is independent of the coordinates (s, ν) but is, generally speaking, a function of t and ω. With these remarks in mind, the formulas for α_m and β_m from Chap. 7 take the form

$$\alpha_{-3}(s, \nu; t) = \int_0^s \frac{d\tau}{c_0(\tau)} \quad ;$$

$$\alpha_{-2}(s, \nu; t) = 0 \quad ;$$

$$\alpha_{-1}(s, \nu; t) = \frac{t}{2^{1/3}} \int_0^s \frac{d\tau}{c_0^{1/3}(\tau) P^{2/3}(\tau)} \quad ;$$

$$\alpha_0(s, \nu; t) = \frac{i}{2} \ln \frac{P^{1/3}(s) c_0^{1/3}(0)}{P^{1/3}(0) c_0^{1/3}(s)} \quad ;$$

$$\alpha_1(s, \nu; t) = \frac{1}{6} \frac{1}{c_0^2(s)} \left[2 \frac{c_0'(s)}{c_0(s)} + \frac{P'(s)}{P(s)} \right] \nu^2 - \frac{t^2}{2^{5/3}} \int_0^s \frac{c_0^{1/3}(\tau)}{P^{4/3}(\tau)}$$

$$\times \left[1 + \frac{4}{3} P(\tau) \left(\frac{2}{5} P(\tau) Q(\tau) - \frac{2}{\varrho(\tau)} \right) \right] d\tau \quad ;$$

$$\alpha_2(s, \nu; t) = \frac{i}{2} \left[\frac{1}{\varrho(s)} - \frac{1}{5} P(s) Q(s) \right] \nu - it[R(s) - R(0)] \quad ;$$

$$\ldots \tag{12.5.4}$$

and

$$\beta_0(s, \nu; t) = -\frac{2^{1/3}}{c_0^{2/3}(s) P^{1/3}(s)} \nu + t \quad ;$$

$$\beta_1(s, \nu; t) = 0 \quad ;$$

$$\beta_2(s, \nu; t) = \frac{2^{1/3}}{10} \frac{P^{2/3}(s)}{c_0^{2/3}(s)} Q(s) \nu^2$$

$$- \frac{2t}{3} \left[\frac{1}{\varrho(s)} - \frac{1}{5} P(s) Q(s) \right] \nu \quad ;$$

$$\ldots \quad . \tag{12.5.5}$$

In these equations the effective radius of curvature $P(s)$ and the function $Q(s)$ are defined by (7.2.1) and (7.3.37), and

$$R(s) = \frac{1}{2^{1/3}} c_0^{2/3}(s) P^{1/3}(s) \left[\frac{1}{\varrho(s)} - \frac{1}{5} P(s) Q(s) - \frac{1}{2P(s)} \right] \quad .$$

Our next job is to find an asymptotic expression for the high-frequency behavior of g in the solution of (12.1.1–2). By analogy with the case of a circle we will look for this asymptotic expression in the form of a superposition of solutions of (12.5.1) that would be analogous to (12.4.28):

$$
g = \int_{C_{t_0}} c(\omega, t) \left\{ w_1 \left[\sum_{m=0}^{M-1} \beta_m(s, \nu; t) \omega^{-m/3} \right] \right.
$$
$$
- \frac{w_1(t)}{w_2(t)} w_2 \left[\sum_{m=0}^{M-1} \beta_m(s, \nu; t) \omega^{-m/3} \right] \Bigg\}
$$
$$
\left. \times \exp \left[i \sum_{m=-3}^{M-1} \alpha_m(s, \nu; t) \omega^{-m/3} \right] \right. dt \quad . \tag{12.5.6}
$$

Here the contour of integration C_{t_0} is the straight line segment joing the points $t_0 e^{i4\pi/3}$ and $t_0 e^{i\pi/3}$, $t_0 > 0$. The weighting function $c(\omega, t)$ (which is independent of s and ν) and the constant t_0 are to be determined.

The integrand in (12.5.6) satisfies the Helmholtz equation to $O(\omega^{-M/3+4/3})$ and vanishes when $\nu = 0$. This occurs by virtue of conditions (12.5.2) and (12.5.3).

The weighting function $c(\omega, t)$ and constant t_0 must be chosen so that:

1) the integrand of g will be an exponentially small in ω of order $O[\exp(-A\omega^\varepsilon)]$, $A > 0$, $\varepsilon > 0$ at the ends of the integration interval[6];
2) the integral g will give a valid description of the field of a source of the given strength;
3) the integral g will satisfy the reciprocity principle.

Using (12.5.4), (12.5.5) and the previously introduced quantity

$$\gamma(s) = \left(\frac{\omega}{2} \right)^{1/3} \int_0^s \frac{d\tau}{c_0^{1/3}(\tau) P^{2/3}(\tau)} \quad ,$$

which was called the *reduced arc length*, we rewrite (12.5.6) in the following form:

$$g = \sqrt[6]{\frac{c_0(s) P(0)}{c_0(0) P(s)}} \exp \left[i\omega \int_0^s \frac{d\tau}{c_0(\tau)} \right] \int_{C_{t_0}} c(\omega, t)$$

---

[6] If this condition is satisfied, the power terms in the asymptotic expansion of g in fractional powers of ω^{-1} will not depend on t_0.

$$\times \left\{ w_1 \left[t - \frac{2^{1/3}}{c_0^{2/3}(s)P^{1/3}(s)} \nu + (\beta_{22}(s)\nu^2 + \beta_{21}(s)\nu)\omega^{-2/3} + \dots \right] \right.$$

$$- \frac{w_1(t)}{w_2(t)} w_2 \left[t - \frac{2^{1/3}}{c_0^{2/3}(s)P^{1/3}(s)} \nu \right.$$

$$\left. \left. + (\beta_{22}(s)\nu^2 + \beta_{21}(s)\nu)\omega^{-2/3} + \dots \right] \right\}$$

$$\times \exp[i\gamma(s)t + i\alpha_1(s,\nu;t)\omega^{-1/3} + i\alpha_2(s,\nu;t)\omega^{-2/3} + \dots]\,dt \quad , \tag{12.5.7}$$

where the ellipsis indicates terms that are of higher order in $\omega^{-1/3}$. Let us go on to selection of the weighting function $c(\omega,t)$. First we satisfy the condition of reciprocity. We calculate the normal derivative $\partial g/\partial n$ when $n = 0$. Differentiating (12.5.7) and taking into consideration that $\partial/\partial n = \omega^{2/3}(\partial/\partial\nu)$ and $w_1 w_2' - w_1' w_2 = 2i$, we get

$$\left. \frac{\partial g}{\partial n} \right|_{n=0} = i\omega^{2/3} \frac{2^{4/3}}{c_0^{2/3}(s)P^{1/3}(s)} \sqrt[6]{\frac{c_0(s)P(0)}{c_0(0)P(s)}}$$

$$\times \exp\left[i\omega \int_0^s \frac{d\tau}{c_0(\tau)} \right] \int_{C_{t_0}} \frac{c(\omega,t)}{w_2(t)} \left[1 - \frac{c_0^{2/3}(s)P^{1/3}(s)}{2^{1/3}} \right.$$

$$\times \beta_{21}(s;t)\omega^{-2/3} + \dots \left] \exp[i\gamma(s)t + i\alpha_{10}(s;t)\omega^{-1/3} \right.$$

$$+ t(R(s) - R(0))\omega^{-2/3} + \dots]\,dt \quad . \tag{12.5.8}$$

The derivative $(\partial g/\partial n)|_{n=0}$ must satisfy the usual reciprocity principle, i.e. it must be a quantity that is symmetric with respect to the observation point and the source. Such symmetry will occur if we set

$$c(\omega,t) = \frac{\omega^{1/3}}{c_0^{1/3}(0)P^{2/3}(0)} \left[1 - \frac{c_0^{2/3}(0)P^{1/3}(0)}{2^{1/3}} \beta_{21}(0;t)\omega^{-2/3} + \dots \right]$$

$$\times \exp[2tR(0)\omega^{-2/3} + \dots]p(t) \quad . \tag{12.5.9}$$

The factor $\omega^{1/3}$ in (12.5.9) is chosen from dimensional considerations. The higher-order terms in the expansions drop off at least as fast as $1/\omega$ and are readily determined from the functions $\beta_{m1}(s;t)$ and $\alpha_{m0}(s;t)$, $m \geq 3$. The new dimensionless weighting function $p(t)$ is no longer dependent on the coordinates of the source. We will assume that $p(t)$ is also independent of ω. This assumption is justified as follows. The frequency ω would have entered into the function $p(t)$ only in dimensionless combinations with other dimensioned parameters of the problem. Since dependence on the coordinates of the source and observation point has already been ruled out, such parameters could only be certain integral characteristics of the problem (for instance the length of boundary S for closed regions). However, if we permitted the function $p(t)$ to depend on the integral characteristics of the problem we would arrive at a contradiction with the principle of localizability, which states that the high-frequency asymptotic behavior of the wave field depends only on the properties of the path between the source and the observation point.

We now substitute this expansion for $c(\omega, t)$ into (12.5.7):

$$
g = \frac{\omega^{1/3}}{\sqrt{c_0(0)P(0)}} \sqrt[6]{\frac{c_0(s)}{P(s)}} \exp\left[i\omega \int_0^s \frac{d\tau}{c_0(\tau)} \right]
$$

$$
\times \int_{C_{t_0}} p(t) \left[1 - \frac{c_0^{2/3}(0)P^{1/3}(0)}{2^{1/3}} \beta_{21}(0; t)\omega^{-2/3} + \dots \right]
$$

$$
\times \left\{ w_1[t - \tilde{\nu}(s) + \beta_2(s, \nu; t)\omega^{-2/3} + \dots] \right.
$$

$$
\left. - \frac{w_1(t)}{w_2(t)} w_2[t - \tilde{\nu}(s) + \beta_2(s, \nu; t)\omega^{-2/3} + \dots] \right\}
$$

$$
\times \exp\{i\gamma(s)t + i\alpha_1(s, \nu; t)\omega^{-1/3}
$$

$$
+ [i\alpha_2(s, \nu; t) + 2tR(0)]\omega^{-2/3} + \dots\} \, dt \quad , \tag{12.5.10}
$$

where the quantity

$$
\tilde{\nu}(s) = 2^{1/3} \frac{1}{c_0^{2/3}(s)P^{1/3}(s)} \nu
$$

can be called the *reduced normal*.

Let us note that the rate of convergence of integral (12.5.10) depends on the reduced arc length γ. In order to keep the most important contribution of the integration contour in (12.5.10) from increasing with γ, we will choose the function $p(t)$ differently depending on the value of $\gamma(s)$. We met a similar situation in the preceding section when we were studying the exact solution of (12.1.1, 2) for a circle. Replacing one expression for $p(t)$ by another in the integrand of (12.5.10) results in a step change in the intensity of the wave propagating near S and in the splitting off of a term from (12.5.10) that can be calculated by the method of steepest descents and has a *ray form*[7]. It is easily shown that this term gives a wave that has been reflected a certain number of times from the boundary. Thus, the step changes in intensity for this boundary wave are associated with the splitting off of waves from it that have undergone various numbers of reflections from the boundary. It is evident from our discussion that the replacement of one expression for $p(t)$ by another as $\gamma(s)$ increases, which

---

[7] We say that a wave has a *ray form* if it can be described by formulas of the ray method.

is necessary to assure the best convergence of the integral (12.5.10), agrees with the physical picture of wave propagation near a boundary (see Sect. 12.1).

Let the reduced arc length $\gamma(s)$ satisfy the condition

$$\frac{2}{\sqrt{3\Delta}} \leq \gamma(s) < 2\sqrt{\Delta}\left(\frac{\omega P(s)}{2c_0(s)}\right)^{\delta} , \qquad (12.5.11)$$

where $\Delta = O(1)$ and $0 < \delta < 1$.

Let us choose the function $p(t)$ so that (12.5.10) describes a field corresponding to a point source of given intensity. Let us turn for a moment to the solution (12.4.28) that we got for the etalon problem in Sect. 12.4. When $c_0(s)$ and $P(s)$ become constants, condition (12.5.11) reduces to (12.4.26), and consequently expression (12.5.10) should reduce to the solution (12.4.28) of the etalon problem for a circle.

When $c(s,\nu) = \text{const}$ and $\varrho(s) = \varrho = \text{const}$, we have $P(s) = \varrho$, $Q(s) = 3\varrho^{-2}$, the reduced normal $\tilde{\nu}(s)$ is the same as the reduced normal $\tilde{\nu}$ (see Sect. 12.4) in the case of a circle, and by virtue of (12.5.4) and (12.5.5) the relations

$$\frac{c_0^{2/3}(0)P^{1/3}(0)}{2^{1/3}}\beta_{21}(0;t)\omega^{-2/3} = -\frac{2}{15}t\left(\frac{2}{k\varrho}\right)^{2/3} , \quad k = \frac{\omega}{c} ,$$

$$\beta_2(s,\nu;t) = \left(\frac{3}{20}\tilde{\nu}^2 - \frac{2}{15}t\tilde{\nu}\right)\left(\frac{2}{k\varrho}\right)^{2/3} ,$$

$$i\alpha_1(s,\nu;t) = i\frac{t^2}{60}\frac{s}{\varrho}\left(\frac{2}{k\varrho}\right)^{1/3} ,$$

$$[i\alpha_2(s,\nu;t) + 2tR(0)]\omega^{-2/3} = \left(-\frac{1}{10}\tilde{\nu} + \frac{t}{10}\right)\left(\frac{2}{k\varrho}\right)^{2/3} \qquad (12.5.12)$$

are satisfied. Thus (12.5.10) will reduce to the solution of the etalon problem if we set

$$p(t) = -\frac{i}{4\pi 2^{1/3}}\frac{1}{v(t)} , \qquad (12.5.13)$$

(see footnote[8]). Later on when we study the way that the formula for a wave propagating in the boundary layer merges into the formulas of geometric optics for waves that have been reflected different numbers of times, we will give further justification for the choice of $p(t)$ in the form (12.5.13).

Investigation of the convergence of the integral (12.5.10) under the condition (12.5.11) and with the choice (12.5.13) for $p(t)$ is similar to the study of the convergence of (12.4.28) for a circle. For sufficiently small values of the reduced normal $\tilde{\nu}(s)$, the quantity t_0 that determines the dimensions of the integration contour in (12.5.10) must be set equal to

---

[8] We can state on the basis of (12.5.12) that when (12.5.13) is satisfied, the general expression (12.5.10) reduces to the solution (12.4.28) of the etalon problem up to terms of order $O(\omega^{-1})$. Calculation of further terms of the expansions would obviously show that this happens for all higher orders as well.

$$t_0 \sim 2\Delta \left(\frac{\omega P(s)}{2c_0(s)}\right)^{2\delta} . \tag{12.5.14}$$

The constant δ must be chosen small enough so that the terms of the expansions in (12.5.10) when evaluated at the ends of the integration interval contain negative powers of the large parameter ω as before. Since $\alpha_{10}(s;t) \sim t^2$ and $t_0 \sim \omega^{2\delta}$, the constant δ must in any event be less than 1/12. It can be shown that when $\delta < 1/12$ the leading terms of the expansions will also contain increasing negative powers of ω. {See the footnote following (12.4.26) concerning the value of the factor $[\omega P(s)/2c_0(s)]^\delta$, and the selection of Δ.}

Let us determine the form of $p(t)$ when $\gamma(s) \geq 2\sqrt{\Delta}[\omega P(s)/2c_0(s)]^\delta$. For the sake of simplicity, we will consider $\partial g/\partial n$ at points on the boundary surface, i.e. at $n = 0$, rather than the expression for g [formula (12.5.10)].

Substituting (12.5.9) and (12.5.13) into (12.5.8) in place of $c(\omega, t)$, and writing out only the principal terms, we get

$$\left.\frac{\partial g}{\partial n}\right|_{n=0} = \frac{\omega}{2\pi} \frac{1}{\sqrt{c_0(0)c_0(s)P(0)P(s)}} \exp\left[i\omega \int_0^s \frac{d\tau}{c_0(\tau)}\right]$$

$$\times \int_{C_{t_0}} \frac{1}{v(t)w_2(t)} \exp[i\gamma(s)t]\{1 + O[(t^2 + 1)\omega^{-1/3}]\} \, dt .$$

$$\tag{12.5.15}$$

For large values of $\gamma(s)$, the factor

$$\frac{1}{v(t)w_2(t)}\exp[i\gamma(s)t] ,$$

appearing in the integrand begins to decrease on the lower part of the integration contour only for t sufficiently large. For large $\gamma(s)$, the quantity t_0 would have to be taken so large that the correction terms would be comparable to the principal term. This implies, as has already been mentioned, that the function $p(t)$ must be selected differently for different values of $\gamma(s)$, and in such a way that:

1) the quantity t_0 does not increase with increasing $\gamma(s)$;
2) the values of $(\partial g/\partial n)|_{n=0}$ change continuously as each new multiply reflected wave emerges in ray form.

Expanding the function $1/v(t)$ in the geometric progression

$$\frac{1}{v(t)} = \frac{2i}{w_1(t) - w_2(t)}$$

$$= -2i \sum_{m=0}^{M-1} \frac{1}{w_2(t)} \left[\frac{w_1(t)}{w_2(t)}\right]^m + \frac{1}{v(t)} \left[\frac{w_1(t)}{w_2(t)}\right]^M ,$$

we transform the integral in (12.5.15) for $(\partial g_n/\partial n)|_{n=0}$ to the sum of integrals

$$\int_{C_{t_0}} \frac{1}{v(t)w_2(t)} \exp[i\gamma(s)t]\{1 + O[(t^2 + 1)\omega^{-1/3}]\}\, dt$$

$$= \sum_{m=0}^{M-1} g_m + \tilde{G}_M \quad , \tag{12.5.16}$$

where

$$g_m = -2i \int_{C_{t_0}} \frac{1}{w_2^2(t)} \left[\frac{w_1(t)}{w_2(t)}\right]^m e^{i\gamma t}\{1 + O[(t^2 + 1)\omega^{-1/3}]\}\, dt$$

and

$$\tilde{G}_M = \int_{C_{t_0}} \frac{1}{v(t)w_2(t)} \left[\frac{w_1(t)}{w_2(t)}\right]^M e^{i\gamma t}\{1 + O[(t^2 + 1)\omega^{-1/3}]\}\, dt \quad .$$

Formula (12.5.16) enables us to write (12.5.15) for $(\partial g/\partial n)|_{n=0}$ as

$$\left.\frac{\partial g}{\partial n}\right|_{n=0} = \sum_{m=0}^{M-1} \left.\frac{\partial u_m}{\partial n}\right|_{n=0} + \left.\frac{\partial u_M}{\partial n}\right|_{n=0} \quad , \tag{12.5.17}$$

where by definition

$$\left.\frac{\partial u_m}{\partial n}\right|_{n=0} = \frac{\omega}{2\pi} \frac{1}{\sqrt{c_0(0)c_0(s)P(0)P(s)}} \exp\left[i\omega \int_0^s \frac{d\tau}{c_0(\tau)}\right] g_m$$

and

$$\left.\frac{\partial u_M}{\partial n}\right|_{n=0} = \frac{\omega}{2\pi} \frac{1}{\sqrt{c_0(0)c_0(s)P(0)P(s)}} \exp\left[i\omega \int_0^s \frac{d\tau}{c_0(\tau)}\right] \tilde{G}_M . \tag{12.5.18}$$

Let us evaluate the integrand of $\tilde{G}_M$ on the integration contour C_{t_0} for $|t| \gg 1$. Using asymptotic formulas for the Airy functions, we get

$$\left|\frac{1}{v(t)w_2(t)} \left[\frac{w_1(t)}{w_2(t)}\right]^M e^{i\gamma t}\right|$$

$$< \begin{cases} \text{const}|t|^{1/2} \exp\left[-\dfrac{2}{\sqrt{3}}\gamma|t|\right] & , \quad \arg t = \dfrac{\pi}{3} \quad , \\[3mm] \text{const}|t|^{1/2} \exp\left[-\dfrac{4}{3}|t|^{3/2}(M+1) + \dfrac{2}{\sqrt{3}}\gamma|t|\right] & , \quad \arg t = \dfrac{4\pi}{3} \quad . \end{cases}$$

The given estimate implies that if the number M is appropriately increased with increasing γ, then there will be no appreciable change in the magnitude of the integrand of $\tilde{G}_M$ along the contour C_{t_0} of integration. From now on we will choose the number M so that it satisfies the inequality

$$(M + 1)2\sqrt{\Delta} \left(\frac{\omega P(s)}{2c_0(s)}\right)^\delta > \gamma(s) \quad . \tag{12.5.19}$$

Then the quantity t_0 that determines the dimensions of the integration contour will satisfy (12.5.14) for all $\gamma(s) \geq (2/\Delta\sqrt{3})$, and (12.5.18) can be rewritten as

$$\left.\frac{\partial u_M}{\partial n}\right|_{n=0} = \frac{\omega}{2\pi} \frac{1}{\sqrt{c_0(0)c_0(s)P(0)P(s)}}$$

$$\times \exp[i\omega \int\limits_0^s \frac{d\tau}{c_0(\tau)}]G_M(\gamma)[1 + O(\omega^{-1/3+4\delta})] \quad , \qquad (12.5.20)$$

where

$$G_M(\gamma) = \int\limits_{C_{t_0}} \frac{1}{v(t)w_2(t)} \left[\frac{w_1(t)}{w_2(t)}\right]^M e^{i\gamma t} \, dt \quad .$$

Next, we study the terms $(\partial u_m/\partial n)|_{n=0}$ in (12.5.17), and show that they describe waves that have been reflected a certain number of times from the boundary surface. Obviously this study boils down to calculating the integrals g_m [see (12.5.16) and below].

As in Sect. 12.4, we introduce the functions

$$A_j(t) = \exp\left[i\frac{\pi}{2}(j-2)\right] t^{1/4} \exp\left[(-1)^{j-1}\frac{2}{3}t^{3/2}\right] w_j(t) \quad , \quad j = 1, 2 \quad .$$

For large $|t|$ the functions $A_j(t)$ in the sectors

$$\frac{\pi}{3}(3-2j) + \varepsilon < \arg t < \frac{\pi}{3}(9-2j) - \varepsilon \quad , \quad \varepsilon > 0 \quad ,$$

can be expanded in the asymptotic series (see Appendix A.1)

$$A_j(t) = 1 + \sum_{n=1}^\infty (-1)^{(2-j)} \frac{a_n}{t^{3/2}} \quad ,$$

$$\left(a_1 = \frac{5}{48} \quad , \quad a_n = \frac{(6n-1)!!}{144^n(2n-1)!!n!}\right) \quad .$$

In terms of the functions $A_j(t)$, the integrands in integrals g_m are written as

$$-2i\frac{1}{w_2^2(t)} \left[\frac{w_1(t)}{w_2(t)}\right]^m e^{i\gamma t} = -\exp\left[i\frac{\pi}{2}(m+1)\right] 2t^{1/2}\frac{1}{A_2^2(t)}\left[\frac{A_1(t)}{A_2(t)}\right]^m$$

$$\times \exp\left\{i\gamma t - \tfrac{4}{3}(m+1)t^{3/2}\right\} \quad ,$$

and when $|t| \gg 1$ and $\pi/3 + \varepsilon < \arg t < 5\pi/3 - \varepsilon$,

$$\frac{1}{A_2^2(t)}\left[\frac{A_1(t)}{A_2(t)}\right]^m = \exp\{-2a_1(m+1)t^{-3/2}[1 + O(t^{-3})]\}$$

$$\times [1 + O(t^{-3})] \quad . \qquad (12.5.21)$$

Let us set

$$\gamma(s) = (m+1)\left(\frac{\omega P(s)}{2c_0(s)}\right)^\delta \tilde{\gamma}(s)$$

and change variables of integration in g_m from t to z, with

$$t = e^{i\pi} \left(\frac{\omega P(s)}{2c_0(s)}\right)^{2\delta} z \quad.$$

As a result we get

$$g_m = -2\exp\left[i\frac{\pi}{2}(m+2)\right]\theta \int_{C_{t_0}} F(z)\exp[i(m+1)\theta f(z)]$$

$$\times \{1 + O[(z^2\theta^{4/3}+1)\omega^{-1/3}]\}\, dz \quad, \tag{12.5.22}$$

where

$$\theta = \left(\frac{\omega P(s)}{2c_0(s)}\right)^{3\delta} \quad, \qquad f(z) = \frac{4}{3}z^{3/2} - \tilde{\gamma}z \quad \text{and}$$

$$F(z) = z^{1/2}\frac{1}{A_2^2(-\theta^{2/3}z)}\left[\frac{A_1(-\theta^{2/3}z)}{A_2(-\theta^{2/3}z)}\right]^m \quad.$$

In defining the functions $f(z)$ and $F(z)$, we assume that the branch cut goes along the negative part of the real axis, and that $z^{3/2} > 0$ and $z^{1/2} > 0$ when $z > 0$. In the sector $-2\pi/3 + \varepsilon < \arg z < 2\pi/3 - \varepsilon$ the function $F(z)$ allows the asymptotic representation implied by (12.5.21):

$$F(z) = z^{1/2}\exp\left\{-2ia_1\frac{m+1}{\theta}z^{-3/2}[1 + O(\theta^{-2}z^{-3})]\right\}$$

$$\times [1 + O(\theta^{-2}z^{-3})] \quad. \tag{12.5.23}$$

Obviously when $\theta \gg 1$, the integrals (12.5.22) can be calculated by the method of steepest descents. The phase function $f(z)$ has the single saddle point $z_0 = [\frac{1}{2}\tilde{\gamma}(s)]^2$. The constant-amplitude contours of $\text{Re}\{f(z)\}$ are shown in Fig. 12.12.

The usual method of steepest descents is applicable to (12.5.22) only when the saddle point z_0 is sufficiently far from the point $z = 0$, which is a branch point of $f(z)$ and $F(z)$. We will assume henceforth that

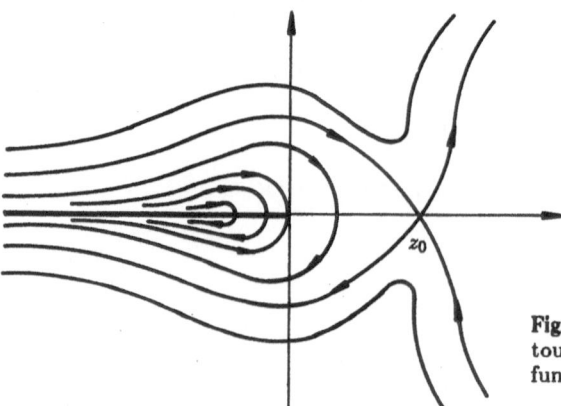

Fig. 12.12. Constant-amplitude contours of the real part of the phase function $f(z)$

$$\tilde{\gamma}(s) \geq 2\sqrt{\Delta} \qquad (12.5.24)$$

and consequently, $z_0 > \Delta$. Inequality (12.5.24) is equivalent to the condition

$$\gamma(s) \geq (m+1) \left(\frac{\omega P(s)}{2c_0(s)} \right)^{\delta} 2\sqrt{\Delta} \quad,$$

which will hold for all m if

$$\gamma(s) \geq 2\sqrt{\Delta} M \left(\frac{\omega P(s)}{2c_0(s)} \right)^{\delta} \quad. \qquad (12.5.25)$$

Inequality (12.5.25) together with (12.5.19) defines the integer M for a given $\gamma(s)$:

$$\frac{\gamma(s)}{2\sqrt{\Delta}} \left(\frac{\omega P(s)}{2c_0(s)} \right)^{-\delta} - 1 < M \leq \frac{\gamma(s)}{2\sqrt{\Delta}} \left(\frac{\omega P(s)}{2c_0(s)} \right)^{-\delta} \quad. \qquad (12.5.26)$$

For finite distances between the source and observation points, $\gamma(s) \sim \omega^{1/3}$ and consequently $M \sim \omega^{1/3-\delta}$. For such values of M the argument of the exponent in (12.5.23) is proportional to $\omega^{1/3-4\delta}$, and therefore when $\delta < 1/12$ the function $F(z)$ cannot, generally speaking, be considered a function that varies slowly in the sense generally accepted for the method of steepest descents. Let us evaluate the contribution that derivatives of the function $F(z)$ make to the correction term in the method of steepest descents. Since in effect the product $(m+1)\theta$ in the integrals g_m is a large parameter, the correction term in the method of steepest descents that is attributable to the function $F(z)$ will be proportional to

$$\left| \frac{1}{2} \frac{F''(z_0)}{F(z_0)} \frac{1}{f''(z_0)(m+1)\theta} \right| \sim \frac{1}{5} \frac{1}{z_0^{9/2}} \frac{m+1}{\theta^3} = \frac{1}{5}(m+1)^{10} \left(\frac{2}{\gamma} \right)^9 \quad. \qquad (12.5.27)$$

The quantity (12.5.27) is $O(\omega^{1/3-10\delta})$ and consequently when $\delta > 1/30$ it approaches zero along with $1/\omega$.

Deforming the integration contour C_{t_0} in (12.5.22) in the vicinity of the saddle point z_0 into the $\mathrm{Re}\{f(z)\} = $ constant line (steepest-descent path), carrying out the usual calculations for the method of steepest descents and combining correction terms of the same order, we get

$$\begin{aligned}
g_m = \sqrt{\pi} \exp\left[i \left(\frac{\pi}{4} + \frac{\pi}{2}m \right) \right] & \frac{\gamma^{3/2}}{(m+1)^2} \exp\left\{ -\frac{i}{12} \frac{\gamma^3}{(m+1)^2} \right. \\
& \left. - 16ia_1 \frac{(m+1)^4}{\gamma^3} \right\} \left\{ 1 + O[z_0^2 \omega^{-1/3} Q^{4/3}] \right. \\
& + O[\Delta^{-9/2}(m+1)Q^{-3}] + O[\Delta^{-3}Q^{-2}] \\
& \left. + O\left[\Delta^{-3/2} \frac{Q^{-1}}{m+1} \right] \right\} \quad. \qquad (12.5.28)
\end{aligned}$$

For the integrals g_m the number $m + 1 \leq M$, and consequently by virtue of (12.5.26), $m + 1$ does not exceed $\omega^{1/3-\delta}$ in order of magnitude. Thus the correction terms in (12.5.28) have the following orders of magnitude: $\omega^{-1/3+4\delta}$, $\omega^{1/3-10\delta}$, $\omega^{-6\delta}$ and $\omega^{-3\delta}$. To get the optimum error of order $\omega^{-1/7}$, we must set $\delta = 1/21$.

Formula (12.5.28) is true both for $\gamma(s) \sim \omega^\delta$, and for $\gamma(s) \sim \omega^{1/3}$, $m+1 = M$, $M-1$, ... In other words, (12.5.28) is valid both for small distances from source to observation points ($s \sim \omega^{-1/3+\delta}$), and for distances that are finite, but for fairly large values of m. For $\gamma(s) \sim \omega^{1/3}$ and small values of m, (12.5.28) is inapplicable, since $|z_0| \sim \omega^{1/3-\delta}$ when m is small. This is because for small m the integrands of g_m do not satisfy the Helmholtz equation with sufficient accuracy in the neighborhood of the saddle point. Using the values found for the integrals g_m and substituting (12.3.34) for $\gamma(s)$, we get

$$
\left. \frac{\partial u_m}{\partial n} \right|_{n=0} = \frac{\omega^{3/2}}{2\sqrt{2\pi}} \exp\left[i\left(\frac{\pi}{4} + \frac{\pi}{2}m \right) \right] \frac{1}{\sqrt{c_0(0)c_0(s)P(0)P(s)}}
$$

$$
\times \frac{1}{(m+1)^2} \left[\int_0^s \frac{d\tau}{c_0^{1/3}(\tau)P^{2/3}(\tau)} \right]^{3/2}
$$

$$
\times \exp\left\{ i\left[\omega \int_0^s \frac{d\tau}{c_0(\tau)} - \frac{1}{24} \frac{\omega}{(m+1)^2} \left(\int_0^s \frac{d\tau}{c_0^{1/3}(\tau)P^{2/3}(\tau)} \right)^3 \right. \right.
$$

$$
\left. \left. - \frac{10}{3} \frac{(m+1)^4}{\omega} \left(\int_0^s \frac{d\tau}{c_0^{1/3}(\tau)P^{2/3}(\tau)} \right)^{-3} \right] \right\} [1 + O(\omega^{-1/7})],
$$

$$
\tag{12.5.29}
$$

for the quantities $\partial u_m/\partial n|_{n=0}$ appearing in (12.5.17). Expression (12.5.29) agrees completely with (12.3.33) obtained in Sect. 12.3 for the normal derivative of the field of a wave that has experienced m reflections and propagates near S.

Let us summarize the results of this section. From the fact that $\partial u_m/\partial n|_{n=0}$ in (12.5.17) describes a wave reflected m times and propagating near S we see that the function $\partial u_M/\partial n|_{n=0}$ of (12.5.18) describes the normal derivative of an interference field generated by waves whose number of reflections exceeds $M-1$. In other words $\partial u_M/\partial n|_{n=0}$ describes the normal derivative of a surface wave. The agreement between (12.5.29) and (12.3.33) can be taken as more evidence of the correctness of the choice of contour C_{t_0} and weighting function $c(\omega, t)$ in the construction of the superposition (12.5.6).

And so to calculate the normal derivative of the field on the boundary S at a fixed observation point $N(s, 0)$ we must first determine the integer M which satisfies (12.5.26). The number M indicates the maximum number of waves in geometric optics that can be observed separately at N. The fields of these waves are described by formulas of the ray method; for waves with larger m, (12.5.29) is valid, and can be obtained either by refining the formulas of the ray method or by separating the corresponding integrals from (12.5.15). The normal derivative of the surface wave, i.e. of the overall effect of waves whose number of reflections exceeds $(M-1)$ is described by (12.5.20):

$$\frac{\partial u_M}{\partial n}\bigg|_{n=0} = \frac{\omega}{2\pi} \frac{1}{\sqrt{c_0(0)c_0(s)P(0)P(s)}} \exp\left[i\omega \int_0^s \frac{d\tau}{c_0(\tau)}\right]$$

$$\times G_M(\gamma)[1 + O(\omega^{-1/7})] \quad,$$

$$G_M(\gamma) = \int_{C_{t_0}} \frac{1}{v(t)w_2(t)} \left[\frac{w_1(t)}{w_2(t)}\right]^M e^{i\gamma(s)t} dt \quad. \tag{12.5.30}$$

The modulus of the integrand of $G_M(\gamma)$ at the selected value of M has a fairly sharp maximum on the integration contour near $t = 0$. This makes tabulation of the integral $G_M(\gamma)$ fairly simple. Tables of the special function $G_M(\gamma)$ are given in Appendix A.4. With an increase in frequency ω or in the coordinate of the observation point s there is an increase in the reduced arc length $\gamma(s)$. At values of $\gamma(s)$ for which the quantity $\gamma(s)/2\sqrt{\Delta}[\omega P(s)/2c_0(s)]^\delta$ becomes an integer, the integrand in (12.5.30) changes stepwise. The former expression for the field of the surface wave becomes equal to the sum of the fields of the new surface wave and a wave that has had $\gamma(s)/2\sqrt{\Delta}[\omega P(s)/2c_0(s)]^{-\delta} - 1$ reflections.

Thus for integer values of $\gamma(s)/2\sqrt{\Delta}[\omega P(s)/2c_0(s)]^{-\delta}$ a geometric optics wave splits off from the surface wave. With increasing $\gamma(s)$ the path of this split-off wave moves further and further from S, and as a result it is no longer described by (12.5.29).

Formula (12.5.30) gives only the principal term of the expansion of the surface wave in negative fractional powers of ω. To get the subsequent terms of the expansion we must refer to (12.5.8), substitute (12.5.9) for the function $c(\omega, t)$, and take the dimensionless function in the form

$$p(t) = -\frac{i}{4\pi 2^{1/3}} \frac{1}{v(t)} \left[\frac{w_1(t)}{w_2(t)}\right]^M \quad.$$

12.6 Notes on the Literature

Buldyrev [12.4] studied the asymptotic expansion of the Green's function of a surface source under conditions where the whispering gallery effect arises. It was assumed in this paper that the impedance condition

$$\frac{\partial u}{\partial n} - i\omega g(s)u\bigg|_S = 0, \quad g(s) > 0,$$

is satisfied on the boundary of the region, and an expression for the surface wave analogous to (12.5.30) was derived. This chapter is more detailed than [12.4]: the ray formulas for noninterfering waves have been refined, and the necessary estimates have been presented.

An SH surface wave on an elastic sphere was studied in the paper of *Buldyrev* and *Lanin* [12.5]. Condition (12.2.18) specifying which reflected waves can be described using ray formulas was derived by *Brekhovskikh* and *Ivanov* [12.6] (see also [12.7]). The etalon problem of Sect. 12.4 has also been studied in somewhat more generality by *Ishihara* et al. [12.8], *Ishihara* and *Felsen* [12.9], and *Topuz* et al. [12.10].

In an analogous way we can derive and study the interference head wave which arises when waves propagate in the presence of two media separated by an interface whose effective radius of curvature $P(s) > 0$. The simplest problem of this kind is that of a line source near the interface of two homogeneous media separated by a circular cylindrical surface. For an interference head wave to exist, the propagation velocity in the inner medium must be greater than that of the outer medium, and the source must be located in the outer medium. Under these conditions, whispering gallery waves propagate in the inner medium near the surface of the cylinder, and refraction of these waves into the outer medium forms a head wave of interference type. This problem, which has been studied by *Buldyrev* [12.11], and by *Buldyrev* and *Lanin* [12.12], can serve as an etalon problem for an entire class of problems in which head waves can arise.

The interference head wave which arises at an arbitrary interface between two elastic media was studied in the three-dimensional case by *Buldyrev* and *Grikurov* [12.13, 14], and by *Grikurov* [12.15]. An estimate of the error in the asymptotic formula for the surface wave was analyzed for a number of concrete problems by *Lanin* [12.3] and by *Buldyrev* and *Lanin* [12.16]. The table of the function $G_M(\gamma)$ in Appendix A.4 was compiled by *Lanin* (see also [12.17]).

Another approach to the construction of the field near a concave boundary based upon matching to ray expansions has been used in the papers by *Babič* [12.18–20].

13. The High-Frequency Asymptotics of the Field Scattered by a Smooth Body

Our goal in this chapter is to construct the high-frequency asymptotic solution for the problem of diffraction of a wave, given as a ray expansion, by a smooth surface S on which the Dirichlet condition $u|_S = 0$ is satisfied. We will assume that the rays of the incident wave have points of tangency with S of first order. This will happen, for example, when the velocity $c(M)$ is constant and S is a convex surface whose Gaussian curvature is positive at each point.

We denote the set of points of tangency between S and rays of the incident wave as the curve C. Since this curve separates the shadow side of S from the illuminated side, we will call it by the astronomical name of "terminator". The asymptotic formulas will be constructed in a certain neighborhood Ω_n of C. They can be continued into the penumbra and into the deep shadow zone. In the illuminated region, they will reduce to the formulas of the ray method to any order in $1/\omega$ as $\omega \to \infty$.

13.1 The Etalon Problem

The formulas to be obtained for the general case are analogous to asymptotic formulas for the solution of the classical diffraction problem of a plane wave by a circle. Considering that the results of this section are to be used merely as guidelines, then, we will not worry about the detailed substantiation of the formulas we obtain.

Let a plane wave $u_{\text{inc}} = e^{ikx}$ be incident on the circle $r \equiv \sqrt{x^2 + y^2} \le a$. The reflected wave u_{refl} must satisfy the Helmholtz equation,

$$(\Delta + k^2)u_{\text{refl}} = 0 \quad (r \ge a) \tag{13.1.1}$$

the boundary condition,

$$u_{\text{inc}} + u_{\text{refl}}\bigg|_{r=a} = 0 \tag{13.1.2}$$

and the radiation condition

$$\sqrt{r}\left(\frac{\partial u_{\text{refl}}}{\partial r} - iku_{\text{refl}}\right) \to 0 \quad \text{as} \quad r \to \infty \quad . \tag{13.1.3}$$

The solution to (13.1.1–3) is well known (cf., for example, [13.1]):

$$u = u_{\text{inc}} + u_{\text{refl}}$$

$$= \sum_{n=0}^{\infty} 2\delta_n \left[J_n(kr) - \frac{H_n^{(1)}(kr)J_n(ka)}{H_n^{(1)}(ka)} \right] e^{i\pi n/2} \cos n\phi \quad .$$

(13.1.4)

Here, (r, ϕ) are polar coordinates, $\delta_0 = \frac{1}{2}$, and $\delta_n = 1$ for $n \geq 1$. The solution (13.1.4) can also be expressed as a contour integral:

$$u = ik \int_{\mathcal{L}} \left[J_{k\zeta}(kr) - \frac{H_{k\zeta}^{(1)}(kr)J_{k\zeta}(ka)}{H_{k\zeta}^{(1)}(ka)} \right] e^{-ik\zeta\pi/2} \frac{\cos k\zeta\phi}{\sin k\zeta\pi} d\zeta \quad (13.1.5)$$

wherein

$$ik \int_{\mathcal{L}} J_{k\zeta}(kr) e^{-i\pi k\zeta/2} \frac{\cos k\zeta\phi}{\sin k\zeta\pi} d\zeta = e^{ikx} \quad . \quad (13.1.6)$$

The contour $\mathcal{L}$ consists of the lines $(+\infty - i\beta, -i\beta)$ and $(i\beta, i\beta + \infty)$ joined by the segment $[-i\beta, i\beta]$ in the complex ζ-plane, where β is a sufficiently small positive number. The direction of integration along this path is clockwise. The integrals (13.1.5) and (13.1.6) must be understood as Cauchy principal values since the integrands have poles at $\zeta = 0$.

We now set out to investigate the integral (13.1.5) in the penumbra zone Ω_p, specifically in a neighborhood of the limiting ray $y = a$, $x \geq 0$ which is independent of k. We must find the section of the integration path which makes the primary contribution to (13.1.5) when the observation point $M(r, \phi) \in \Omega_p$.

It is not difficult to see that the result of integrating along the lower part of $\mathcal{L}$ ($\text{Im } \zeta = -\beta$; $\text{Re } \zeta \geq 0$) in (13.1.5) and (13.1.6) will be exponentially small. To see this, replace the Hankel and Bessel functions by their asymptotic expansions (cf. Chap. 3) and use the fact that

$$\frac{\cos k\zeta\phi}{\sin k\zeta\pi} = i \exp[ik\zeta(\phi - \pi)]\{1 + O[\exp(-2k\beta\phi)]\} \quad .$$

On the upper part $\mathcal{L}'$ of $\mathcal{L}$: $\text{Im } \zeta = \beta$, $0 \leq \text{Re } \zeta < +\infty$, we put

$$\frac{\cos k\zeta\phi}{\sin k\zeta\pi} = -i \exp[ik\zeta(\pi - \phi)]\{1 + O[\exp(-2k\beta\phi)]\} \quad .$$

Dropping exponentially small terms of order $e^{-k\delta}$, $\delta > 0$, we obtain

$$u \sim k \int_{\mathcal{L}'} \left[J_{k\zeta}(kr) - \frac{H_{k\zeta}^{(1)}(kr)J_{k\zeta}(ka)}{H_{k\zeta}^{(1)}(ka)} \right] \exp\left[ik\zeta\left(\frac{\pi}{2} - \phi\right)\right] d\zeta$$

$$+ ik \int_{-i\beta}^{i\beta} \left[J_{k\zeta}(kr) - \frac{H_{k\zeta}^{(1)}(kr)J_{k\zeta}(ka)}{H_{k\zeta}^{(1)}(ka)} \right] e^{-ik\zeta\pi/2} \frac{\cos k\zeta\phi}{\sin k\zeta\pi} d\zeta$$

(13.1.7)

and

$$u_{\text{inc}} = e^{ikx} \sim k \int\limits_{\mathcal{L}'} J_{k\zeta}(kr) \exp\left[ik\zeta\left(\frac{\pi}{2} - \phi\right)\right] d\zeta$$

$$+ ik \int\limits_{-i\beta}^{i\beta} J_{k\zeta}(kr) e^{-ik\zeta\pi/2} \frac{\cos k\zeta\phi}{\sin k\zeta\pi} d\zeta \quad . \tag{13.1.8}$$

Let us take a look at the behavior of the integrand of the first term of (13.1.8), using the asymptotic forms of the Bessel and Hankel functions. If $J_{k\zeta}$ is replaced by $[H_{k\zeta}^{(1)} + H_{k\zeta}^{(2)}]/2$, then the integrand of (13.1.8) splits into two terms whose phase factors are $\exp[ikf_0(r, \phi; \zeta)]$ and $\exp[ikf_1(r, \phi; \zeta)]$, where

$$f_j(r, \phi; \zeta) = (-1)^j\left[\sqrt{r^2 - \zeta^2} - \zeta\arccos\frac{\zeta}{r}\right] + \left(\frac{\pi}{2} - \phi\right)\zeta \quad (j = 0, 1) \quad .$$

From the Cauchy-Riemann equations it follows that the imaginary part $\text{Im}\{f_j\}$ will increase or decrease with increasing $\text{Im}\{\zeta\}$ accordingly as $df_j/d\zeta$ on $\text{Im}\{\zeta\} = 0$ is positive or negative. Taking this into account, it is easy to verify that for $0 < \phi < \pi$, the term containing $H_{k\zeta}^{(1)}$ [or $H_{k\zeta}^{(2)}$] is exponentially large (or small, respectively) on the segment $0 < \text{Im}\{\zeta\} \leq \beta$, $\text{Re}\{\zeta\} = 0$, and also on the segment $\text{Im}\{\zeta\} = \beta$, $0 \leq \text{Re}\{\zeta\} \leq r$, or on some part of it adjacent to the point $\zeta = i\beta$. One and only one of the phase factors $\exp(ikf_j)$, $j = 0, 1$, has a saddle point, located at $\zeta_0 = r\sin\phi$. If the point $M = M(r, \phi)$ is near the limiting ray (i.e., $r \simeq a/\sin\phi$), then the saddle point is near a.

Having established the behavior of the integrands, we will deform the contour of integration $\mathcal{L}'$ (differently for each term). Our immediate goal is to transform the expressions for u_{inc} and $u_{\text{inc}} + u_{\text{refl}}$ into a form suitable for finding their shortwave asymptotics.

We replace $\mathcal{L}'$ by the segment $[i\beta, \zeta_1]$ joined to the line $(\zeta_1, +\infty)$, where ζ_1 is an arbitrary positive number. After doing this, the terms containing $H_{k\zeta}^{(1)}$ can be integrated along the broken line $(i\beta, -i\beta, \zeta_1)$ instead of the segment $[i\beta, \zeta_1]$. We have thus obtained two integrals over the segment $[i\beta, -i\beta]$: one arising from the indicated contour deformation and the other from the second term of (13.1.8). Combining these integrals gives an integral which vanishes because its integrand is an odd function.

We thus arrive at an integral over the contour $\mathcal{L}_1$ which has a distinctive "forked" shape (Fig. 13.1). On the segment $[i\beta, \zeta_1]$ (respectively, $[-i\beta, \zeta_1]$) a

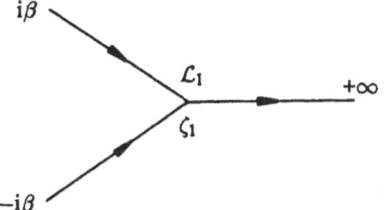

Fig. 13.1. Contour of integration for (13.1.9)

359

term containing $H_{k\zeta}^{(2)}/2$ [respectively, $H_{k\zeta}^{(1)}/2$] is integrated; along the half-line $(\zeta_1, +\infty)$ the integrand is the sum of both terms.

We can now transform the expression for $u = u_{\text{inc}} + u_{\text{refl}}$. We have already dealt with the first term u_{inc}. In the term $H_{k\zeta}^{(1)}(kr)J_{k\zeta}(ka)/H_{k\zeta}^{(1)}(ka)$ we replace the Bessel function by half the sum of the two Hankel functions. The integrand splits accordingly into the sum of two terms of which one contains the factor $H_{k\zeta}^{(2)}(ka)/H_{k\zeta}^{(1)}(ka)$, while the other contains neither $H_{k\zeta}^{(1)}(ka)$ nor $H_{k\zeta}^{(2)}(ka)$. The second term is integrated over the broken line $(i\beta, -i\beta, \zeta_1)$. The integral over the segment $[-i\beta, i\beta]$ cancels the corresponding term in (13.1.7), since when both integrals are combined into a single one, the integrand of the resulting expression becomes an odd function.

We come now to the final result. We have the following representation for the field:

$$u = u_{\text{inc}} + u_{\text{refl}} \simeq k \int_{\mathcal{L}_1} \left\{ J_{k\zeta}(kr) \exp\left[ik\zeta\left(\frac{\pi}{2} - \phi\right)\right]\right.$$

$$\left. - H_{k\zeta}^{(1)}(kr)\exp\left[ik\zeta\left(\frac{\pi}{2} - \phi\right)\right] J_{k\zeta}(ka)/H_{k\zeta}^{(1)}(ka)\right\} d\zeta \qquad (13.1.9)$$

where

$$u_{\text{inc}} \simeq k \int_{\mathcal{L}_1} J_{k\zeta}(kr) \exp\left[ik\zeta\left(\frac{\pi}{2} - \phi\right)\right] d\zeta \quad . \qquad (13.1.10)$$

The contour $\mathcal{L}_1$ is shown in Fig. 13.1. The integration is to be carried out as follows: the Bessel functions in (13.1.9) and (13.1.10) are written as $[H_{k\zeta}^{(1)} + H_{k\zeta}^{(2)}]/2$. The integrands then split into sums of two terms $F_1 + F_2$. The term F_2, containing the functions $H_{k\zeta}^{(2)}$, should be integrated from left to right along the upper leg of the forked path, and the remaining term F_1 should be integrated along the lower leg. On the half-line $(\zeta_1, +\infty)$ need not be split up, and the integration is carried out in the usual way.

It is not difficult to see that the choice of the point ζ_1 (see Fig. 13.1) has no effect on the values of the integrals (13.1.9) and (13.1.10). If the observation point $M(r, \phi)$ at which the field is being calculated is located near a limiting ray, then the important part of the range of integration is located near the point $\zeta = a$. Within an accuracy of terms which are exponentially small as $k \to \infty$, we can replace the contour $\mathcal{L}_1$ by a "small" contour $\mathcal{L}_2$ located near $\zeta = a$ (cf. Fig. 13.2). The endpoints of $\mathcal{L}_2$ are located at points where the integrands are

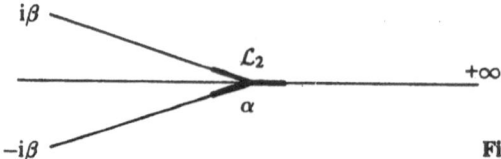

Fig. 13.2. Modified integration contour

exponentially small. The unusual form of this contour does not keep us from being able to asymptotically evaluate (13.1.9) and (13.1.10), and any convenient method (e.g., the saddle-point method or the residue theorem) can be used for this purpose.

In Sect. 3.2 it was shown that the field described by the function $J_{k\zeta}(kr)$ $\exp[ik\zeta(\pi/2 - \phi)]$ has a caustic $\mathcal{K}$ which is a circle of radius ζ centered at the origin. The function $H_{k\zeta}^{(1)}(kr)\exp[ik\zeta(\pi/2 - \phi)]$ is also naturally associated with the caustic $\mathcal{K}$. It describes waves leaving the caustic and satisfying the radiation condition.

Thus, asymptotic expressions for the incident and reflected waves can be obtained in the form of superpositions of solutions of the Helmholtz equation which have caustics. The superposition of solutions containing the Bessel function $J_{k\zeta}(kr)$ describes the incident wave, and that containing the Hankel function $H_{k\zeta}^{(1)}(kr)$ describes the reflected wave.

We now replace the functions $J_{k\zeta}(ka)$ and $H_{k\zeta}^{(1)}(ka)$ in (13.1.9) by Cherry's form (cf. [13.2] and also Sect. 7.2) of their asymptotic expansions. Then

$$\frac{J_{k\zeta}(ka)}{H_{k\zeta}^{(1)}(ka)} \sim \frac{iv(h)}{w_1(h)} \quad ; \quad h = T_a + \frac{P_1(z)}{(ka)^{4/3}} + \ldots + \frac{P_N(z)}{(ka)^{2N-2/3}} \qquad (13.1.11)$$

where $v(h)$ and $w_1(h)$ are Airy functions, and T_a and z are given by the relations

$$T_a = (k\zeta)^{2/3} z = \left[\frac{3}{2} ika \int\limits_{1}^{\zeta/a} \arccos x \, dx \right]^{2/3}$$

$$= 2\left(\frac{ka}{2} \right)^{2/3} \left(\frac{\zeta}{a} - 1 \right) \left[1 - \frac{1}{30}\left(\frac{\zeta}{a} - 1 \right) + \ldots \right] \quad .$$

The functions $P_j(z)$, $j = 1, \ldots, N$ are regular in a neighborhood of $z = 0$, and N is a sufficiently large positive integer. The functions $J_{k\zeta}(kr)$ and $H_{k\zeta}^{(1)}(kr)$ are replaced by Olver's form (cf. [13.3,4] as well as Sect. 3.2) of their asymptotic expressions. Having made these replacements, we change to a new variable of integration γ according to

$$k^{2/3}\gamma = T_a + \sum_{j=1}^{N} \frac{P_j(z)}{(ka)^{2j-2/3}} \quad . \qquad (13.1.12)$$

It is clear that

$$\zeta = a + \gamma \left(\frac{a}{2} \right)^{1/3} + O(\gamma^2 + k^{-2})$$

and that the integration in the plane of the new variable γ will be carried out over a forked contour $\mathcal{L}_3$ analogous to $\mathcal{L}_2$. The point $\zeta = a$ will correspond to $\gamma = 0$ (cf. Fig. 13.3). We can take the segments making up $\mathcal{L}_3$ to be straight lines, and take $|\gamma_1| = |\gamma_2| = \gamma_0$ where γ_0 is a small positive number. It is important

361

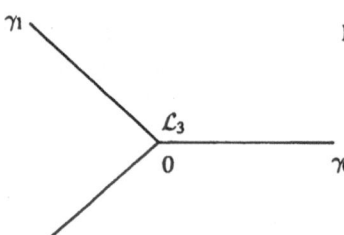

Fig. 13.3. Integration contour for (13.1.13)

to note that for all the forked contours occurring here, the endpoints lie in a region where the integrand is exponentially small as $k \to \infty$, and thus the exact locations of these endpoints do not affect the terms of order k^{-j} for any j in the asymptotic expansions of these integrals.

After the indicated transformations, we can obtain from (13.1.9) the following expression, accurate to any desired order as $k \to \infty$:

$$u \simeq k \int_{\mathcal{L}_3} \left[J_c(M, \gamma) - H_c(M, \gamma) \frac{v(k^{2/3}\gamma)}{w_1(k^{2/3}\gamma)} \right] e^{ik\xi_c(M,\gamma)}\, d\gamma \qquad (13.1.13)$$

where

$$J_c(M, \gamma) e^{ik\xi_c(M,\gamma)} = k^{-1/3} \left[\sum_{n=0}^{N} A_n^c(M, \gamma) k^{-n} v \left(- k^{2/3}\mu_c(M, \gamma) \right) \right.$$

$$\left. + ik^{-1/3} \sum_{n=0}^{N} B_n^c(M, \gamma) k^{-n} v' \left(- k^{2/3}\mu_c(M, \gamma) \right) \right]$$

$$\times e^{ik\xi_c(M,\gamma)} \qquad (13.1.14)$$

and

$$H_c(M, \gamma) e^{ik\xi_c(M,\gamma)} = k^{-1/3} \left[\sum_{n=0}^{N} A_n^c(M, \gamma) k^{-n} w_1 \left(- k^{2/3}\mu_c(M, \gamma) \right) \right.$$

$$\left. + ik^{-1/3} \sum_{n=0}^{N} B_n^c(M, \gamma) k^{-n} w_1' \left(- k^{2/3}\mu_c(M, \gamma) \right) \right]$$

$$\times e^{ik\xi_c(M,\gamma)} \qquad (13.1.15)$$

are the asymptotics of the functions

$$J_{k\zeta}(kr) \exp\left[ik\zeta \left(\frac{\pi}{2} - \phi \right) \right] \frac{d\zeta}{d\gamma}$$

and

$$-i H_{k\zeta}^{(1)}(kr) \exp\left[ik\zeta \left(\frac{\pi}{2} - \phi \right) \right] \frac{d\zeta}{d\gamma}$$

after the change of variable (13.1.12) has been made. The point $M = M(r, \phi)$ is the observation point. In (13.1.13–15), the index c indicates that the quantities involved pertain to the case of a circle. Evidently, the functions (13.1.14) and (13.1.15) will have caustics corresponding to those of the functions which they approximate, i.e., circles.

The functions $J_{k\zeta}(kr)\exp[ik\zeta(\pi/2 - \phi)]$ and $H_{k\zeta}^{(1)}(kr)\exp[ik\zeta(\pi/2 - \phi)]$ satisfy the Helmholtz equation. Therefore after our sequence of asymptotic approximation and change of variable, we will still have solutions of Helmholtz equation, albeit now merely a formal one. The finite sum in (13.1.14) satisfies the Helmholtz equation up to terms of order $O(k^{-N})$, and thus the coefficients A_n^c and B_n^c must satisfy the recurrence relations of Sect. 3.4.

The integrand of (13.1.13) satisfies the boundary condition $u|_S = 0$, since this is also true of (13.1.9).

In the following, we will use the term *caustic sum* to denote an expression obtained by replacing the function v by an arbitrary solution w of Airy's equation in the formal expansion (3.4.2) and then discarding terms of order $O(\omega^{-N-1})$ and higher. Expressions (13.1.14) and (13.1.15) are special cases of caustic sums. Taking as our point of departure eq. (13.1.13), which provides us with the asymptotic field in the penumbra for the etalon problem, we pass on to the general case.

Let a wave given by its ray expansion

$$u_{\text{inc}} \sim e^{i\omega\tau^{\text{inc}}(M)} \sum_{j=0}^{N} \frac{u_j^{\text{inc}}(M)}{(-i\omega)^j} \quad ; \quad M = M(x, y, z) \qquad (13.1.16)$$

be incident at a smooth surface S. The total field $u = u_{\text{inc}} + u_{\text{refl}}$ in the illuminated region can be found easily by the ray method. The rays tangent to S separate the shadow zone from the illuminated region (Fig. 13.4). A neighborhood Ω_p of these rays is the penumbra. The construction of the asymptotic field u in the penumbra and the deep shadow both is a considerably more difficult problem.

The foregoing considerations suggest that we seek the diffracted field in Ω_p in the form

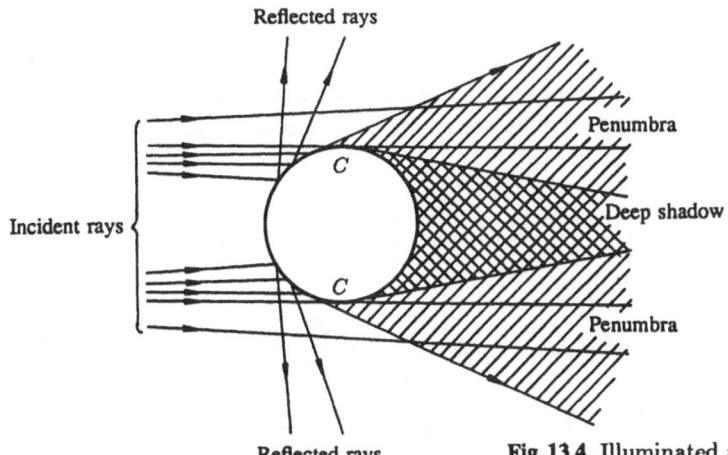

Fig. 13.4. Illuminated and shadow zones

$$u = u_{\text{inc}} + u_{\text{refl}}$$

$$\simeq \omega \int_{\mathcal{L}_3} \left[J(M,\gamma) - H(M,\gamma) \frac{v(\omega^{2/3}\gamma)}{w_1(\omega^{2/3}\gamma)} \right] e^{i\omega\xi(M,\gamma)} \, d\gamma \qquad (13.1.17)$$

where

$$J(M,\gamma) e^{i\omega\xi(M,\gamma)} = \omega^{-1/3} \left[v\left(-\omega^{2/3}\mu(M,\gamma) \right) \sum_{n=0}^{N} A_n(M,\gamma)\omega^{-n} \right.$$

$$\left. + i\omega^{-1/3} v'\left(-\omega^{2/3}\mu(M,\gamma) \right) \sum_{n=0}^{N} B_n(M,\gamma)\omega^{-n} \right] e^{i\omega\xi(M,\gamma)} \qquad (13.1.18)$$

and

$$H(M,\gamma) e^{i\omega\xi(M,\gamma)} = \omega^{-1/3} \left[w_1\left(-\omega^{2/3}\mu(M,\gamma) \right) \sum_{n=0}^{N} A_n(M,\gamma)\omega^{-n} \right.$$

$$\left. + i\omega^{-1/3} w_1'\left(-\omega^{2/3}\mu(M,\gamma) \right) \sum_{n=0}^{N} B_n(M,\gamma)\omega^{-n} \right] e^{i\omega\xi(M,\gamma)} \qquad (13.1.19)$$

are caustic sums analogous to those of (13.1.14) and (13.1.15). The contour $\mathcal{L}_3$ is shown in Fig. 13.3. The integration over $\mathcal{L}_3$ is carried out just the same way as the corresponding integration for the case of a circle: the function v appearing in the integrand is replaced by $v = (w_1 - w_2)/2i$, and the expressions (13.1.18) and (13.1.19) are split into two terms. The expressions containing w_2 are integrated from left to right along the upper branch $\gamma_1 0$ of the fork (Fig. 13.3), and the other term along the branch $\gamma_2 0$. On the segment $0\gamma_0$ the function v need not be so split up, and the integral can be carried out in the usual manner.

13.2 Construction of Approximate Caustic Sums – Equations for the Expansion Coefficients

Since we are only interested in the asymptotics of $u = u_{\text{inc}} + u_{\text{refl}}$ near the light-shadow boundary, the contour $\mathcal{L}_3$ can be reckoned to be located in a small neighborhood of the origin. It thus seems natural to seek ξ, μ, A_n and B_n as power series expansions in γ:

$$\xi(M,\gamma) \sim \sum_{j=0}^{\infty} l_j(M)\gamma^j \quad ; \quad \mu(M,\gamma) \sim \sum_{j=0}^{\infty} m_j(M)\gamma^j \qquad (13.2.1)$$

$$A_n(M,\gamma) \sim \sum_{j=0}^{\infty} A_{nj}(M)\gamma^j \quad ; \quad B_n(M,\gamma) \sim \sum_{j=0}^{\infty} B_{nj}(M)\gamma^j \quad . \qquad (13.2.2)$$

Now the series expansions (13.2.1) and (13.2.2) are in general divergent (cf. [13.5]). We will obtain approximate expressions for the caustic sums if ξ, μ, A_n

and B_n are represented by finite portions of the series in (13.2.1) and (13.2.2). These approximate expressions will satisfy the relationships of Sect. 3.4 up to an appropriate power of γ. It turns out that this sort of approximate construction for the caustic sums is entirely adequate for our purposes. Thus, the expressions

$$
\mathcal{J}(M,\gamma)\, e^{i\omega l(M,\gamma)} = \omega^{-1/3}\left[\sum_{n=0}^{N} \mathcal{A}_n(M,\gamma)\omega^{-n}v\left(-\omega^{2/3}m(M,\gamma)\right)\right.
$$

$$
\left. + i\omega^{-1/3}\sum_{n=0}^{N} \mathcal{B}_n(M,\gamma)\omega^{-n}v'\left(-\omega^{2/3}m(M,\gamma)\right)\right] e^{i\omega l(M,\gamma)} \qquad (13.2.3)
$$

and

$$
\mathcal{H}(M,\gamma)\, e^{i\omega l(M,\gamma)} = \omega^{-1/3}\left[\sum_{n=0}^{N} \mathcal{A}_n(M,\gamma)\omega^{-n}w_1\left(-\omega^{2/3}m(M,\gamma)\right)\right.
$$

$$
\left. + i\omega^{-1/3}\sum_{n=0}^{N} \mathcal{B}_n(M,\gamma)\omega^{-n}w_1'\left(-\omega^{2/3}m(M,\gamma)\right)\right] e^{i\omega l(M,\gamma)}
$$

$$
\qquad (13.2.4)
$$

in which l, m, $\mathcal{A}_n$ and $\mathcal{B}_n$ are partial sums of the series (13.2.1) and (13.2.2):

$$
l(M,\gamma) = \sum_{j=1}^{N_1} l_j(M)\gamma^j \quad ; \quad m(M,\gamma) = \sum_{j=0}^{N_1} m_j(M)\gamma^j \qquad (13.2.5)
$$

$$
\mathcal{A}_n(M,\gamma) = \sum_{j=0}^{N_1} \mathcal{A}_{nj}(M)\gamma^j \quad ; \quad \mathcal{B}_n(M,\gamma) = \sum_{j=0}^{N_1} \mathcal{B}_{nj}(M)\gamma^j \qquad (13.2.6)
$$

will be called *approximate caustic sums* (AC sums for short).

In what follows, the diffracted field u will be constructed in the form of an integral:

$$
u = u_{\mathrm{inc}} + u_{\mathrm{refl}} \simeq I = I_1 + I_2
$$

$$
= \omega \int_{\mathcal{L}_3}\left[\mathcal{J}(M,\gamma) - \mathcal{H}(M,\gamma)\frac{v(\omega^{2/3}\gamma)}{w_1(\omega^{2/3}\gamma)}\right] e^{i\omega l(M,\gamma)}\, d\gamma \qquad (13.2.7)
$$

which represents a superposition of AC sums like (13.2.3) and (13.2.4). In (13.2.7) the integrals

$$
I_1 = \omega \int_{\mathcal{L}_3} \mathcal{J}(M,\gamma)\, e^{i\omega l(M,\gamma)}\, d\gamma \qquad (13.2.8)
$$

and

$$
I_2 = -\omega \int_{\mathcal{L}_3} \mathcal{H}(M,\gamma)\frac{v(\omega^{2/3}\gamma)}{w_1(\omega^{2/3}\gamma)}\, e^{i\omega l(M,\gamma)}\, d\gamma \qquad (13.2.9)
$$

describe the incident and reflected waves respectively. The quantity γ_0 which determines the length of the branches of $\mathcal{L}_3$ will be taken to be a small positive number. In the following derivation it will come out that likewise the part of $\mathcal{L}_3$ lying outside the circle $|\gamma| \leq \text{const} \times \omega^{-\varepsilon}$, $0 < \varepsilon < \frac{1}{3}$, contributes only an exponentially small amount to the asymptotics of u_{inc} and $u_{\text{inc}} + u_{\text{refl}}$. In view of the fact that the AC sums formally satisfy the equation

$$\left(\Delta + \frac{\omega^2}{c^2} \right) u = 0$$

up to terms of order $O[(|\gamma|^{N_1} + \omega^{-N})\omega^{2/3}]$, the integral (13.2.7) can be made to satisfy this same equation up to any desired order in ω by choosing N_1 and N large enough.

The method of constructing the AC sums (13.2.3) and (13.2.4) is briefly as follows: the formal series (13.2.1) and (13.2.2) are substituted into (3.4.5, 8, 9), and we require that the coefficients of successive powers of ω^{-1} on both sides of these equations be identical. This leads to recurrence relations between l_j, m_j, $\mathcal{A}_{nj}$ and $\mathcal{B}_{nj}$. These functions are found using the aforementioned recurrence relations and certain additional reasonable requirements.

And so we will construct the AC sums. We require that the series (13.2.1) and (13.2.2) satisfy

$$(\nabla \xi)^2 + \mu(\nabla \mu)^2 = 1/c^2(M) \tag{13.2.10}$$

$$\nabla \xi \cdot \nabla \mu = 0 \tag{13.2.11}$$

$$2\nabla \xi \cdot \nabla A_n + A_n \Delta \xi + 2\mu \nabla \mu \cdot \nabla B_n + B_n \mu \Delta \mu + (\nabla \mu)^2 B_n$$
$$= \Delta A_{n-1} \quad . \tag{13.2.12}$$

$$2\nabla \mu \cdot \nabla A_n + A_n \Delta \mu + 2\nabla \xi \cdot \nabla B_n + B_n \Delta \xi = \Delta B_{n-1}$$
$$(n = 0, 1, 2, \ldots)$$
$$A_{-1} = B_{-1} = 0 \quad . \tag{13.2.13}$$

Equations (13.2.10) and (13.2.11) follow from (3.4.5), while the recurrence relations (13.2.12) and (13.2.13) are the result of (3.4.8) and (3.4.9). We shall further require that

$$\mu|_S = -\gamma \quad . \tag{13.2.14}$$

This condition arises because the boundary condition $I_1 + I_2|_S = 0$ must be satisfied. When (13.2.14) holds, we have

$$m_1|_S = -1 \quad ; \quad m_j|_S = 0 \quad (j \neq 1) \tag{13.2.15}$$

and consequently for arbitrary N_1 the partial sums m of the series $\mu(M, \gamma)$ satisfy

$$m|_S = -\gamma \quad . \tag{13.2.16}$$

366

The use of (13.2.16) and the condition $B_n(M, \gamma) = 0$ for $M \in S$ will guarantee that the integrand of (13.2.7) vanishes on S.

From the development of Chap. 3 it follows that the surface on which the argument of the Airy function v is equal to zero is a caustic of the corresponding wave. Thus for expression (13.2.3) the caustic is the surface described by $m(M, \gamma) = 0$. We can see from (13.2.4) that when $\gamma = 0$ the caustic coincides with S. For small γ the equation $m(M, \gamma) = 0$ describes a surface situated near S. It seems natural to call these surfaces approximate caustics, since the function $m(M, \gamma)$ will depend on the choice of N_1.

We now return to the construction of the coefficients in (13.2.1). We substitute (13.2.1) into (13.2.10, 11) and (13.2.14) and, setting the coefficient of each power of γ equal to zero, we obtain a recurrent system of equations for the coefficients l_j and m_j for $j \geq 0$;

$$(\nabla l_0)^2 + m_0(\nabla m_0)^2 = \frac{1}{c^2(M)} \tag{13.2.17}$$

$$\nabla l_0 \cdot \nabla m_0 = 0 \tag{13.2.18}$$

$$m_0|_S = 0 \tag{13.2.19}$$

$$2\nabla l_0 \cdot \nabla l_1 + 2m_0 \nabla m_0 \cdot \nabla m_1 + m_1(\nabla m_0)^2 = 0 \tag{13.2.20}$$

$$\nabla m_0 \cdot \nabla l_1 + \nabla m_1 \cdot \nabla l_0 = 0 \tag{13.2.21}$$

$$m_1|_S = -1 \tag{13.2.22}$$

and so on.

Consider (13.2.17–19) when $M \in S$. From (13.2.18) and (13.2.19) it follows that ∇m_0 is a vector normal to S, and that ∇l_0 lies in the plane tangent to S at M. From (13.2.17) we then have

$$(\nabla l_0)^2 = \frac{1}{c^2(M)} \quad ; \quad M \in S \quad . \tag{13.2.23}$$

Equation (13.2.23) is the eikonal equation for surface rays — those which are determined from Fermats principle

$$\delta \int \frac{ds}{c} = 0$$

with the variation taken only over curves which lie on the surface S. Equation (13.2.23) will be called the surface eikonal equation. The fundamental results from Sects. 2.2 and 2.3 are readily adapted to apply to the surface eikonal equation.

From here on (in this and in later sections) all our constructions will take place in a neighborhood of the terminator C, that is, the geometrical locus of the points of tangency between the incident rays and S. This is, in particular, related to the fact that the field of the surface rays determined by $l_0|_S$ can lose its regularity if we wander too far from C.

In some neighborhood of C, let a function $l_0(M)$ satisfying (13.2.23) be constructed over some field of surface rays on S. In what follows this field must be identical with that of the spatial rays of the incident wave, and initial values of the function $l_0(M)$ on C must be given (see Sect. 13.4).

Once the function $l_0(M)$ is constructed on S, it is not difficult to find $l_0(M)$ and $m_0(M)$ outside of S. Equations (13.2.17) and (13.2.18) are equivalent to the system of equations (cf. Sect. 3.4):

$$\left[\nabla \left(l_0 \pm \frac{2}{3} m_0^{3/2} \right) \right]^2 = \frac{1}{c^2(M)} \quad \text{or}$$

$$(\nabla \tau_0^\pm)^2 = \frac{1}{c^2(M)} \quad ; \quad \tau_0^\pm = l_0 \pm \frac{2}{3} m_0^{3/2} \quad . \tag{13.2.24}$$

Since $m_0|_S = 0$, it follows that $\tau_0^\pm$ satisfies the conditions

$$\nabla \tau_0^\pm(M) = \nabla l_0(M)$$
$$\tau_0^\pm(M) = l_0(M) \quad M \in S \tag{13.2.25}$$

on S.

Thus, $l_0(M)$ and $m_0(M)$ can be constructed outside S, if the eikonals $\tau_0^\pm(M)$ are known there. These eikonals are constructed along spatial rays (extremals of the functional $\int ds/c$) tangential to surface rays on S, in view of the first of conditions (13.2.25).

We now determine the coefficients $l_1(M)$ and $m_1(M)$. We consider first (13.2.20) on the surface S. Taking (13.2.22) into account, we have

$$2\nabla l_0 \cdot \nabla l_1 = (\nabla m_0)^2 \quad .$$

This is equivalent to the following ordinary differential equation along a surface ray:

$$\frac{2}{c} \frac{dl_1}{ds} = (\nabla m_0)^2 \tag{13.2.26}$$

where ds is the differential arc length along the surface ray. Since S is a caustic for rays corresponding to the eikonals $\tau_0^\pm = l_0 \pm \frac{2}{3} m_0^{3/2}$, we obtain from (3.3.21) that

$$(\nabla m_0)^2 = \left(\frac{2}{c^2 P(M)} \right)^{2/3} \quad M \in S \tag{13.2.27}$$

where $P(M)$ is the effective radius of curvature of a cross-section normal to S, measured along the surface ray at the point $M \in S$. Equation (13.2.26) must be supplemented with initial data. This data will be obtained in Sect. 13.4 by matching the asymptotics of the integral (13.2.7) to the ray expansion of the incident wave in the illuminated region.

Having determined l_1 on S, we construct the functions l_1 and m_1 outside of S. Multiplying (13.2.21) by $\pm 2\sqrt{m_0}$ and adding to (13.2.20), we have

$$2\nabla \left(l_0 \pm \tfrac{2}{3} m_0^{3/2} \right) \cdot \nabla (l_1 \pm m_1 \sqrt{m_0}) = 0 \tag{13.2.28}$$

from which it follows that, along a ray corresponding to the eikonal $l_0 \pm \tfrac{2}{3} m_0^{3/2}$, we have

$$l_1 \pm m_1 \sqrt{m_0} = \text{const} \quad . \tag{13.2.29}$$

From (13.2.29) it follows that l_1 and m_1 outside of S can be determined uniquely from $l_1|_S$.

In the same way, we find l_j and m_j for $j > l$. The functions $l_j|_S$ satisfy differential equations analogous to (13.2.26) along surface rays. The initial data for these equations are also obtained in Sect. 13.4 by matching the asymptotics of (13.2.7) to the ray expansion in the illuminated region.

We turn now to the determination of the coefficients of the formal expansions (13.2.6). We demand that the integrand in (13.2.7) vanish at the surface S. For $M \in S$, the function $m(M, \gamma)$ must take the value of $-\gamma$. Then from (13.2.3) and (13.2.4) it easily follows that the integrand vanishes on S if the coefficients in the expansions for B_n in (13.2.6) are equal to zero on S:

$$B_{nj}|_S = 0 \quad . \tag{13.2.30}$$

To construct $\mathcal{A}_0$ and $\mathcal{B}_0$ we return to (13.2.12) and (13.2.13), setting $n = 0$. Let $\gamma = 0$ and $M \in S$: then (13.2.12) reduces to

$$2\nabla l_0 \cdot \nabla \mathcal{A}_{00} + \mathcal{A}_{00} \Delta l_0 = 0 \tag{13.2.31}$$

which is a linear differential equation along a surface ray.

The functions $\mathcal{A}_{00}(M)$ and $\mathcal{B}_{00}(M)$ outside S satisfy the same equations as $\mathcal{A}_0$ and $\mathcal{B}_0$ in Sect. 3.4, and can be constructed using (3.5.6). Given the function $\mathcal{A}_{00}$ on S, we uniquely determine χ_{00} that appears in these formulas, and as a result, we uniquely determine $\mathcal{A}_{00}(M)$ and $\mathcal{B}_{00}(M)$ outside S. Differentiating (13.2.12) and (13.2.13) with respect to γ for $n = 0$, and putting $\gamma = 0$, we find in a similar fashion $\mathcal{A}_{0j}$ and $\mathcal{B}_{0j}$ for $j \geq 0$, and so on.

The initial data for (13.2.31) and its analogs for $\mathcal{A}_{nj}$ are (as in the case of (13.2.23) and (13.2.26)) found by requiring that (13.2.7) be asymptotically equivalent to the ray expansion of the incident and reflected waves as we move from the penumbra to the illuminated region.

Let us emphasize in concluding this section that from condition (13.2.30) and the equations

$$m_0|_S = 0 \quad ; \quad m_1|_S = -1 \quad ; \quad m_j|_S = 0 \quad , \quad (j \geq 2)$$

follow the fact that the integral (13.2.7) must satisfy the boundary conditions $I|_S = 0$ exactly.

13.3 Asymptotic Evaluation of the Integral I_1, in the Vicinity of the Terminator C

As we have already noted, our problem is to construct expressions for l_j, m_j, $\mathcal{A}_{nj}$ and $\mathcal{B}_{nj}$ such that the asymptotic expression of the integral (13.2.7) in the illuminated region matches the formulas of the ray method. The integral I_1 in (13.2.8) should thus have the expansion (13.1.16), and the integral I_2 of (13.2.9) will have the ray expansion of the reflected wave. Since the ray expansion of the reflected wave is determined uniquely from that of the incident wave (Sect. 2.6), it is natural to expect that it will be sufficient to match only the asymptotic expansion of the incident wave (13.1.16) and I_1. The matching of the asymptotic expansion of I_2 with the ray expansion of the reflected wave will then be automatic. We will show that I_1 has an asymptotic expansion of the ray type. Naturally, we try to find it by the method of steepest descent or stationary phase, replacing the Airy function v by its asymptotic representation (cf., for instance *Fedoryuk* [13.6] or *Bleistein* and *Handelsman* [13.7]).

But this route is fraught with difficulties. First of all, we have to justify the application of the stationary phase method to the forked contour $\mathcal{L}_3$; secondly, the stationary point might show up in a region where the argument of the Airy function is $O(1)$, and hence for which the use of the asymptotic representation of the Airy function is not allowed. To overcome the first of these difficulties, we deform the forked contour $\mathcal{L}_3$ into a new forked contour $\mathcal{L}_4$ (Fig. 13.5). We choose a point $-\gamma_0/2$ on the real axis and deform the upper leg of the forked path into the broken line going from 0 to $-\gamma_0/2$ to γ_1, and the lower leg into the broken line from 0 to $-\gamma_0/2$ to γ_2. On the segment $[-\gamma_0/2, 0]$, both integrals are absorbed into the integral of the function $\mathcal{J}(M, \gamma)$.[1] We introduce a neutralizing function $\Upsilon(\gamma)$ which is infinitely differentiable on $-\infty < \gamma < +\infty$, identically equal to one for $|\gamma| \le \gamma_0/4$ (γ_0 was the length of one of the legs of contour $\mathcal{L}_3$), equal to zero for $|\gamma| > \gamma_0/2$, and equal to zero on the upper and lower legs of $\mathcal{L}_4$. Carrying out the indicated deformation of $\mathcal{L}_3$ and using the neutralizing function allows us to express the integral I_1 as

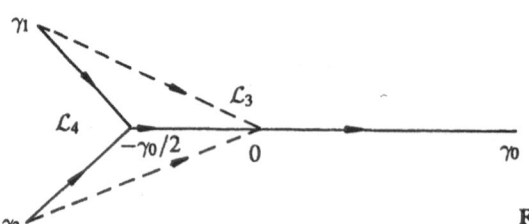

Fig. 13.5. Forked contour for I_1

---

[1] This reasoning demonstrates that the value of the integral along the forked contour does not depend on the choice of the point where the legs join.

$$I_1 = \omega \int_{-\gamma_0/2}^{\gamma_0/2} \mathcal{J}(M,\gamma) e^{i\omega l(M,\gamma)} \Upsilon(\gamma) d\gamma$$

$$+ \omega \int_{\mathcal{L}_4} \mathcal{J}(M,\gamma) e^{i\omega l(M,\gamma)} [1 - \Upsilon(\gamma)] d\gamma \quad .$$

(On the legs $[\gamma_1, -\gamma_0/2]$ and $[\gamma_2, -\gamma_0/2]$ of $\mathcal{L}_4$ – Fig. 13.5 – we put $\Upsilon(\gamma) = 0$). Integrating by parts and taking into account the exponentially small value of the integrand at the finite points γ_0, γ_1 and γ_2, we can state that

$$\omega \int_{\mathcal{L}_4} \mathcal{J}(M,\gamma) e^{i\omega l(M,\gamma)} [1 - \Upsilon(\gamma)] d\gamma = O(\omega^{-\tilde{N}})$$

where $\tilde{N}$ is an arbitrary positive number.

To asymptotically evaluate the integral over the interval $(-\gamma_0/2, \gamma_0/2)$, we must deal with the difficulty mentioned earlier: the formally calculated stationary point turns up in a region where the asymptotic expansion of v appearing in (13.1.18) is not valid. As we will see shortly this difficulty can be overcome by replacing the Airy function by its integral representation (see Appendix A.1):

$$v(-\omega^{2/3} m) = \frac{\omega^{1/3}}{2\sqrt{\pi}} \int_{-\infty}^{\infty} e^{i\omega(m\nu - \nu^3/3)} d\nu$$

and passing to a double integral. This double integral has the form

$$I_1 = \frac{\omega}{2\sqrt{\pi}} \int_{-\gamma_0/2}^{\gamma_0/2} \int_{-\infty}^{\infty} e^{i\omega(l+m\nu-\nu^3/3)} (A + \nu B) \Upsilon(\gamma) d\gamma \, d\nu \qquad (13.3.1)$$

where

$$A = \sum_{n=0}^{N} \frac{A_n}{\omega^n} \quad ; \quad B = \sum_{n=0}^{\infty} \frac{B_n}{\omega^n} \quad .$$

The asymptotic expansion of the double integral (13.3.1) is carried out using the method of stationary phase for multiple integrals (see [13.6] or [13.7]). The equations determining the stationary point are

$$\frac{\partial \Phi}{\partial \gamma} = \frac{\partial l}{\partial \gamma} + \frac{\partial m}{\partial \gamma} \nu = 0 \quad ; \quad \frac{\partial \Phi}{\partial \nu} = m - \nu^2 = 0 \qquad (13.3.2)$$

where $\Phi = l + m\nu - \nu^3/3$. We denote the solution of (13.3.2), i.e., the coordinates of the stationary point in the (γ, ν) plane, by $\gamma_c(M)$ and $\nu_c(M)$. If the Hessian

$$\Delta = \begin{vmatrix} \Phi_{\gamma\gamma} & \Phi_{\gamma\nu} \\ \Phi_{\nu\gamma} & \Phi_{\nu\nu} \end{vmatrix}$$

is nonzero at the stationary point $(\gamma_c, \nu_c)^2$, then the integral I_1 has the asymptotic expansion

$$I_1 \sim e^{i\omega\tau^{(1)}(M)} \sum_{j=0}^{\infty} \frac{u_j^{(1)}(M)}{(-i\omega)^j} \qquad (13.3.3)$$

where

$$\tau^{(1)}(M) = l(M, \gamma_c) + m(M, \gamma_c)\nu_c - \nu_c^3/3 \quad . \qquad (13.3.4)$$

Expression (13.3.3) has the form of a ray expansion, and consequently, the functions $\tau^{(1)}(M)$ and $u_j^{(1)}(M)$ must satisfy the eikonal equation and the transport equations respectively. Expansion (13.3.3) must also be compatible with expansion (13.1.16) of the incident wave. This latter requirement allows us to determine the initial data for the equations, from which l_0 and l_j for $j \geq 1$ can be found. For our purposes it will be sufficient to make τ^{inc} and $\tau^{(1)}$ coincide only in a neighborhood of the surface $\sum$ formed by rays tangent to S at points of C. We now proceed to carry out the necessary constructions.

We assume that $\sum$ is smooth. In the neighborhood of $\sum$ we introduce a coordinate system $M(\alpha, \beta, n)$: the quantity n is the distance from M to $\sum$, and α and β are parameters specifying the location of the orthogonal projection M' of M onto $\sum$. For α we choose the parameter α of the rays forming $\sum$, while for β we choose the eikonal τ^{inc} along these rays. For points M lying on the same side of $\sum$ as the surface S, we take $n < 0$; on the other side $n > 0$. The coordinates (α, β, n) are regular in a neighborhood of $\sum$, and are orthogonal at $\sum$. We note that (α, β, n) are analogous to the coordinates introduced in Chap. 3 in the neighborhood of a caustic.

We expand τ^{inc}, u_j^{inc} and $\tau^{(1)}$, $u_j^{(1)}$ in powers of n in a neighborhood of $\sum$:

$$\tau^{\text{inc}}(M) = \sum_{r=0}^{N} \frac{1}{r!} \tau_r^{\text{inc}}(M')n^r + O(n^{N+1}) \qquad (13.3.5)$$

$$u_j^{\text{inc}}(M) = \sum_{r=0}^{N} \frac{1}{r!} u_{jr}^{\text{inc}}(M')n^r + O(n^{N+1}) \qquad (13.3.6)$$

$$\tau^{(1)}(M) = \sum_{r=0}^{N} \frac{1}{r!} \tau_r^{(1)}(M')n^r + O(n^{N+1}) \qquad (13.3.7)$$

$$u_j^{(1)}(M) = \sum_{r=0}^{N} \frac{1}{r!} u_{jr}^{(1)}(M')n^r + O(n^{N+1}) \quad . \qquad (13.3.8)$$

Here and in what follows, M' is the projection of M on $\sum$. We require that

---

[2] The calculation of the Hessian at the stationary point and the verification that $\Delta \neq 0$ will be carried out in Sect. 13.4.

$$\tau_r^{\text{inc}}(M') = \tau_r^{(1)}(M') \equiv \tau_r(M') \quad , \quad \text{and}$$
$$u_{jr}^{\text{inc}}(M') = u_{jr}^{(1)}(M') \equiv u_{jr}(M') \tag{13.3.9}$$

Having enunciated these conditions, we will in fact regard then as being satisfied. In (α, β, n) coordinates, by the very manner in which they were defined, we have

$$\left.\frac{\partial \tau^{\text{inc}}}{\partial \alpha}\right|_{n=0} = \left.\frac{\partial \tau^{\text{inc}}}{\partial n}\right|_{n=0} = 0 \quad \text{and}$$
$$\left.\frac{\partial \tau^{\text{inc}}}{\partial \beta}\right|_{n=0} = 1 \tag{13.3.10}$$

From (13.3.9) and (13.3.10) it follows that

$$\frac{\partial \tau_0(M')}{\partial \alpha} = 0 \quad ; \quad \tau_1(M') = 0 \quad ; \quad \frac{\partial \tau_0(M')}{\partial \beta} = 1 \quad . \tag{13.3.11}$$

We next obtain the coefficients $\tau_r(M')$ and $u_{jr}(M')$ of the expansions (13.3.5–8). We write the eikonal equation, which must be satisfied by $\tau^{\text{inc}}(M)$ and $\tau^{(1)}(M)$ alike, in (α, β, n) coordinates:

$$(\nabla \tau)^2 = g^{\alpha\alpha} \left(\frac{\partial \tau}{\partial \alpha}\right)^2 + 2g^{\alpha\beta} \frac{\partial \tau}{\partial \alpha} \frac{\partial \tau}{\partial \beta} + g^{\beta\beta} \left(\frac{\partial \tau}{\partial \beta}\right)^2 + \left(\frac{\partial \tau}{\partial n}\right)^2$$
$$= \frac{1}{c^2(M)} \tag{13.3.12}$$

where

$$\left.g^{\alpha\alpha}\right|_{n=0} > 0 \quad ; \quad g^{\alpha\beta} = O(n) \quad ;$$
$$g^{\beta\beta} = \frac{1}{c^2(\alpha, \beta)} - \frac{2n}{\varrho(\alpha, \beta)c^2(\alpha, \beta)} + O(n^2)$$

and $1/\varrho(\alpha, \beta)$ is the curvature of a cross-section normal to $\sum$, measured along the ray $\alpha = \text{const}$ (see Appendix A.2 and Sect. 3.3). Substituting the expansion

$$\tau(M) = \sum_{r=0}^{N} \frac{1}{r!} \tau_r(M') n^r + O(n^{N+1})$$

into (13.3.12), taking (13.3.11) into account, and setting the coefficients of n and n^2 equal to 0, we obtain

$$\frac{2}{c^2(\alpha, \beta)} \left[\frac{c_n(\alpha, \beta)}{c(\alpha, \beta)} - \frac{1}{\varrho(\alpha, \beta)}\right] = 0 \tag{13.3.13}$$

$$\frac{1}{c^2(\alpha, \beta)} \frac{\partial \tau_2}{\partial \beta} + \tau_2^2 + \Psi(\alpha, \beta) = 0 \tag{13.3.14}$$

where $c_n(\alpha, \beta) = \partial c(M)/\partial n|_{n=0}$ and

$$\Psi(\alpha, \beta) = \tfrac{1}{2} \left(\frac{\partial^2 g^{\beta\beta}}{\partial n^2} - \frac{\partial^2 (1/c^2)}{\partial n^2} \right)_{n=0} .$$

Equation (13.3.13) is merely the identity $0 = 0$, since the effective radius of curvature $P = (1/\varrho - c_n/c)^{-1}$ of the surface $\sum$ along the ray $\alpha = $ const is infinite. This is because the surface was formed by rays. Equation (13.3.14) is a Riccati equation for $\tau_2(M')$ along a ray $\alpha = $ const. Equating the coefficients of n^j, for $j \geq 3$, to zero leads once more to linear first-order differential equations for $\tau_j(M')$ along the ray $\alpha = $ const.

Now consider the transport equations:

$$2\nabla\tau \cdot \nabla u_j + u_j \Delta\tau \equiv 2 \left[g^{\alpha\alpha} \frac{\partial\tau}{\partial\alpha} \frac{\partial u_j}{\partial\alpha} + g^{\alpha\beta} \frac{\partial\tau}{\partial\alpha} \frac{\partial u_j}{\partial\beta} + g^{\alpha\beta} \frac{\partial\tau}{\partial\beta} \frac{\partial u_j}{\partial\alpha} \right.$$
$$\left. + g^{\beta\beta} \frac{\partial\tau}{\partial\beta} \frac{\partial u_j}{\partial\beta} + \frac{\partial\tau}{\partial n} \frac{\partial u_j}{\partial n} \right] + u_j \Delta\tau$$
$$= \Delta u_{j-1} \quad (j = 0, 1, \ldots) \quad (u_{-1} \equiv 0) . \qquad (13.3.15)$$

Substituting the expansion

$$u_j(M) = \sum_{r=0}^{N} \frac{1}{r!} u_{jr}(M') n^r + O(n^{N+1})$$

into (13.3.15) and equating coefficients of various powers of n, we obtain a linear differential equation of first order for $u_{jr}(M')$ along the ray $\alpha = $ const.

Since the coefficients $\tau_r(M')$ and $u_{jr}(M')$ in (13.3.5–8) satisfy first-order differential equations along $\alpha = $ const, they are completely determined by their values at one point of this ray. Thus, we come to the following assertion which will be very important in what follows:

Equations (13.3.9) will be satisfied on $\sum$ if and only if they are satisfied on the terminator C, i.e.,

$$\tau_r^{\text{inc}}(M')|_C = \tau_r^{(1)}(M')|_C \; ; \quad (r = 0, 1, 2, \ldots) \qquad (13.3.16)$$

$$u_{jr}^{\text{inc}}(M')|_C = u_{jr}^{(1)}(M')|_C \; ; \quad \begin{pmatrix} r = 0, 1, 2, \cdots \\ j = 0, 1, 2, \cdots \end{pmatrix} . \qquad (13.3.17)$$

Equations (13.3.16) and (13.3.17), which guarantee that the asymptotic expansion (13.3.3) of I_1 is the same as the ray expansion of the incident wave (13.1.16) in a neighborhood of $\sum$, also serve to determine the initial data which must be appended to the equations for $l_j(M)|_S$; $j \geq 0$ and $\mathcal{A}_{nj}(M)|_S$; $n, j \geq 0$.

The equation for $l_0(M)|_S$ is nonlinear in $\nabla l_0(M)|_S$. To determine $l_0(M)|_S$ uniquely, we must specify not only the value of $l_0(M)|_C$, but also the value of $\nabla l_0(M)|_C$. We thus supplement (13.3.16) and (13.3.17) by the equation

$$\nabla\tau^{\text{inc}}(M)|_C = \nabla\tau^{(1)}(M)|_C \qquad (13.3.18)$$

which follows from $\tau^{\text{inc}}(M) = \tau^{(1)}(M)$.

13.4 Choice of the Initial Data; Fock's Formula

The initial data for the equations considered in Sect. 13.2 that determine the functions l, m, A_n and B_n appearing in the integrals I_1 and I_2 must be such that on C equations (13.3.16–18) are satisfied. Having noted that

$$\tau_0^{\mathrm{inc}}(M^1)|_C = \tau^{\mathrm{inc}}(M)|_C \quad \text{and}$$

$$\tau_0^{(1)}(M)|_C = l(M, \gamma_c) + m(M, \gamma_c)\nu_c - \tfrac{1}{3}\nu_c^3|_C$$

we write (13.3.16) for $r = 0$ together with the equations for the stationary point (γ_c, ν_c):

$$\tau^{\mathrm{inc}}(M)|_C = l(M, \gamma_c) + m(M, \gamma_c)\nu_c - \tfrac{1}{3}\nu_c^3|_C \tag{13.4.1}$$

$$\frac{\partial l(M, \gamma_c)}{\partial \gamma} + \frac{\partial m(M, \gamma_c)}{\partial \gamma}\nu_c \bigg|_C = 0 \tag{13.4.2}$$

$$m(M, \gamma_c) - \nu_c^2|_C = 0 \quad . \tag{13.4.3}$$

We supplement (13.4.1–3) by (13.2.16), taken on C:

$$m(M, \gamma_c)|_C = -\gamma_c|_C \quad . \tag{13.4.4}$$

From the system (13.4.1–4) we must find the two functions l and m and the two numbers γ_c and ν_c. It is easy to see that this system will be satisfied if we put[3]

$$\gamma_c = \nu_c = 0 \quad , \quad m(M, \gamma_c) = 0 \quad ,$$
$$l(M, \gamma_c) = \tau^{\mathrm{inc}}(M)|_C$$

and from the last of these it follows that

$$l_0(M)|_C \equiv l(M, 0)|_C = \tau^{\mathrm{inc}}(M)|_C \quad . \tag{13.4.5}$$

Now we turn to (13.3.18). Since the derivatives $\partial l/\partial \gamma$ and $\partial m/\partial \gamma$ satisfy (13.4.2), we have

$$\nabla \tau^{(1)}(M)|_C = \nabla l + \nabla m \nu_c - \nu_c^2 \nabla \nu_c|_C = \nabla l(M, 0)|_C$$
$$= \nabla l_0(M)|_C$$

and consequently

$$\nabla l_0(M)|_C = \nabla \tau^{\mathrm{inc}}(M)|_C \quad . \tag{13.4.6}$$

Moreover, from (13.4.2) there also results

---

[3] As a guide in our search for solutions of (13.4.1–4), we can make use of the following consideration: in the etalon problem for the circle, when the observation point lies on a limiting ray the stationary point of the integrand in (13.1.14) lies at the origin.

$$\left.\frac{\partial l(M,\gamma_c)}{\partial\gamma}\right|_C = 0$$

which means that

$$l_1(M)|_C \equiv \left\{\left.\frac{\partial l(M,\gamma)}{\partial\gamma}\right|_{\gamma=0}\right\}\Bigg|_C = 0 \quad . \tag{13.4.7}$$

Equations (13.4.5–7) give the initial conditions for (13.2.23) and (13.2.26). The functions $l_0(M)$ and $l_1(M)$ are uniquely determined on and outside of S. All the foregoing calculations make sense if at the stationary point,

$$\Delta = \begin{vmatrix} \Phi_{\gamma\gamma} & \Phi_{\gamma\nu} \\ \Phi_{\nu\gamma} & \Phi_{\nu\nu} \end{vmatrix}_{M\in C} \neq 0 \quad ; \quad \Phi \equiv l + m\nu - \tfrac{1}{3}\nu^3$$

is satisfied, which is the condition under which the stationary phase method for I_1 is applicable. The determinant Δ can easily be calculated with the help of (13.2.15):

$$\Delta = \begin{vmatrix} \Phi_{\gamma\gamma}|_{M\in C} & -1 \\ -1 & 0 \end{vmatrix} = -1 \quad .$$

Thus, the stationary phase method for I_1 is in fact valid. From the condition $\Delta \neq 0$ it also follows that (13.4.2) and (13.4.3) have a unique solution for points M lying in a sufficiently small neighborhood of the curve C, no matter what $\tau^{\mathrm{inc}}(M)$ is.

We next find the initial conditions for the equation which determines

$$l_2(M)|_S = \frac{1}{2}\left\{\left.\frac{\partial^2 l(M,\gamma)}{\partial\gamma^2}\right|_{\gamma=0}\right\}_S \quad .$$

We evaluate the quantity $\tau_2^{(1)}(M^1)|_C$ appearing in (13.3.16) for $r = 2$. Differentiating (13.3.4) with respect to n, and taking (13.4.2) into account, we obtain:

$$\frac{\partial\tau^{(1)}(M)}{\partial n} = \frac{\partial l(M,\gamma_c)}{\partial n} + \frac{\partial m(M,\gamma_c)}{\partial n}\nu_c \quad .$$

A second differentiation evaluated at a point M lying on C gives

$$\left.\frac{\partial^2\tau^{(1)}(M)}{\partial n^2}\right|_C = \left.\frac{\partial^2 l(M,0)}{\partial n^2}\right|_C + \left.\frac{\partial^2 l(M,0)}{\partial n\partial\gamma}\frac{\partial\gamma_c}{\partial n}\right|_C$$
$$+ \left.\frac{\partial m(M,0)}{\partial n}\frac{\partial\nu_c}{\partial n}\right|_C \quad . \tag{13.4.8}$$

We can rewrite the terms on the right side of (13.4.8). Since $\nabla l \cdot \nabla m = 0$ and $\nabla m|_C$ is directed normally to the surface $\sum$, then $\partial l(M,\gamma)/\partial n|_C = 0$, and so

$$\left.\frac{\partial^2 l(M,0)}{\partial n\partial\gamma}\right|_C = 0 \quad . \tag{13.4.9}$$

We differentiate (13.4.2) for the stationary point with respect to n. After that, we take $M \in C$ and obtain, in view of (13.4.9) and

$$\left.\frac{\partial m(M,0)}{\partial \gamma}\right|_C = -1$$

the equations

$$\frac{\partial^2 l(M,0)}{\partial \gamma^2}\left.\frac{\partial \gamma_c}{\partial n}\right|_C - \left.\frac{\partial \nu_c}{\partial n}\right|_C = 0 \quad ,$$

$$\left.\frac{\partial m(M,0)}{\partial n}\right|_C - \left.\frac{\partial \gamma_c}{\partial n}\right|_C = 0 \quad .$$

Hence,

$$\left.\frac{\partial \nu_c}{\partial n}\right|_C = \frac{\partial^2 l(M,0)}{\partial \gamma^2}\left.\frac{\partial m(M,0)}{\partial n}\right|_C \quad . \tag{13.4.10}$$

Finally, we differentiate the equality

$$\nabla l_0 \cdot \nabla m_0 = \frac{\partial l_0}{\partial n}\frac{\partial m_0}{\partial n} + g^{\alpha\alpha}\frac{\partial l_0}{\partial \alpha}\frac{\partial m_0}{\partial \alpha} + g^{\alpha\beta}\left(\frac{\partial l_0}{\partial \alpha}\frac{\partial m_0}{\partial \beta} + \frac{\partial l_0}{\partial \beta}\frac{\partial m_0}{\partial \alpha}\right)$$

$$+ g^{\beta\beta}\frac{\partial l_0}{\partial \beta}\frac{\partial m_0}{\partial \beta} = 0$$

by n and take $M \in C$. Since

$$\nabla l_0(M)|_C = \nabla \tau^{\text{inc}}(M)|_C$$

we have

$$\left.\frac{\partial l_0}{\partial \beta}\right|_C = 1 \quad ; \quad \left.\frac{\partial l_0}{\partial \alpha}\right|_C = \left.\frac{\partial l_0}{\partial n}\right|_C = 0 \quad .$$

Moreover, $\partial m_0/\partial \alpha|_C = \partial m_0/\partial \beta|_C = 0$, since $\nabla m_0|_C$ is directed normally to Σ. In view of these relations, we get

$$\frac{\partial^2 l_0}{\partial n^2}\left.\frac{\partial m_0}{\partial n}\right|_C + \frac{1}{c^2(M)}\left.\frac{\partial^2 m_0}{\partial \beta \partial n}\right|_C = 0 \quad . \tag{13.4.11}$$

Equations (13.4.9–11) allow formula (13.4.8) to be written in the form

$$\tau_2^{(1)}(M)|_C = -\frac{1}{c^2(M)}\frac{\partial}{\partial \beta}\left.\left(\ln\frac{\partial m_0}{\partial n}\right)\right|_C + \frac{\partial^2 l(M,0)}{\partial \gamma^2}\left.\left(\frac{\partial m_0}{\partial n}\right)^2\right|_C \quad .$$

Here $(\partial m_0/\partial n)^2|_C$ can be replaced by the right side of (13.2.27), since on C it is equal to $(\nabla m_0)^2$. Substituting this value of $\tau_2^{(1)}(M)|_C$ into (13.3.16), we obtain

$$\tau_2^{\text{inc}}(M)|_C = \frac{1}{3c^2}\left.\frac{\partial(c^2 P_0)/\partial \beta}{c^2 P_0}\right|_C + 2\left(\frac{2}{c^2 P_0}\right)^{2/3} l_2(M)|_C \tag{13.4.12}$$

where P_0 is the effective radius of curvature of the surface S at the point $M \in C$, measured along an incident ray tangent to S at M. From (13.4.12) we can uniquely determine $l_2(M)|_C$.

For $r \geq 3$, the values of $l_r(M)|_C$; i.e., the initial data for the equations which determine $l_r(M)$, are obtained in an analogous fashion.

From (13.3.17), the initial data for (13.2.31) and its analogs for $\mathcal{A}_{nj}|_S$ are determined uniquely. We present here only the value of

$$\mathcal{A}_{00}(M)|_C = \frac{1}{\sqrt{\pi}} u_0^{\text{inc}}(M)|_C \tag{13.4.13}$$

which serves as the initial data for the equation determining the principal approximation. Thus, all the functions appearing in the caustic sums are determined in a neighborhood of the surface $\sum$ with the help of the differential equations and the given initial conditions. By the same token, integrands are constructed for I_1 and I_2 – completely and uniquely in a neighborhood of $\sum$.

Let us remind the reader of an important point. The integrands of I_1 and I_2 are expressed in terms of solutions of differential equations, and can be constructed only so far as we can construct the solutions to these differential equations. We have only guaranteed the existence and uniqueness of solutions to the differential equations for l, m, A_n and B_n in a sufficiently small neighborhood of C, and therefore the existence and uniqueness of the caustic sums can only be guaranteed in this neighborhood as well. In the next section the integrands of I_1 and I_2 will be manipulated into new forms. It turns out that I_1 and I_2 will be superpositions of ray sums. This will allow us to calculate the field in a neighborhood of the surface $\sum^+$ at finite distances from C. Here $\sum^+$ is that part of $\sum$ formed by the extensions of incident rays tangent to S from the terminator C; $\sum^+$ is evidently the boundary between the illuminated and shadow regions.

The integral I_1, as follows from our constructions is the incident wave near C. We will show that in the illuminated region near C the integral I_2 describes the reflected wave. We consider I_2 in the portion of a neighborhood of C which is covered by rays reflected from S at points $M \in S$ which are at distances d from C, such that

$$(\text{const})\omega^{-\varepsilon} > d > (\text{const})\omega^{-\varepsilon_1} \quad ; \quad 0 < \varepsilon < \varepsilon_1 < \tfrac{1}{3} \quad .$$

In this region the Airy functions in I_2 can be replaced by their asymptotic expansions, and the integral I_2 calculated by the method of stationary phase. Performing the indicated calculations, we obtain an asymptotic expansion for I_2 which has the form of a ray expansion. Because the phase function and the coefficients of this expansion (a) satisfy the eikonal and transport equations of the ray method, and (b) are such that the boundary condition $I_1 + I_2|_S = 0$ is satisfied, it follows that this expansion must be the ray expansion of the reflected wave. Thus, outside of some neighborhood of the light-shadow boundary, the asymptotic value of $I_1 + I_2$ on the illuminated side is just the asymptotic value for the field that would have been obtained by the ordinary ray method; i.e., the geometric-optics expansion of the field. We will omit the corresponding computations.

378

Starting with the formulas for I_1 and I_2 we can obtain expressions for the field in various physically interesting regions. Here we will consider a small neighborhood of the terminator C. Returning to (13.2.17–19), we replace all the functions appearing in (13.1.17) by their first approximations, reckoning γ and the distance to C to be small quantities. We will characterize surface rays by the parameter α_1, and a point on a ray by the value of the eikonal,

$$\beta_1 = \tau^{\text{inc}}(M_0) + \int_{M_0}^{M} \frac{ds}{c} \quad ; \quad M \in S \quad ; \quad M_0 \in C \quad .$$

Thus, a coordinate system (α_1, β_1) is defined on S near C, analogous to (α, β) on $\sum$; see Sect. 13.3. For the third coordinate we choose the normal distance n_1 from S.

Taking formulas (13.2.17–19), (13.2.22) and (13.4.7, 9, 13) into account, we find that near an arbitrary point $M_0(\alpha_1, \beta_1) \in C$,

$$u_{\text{inc}} + u_{\text{refl}} \simeq \frac{u_0(M_0)}{\sqrt{\pi}} e^{i\omega\beta_1} \int_{\mathcal{L}_3} \exp\left[\frac{i\omega c_0^2}{2}\left(\frac{2}{c_0^2 P_0}\right)^{2/3}(\beta_1 - \beta_1^0)\gamma\right]$$

$$\times \left\{ v\left[\omega^{2/3}\left(\gamma - \sqrt[3]{\frac{2}{c_0^2 P_0}}n_1\right)\right] - \frac{v(\omega^{2/3}\gamma)}{w_1(\omega^{2/3}\gamma)} \right.$$

$$\left. \times w_1\left[\omega^{2/3}\left(\gamma - \sqrt[3]{\frac{2}{c_0^2 P_0}}n_1\right)\right] \right\} d\gamma \quad . \qquad (13.4.14)$$

Here

$$M_0 \in C \quad ; \quad \beta_1^0 = \beta_1(M_0) \quad ; \quad c_0 = c(M_0) \quad ;$$
$$P_0^{-1} = \frac{1}{\varrho_0(M_0)} - \frac{c_{n_1}(M_0)}{c(M_0)} \quad ,$$

$1/\varrho_0$ is the curvature of a normal cross-section of S, measured along the ray tangent to S at M_0; and $\mathcal{L}_3$ is the same contour as in Sect. 13.1. The endpoints of $\mathcal{L}_3$ can be extended to infinity, with only an exponentially small error. We locate the left leg of $\mathcal{L}_3$ (now infinitely long) on the negative real axis, without affecting the value of the integral. Making the change of variable $\omega^{2/3}\gamma = \zeta$ in (13.4.14) and introducing the notation

$$\sigma = \left(\frac{\omega}{2}\right)^{1/3}\left(\frac{2c_0}{P_0}\right)^{2/3}(\beta_1 - \beta_1^0) \quad ; \quad \nu = \omega^{2/3}\left(\frac{2}{c_0^2 P_0}\right)^{1/3} \qquad (13.4.15)$$

we obtain the formula

$$u_{\text{inc}} + u_{\text{refl}} \simeq \frac{u_0(M_0)}{\sqrt{\pi}} e^{i\omega\beta_1}$$

$$\times \int_{-\infty}^{\infty} e^{i\sigma\zeta}\left[v(\zeta - \nu) - \frac{v(\zeta)}{w_1(\zeta)}w_1(\zeta - \nu)\right] d\zeta \quad . \qquad (13.4.16)$$

Formula (13.4.16) is called Fock's formula [13.8] because an analogous formula was obtained by rather different means in the 1940's by V.A. Fock.

13.5 Transformation of the Integrals I_1 and I_2 in the Neighborhood of the Light-Shadow Boundary

We have found that I_1 represents the incident wave in the neighborhood of C, since there the determinant $\Delta \neq 0$. Beyond the limits of this neighborhood the behavior of the integrals I_1 and I_2 is not evident. Of immediate interest to us is a neighborhood of that part of the surface $\sum$ (cf. Sect. 13.4) which separates the illuminated region from the shadow zone. This surface, which will be denoted $\sum^+$, is formed by the continuations of incident rays tangent to S beyond the terminator C.

Outside our neighborhood of C we have $|m| \geq \text{const} > 0$, and the Airy functions $w_j(-\omega^{2/3}m)$ and $v(-\omega^{2/3}m)$ can be replaced by their asymptotics. If this is done, then the integrals representing the field reduce not to a superposition of AC sums, but to a superposition of similarly approximate, but now *ray* sums. It turns out that such a representation of the solution is more to the point and makes it possible for us to obtain a number of important formulas for the field.

Let us carry out the relevant derivations. Let the observation point be located near $\sum^+$ and not too far from C, so that $\Delta \neq 0$. The integrand of I_1 contains the Airy functions v and v' on the right prong of the contour, and w_j and w'_j ($j = 1, 2$) on the left prongs. Replacing all these functions by their asymptotics we can write I_1 as the sum of two integrals

$$I_1 = \frac{1}{\sqrt{\omega}} \int_{\mathcal{L}_-} u^+ \exp[i\omega\tau^+(M,\gamma)]d\gamma$$

$$+ \frac{1}{\sqrt{\omega}} \int_{\mathcal{L}_+} u^- \exp[i\omega\tau^-(M,\gamma)]d\gamma \tag{13.5.1}$$

along the contours $\mathcal{L}_-$: $(\gamma_0 e^{-2\pi i/3}, 0, \gamma_0)$ and $\mathcal{L}_+$: $(\gamma_0 e^{2\pi i/3}, 0, \gamma_0)$ shown in Fig. 13.6. The functions $u^{\pm}$ have the form

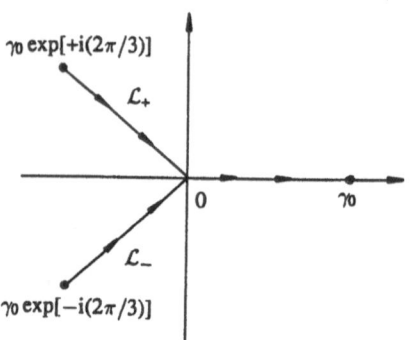

$\gamma_0 \exp[+i(2\pi/3)]$

$\mathcal{L}_+$

0

γ_0

$\mathcal{L}_-$

$\gamma_0 \exp[-i(2\pi/3)]$

Fig. 13.6. Integration contours for (13.5.1)

$$u^{\pm} = \sum_{j=0}^{N} \frac{u_j^{\pm}}{(-i\omega)^j} \quad \text{where} \quad u_j^{\pm} = \sum_{r=0}^{N_1(j)} u_{jr}^{\pm} \gamma^r \quad ;$$

$$\tau^{\pm}(M,\gamma) = l(M,\gamma) \pm \tfrac{2}{3} m^{3/2}(M,\gamma) \quad . \tag{13.5.2}$$

The integrand

$$u^{\pm} \exp[i\omega\tau^{\pm}(M,\gamma)]$$

is, to the highest-order terms in γ and ω, a solution to the equation

$$\left(\Delta + \frac{\omega^2}{c^2(M)}\right) u = 0$$

in which $\tau^{\pm}(M,\gamma)$ serves as the eikonal, likewise within terms of highest order in γ satisfying the eikonal equation

$$[\nabla\tau(M,\gamma)]^2 = 1/c^2 \quad . \tag{13.5.3}$$

It is natural to try to evaluate the integrals in (13.5.1) by the saddle-point method. To do this, we first of all clarify the existence of a stationary point for the functions $\tau^{\pm}(M,\gamma)$. Differentiating the eikonal equation (13.5.3) with respect to γ, and putting $\gamma = 0$, we obtain

$$\nabla\tau^{\pm}(M,0) \cdot \nabla\frac{\partial\tau^{\pm}(M,\gamma)}{\partial\gamma}\bigg|_{\gamma=0} = 0 \tag{13.5.4}$$

that is, $\partial\tau^{\pm}/\partial\gamma = \text{const.}$ along rays corresponding to the eikonal $\tau^{\pm}(M,0) \equiv \tau_0^{\pm}$. Note that in accordance with the constructions of Chap. 3, rays corresponding to the eikonal $\tau_0^{+} = l(M,0) + \tfrac{2}{3} m^{3/2}(M,0)$ [respectively $\tau_0^{-} = l(M,0) - \tfrac{2}{3} m^{3/2}(M,0)$] are rays leaving (respectively, approaching) the caustic. The role of the caustic $m(M,0) = 0$ is played here by the surface S. In view of the fact that, on C, $\nabla\tau^{\text{inc}} = \nabla\tau_0^{+}$, the rays forming $\sum^{+}$ enter not only into the rays of the eikonal τ^{inc}, but also those of τ^{+}. For $M \in C$ and $\gamma = 0$ the equality

$$\tau_\gamma^{+}(M) = l_\gamma(M,0) + \sqrt{m(M,0)} m_\gamma(M,0) = l_\gamma(M,0) = 0$$

is satisfied, where $\tau_\gamma^{\pm}(M) \equiv \tau_\gamma^{\pm}(M,\gamma)|_{\gamma=0} \equiv \partial l(M,\gamma)/\partial\gamma|_{\gamma=0}$ and so on. Thus, the phase factor $\exp[i\omega\tau^{+}(M,\gamma)]$ has $\gamma = 0$ as its stationary point for all $M \in \sum^{+}$. Calculating the Hessian

$$\Delta = \begin{vmatrix} \Phi_{\gamma\gamma} & \Phi_{\gamma\nu} \\ \Phi_{\nu\gamma} & \Phi_{\nu\nu} \end{vmatrix} \quad ; \quad \Phi = l + m\nu - \frac{\nu^3}{3}$$

at the stationary point (γ_c, ν_c), where $\nu_c = \sqrt{m}$ and $l_\gamma + m_\gamma \nu_c = 0$, it is not difficult to obtain that

$$\Delta = -2m\frac{\partial^2\tau^{+}}{\partial\gamma^2} \quad .$$

In a neighborhood of C we have $\Delta < 0$ (cf. Sect. 13.4), and therefore $\partial^2\tau^{+}/\partial\gamma^2$

> 0. In view of the fact that $\tau_\gamma = 0$ on $\sum^+$, the positivity of $\tau_{\gamma\gamma}^+$ and the implicit function theorem for a point M located near $\sum^+$ and not too far from C, the equation $\tau_\gamma^+(M, \gamma) = 0$ will have the root $\gamma = \gamma_c(M)$ near $\gamma = 0$. In the next section we will obtain a formula for $\tau_\gamma^+(M)$ for $\gamma = 0$ and $M \in \sum^+$, allowing us to specify the region where $\tau_{\gamma\gamma}^+(M) > 0$.

We will now show that the function $\tau^-(M, \gamma)$ for M near $\sum^+$, has no stationary point. Since

$$\frac{\partial \tau^-}{\partial \gamma} - \frac{\partial \tau^+}{\partial \gamma} = m_\gamma \sqrt{m}$$

and for $M \in \sum^+$ we have $m_\gamma > 0$ and $m > 0$ while $\partial \tau^+/\partial \gamma$ vanishes, then $\partial \tau^-/\partial \gamma|_{M \in \sum^+} > 0$. Clearly the derivative $\partial \tau^-/\partial \gamma$ remains positive for M located sufficiently close to $\sum^+$.

Therefore the second integral in (13.5.1) has no stationary point and thus makes no contribution to the asymptotics of I_1 and can be neglected. We next deform the contour of integration of the first integral. We note first of all that near C, because $\tau_{\gamma\gamma}^+(M) > 0$ for $M \in \sum^+$ and also that $\tau_\gamma^+(M) = 0$ for $M \in \sum^+$, $\gamma = 0$, we have $\tau_\gamma^+ > 0$ for $\gamma > 0$. It will be shown in the next section that the derivative $\partial \tau^+/\partial \gamma$ is positive for $\gamma > 0$ not just near C, but in some neighborhood of $\sum^+$ as well. This fact allows us to replace the horizontal segment $(0, \gamma_0)$ of the contour $\mathcal{L}_-$ by the inclined segment $(0, \gamma_0 e^{i\pi/3})$ with only an exponentially small error as $\omega \to \infty$. As a result, the expression for $u_{\text{inc}} \simeq I_1$ reduces to

$$u_{\text{inc}} \simeq I_1 \simeq \omega \int_{\mathcal{L}_5} u^+ e^{i\omega\tau^+(M,\gamma)} \, d\gamma \tag{13.5.5}$$

where the contour of integration is a line segment of length $2\gamma_0$ passing through $\gamma = 0$ at a nonzero angle (Fig. 13.7). The value of γ_0 is such that the integrand in (13.5.5) is exponentially small at the endpoints of $\mathcal{L}_5$ (the exponential smallness of the integrand at the lower endpoint of $\mathcal{L}_5$ is guaranteed by virtue of $\partial^2\tau^+/\partial\gamma^2 > 0$).

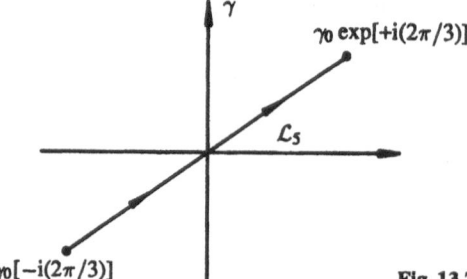

Fig. 13.7. Integration contour $\mathcal{L}_5$ for u_{inc}

Analogous transformations of the integral $u_{\text{refl}} \simeq I_2$ lead to the expression

$$u_{\text{refl}} \simeq I_2 \simeq - \sqrt{\omega} 2\mathrm{i} \int\limits_{\mathcal{L}_3} u^+ \frac{v(\omega^{2/3}\gamma)}{w_1(\omega^{2/3}\gamma)} \, \mathrm{e}^{\mathrm{i}\omega\tau^+(M,\gamma)} \, d\gamma \qquad (13.5.6)$$

where, however, the contour $\mathcal{L}_3$ is a forked one (cf. Fig. 13.3). The integration over $\mathcal{L}_3$ is carried out as indicated in Sect. 13.1.

The integrals (13.5.5) and (13.5.6) give us new expressions for the field. They represent superpositions of approximate ray sums in contrast to the integrals (13.2.8) and (13.2.9), which are superpositions of approximate caustic sums. In the next section we show that in order to determine the coefficients in the expansions of $\tau^+(M,\gamma)$ and $u^+(M,\gamma)$ in powers of γ, we must solve ordinary differential equations along the ray corresponding to the eikonal $\tau_0^+(M)$. These equations can be called transport equations for the corresponding coefficients. Thus, to construct the functions $\tau^+(M,\gamma)$ and $u^+(M,\gamma)$ it is only necessary that the ray field corresponding to the eikonal $\tau_0^+(M)$ be regular. The ray field corresponding to the eikonal $\tau_0^-(M)$ plays no part in the description of $\tau^+(M,\gamma)$ and $u^+(M,\gamma)$ and the regularity of the ray field pertaining to $\tau^-(M,\gamma)$ is, by the same token, not essential. We note that for the construction of the approximate caustic sums in terms of which the field near the terminator C is described, it is necessary that the ray fields corresponding to both $\tau_0^+(M)$ and $\tau_0^-(M)$ be regular.

13.6 Calculation of the Derivatives of $\tau^+(M, \gamma)$ and $u^+(M, \gamma)$ on Σ^+

Our task in this section is to calculate the derivative $\partial^2 \tau^+/\partial \gamma^2$ at a point M located near Σ^+, for $\gamma = 0$, and to show that $\partial^2 \tau^+/\partial \gamma^2 > 0$ near Σ^+ for $\gamma = 0$ if the incident ray field is regular near Σ^+. In doing so we will have not only proved that the contour $\mathcal{L}_-$ can be deformed into $\mathcal{L}_5$ for the first integral in (13.5.1), but also validated our use of the standard saddle-point method for calculating this integral near Σ^+. Having calculated the integral (13.5.5) by the saddle-point method and using the value we have found for the derivative, it will not be difficult to convince ourselves that I_1 represents the incident wave not just near C, but in a neighborhood of Σ^+ as well. In fact, the asymptotics of both the incident wave and I_1 as $\omega \to \infty$ are ray series. The corresponding expansion coefficients from (13.3.5–8) satisfy the same recurrent ordinary differential equations along the rays which form Σ^+. In the neighborhood of C these coefficients are the same, so consequently (by the uniqueness theorem for the solution of the Cauchy problem for ordinary differential equations) these coefficients are the same everywhere on Σ^+.

The formulas obtained in this section will also play an important role in the next section in the derivation of the Fresnel-Fock formula for the field in the penumbra. It is not out of place to draw the reader's attention at the beginning of this section to the fact that the method used to obtain the formula for $\partial^2\tau + /\partial\gamma^2$ will also indicate the conditions under which the asymptotic value of I_1 found by the saddle-point method is the same as the original expansion (13.1.16) of the incident wave.

We turn again to the (α, β, n) coordinate system introduced in Sect. 13.3. By the same method as in Sect. 13.3, it is not difficult to verify that the derivatives $\partial^2\tau^{\mathrm{inc}}/\partial n^2$ and $\partial^2\tau_0^+/\partial n^2$ satisfy a Riccati equation:

$$\frac{1}{c^2}\frac{dy}{d\beta} + y^2 + \Psi = 0$$

where

$$\Psi = \frac{1}{2}\left(\frac{\partial^2 g^{\beta\beta}}{\partial n^2} - \frac{\partial^2 c^{-2}}{\partial n^2}\right)\Bigg|_{n=0} = c^{-4}\left(c_{nn}c - 3c_n^2 + \frac{c^4}{2}\frac{\partial^2 g^{\beta\beta}}{\partial n^2}\right)_{n=0} .$$

Putting

$$\frac{\partial^2\tau^{\mathrm{inc}}}{\partial n^2} \equiv \tau_{nn}^{\mathrm{inc}} = \frac{1}{c^2}\frac{dZ_1/d\beta}{Z_1}$$

$$\frac{\partial^2\tau^+}{\partial n^2} \equiv \tau_{0nn}^+ = \frac{1}{c^2}\frac{dZ_2/d\beta}{Z_2} \tag{13.6.1}$$

we find that the functions $Z_1(\beta)$ and $Z_2(\beta)$ are solutions of the second-order differential equation

$$\frac{d^2 Z}{d\beta^2} - \frac{2c_\beta'}{c}\frac{dZ}{d\beta} + \Psi Z = 0 \quad . \tag{13.6.2}$$

We identify the solution Z_1 of (13.6.2) through the properties of the incident wave. We obtain the requisite formula by evaluating the Laplacian $\Delta\tau^{\mathrm{inc}}$ on Σ^+ by two different methods. In (α, β, n) coordinates, the Laplacian has the form

$$\Delta\tau = \frac{1}{\sqrt{g}}\left[\frac{\partial}{\partial n}\left(\sqrt{g}\frac{\partial\tau}{\partial n}\right) + \frac{\partial}{\partial\alpha}\left(\sqrt{g}g^{\alpha\alpha}\frac{\partial\tau}{\partial\alpha}\right) + \frac{\partial}{\partial\alpha}\left(\sqrt{g}g^{\alpha\beta}\frac{\partial\tau}{\partial\beta}\right)\right.$$
$$\left. + \frac{\partial}{\partial\beta}\left(\sqrt{g}g^{\alpha\beta}\frac{\partial\tau}{\partial\alpha}\right) + \frac{\partial}{\partial\beta}\left(\sqrt{g}g^{\beta\beta}\frac{\partial\tau}{\partial\beta}\right)\right] \quad .$$

Hence

$$\Delta\tau^{\mathrm{inc}}\big|_{M\in\Sigma^+} = \frac{\partial^2\tau^{\mathrm{inc}}}{\partial n^2}\bigg|_{n=0} + \frac{1}{c|\vec{r}_\alpha|}\frac{\partial}{\partial\beta}\left(\frac{|\vec{r}_\alpha|}{c}\right)\bigg|_{M\in\Sigma^+} \tag{13.6.3}$$

where $\vec{r} = \vec{r}(\alpha, \beta, n) = (x, y, z)$ is the vector function expressing the conversion from (α, β, n) to cartesian coordinates. On the other hand, we obtained in Chap. 2 the expression

$$\Delta\tau = \frac{1}{Jc}\frac{d}{d\tau}\left(\frac{J}{c}\right)$$

where J is the geometrical divergence of the ray field corresponding to the eikonal τ. Therefore, for $\Delta\tau^{\text{inc}}$ on $\sum^+$ we can write

$$\Delta\tau^{\text{inc}} = \frac{1}{J^{\text{inc}}c} \frac{\partial}{\partial\beta}\left(\frac{J^{\text{inc}}}{c}\right) \qquad (13.6.4)$$

where J^{inc} is the divergence of the incident wave ray field on $\sum^+$. Comparing (13.6.3) and (13.6.4) and recalling the definition (13.6.1) of the function Z_1, we obtain

$$\frac{1}{Z_1}\frac{dZ_1}{d\beta} + \frac{c}{|\vec{r}_\alpha|}\frac{\partial}{\partial\beta}\left(\frac{|\vec{r}_\alpha|}{c}\right) = \frac{c}{J^{\text{inc}}}\frac{\partial}{\partial\beta}\left(\frac{J^{\text{inc}}}{c}\right) \quad .$$

Thus,

$$\left.\frac{Z_1|\vec{r}_\alpha|}{J^{\text{inc}}}\right|_{M\in\sum^+} = \text{const.} \qquad (13.6.5)$$

We next make precise the choice of the solution Z_2 of (13.6.2), which appears as the second line of (13.6.1). In view of the fact that $\tau_0^+ = l_0 + \frac{2}{3}m_0^{3/2}$ and $m_0|_S = 0$, $(\partial m_0/\partial n)|_S = -1$, the derivative τ_{0nn}^+ on the surface S, and particularly at points on the terminator C, becomes infinite. Therefore, we must satisfy the condition

$$Z_2|_C = 0 \quad (Z_2 \not\equiv 0) \quad . \qquad (13.6.6)$$

The solution Z_2 of (13.6.2) will be determined uniquely if we supplement (13.6.6) by a second initial condition

$$\left.\frac{dZ_2}{d\beta}\right|_C = 2^{1/3}c_0^{4/3}|_C \qquad (13.6.7)$$

where $c_0|_C$ is the value of the wave velocity at the terminator.[4] Let us note the following important point: the solution Z_2 determined by (13.6.6) and (13.6.7) depends in no way on the properties of S at points of C. We also note that the formula

$$\left.Z_2\frac{|\vec{r}_\alpha|}{J^+}\right|_{M\in\sum^+} = \text{const} \qquad (13.6.8)$$

holds, where J^+ is the divergence of the ray field corresponding to the eikonal τ_0^+. The derivation of (13.6.8) is completely analogous to that of (13.6.5). To calculate the derivative $\partial^2\tau^+/\partial\gamma^2$ we will require the value of the derivative

$$\frac{\partial^2\tau_0^+}{\partial n\partial\gamma} \equiv \tau_{0n\gamma}^+$$

which we will also determine shortly. Differentiating the eikonal equation

---

[4] The somewhat exotic choice of condition (13.6.7) leads to certain simplifications in the formulas of the next section.

$$[\nabla \tau^+(M,\gamma)]^2 = \frac{1}{c^2}$$

written in (α, β, n) coordinates by γ and n, and then putting $\gamma = n = 0$, we obtain

$$\tau^+_{0nn}\tau^+_{0n\gamma} + \frac{1}{c^2}\frac{d}{d\beta}\tau^+_{0n\gamma} = 0 \quad .$$

We replace τ^+_{0nn} by $(dZ_2/d\beta)/c^2 Z_2$ and integrate which results in

$$\tau^+_{0n\gamma}\big|_{\Sigma^+} = \frac{\text{const}}{Z_2} \quad . \tag{13.6.9}$$

By the formula $\tau^+(M,\gamma) = l + \frac{2}{3}m^{3/2}$ we have

$$\tau^+_{n\gamma} = l_{n\gamma} + \sqrt{m}\,m_{n\gamma} + \frac{m_\gamma m_n}{2\sqrt{m}} \quad .$$

Comparing the singularities of both sides of (13.6.9) as $M \to C$ and using the last formula above together with (13.6.6) and (13.6.7), we find that const. $= -P_0^{1/3}$, where P_0 is the effective radius of curvature of the surface S at the appropriate point on the terminator C. Thus,

$$\tau^+_{0\gamma n}\big|_{\Sigma^+} = -\frac{P_0^{1/3}}{Z_2} \quad . \tag{13.6.10}$$

We come now directly to the evaluation of $\partial^2\tau^+/\partial\gamma^2$ at points of Σ^+ for $\gamma = 0$. From Sect. 13.4 it follows that near C (and hence also in a neighborhood of Σ – see the beginning of this section)

$$\tau^{\text{inc}}(M) = \tau^+(M, \gamma_c) \tag{13.6.11}$$

holds up to terms of sufficiently high order in n as $n \to 0$. The function $\gamma_c = \gamma_c(M)$ in (13.6.11) is the root of $\tau^+_\gamma(M, \gamma) = 0$. Differentiating (13.6.11) twice by n and recalling that $\tau^+_\gamma(M, \gamma_c) = 0$, we obtain

$$\frac{\partial^2\tau^{\text{inc}}}{\partial n^2} = \frac{\partial^2\tau^+}{\partial n^2} + \frac{\partial^2\tau^+}{\partial n\partial\gamma}\frac{\partial\gamma}{\partial n} = \frac{\partial^2\tau^+}{\partial n^2} - \frac{(\partial^2\tau^+/\partial n\partial\gamma)^2}{\partial^2\tau^+/\partial\gamma^2} \quad .$$

Hence,

$$\frac{\partial^2\tau^+}{\partial\gamma^2}\bigg|_{\substack{M \in \Sigma^+ \\ \gamma=0}} = \frac{(\partial^2\tau^+/\partial n\partial\gamma)^2}{(\partial^2\tau^+/\partial n^2) - (\partial^2\tau^{\text{inc}}/\partial n^2)}\bigg|_{\substack{M \in \Sigma^+ \\ \gamma=0}} \quad .$$

The second derivatives can be replaced by their expressions from (13.6.1) and (13.6.10), resulting in

$$\frac{\partial^2\tau^+}{\partial\gamma^2}\bigg|_{\substack{M \in \Sigma^+ \\ \gamma=0}} = c^2 P_0^{2/3}\frac{Z_1}{Z_2}\frac{1}{Z_1 Z_2' - Z_1' Z_2} \quad .$$

The Wronskian $W = Z_1 Z_2' - Z_1' Z_2$ of two solutions of (13.6.2) satisfies

$$\frac{dW}{d\beta} = \frac{2c'_\beta}{c} W$$

and consequently

$$W = (\text{const})c^2 \quad .$$

We choose the constant in (13.6.5) so that $W = c^2$; in other words, we normalize Z_1 so that this will happen. Then,

$$\left.\frac{\partial^2 \tau^+}{\partial \gamma^2}\right|_{\substack{M \in \sum^+ \\ \gamma=0}} = P_0^{2/3} \frac{Z_1}{Z_2} \quad . \tag{13.6.12}$$

Clearly neither Z_1 nor Z_2 depends on the properties of the reflecting surface. This follows from the fact that the constant in (13.6.5) was not connected with S in any way; it was chosen to normalize the Wronskian which itself is independent of S.

There is an important consequence of (13.6.12); if a caustic of the incident wave does not pass through $\sum^+$, then the integral I_1 can be calculated by the standard version of the saddle-point method. In fact, the absence of common points between $\sum^+$ and the caustic is equivalent to $J^{\text{inc}} \neq 0$, while J^{inc} is proportional to Z_1 [cf. (13.6.5)]. The inequality $Z_1 \neq 0$ is equivalent to $\partial^2 \tau^+/\partial \gamma^2 \neq 0$ [cf. (13.6.12)], and this indicates that the standard version of the saddle-point method applies. Its use here leads to an expansion of the form (13.1.16). We note that the first zero of Z_1 (if one exists in the general case) is located on an incident ray nearer the terminator C than the first zero of Z_2. This follows from the alternating of zeroes between solutions of second-order equations and the equality $Z_2|_C = 0$.

We come now to the calculation of the coefficients $u_{jr}^+(M)$ in the expansion of $u^+(M, \gamma, \omega)$ in powers of ω^{-1} and γ [cf. (13.5.2)]. Since $u^+(M, \gamma, \omega)$ satisfies, to terms of sufficiently high order in ω^{-1}, the equation $(\Delta + \omega^2/c^2)u^+ = 0$, the coefficients $u_j^+(M, \gamma)$, $j = 0, 1, 2, \ldots$ must satisfy the transport equations

$$2\nabla \tau^+ \cdot \nabla u_j^+ + u_j^+ \Delta \tau^+ = \Delta u_{j-1}^+; \quad j = 0, 1, 2, \ldots, \quad u_{-1}^+ \equiv 0. \tag{13.6.13}$$

Successively differentiating (13.6.13) by γ and then putting $\gamma = 0$, we obtain transport equations for the $u_{jr}^+(M)$. Near $\sum^+$ the coefficients $u_{jr}^+(M)$ can be expanded in powers of n. The derivatives $\partial^s u_{jr}^+/\partial n^s|_{M \in \sum^+}$ which appear in the coefficients of this expansion satisfy transport equations in their own right, which are found by differentiating those for u_{jr}^+ by n and setting $n = 0$. The transport equations for $\partial^s u_{jr}^+/\partial n^s|_{M \in \sum^+}$ reduce in the usual way to ordinary differential equations along the rays covering $\sum^+$. The initial data are determined by requiring the asymptotic expansion of I_1 obtained by the saddle-point method to be the same as expansion (13.1.16) for the incident wave. We will carry out a detailed calculation of $u_{00}^+(M)|_{M \in \sum^+}$. The resulting formula will be used in the next section in the derivation of the Fresnel-Fock formula for the field in the penumbra.

Let M' and M'' be two points along the same ray (corresponding to the eikonal τ_0^+) and lying in the surface $\sum^+$. Since $u_{00}^+(M)$ satisfies the transport equation along the rays making up $\sum^+$, we have [cf. (2.4.5)]

$$u_{00}^+(M') = \sqrt{\frac{c(M')}{c(M'')} \frac{J^+(M'')}{J^+(M')}} u_{00}^+(M'')$$

which thanks to (13.6.8) can be rewritten as

$$u_{00}^+(M') = \sqrt{\frac{c(M')}{c(M'')}} \sqrt{\frac{Z_2(M'')|\vec{r}_\alpha(M'')|}{Z_2(M')|\vec{r}_\alpha(M')|}} u_{00}^+(M'') \quad .$$

We now let M'' tend to a point M_1, on the terminator C. For small values of $|M'' M_1|$, the function $u_{00}^+(M'')$ can be replaced by the expression

$$\frac{e^{i\pi/4}}{2i} \left(A_{00} m_0^{-1/4} + B_{00} m^{1/4} \right)$$

[cf. (13.2.3, 6) and the asymptotic formulas for v and v' from Appendix A.1], and then, letting $M'' \to M_1$ and using the original values of A_{00} and B_{00}, we get

$$u_{00}^+(M') = \frac{e^{i\pi/4}}{2\sqrt{\pi}} u_0^{inc}(M_1) \sqrt{\frac{c(M')|\vec{r}_\alpha(M_1)|}{c(M_1)|\vec{r}_\alpha(M')|}}$$

$$\times \lim_{M'' \to M_1} \sqrt{\frac{Z_2(M'')}{\sqrt{m_0(M'')}}} \quad . \tag{13.6.14}$$

The limit in (13.6.14) is easily found if we take into account (13.2.27), (13.6.6, 7) and (13.3.17). Straightforward calculations lead us to the desired formula:

$$u_0^+(M') = \frac{e^{-i\pi/4}}{2^{2/3}\sqrt{\pi}} u_0^{inc}(M_1) c^{1/3}(M_1) P_0^{1/3}(M_1)$$

$$\times \sqrt{\frac{|\vec{r}_\alpha(M_1)|}{c(M_1)}} \sqrt{\frac{c(M')}{|\vec{r}_\alpha(M')| Z_2(M')}} \quad . \tag{13.6.15}$$

13.7 The Fresnel-Fock Formula in the Neighborhood of the Light-Shadow Boundary

In the previous section it was shown that the asymptotic expansion of the integral I_1 as $\omega \to \infty$ can be obtained by the standard version of the saddle-point method, and that it is identical with the expansion of the incident wave u_{inc}.

The asymptotic expansion of I_2 can be found in a similar way only for sufficiently large positive values of n. The fact is that the usual saddle-point method can be applied to I_2 only if on the important part of the integration the Airy functions $v(\omega^{2/3}\gamma)$ and $w_1(\omega^{2/3}\gamma)$ appearing in the integrand can be replaced by their asymptotics. The replacement of the Airy functions by their asymptotics is possible only if the stationary phase point $\tilde{\gamma}_c$ of the phase function

$$\tau^+(M,\gamma) - \tfrac{4}{3}(-\gamma)^{3/2}$$

which will arise in one of the terms of the integrand of I_2 after this replacement, satisfies the inequality

$$-\omega^{2/3}\tilde{\gamma}_c \geq \text{const}\,\omega^\varepsilon \quad ;$$
$$\text{const} > 0 \quad , \quad \varepsilon > 0 \quad . \tag{13.7.1}$$

For small n and $\tau^+_{on\gamma} \neq 0$, we have $-\tilde{\gamma}_c \sim n^2$ and consequently (13.7.1) is fulfilled when

$$n > \text{const}\,\omega^{-1/3+\varepsilon/2} \quad . \tag{13.7.2}$$

Evaluating I_2 when (13.7.2) holds by the saddle-point method, we obtain an asymptotic expansion of the reflected wave (cf. also in this regard Sect. 13.4).

The neighborhood of the light-shadow boundary $\sum^+$ described by the inequality

$$|n| < \text{const}\,\omega^{-1/3} \tag{13.7.3}$$

is usually called the penumbra. In the penumbra the standard saddle-point method cannot be applied to I_2, and consequently we must devise a new way to calculate the asymptotics of the field.

In the penumbra (13.7.3), the integrals I_1 and I_2 must be treated simultaneously. The physical reason for this is as follows. Near the light-shadow boundary $\sum^+$ the phases of the incident and reflected waves are close to each other. Because of this, the direct and reflected waves interfere, and as a result they lose their distinct identities. There arises instead a field whose description requires a special function – in the present case a Fresnel integral, as we shall see below.

If we combine the integrals I_1 and I_2, then the integral over the lower left fork cancels, and we get

$$u \simeq I_1 + I_2 \simeq u^{(1)} + u^{(2)} + u^{(3)}$$

where

$$u^{(1)} = \sqrt{\omega} \int_{\mathcal{L}'_6} u^+ e^{i\omega\tau^+}\, d\gamma$$

$$u^{(2)} = \sqrt{\omega} \int_{\mathcal{L}''_6} \frac{w_2(\omega^{2/3}\gamma)}{w_1(\omega^{2/3}\gamma)} u^+ e^{i\omega\tau^+}\, d\gamma$$

$$u^{(3)} = -2i\sqrt{\omega} \int_{\mathcal{L}'''_6} \frac{v(\omega^{2/3}\gamma)}{w_1(\omega^{2/3}\gamma)} u^+ e^{i\omega\tau^+}\, d\gamma \quad .$$

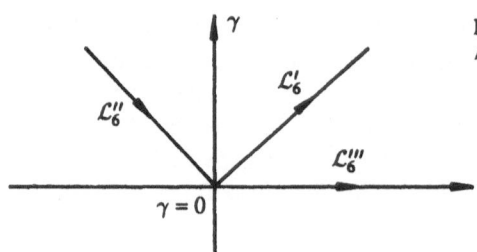

Fig. 13.8. Integration contours $\mathcal{L}_6'$, $\mathcal{L}_6''$ and $\mathcal{L}_6'''$ for the penumbra field

Here $\mathcal{L}_6'$, $\mathcal{L}_6''$ and $\mathcal{L}_6'''$ are the segments shown in Fig. 13.8. The integral $u^{(1)}$, as will be shown below, can, in the first approximation, be expressed as a Fresnel integral and gives the major contribution to the field in the penumbra, which does not depend on the properties of the surface S. The integrals $u^{(2)}$ and $u^{(3)}$ are correction terms, describing (in the terminology of V.A. Fock) the wave background.

If we limit ourselves to the evaluation of only the dominant terms in the asymptotic expansions of $u^{(j)}$, $j = 1, 2, 3$, as $\omega \to \infty$, then we can replace the functions u^+ and τ^+ by the first terms of their power series in γ. After this has been done, the segments $\mathcal{L}_6'$, $\mathcal{L}_6''$ and $\mathcal{L}_6'''$ over which the integrals are carried out can be extended out to infinity. (This will only introduce an exponentially small error into the values of $u^{(j)}$.) When all this has been done, we obtain

$$u^{(1)} \simeq \sqrt{\omega} \int_{\mathcal{L}_7'} \exp\left\{ i\omega[\tau_0^+(M) + \gamma\tau_{0\gamma}^+(M) + \frac{\gamma^2}{2}\tau_{0\gamma\gamma}^+(M)] \right\} u_{00}^+(M)d\gamma$$

(13.7.4)

$$u^{(2)} \simeq \sqrt{\omega} \int_{\mathcal{L}_7''} \frac{w_2(\omega^{2/3}\gamma)}{w_1(\omega^{2/3}\gamma)} \exp\{i\omega[\tau_0^+(M) + \gamma\tau_{0\gamma}^+(M)]\} u_{00}^+(M) \, d\gamma$$

(13.7.5)

$$u^{(3)} \simeq - \sqrt{\omega} \int_{\mathcal{L}_7'''} \frac{2iv(\omega^{2/3}\gamma)}{w_1(\omega^{2/3}\gamma)} \exp\{i\omega[\tau_0^+(M) + \gamma\tau_{00}^+(M)]\} u_{00}^+(M) \, d\gamma \quad .$$

(13.7.6)

Here $\mathcal{L}_7'$, $\mathcal{L}_7''$ and $\mathcal{L}_7'''$ are half-lines leaving the point $\gamma = 0$, the result of extending the segments $\mathcal{L}_6'$, $\mathcal{L}_6''$ and $\mathcal{L}_6'''$ to infinity. Recall that $\tau_0^+(M) = \tau^+(M, \gamma)|_{\gamma=0}$, $\tau_{0\gamma}^+(M) = \partial\tau^+(M, \gamma)/\partial\gamma|_{\gamma=0}$, and $\tau_{0\gamma\gamma}^+(M) = \partial^2\tau^+(M, \gamma)/\partial\gamma^2|_{\gamma=0}$. The term $\gamma^2\tau_{0\gamma\gamma}^+/2$ was retained in the expansion of the phase function in $u^{(1)}$ because this term will guarantee the exponentially small behavior of the integrand at the end of $\mathcal{L}_6'$ if $\tau_{0\gamma}^+(M) < 0$, and guarantee the convergence at infinity of (13.7.4).

We make some further simplifications in (13.7.4–6) which do not affect the values of $u^{(j)}$, $j = 1, 2, 3$, in the dominant terms. We replace

$$\tau_{0\gamma}^+(M) \to \tau_{0\gamma n}^+(M')n$$

$$\tau_{0\gamma\gamma}^+(M) \to \tau_{0\gamma\gamma}^+(M') \quad \text{and} \quad u_{00}^+(M) \to u_{00}^+(M')$$

where M' is the projection of the point M on the surface $\sum^+$. Having done this, we perform a change of variable in (13.7.4) according to

$$\zeta = \sqrt{\frac{\omega}{2}} \left[\gamma \sqrt{\tau_{0\gamma\gamma}^+(M')} + \frac{\tau_{0\gamma}(M')}{\sqrt{\tau_{0\gamma\gamma}^+(M')}} \right]$$

and in (13.7.5) and (13.7.6), by

$$\zeta = \omega^{2/3}\gamma \quad .$$

Finally, we replace $\tau_{0\gamma n}^+(M')$, $\tau_{0\gamma\gamma}^+(M')$ and $u_{00}^+(M')$ by their values given in (13.6.10, 12) and (13.6.15). As a result we obtain

$$u^{(1)}(M) \simeq u_0^{\text{inc}}(M_1) e^{i\omega\tau_0^+(M)} \sqrt{\frac{c(M')}{c(M_1)} \frac{J^{\text{inc}}(M_1)}{J^{\text{inc}}(M')}} F(\xi) \tag{13.7.7}$$

where

$$F(\xi) = \frac{e^{-i\pi/4 - i\xi^2}}{\sqrt{\pi}} \int_\xi^\infty e^{i\zeta^2} d\zeta \tag{13.7.8}$$

$$\xi = -\frac{\sqrt{\omega}n}{\sqrt{2Z_1(M')Z_2(M')}} \tag{13.7.9}$$

and

$$u^{(2)} \simeq \frac{u_0^{\text{inc}}(M_1) e^{i\omega\tau_0^+(M)}}{\omega^{1/6}} \int_\infty^0 e^{i2\pi/3} \frac{w_2(\zeta)}{w_1(\zeta)} e^{i\eta\zeta} d\zeta \tag{13.7.10}$$

$$u^{(3)} \simeq -\frac{u_0^{\text{inc}}(M_1) e^{i\omega\tau_0^+(M)}}{\omega^{1/6}} \int_0^\infty \frac{2iv(\zeta)}{w_1(\zeta)} e^{i\eta\zeta} d\zeta \tag{13.7.11}$$

where

$$\eta = -\frac{P_0^{1/3}(M_1)}{Z_2(M')}\omega^{1/3}n \quad . \tag{13.7.12}$$

In these formulas, M_1 (Sect. 13.6) is the point of intersection of the incident ray lying in $\sum^+$ and passing through the point M' (that is, through the projection of the observation point M onto $\sum^+$), with the terminator C.

Let us analyze the content of (13.7.7–12). In the penumbra, where (13.7.3) holds, the variable η remains bounded as $\omega \to \infty$, while ξ can grow, but no faster than $\omega^{1/6}$. Therefore, the integrals $u^{(2)}$ and $u^{(3)}$ are proportional to $\omega^{-1/6}$ in the penumbra. The function $F(\xi)$ as $\xi \to \infty$ has the asymptotic value

$$F(\xi) \sim \frac{e^{i\pi/4}}{2\sqrt{\pi\xi}} \quad .$$

From (13.7.9), $F(\xi) \sim \omega^{-1/6+\varepsilon}$, i.e., the integral $u^{(1)}$ decays more slowly than $\omega^{-1/6}$. In the neighborhood of the surface $\sum^+$ defined by

$$|n| \le \text{const.}\, \omega^{-1/2}$$

we have $u^{(1)} = O(1)$ as $\omega \to \infty$. Thus, as was mentioned above, the integral $u^{(1)}$ is the main contributor to the field, while the integrals $u^{(2)}$ and $u^{(3)}$ are more in the nature of corrections. The function $F(\xi)$ clearly reduces to the Fresnel integrals

$$S(\xi) = \sqrt{\frac{2}{\pi}} \int\limits_0^\xi \sin x^2\, dx \quad \text{and} \quad C(\xi) = \sqrt{\frac{2}{\pi}} \int\limits_0^\xi \cos x^2\, dx$$

and consequently, the field in the penumbra is in fact described primarily by Fresnel integrals. Finally, since $Z_1(M')$ and $Z_2(M')$ do not depend on the properties of S (cf. Sect. 13.6), the dominant part of the field is likewise independent of the properties of S. The surface S affects only the wave background, given by (13.7.10) and (13.7.11). This influence shows up via the quantity $P_0(M_1)$, i.e., the effective radius of curvature of S at points on the terminator C [see (13.7.12)].

13.8 Asymptotics of the Field in the Deep Shadow

The analysis of (13.7.7) carried out at the end of the last section shows that as the observation point M sinks into the shadow zone, the field decreases, and at the boundary of the penumbra (where $n = -\text{const.}\,\omega^{-1/3}$; const. > 0) it becomes proportional to $\omega^{-1/6}$. In the deep shadow (i.e., for $n < -\text{const.}\,\omega^{-1/3+\varepsilon}$, $\varepsilon > 0$), the field, as we will show in this section, is exponentially small as $\omega \to \infty$.

The asymptotics of the field in the deep shadow must be obtained using the residue theorem, applied to the integral

$$u \cong \omega \int\limits_{\mathcal{L}_3} \left[\mathcal{J}(M,\gamma) - \mathcal{H}(M,\gamma) \frac{v(\omega^{2/3}\gamma)}{w_1(\omega^{2/3}\gamma)} \right] e^{i\omega l(M,\gamma)}\, d\gamma \qquad (13.8.1)$$

which describes the total field $u = u_{\text{inc}} + u_{\text{refl}}$ [cf. (13.2.7)]. On the lower left fork of $\mathcal{L}_3$ the term in brackets in the integrand of (13.8.1) vanishes, and (13.8.1) reduces to an integral over the broken line $(\gamma_1, 0, \gamma_0)$; $|\gamma_1| = \gamma_0$ (cf. Fig. 13.9). The crosses in this figure denote poles of the integrand, which occur at the points

$$\gamma_h = \omega^{-2/3}\xi_h$$
$$\xi_h = |\xi_h|\, e^{i\pi/3} \qquad\qquad (13.8.2)$$

where ξ_h is a root of the Airy function $w_1(\xi)$.

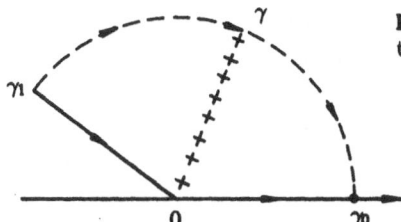

Fig. 13.9. Integration contours and pole locations for the deep shadow field

The integral over the broken line $(\gamma_1, 0, \gamma_0)$ is replaced by one over the arc of a circle between γ_1 and γ_0, centered at $\gamma = 0$ (the dashed curve in Fig. 13.9), plus the sum of the residues at the poles γ_h of the integrand. Neglecting the exponentially small integral over the arc (γ_1, γ_0), we obtain

$$u \simeq \sum_h u_h \tag{13.8.3}$$

where u_h is the residue of the integrand:

$$
u_h = \frac{\pi w_2(\xi_h)}{w_1'(\xi_h)} \left[\sum_{r=0}^{N} \sum_{j=0}^{N_1} \mathcal{A}_{rj} \gamma_h^j (-i\omega)^{-r} w_1(-\omega^{2/3} m(M, \gamma_h)) \right.
$$
$$
\left. + i\omega^{-1/3} \sum_{r=0}^{N} \sum_{j=0}^{N_1} \mathcal{B}_{rj} \gamma_h^j (-i\omega)^{-r} w_1'(-\omega^{2/3} m(M, \gamma_h)) \right]
$$
$$
\times e^{i\omega l(M, \gamma_h)}. \tag{13.8.4}
$$

Contrary to the formal solution constructed in Sect. 7.5, expression (13.8.4) is defined not only in a boundary layer, but also at points more distant from S than a constant not depending on ω. Replacing l and m in (13.8.4) by the sums (13.2.5), we arrive at an expression of the form

$$
u_h = \left[\left(\sum_{n=0}^{N_2} F_n \omega^{-n/3} \right) w_1 \left(-\omega^{2/3} \sum_{j=0}^{N_3} \theta_j(M) \omega^{-2j/3} \right) \right.
$$
$$
\left. + i\omega^{-1/3} \left(\sum_{n=0}^{N_2} G_n \omega^{-n/3} \right) w_1' \left(-\omega^{2/3} \sum_{j=0}^{N_3} \theta_j(M) \omega^{-2j/3} \right) \right]
$$
$$
\times \exp\left(i\omega \sum_{j=0}^{N_4} \psi_j(M) \omega^{-2j/3} \right) . \tag{13.8.5}
$$

Here N_2, N_3, N_4 are expressed in terms of N and N_1; F_n, G_n, θ_j and ψ_j are certain functions of M, in which

$$
\psi_0(M) = l_0(M) \quad ; \quad \psi_1(M) = \xi_h l_1(M)
$$
$$
\theta_0(M) = m_0(M) \quad ; \quad \theta_1(M) = \xi_h m_1(M)
$$
$$
F_0 = \frac{\pi w_2(\xi_h)}{w_1'(\xi_h)} \mathcal{A}_{00} \quad ; \quad G_0 = \frac{\pi w_2(\xi_h)}{w_1'(\xi_h)} \mathcal{B}_{00} \quad .
$$

We will carry out a detailed study of the leading terms of (13.8.5). Using the relation $w_2(\xi_h) = -2i/w_1'(\xi_h)$, which follows from the Wronskian relation since ξ_h is a root of w_1, we get

$$
\begin{aligned}
u_h \simeq \; & -\frac{2\pi i}{[w_1'(\xi_h)]^2} \exp\left[i\omega\left(l_0(M) + \xi_h\omega^{-2/3}l_1(M_1)\right)\right] \\
& \times \left[\mathcal{A}_{00}(M)w_1\left(-\omega^{2/3}m_0(M) - \xi_h m_1(M)\right)\right. \\
& \left. + i\omega^{-1/3}\mathcal{B}_{00}(M)w_1'\left(-\omega^{2/3}m_0(M) - \xi_h m_1(M)\right)\right] \; . \quad (13.8.6)
\end{aligned}
$$

If the point M is located within a distance of $O(\omega^{-2/3})$ from S, then (13.8.6) can be reduced to the same form as the expressions of Sect. 7.5. Of more immediate interest to us is the opposite case, when M is sufficiently far from S, and the functions w_1 and w_1' in (13.8.6) can be replaced by their asymptotics. Doing this results in

$$
\begin{aligned}
u_h \simeq \; & -\frac{2\pi i\, e^{i\pi/4}}{\omega^{1/6}[w_1'(\xi_h)]^2} \exp\left[i\omega\left(l_0 + \tfrac{2}{3}m_0^{3/2}\right) + i\omega^{1/3}\xi_h(l_1 + \sqrt{m_0}\,m_1)\right] \\
& \times \left(\mathcal{A}_{00}m_0^{-1/4} + \mathcal{B}_{00}m_0^{1/4}\right) \; . \quad (13.8.7)
\end{aligned}
$$

We can transform (13.8.7) into a form possessing an interesting physical interpretation. First of all, it is not difficult to observe the phase factor $\exp[i\omega(l_0(M) + \tfrac{2}{3}m_0^{3/2}(M))]$. In fact, the function $l_0(M) + \tfrac{2}{3}m_0^{3/2}(M)$ is equal to the eikonal τ_0^+. We can explain, on the basis of τ_0^+, how a wave disturbance reaches an observation point M in the shadow zone. Recall (Sect. 13.3) that at points on the terminator we have $\tau_0^+(M_1) = \tau^{\mathrm{inc}}(M_1)$, and $\nabla\tau_0^+(M_1) = \nabla l_0(M_1) = \nabla\tau^{\mathrm{inc}}(M_1)$ when $M_1 \in C$; and that along the surface and space rays $\partial\tau_0^+/\partial s = 1/c$, so that a surface ray is transformed into a space ray such that at their common point on S, they have the same tangent. Therefore, we can write

$$
\tau_0^+(M) = \tau^{\mathrm{inc}}(M_1) + \int\limits_{M_1}^{M} \frac{ds}{c} \; .
$$

The integral $\int_{M_1}^{M} ds/c$ is carried out initially over the surface ray M_1M_2 which is tangent to the incident ray at $M_1 \in C$. From M_2 the integral is carried out along the space ray M_2M which is shed from S at M_2 and proceeds toward the observation point M (Fig. 13.10). At the point M_2 where the ray is shed, the surface ray and the space ray are mutually tangent. From the viewpoint of the geometrical theory of diffraction, this path is the one along which a wave disturbance propagates into the shadow zone.

We turn next to the factor $\exp[i\omega^{1/3}\xi_h(l_1 + \sqrt{m_0}\,m_1)]$. In Sect. 13.2, it was shown [see (13.2.29)] that $l_1 + \sqrt{m_0}\,m_1 = \mathrm{const.}$ along the ray M_2M. Moreover, using (13.2.26–27) we find that

$$l_1 + \sqrt{m_0}\,m_1 = \int_{M_1}^{M_2} \frac{ds}{2^{1/3}c^{1/3}P^{2/3}} \qquad (13.8.8)$$

where $P(s)$ is the effective radius of curvature of the normal cross-section of S, measured at a point s lying on the surface ray $M_1 M_2$, along the tangent to that ray.

The factor $\mathcal{A}_{00}m^{-1/4}+\mathcal{B}_{00}m^{1/4}$ obeys the normal transport equation (2.1.6), and thus

$$\mathcal{A}_{00}(M)m_0^{-1/4}(M) + \mathcal{B}_{00}(M)m_0^{1/4}(M)$$

$$= \left[\mathcal{A}_{00}(M_0)m_0^{-1/4}(M_0) + \mathcal{B}_{00}(M_0)m_0^{1/4}(M_0)\right]\sqrt{\frac{J(M_0)}{J(M)}} \quad . \qquad (13.8.9)$$

Here J is the geometrical divergence and M_0 is an arbitrary point on the ray $M_2 M$. The geometrical divergence is a quantity determined along a ray only up to a factor which is constant on the ray. From the constructions of Sect. 3.3 it follows that as $M_0 \to M_2$, the ratio

$$\frac{J(M_0)}{\tau_0^+(M_0) - \tau_0^+(M_2)}$$

has a finite limit [cf. (3.3.17) and (3.3.25)]. We normalize $J(M)$ so that

$$J(M) = \tau_0^+(M) - \tau_0^+(M_2) + o\big(\tau_0^+(M) - \tau_0^+(M_2)\big) \qquad (13.8.10)$$

holds. Then, taking the limit of (13.8.9) as $M_0 \to M_2$ and using (3.2.27) and (3.3.17), we obtain

$$\mathcal{A}_{00}(M)m_0^{-1/4} + \mathcal{B}_{00}(M)m_0^{1/4}(M)$$

$$= \mathcal{A}_{00}(M_2)2^{1/6}P^{1/3}(M_2)c^{-1/3}(M_2)/\sqrt{J(M)} \qquad (13.8.11)$$

where $P(M_2)$ is the effective radius of curvature of a normal cross-section to S, measured along the tangent to the surface ray $M_1 M_2$ at M_2. To determine $\mathcal{A}_{00}(M_2)$ we use (13.2.31). Equation (13.2.31) is an ordinary differential equation along the surface ray. Introducing the eikonal $\tau \equiv \beta_1$ as a parameter on the surface ray [cf. the introduction of (α_1, β_1, n_1) coordinates at the end of Sect. 13.4] and using the initial conditions (13.4.13), we arrive at:

$$\mathcal{A}_{00}(M_2) = \frac{1}{\sqrt{\pi}}u_0^{\text{inc}}(M_1)\exp\left[-\frac{1}{2}\int_{M_1}^{M_2} c^2\Delta l_0\,d\tau\right] \quad . \qquad (13.8.12)$$

To calculate Δl_0 we use (α_1, β_1, n_1) coordinates and express the Laplacian in this curvilinear coordinate system (cf. Appendix A.2). Using (13.2.17, 18) and (13.2.23), it is not difficult to derive the formula

$$\Delta l_0 = \frac{1}{jc}\frac{d}{d\tau}\left(\frac{j}{c}\right) + \frac{1}{3c^4 P(\tau)}\frac{d}{d\tau}\left(c^2 P(\tau)\right) \quad . \qquad (13.8.13)$$

Here j is the geometrical divergence of a surface ray. From (13.8.12–13) it follows that

$$\mathcal{A}_{00}(M_2) = \frac{1}{\sqrt{\pi}} u_0^{\mathrm{inc}}(M_1) \sqrt{\frac{j(M_1)}{j(M_2)}} \left(\frac{c(M_2)P(M_1)}{c(M_1)P(M_2)} \right)^{1/6} . \tag{13.8.14}$$

Collecting formulas (13.8.5–14) we finally obtain

$$u_h \simeq u_0^{\mathrm{inc}}(M_1) \frac{T(M_1)T(M_2)}{\sqrt{J(M)}} \sqrt{\frac{j(M_1)}{j(M_2)}}$$

$$\times \exp\left[i\omega^{1/3}\xi_h \int_{M_1}^{M_2} \frac{ds}{2^{1/3}P^{2/3}(s)c^{1/3}(s)} + i\omega\tau_0^+(M) \right] . \tag{13.8.15}$$

Here

$$\tau_0^+(M) = l_0(M) + \frac{2}{3}m_0^{3/2}(M) = \tau^{\mathrm{inc}}(M_1) + \int_{M_1}^{M} \frac{ds}{c}$$

is the eikonal of the wave in the shadow zone,

$$T(M_j) = e^{-i\pi/8} \frac{2^{7/12}\pi^{1/4}}{w_1'(\xi_h)}$$

$$\times P^{1/6}(M_j)\omega^{-1/12}c^{-1/6}(M_j) \quad (j = 1, 2) . \tag{13.8.16}$$

And so, the field $u = u_{\mathrm{inc}} + u_{\mathrm{refl}}$, to within negligibly small terms, is equal to the sum of residues $\sum u_h$. Each of these terms in the deep shadow is exponentially small due to the presence of the factor

$$\exp\left[i\omega^{1/3}\xi_h \int_{M_1}^{M_2} \frac{ds}{2^{1/3}P^{2/3}(s)c^{1/3}(s)} \right]$$

(recall that $\arg \xi_h = \pi/3$). Thus, the asymptotics of $u_{\mathrm{inc}} + u_{\mathrm{refl}}$ is given by only the first residue term of the integral (13.8.1).

Equation (13.8.15) can, according to J.B. Keller (see Sect. 13.9), be interpreted in the following way. The incident wave travels along the ray $M'M_1$, and then creeps along the surface on the ray M_1M_2 tangent to S (Fig. 13.10), and finally sheds again as a space ray until it reaches the point M. Thus, the phase factor $\exp[i\omega\tau_0^+(M)]$ is the optical path length $M'M_1M_2M$ traversed by the wave.

The amplitude of the wave at M is clearly proportional to the amplitude at M_1 [that is $u_0^{\mathrm{inc}}(M_1)$]. A wave arriving at M_1 has two possibilities: either it continues its path along M_1M'', or it moves along the surface ray M_1M_2. There are no other possibilities — otherwise Fermat's principle would be violated. How much of the wave proceeds along M_1M'' and how much along the surface ray is determined by the boundary conditions and by the geometrical characteristics of the surface S at M_1. The point M_1 acts as if it were a policeman directing traffic

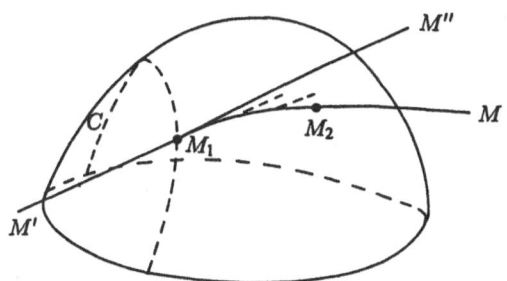

Fig. 13.10. Shedding of space rays from a surface ray

at an intersection: of the incident wave (proportional to u_0^{inc}) it allows only a certain portion – given by the factor $T(M_1)$ – to continue along the geodesic.

Waves moving along the surface S are continuously shedding their energy into waves whose rays are tangential to the arc $M_1 M_2$ (cf. the dashed line in Fig. 13.10). This shedding leads to a decay in amplitude, which is described in (13.8.15) by the factor

$$\exp\left[i\omega^{1/3}\frac{\xi_h}{2^{1/3}}\int\limits_{M_1}^{M_2}\frac{ds}{P^{2/3}(s)c^{1/3}(s)}\right] \; . \tag{13.8.17}$$

The physical content of this factor is explained by the following arguments. Assuming that the decrease in the amplitude of Ψ along a segment ds is proportional to the amplitude and to the segment length ds, we have

$$d\Psi = -\alpha(s)\Psi\,ds \tag{13.8.18}$$

where $\alpha(s)$ is a proportionality coefficient. It is natural to suppose that $\alpha(s)$ depends only on the boundary condition and geometrical properties of S at the point s itself. Then, integrating (13.8.18), we obtain that the attenuation is given by the factor $\exp[-\int_{M_1}^{M_2}\alpha(s)ds]$ which agrees with (13.8.17) if

$$\alpha(s) = \frac{-i\omega^{1/3}\xi_h}{2^{1/3}P^{2/3}(s)c^{1/3}(s)} \tag{13.8.19}$$

or, more precisely, the real part of the right side of (13.8.19).

In propagating along S the amplitude also decays by virtue of the geometrical divergence of the surface rays on S, which accounts for the presence of the factor $\sqrt{j(M_1)/j(M_2)}$ in (13.8.15). Now, how much of the wave moving along $M_1 M_2$ and arriving at M_2 will be shed into the ray $M_2 M$? Clearly, this portion will be proportional to the amplitude of the wave arriving at M_2. In (13.8.15), $T(M_2)$ is this proportionality factor. That $T(M_1)$ and $T(M_2)$ are given by the same function could have been predicted ahead of time from the symmetry of the Green's function in the source and observation points. In its propagation along the ray $M_2 M$, the wave's amplitude will be reduced by the geometrical divergence of the space rays. This accounts for the presence of the factor $J^{-1/2}(M)$ in (13.8.15).

And so, formula (13.8.15) is just what we must find if we start from clear physical reasoning (ingenious, even if it is rather speculative). This can be taken as one more confirmation of its validity.

13.9 Notes on the Literature

The notes to Chap. 11 mentioned papers dealing with asymptotic expansions of the Green's function in exterior problems using the parabolic equation method. We will not go through these again, but only cite papers here that have directly to do with the present chapter.

Ludwig's paper [13.5] gives the uniform asymptotic expansion for the solution of the diffraction of a given incident wave at a smooth convex surface S. Unfortunately, this paper contains a number of errors and inaccuracies. We will note some of them here.

To construct the asymptotic expansion of the reflected field, *Ludwig* used an eikonal which in our notation has the form $\tau^+ = \xi_{inc} + \frac{2}{3}\mu_{inc}^{3/2}$ (Sect. 13.3). We can easily construct an example (in three dimensions) where the direction of $\nabla\tau^+$ is not the same as that of the reflected ray passing through M tangent to S. This essentially invalidates all of *Ludwig*'s subsequent constructions for the three-dimensional case.

The neutralizing factor proposed by *Ludwig* [Ref. 13.5, Sect. 5], has the form $1 + R$ at interior points of the integration interval, where R is bounded above and below by expressions of the form const $\times k^{-1/3}$; const > 0. This neutralizing factor influences the asymptotic expansion of the diffracted field as early as the second term of the series.

We note further that *Ludwig*'s paper makes no mention of the choice of initial data for the equations which (in our notation) are satisfied by the coefficients $A_{n,j}$ in the partial sums (13.2.6).

However, as shown by *Babič*, the errors and inaccuracies in *Ludwig*'s work can be eliminated. Thus the principal importance of *Ludwig*'s work is in the ideas that it contains. These ideas formed the basis of this chapter, which differs to a considerable extent from *Ludwig*'s work in a number of the constructions and in the overall form of the whole development. On the basis of *Ludwig*'s work [13.5], and a priori integral estimates of solutions of the Helmholtz equation, *Morawetz* and *Ludwig* [13.10] rigorously substantiated the ray expansions for the Green's function of the Dirichlet problem in the illuminated region, and the asymptotic formulas for it in the penumbra.

A very important result of this chapter is (13.8.15) describing the field in the shadow behind a three-dimensional obstacle. On the basis of heuristic considerations (partially described in Sect. 13.7), this formula was first derived by *Keller* [13.11] in 1956 for the two-dimensional case, and was later obtained by *Levy* and *Keller* [13.12] for the three-dimensional case.

The derivation of *Fock*'s formula on the basis of *Ludwig*'s work is due to *Leont'ev*. Finding the initial data for the coefficients in the expansions of ξ and A_j, forsaking the use of a neutralizing factor, introducing the forked contours, deriving the formulas containing Fresnel integrals for the penumbra field on the basis of *Ludwig*'s formulas, and several other technical devices are all due to *Babič*. *Buslaev* [13.13] showed that the formula for the Green's function in the penumbra, obtained by the parabolic equation method, is indeed asymptotic. Amending the constructions of this chapter to deal with the case of Neumann boundary conditions presents no difficulty in the first approximation. The treatment of higher-order terms runs into a number of complications which have not yet been overcome. The higher approximations can be constructed by the method of matched asymptotic expansions [13.14, 15].

Appendix

A.1. The Airy Equation and Airy Functions[1]

The Airy equation is

$$w'' - tw = 0 \quad ; \quad w = w(t) \quad . \tag{A.1.1}$$

For linear equations, we know there can be singularities in the solutions only at points where the coefficients have singularities. Therefore solutions of the Airy equation cannot have singularities at any point of the complex plane. And so all solutions of the Airy equation are entire functions of t and can be expanded in powers of t in series that converge for any t.

Let us substitute the power series $w(t) = \sum_{k=0}^{\infty} C_k t^k$ ($C_0 = w(0)$, $C_1 = w'(0)$) with undetermined coefficients into Airy's equation, and set coefficients of like powers of t equal to zero. Then we readily see that any solution of the Airy equation takes the form

$$
\begin{aligned}
w(t) = w(0) &\left(1 + \frac{t^3}{2 \cdot 3} + \frac{t^6}{2 \cdot 3 \cdot 5 \cdot 6} + \cdots \right) \\
&+ w'(0) \left(t + \frac{t^4}{3 \cdot 4} + \frac{t^7}{3 \cdot 4 \cdot 6 \cdot 7} + \cdots \right) \quad .
\end{aligned}
\tag{A.1.2}
$$

The Airy equation can be reduced to Bessel's equation. It is pointed out in many handbooks that if $Z_\nu(x)$ is a solution of Bessel's equation of order ν, then the function

$$u = t^\alpha Z_\nu(\beta t^\gamma)$$

satisfies the equation

$$u''_{tt} + \frac{1 - 2\alpha}{t} u'_t + \left[(\beta \gamma t^{\gamma - 1})^2 + \frac{\alpha^2 - \nu^2 \gamma^2}{t^2} \right] u = 0 \quad .$$

If we require that this latter equation coincide with the Airy equation, we get

$$1 - 2\alpha = 0 \quad , \quad (\beta \gamma)^2 = -1 \quad , \quad 2\gamma - 2 = 1 \quad , \quad \alpha^2 - \nu^2 \gamma^2 = 0 \quad .$$

---

[1] Other expositions of the theory of the Airy equation and Airy functions can be found in the books by *Smirnov* [A.1.1], *Lebedev* [A.1.2], *Fock* [A.1.3], *Yakovleva* [A.1.4] and *Miller* [A.1.5]. The last three books contain tables of Airy functions.

These formulas imply that any solution w of the Airy equation has the form

$$w(t) = t^{1/2} Z_{1/3}\left(\tfrac{2}{3} i t^{3/2}\right) \quad ; \quad i = \sqrt{-1} \quad , \tag{A.1.3}$$

where $Z_{1/3}(x)$ is some solution of the Bessel equation of order 1/3.

Using the theory of Bessel functions and formula (A.1.3), it is easy to get integral representations and asymptotic formulas for solutions of the Airy equation as $|t| \to \infty$. However, it is more convenient to make a direct examination of the Airy equation.

Let us first obtain integral representations for the solutions of the Airy equation. Let us note that this equation belongs to the class of so called Laplace equations, i.e. equations whose coefficients are linear in the independent variables. The solutions of these equations can be represented by Laplace integrals. According to the general rule, we will seek the solution of (A.1.1) in the form

$$w(t) = \int_{\mathcal{L}} e^{tz} s(z) dz \quad ,$$

where $\mathcal{L}$ is some contour in the complex z-plane.

Substituting into the Airy equation and integrating by parts, we get

$$w'' - tw = -[s(z) e^{tz}]_{\mathcal{L}} + \int_{\mathcal{L}} e^{tz} [z^2 s(z) + s'(z)] dz = 0 \quad ,$$

where $[s(z) e^{tz}]_{\mathcal{L}}$ is the increment of $s(z) e^{tz}$ over the length of $\mathcal{L}$. To satisfy the equation, it is sufficient to take $s(z)$ and the contour $\mathcal{L}$ such that the relations

$$s' + z^2 s(z) = 0 \quad ; \quad [s(z) e^{tz}]_{\mathcal{L}} = 0$$

are satisfied.

Whence

$$s(z) = \text{const}\, e^{-z^3/3} \quad ,$$
$$w(t) = \text{const} \int_{\mathcal{L}} e^{tz - z^3/3} dz, \quad [e^{tz - z^3/3}]_{\mathcal{L}} = 0 \quad . \tag{A.1.4}$$

When $z \to \infty$ the function $e^{tz - z^3/3}$ approaches zero in the sectors $|\arg z| < \pi/6$; $\pi/2 < \arg z < 5\pi/6$ and $-5\pi/6 < \arg z < -\pi/2$. The latter two sectors are obtained from the sector $|\arg z| < \pi/6$ by rotations through angles $\pm 2\pi/3$. To satisfy the condition $[\exp(tz - z^3/3)]_{\mathcal{L}} = 0$, (and end up with a nontrivial result) we must take one of the contours which begins at infinity in one of these sectors, passes into another sector and goes out to infinity within this second sector (Fig. A.1.1).

We introduce the contours $\mathcal{L}_j$, $j = 1, 2, 3$, formed by the half-lines

$$z = r e^{2\pi i(j-2)/3} \quad , \quad 0 \leq r < \infty \quad , \quad \text{and}$$
$$z = r e^{2\pi i(j-1)/3} \quad , \quad 0 \leq r < \infty \quad , \quad j = 1, 2, 3 \quad . \tag{A.1.5}$$

The contours $\mathcal{L}_j$ are shown in Fig. A.1.1. The arrows denote the direction of integration. We define the solutions $W_j(t)$ of Airy's equation by

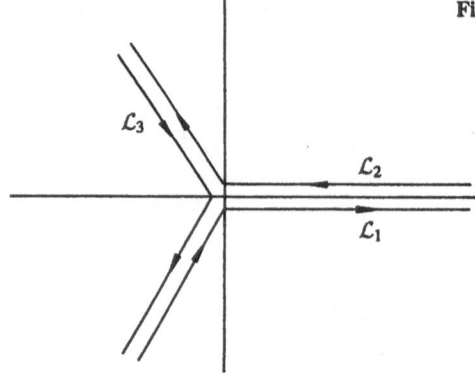

$$W_j(t) = \int\limits_{\mathcal{L}_j} e^{tz - z^3/3}\, dz \quad . \tag{A.1.6}$$

Clearly,

$$W_1(t) + W_2(t) + W_3(t) = 0 \quad . \tag{A.1.7}$$

The set of solutions W_j connected by relation (A.1.7) is a very natural one for Airy's equation from a variety of standpoints. This naturalness shows itself first of all in the remarkable symmetry of the transformation formulas for $W_j(t)$ when t is replaced by $t\exp(\pm 2\pi i/3)$.

Airy's equation has an important property: if a function $w(t)$ is a solution of it, then the functions $w[t\exp(\pm 2\pi i/3)]$ are also solutions. This is easily verified using (A.1.2), or by making the changes of variable $t_1 = t\exp(\pm 2\pi i/3)$ in (A.1.1).

This is how the $W_j(t)$ are transformed when $t \to t\exp(\pm 2\pi i/3)$. For our present purposes, we will reckon that j in the definition of the contour $\mathcal{L}_j$ can be any integer. The rays (A.1.5) are invariant under the replacement of j by $j + 3n$ (n an integer), and so $\mathcal{L}_{j+3n} = \mathcal{L}_j$. Letting the index j in (A.1.6) likewise to be an arbitrary integer, we clearly have $W_{j+3n} = W_j$ for n an integer. We perform the change of variable $z = z_1\exp(2\pi i/3)$ in (A.1.6). Upon doing so, the contour $\mathcal{L}_j$ becomes $\mathcal{L}_{j-1}$, and we obtain

$$W_j(t) = e^{2\pi i/3} \int\limits_{\mathcal{L}_{j-1}} \exp(t\,e^{2\pi i/3}z_1 - z_1^3/3)dz_1$$

$$= e^{2\pi i/3}\, W_{j-1}(t\,e^{2\pi i/3}) \quad .$$

Replacing j by $j + 1$, or t by $t\exp(-2\pi i/3)$, we get the relations

$$W_{j+1}(t) = e^{2\pi i/3}\, W_j(t\,e^{2\pi i/3})$$

$$W_{j-1}(t) = e^{-2\pi i/3}\, W_j(t\,e^{-2\pi i/3}) \quad . \tag{A.1.8}$$

The naturalness of choosing the set of functions W_j becomes apparent not only in the symmetry of formulas (A.1.7–8), but also (as will become clear later) in the behavior of $W_j(t)$ as $t \to \infty$ in the complex plane.

However in practice we use not the functions W_j, but the functions[2] $w_1(t)$, $w_2(t)$ and $v(t)$ which are proportional to the W_j:

$$w_1(t) = \frac{1}{\sqrt{\pi}} W_1(t) = \frac{1}{\sqrt{\pi}} \int_{L_1} e^{tz - z^3/3} \, dz \qquad (A.1.9)$$

$$w_2(t) = -\frac{1}{\sqrt{\pi}} W_2(t) = -\frac{1}{\sqrt{\pi}} \int_{L_2} e^{tz - z^3/3} \, dz \qquad (A.1.10)$$

$$v(t) = \frac{i}{2\sqrt{\pi}} W_3(t) = \frac{i}{2\sqrt{\pi}} \int_{L_3} e^{tz - z^3/3} \, dz \quad . \qquad (A.1.11)$$

The use of w_1, w_2 and v rather than W_j is accounted for by the fact that $v(t)$ is real when t is real (W_3 does not have this property) and by the relations

$$v(t) = \text{Im}(w_1(t)) = -\text{Im}(w_2(t))$$

which are valid for real t. These latter relations as well as the reality of v follow from the readily verified equations

$$\overline{w_1(t)} = w_2(\bar{t}) \qquad (A.1.12)$$

$$v(t) = \frac{w_1(t) - w_2(t)}{2i} \quad . \qquad (A.1.13)$$

We can transform the integral (A.1.11) into a real form. Deforming the contour of integration onto the imaginary axis, we obtain

$$v(t) = \frac{1}{2i\sqrt{\pi}} \int_{-i\infty}^{i\infty} e^{tz - z^3/3} \, dz \quad .$$

Performing the change of variable $z = ix$ and using

$$\exp[i(tx + x^3/3)] = \cos(tx + x^3/3) + i \sin(tx + x^3/3)$$

while observing that the first of these terms is even and the second odd for real t, we arrive at

$$v(t) = \frac{1}{\sqrt{\pi}} \int_0^\infty \cos(tx + x^3/3) \, dx \quad (\text{Im}\{t\} = 0) \quad . \qquad (A.1.14)$$

Often the Airy function $u(t)$ is also used as a linearly independent solution of (A.1.1) from $v(t)$ which is also real for real t. It is defined as

$$u(t) = \frac{w_1(t) + w_2(t)}{2} \quad . \qquad (A.1.15)$$

---

[2] We follow *Fock* [A.1.3, 6] in our definitions and notation for the Airy functions w_1, w_2 and v.

Manipulations similar to those done in the derivation of (A.1.14) lead to

$$u(t) = \frac{1}{\sqrt{\pi}} \int\limits_0^\infty \left[\sin(tx + x^3/3) + \exp(tx - x^3/3)\right]dx \quad (\mathrm{Im}\{t\} = 0) \quad .$$

It is evident that when t is real we also have

$$u(t) = \mathrm{Re}\{w_1(t)\} = \mathrm{Re}\{w_2(t)\} \quad .$$

Relations between the functions w_1, w_2, v and u which are valid for arbitrary complex t follow from (A.1.8) and have the form

$$w_1\left(t\,e^{2\pi i/3}\right) = e^{\pi i/3}\, w_2(t) \quad ; \quad w_1\left(t\,e^{-2\pi i/3}\right) = 2\,e^{\pi i/6}\, v(t)$$

$$w_2\left(t\,e^{2\pi i/3}\right) = 2\,e^{-\pi i/6}\, v(t) \quad ; \quad w_2\left(t\,e^{-2\pi i/3}\right) = e^{-\pi i/3}\, w_1(t)$$

$$v\left(t\,e^{2\pi i/3}\right) = \tfrac{1}{2}e^{-\pi i/6}\, w_1(t) \quad ; \quad v\left(t\,e^{-2\pi i/3}\right) = \tfrac{1}{2}e^{\pi i/6}\, w_2(t). \quad \text{(A.1.16)}$$

Expressing $w_1(t)$ and $w_2(t)$ in terms of $v(t)$ by (A.1.16) and using (A.1.15), we obtain

$$u(t) = e^{i\pi/6}\, v\left(t\,e^{2\pi i/3}\right) + e^{-i\pi/6}\, v\left(t\,e^{-2\pi i/3}\right) \quad . \tag{A.1.17}$$

In the western literature it is more common to use the two linearly independent Airy functions $Ai(t)$ and $Bi(t)$, which differ from the functions $v(t)$ and $u(t)$ only by a constant factor:

$$Ai(t) = \frac{1}{\sqrt{\pi}}v(t) \quad ; \quad Bi(t) = \frac{1}{\sqrt{\pi}}u(t) \quad .$$

From (A.1.9) it is not difficult to find $w_1(0)$ and $w_1'(0)$. These quantities are expressed in terms of Euler's gamma function $\Gamma(x)$:

$$w_1(0) = \frac{2\sqrt{\pi}\,e^{\pi i/6}}{3^{2/3}\Gamma(\frac{2}{3})} \quad ; \quad w_1'(0) = \frac{2\sqrt{\pi}\,e^{-\pi i/6}}{3^{4/3}\Gamma(\frac{4}{3})} \tag{A.1.18}$$

whence

$$w_2(0) = \overline{w_1(0)} = \frac{2\sqrt{\pi}}{3^{2/3}\Gamma(\frac{2}{3})}e^{-i\pi/6} \quad ;$$

$$w_2'(0) = \overline{w_1'(0)} = \frac{2\sqrt{\pi}}{3^{4/3}\Gamma(\frac{4}{3})}e^{i\pi/6} \quad , \tag{A.1.19}$$

$$v(0) = \mathrm{Im}\{w_1(0)\} = \frac{\sqrt{\pi}}{3^{2/3}\Gamma(\frac{2}{3})} \quad ,$$

$$v'(0) = \mathrm{Im}\{w_1'(0)\} = -\frac{\sqrt{\pi}}{3^{4/3}\Gamma(\frac{4}{3})}. \tag{A.1.20}$$

Formula (A.1.18) and the values of $w_k(0)$, $w_k'(0)$, $v(0)$, $v'(0)$ enable us to write out the power series (A.1.2) for $w_k(t)$ and $v(t)$.

In applications we often encounter the Wronskian determinants of various pairs of linearly independent Airy functions. It is not difficult to obtain their values using (A.1.18–20) (or with the help of the asymptotic expansions given below):

$$W[w_1, w_2] = w_1 \frac{dw_2}{dt} - w_2 \frac{dw_1}{dt} = 2$$

$$W[w_1, v] = w_1 \frac{dv}{dt} - v \frac{dw_1}{dt} = i$$

$$W[w_2, v] = w_2 \frac{dv}{dt} - v \frac{dw_2}{dt} = i \qquad (A.1.21)$$

$$W[u, v] = u \frac{dv}{dt} - v \frac{du}{dt} = 1 \quad . \qquad (A.1.22)$$

Asymptotic formulas for the Airy functions are of considerable interest. Before deriving these formulas (using the method of steepest descent) we will write them out in their entirety, and abstract from them some simple consequences that relate to the behavior of the Airy functions and their derivatives in the complex plane, and of their zeros. The structure of the asymptotic formulas for complex t, $|t| \to \infty$, depends on $\arg t$. In (A.1.23–30) we indicate for which sector each formula holds true:

$$w_1(t) \sim \frac{1}{\sqrt{\pi}} \frac{e^{(2/3)t^{3/2}}}{t^{1/4}} \sum_{n=0}^{\infty} \frac{\Gamma(3n + \frac{1}{2})}{(2n)!} (9t^{3/2})^{-n}$$

$$\sim t^{-1/4} e^{(2/3)t^{3/2}} \left(1 + O\left(\frac{1}{t^{3/2}} \right) \right) \quad ,$$

$$(-4\pi/3 \leq \arg t \leq 0) \qquad (A.1.23)$$

$$w_2(t) \sim \frac{1}{\sqrt{\pi}} \frac{e^{(2/3)t^{3/2}}}{t^{1/4}} \sum_{n=0}^{\infty} \frac{\Gamma(3n + \frac{1}{2})}{(2n)!} (9t^{3/2})^{-n}$$

$$\sim t^{-1/4} e^{(2/3)t^{3/2}} \left(1 + O\left(\frac{1}{t^{3/2}} \right) \right) \quad ,$$

$$(0 \leq \arg t \leq 4\pi/3) \qquad (A.1.24)$$

$$v(t) \sim \frac{1}{\sqrt{\pi}} \frac{e^{-(2/3)t^{3/2}}}{2t^{1/4}} \sum_{n=0}^{\infty} \frac{\Gamma(3n + \frac{1}{2})}{(2n)!} (-9t^{3/2})^{-n}$$

$$\sim \frac{1}{2} t^{-1/4} e^{-(2/3)t^{3/2}} \left(1 + O\left(\frac{1}{t^{3/2}} \right) \right) \quad ,$$

$$(-2\pi/3 \leq \arg t \leq 2\pi/3) \qquad (A.1.25)$$

$$v(-t) = \frac{1}{2} e^{-i\pi/6} w_1(e^{i\pi/3} t) = \frac{1}{2} e^{i\pi/6} w_2(e^{-i\pi/3} t)$$

$$\sim \frac{1}{\sqrt{\pi}} t^{-1/4} \sin \left(\frac{2}{3} t^{3/2} + \frac{\pi}{4} \right) \sum_{n=0}^{\infty} (-1)^n \frac{\Gamma(6n + \frac{1}{2})}{(4n)!} (9t^{3/2})^{-2n}$$

$$- \frac{1}{\sqrt{\pi}} t^{-1/4} \cos \left(\frac{2}{3} t^{3/2} + \frac{\pi}{4} \right) \sum_{n=0}^{\infty} (-1)^n \frac{\Gamma(6n + \frac{7}{2})}{(4n + 2)!} (9t^{3/2})^{-2n-1}$$

$$\sim t^{-1/4} \sin\left(\frac{2}{3}t^{3/2} + \frac{\pi}{4}\right)[1 + O(t^{-3})]$$

$$- \frac{5}{48}t^{-7/4} \cos\left(\frac{2}{3}t^{3/2} + \frac{\pi}{4}\right)[1 + O(t^{-3})]$$

$$(-\pi/3 \le \arg t \le \pi/3) \quad . \tag{A.1.26}$$

An asymptotic formula for $u(t)$ can be obtained using (A.1.15) or (A.1.17) and the asymptotic expansions above for w_1, w_2 and v:

$$u(t) \sim \frac{1}{\sqrt{\pi}} \frac{e^{(2/3)t^{3/2}}}{t^{1/4}} \sum_{n=0}^{\infty} \frac{\Gamma(3n + \frac{1}{2})}{(2n)!}(9t^{3/2})^{-n}$$

$$\sim t^{-1/4} e^{(2/3)t^{3/2}} \left(1 + O\left(\frac{1}{t^{3/2}}\right)\right) \quad , \quad \arg t = 0 \quad . \tag{A.1.27}$$

$$u(t) \sim \frac{1}{\sqrt{\pi}} \frac{e^{(2/3)t^{3/2}}}{t^{1/4}} \sum_{n=0}^{\infty} \frac{\Gamma(3n + \frac{1}{2})}{(2n)!}(9t^{3/2})^{-n} + \frac{i}{2\sqrt{\pi}} \frac{e^{-(2/3)t^{3/2}}}{t^{1/4}}$$

$$\times \sum_{n=0}^{\infty} \frac{\Gamma(3n + \frac{1}{2})}{(2n)!}(9t^{3/2})^{-n} \sim t^{-1/4} e^{(2/3)t^{3/2}} \left(1 + O\left(\frac{1}{t^{3/2}}\right)\right)$$

$$+ \frac{i}{2}t^{-1/4} e^{-(2/3)t^{3/2}} \left(1 + O\left(\frac{1}{t^{3/2}}\right)\right) \quad ,$$

$$0 < \arg t \le \frac{2\pi}{3} \quad ; \tag{A.1.28}$$

$$u(t) \sim \frac{1}{\sqrt{\pi}} \frac{e^{(2/3)t^{3/2}}}{t^{1/4}} \sum_{n=0}^{\infty} \frac{\Gamma(3n + \frac{1}{2})}{(2n)!}(9t^{3/2})^{-n} - \frac{i}{2\sqrt{\pi}} \frac{e^{-(2/3)t^{3/2}}}{t^{1/4}}$$

$$\times \sum_{n=0}^{\infty} \frac{\Gamma(3n + \frac{1}{2})}{(2n)!}(9t^{3/2})^{-n} \sim t^{-1/4} e^{(2/3)t^{3/2}} \left(1 + O\left(\frac{1}{t^{3/2}}\right)\right)$$

$$- \frac{i}{2}t^{-1/4} e^{-(2/3)t^{3/2}} \left(1 + O\left(\frac{1}{t^{3/2}}\right)\right) \quad ,$$

$$- \frac{2\pi}{3} \le \arg t < 0 \quad ; \tag{A.1.29}$$

$$u(-t) \sim \frac{1}{\sqrt{\pi}} t^{-1/4} \cos\left(\frac{2}{3}t^{3/2} + \frac{\pi}{4}\right) \sum_{n=0}^{\infty} (-1)^n \frac{\Gamma(6n + \frac{1}{2})}{(4n)!}(9t^{3/2})^{-2n}$$

$$- \frac{1}{\sqrt{\pi}} t^{-1/4} \sin\left(\frac{2}{3}t^{3/2} + \frac{\pi}{4}\right) \sum_{n=0}^{\infty} (-1)^n \frac{\Gamma(6n + \frac{7}{2})}{(4n + 2)!}(9t^{3/2})^{-2n-1}$$

$$\sim t^{-1/4} \cos\left(\frac{2}{3}t^{3/2} + \frac{\pi}{4}\right)[1 + O(t^{-3})]$$

$$+ \frac{5}{48}t^{-7/4} \sin\left(\frac{2}{3}t^{3/2} + \frac{\pi}{4}\right)$$

$$\times [1 + O(t^{-3})] \quad , \quad (-\pi/3 \le \arg t \le \pi/3) \quad . \tag{A.1.30}$$

A few words of explanation are in order for these formulas. The multi-valued functions t^β which enter into them are evaluated on their principal branches, i.e., $t^\beta = |t|^\beta e^{i\beta \arg t}$. At the boundaries of the sector indicated, formula (A.1.26) reproduces the results of formulas (A.1.23–25) to within quantities that are exponentially small as $|t| \to \infty$. We will verify this for $v(t)$ along the rays $\arg t = \pm 2\pi/3$. We put $-t = t e^{i\pi}$ in place of t in (A.1.26) (as a result of which $\arg t$ now varies from $2\pi/3$ to $4\pi/3$) and use Euler's formula. As a result, we obtain

$$v(t) \sim \frac{1}{\sqrt{\pi}} \frac{e^{-(2/3)t^{3/2}}}{2t^{1/4}} \sum_{n=0}^{\infty} \frac{\Gamma(3n + \tfrac{1}{2})}{(2n)!}(-9t^{3/2})^{-n}$$

$$-\frac{1}{\sqrt{\pi}} \frac{e^{(2/3)t^{3/2}}}{2it^{1/4}} \sum_{n=0}^{\infty} \frac{\Gamma(3n + \tfrac{1}{2})}{(2n)!}(9t^{3/2})^{-n} . \tag{A.1.31}$$

For $\arg t = 2\pi/3$ we have $t^{3/2} = -|t|^{3/2}$, and thus the second term of (A.1.31) is exponentially small and can be dropped, after which (A.1.31) is identical with (A.1.25). If $\arg t = 4\pi/3$, then $t^{3/2} = |t|^{3/2}$, and the first term of (A.1.31) can be dropped. Replacing t by $t e^{2\pi i}$ in the second term, we get

$$v(t)\big|_{\arg t = -2\pi/3} \sim \frac{1}{\sqrt{\pi}} \frac{e^{-(2/3)t^{3/2}}}{2t^{1/4}} \sum_{n=0}^{\infty} \frac{\Gamma(3n + \tfrac{1}{2})}{(2n)!}(-9t^{3/2})^{-n}$$

which again is the same as (A.1.25).

From the foregoing considerations it follows that the asymptotic formula (A.1.26) for $v(t)$ in the sector $|\arg t - \pi| \leq \pi/3$ can be obtained in the following way. Analytically continue the expression in (A.1.25) into the sector $|\arg t - \pi| \leq \pi/3$ both from above, through the ray $\arg t = 2\pi/3$, and from below, through the ray $\arg t = 4\pi/3$ or $\arg t = -2\pi/3$. As a result, we have two asymptotic series at each point t in the sector $|\arg t - \pi| \leq \pi/3$, in which we must take $2\pi/3 \leq \arg t \leq 4\pi/3$ in the first and $-4\pi/3 \leq \arg t \leq -2\pi/3$ in the second. *The asymptotic expansion of $v(t)$ in this range is given by their sum.* Replacing t by $t e^{-2\pi i}$ in the second series, we obtain (A.1.31), wherein $2\pi/3 \leq \arg t \leq 4\pi/3$ for both series; and finally, replacing t by $t e^{i\pi}$, we get (A.1.26).

In similar fashion, the analytic continuation of (A.1.23) into the sector $|\arg t - \pi/3| < \pi/3$ through the rays $\arg t = 0$ and $\arg t = -4\pi/3$ leads to formula (A.1.26) for $w_1(t e^{i\pi/3})$. The continuation of (A.1.24) into $|\arg t + \pi/3| < \pi/3$ through the rays $\arg t = 0$ and $\arg t = 4\pi/3$ leads to (A.1.26) for $w_2(t e^{-i\pi/3})$. We also note that the ranges of validity for (A.1.23–25) which contain only one asymptotic series each, can be widened: (A.1.23) is valid in the sector $-5\pi/3 + \delta \leq \arg t \leq \pi/3 - \delta$; (A.1.24) is valid in $-\pi/3 + \delta \leq \arg t \leq 5\pi/3 - \delta$; (A.1.25) is valid in $-\pi + \delta \leq \arg t \leq \pi - \delta$ (in all these formulas, δ is a sufficiently small positive value). These extensions are possible because the replacement of the sines and cosines by exponentials in (A.1.26) results in some terms which are exponentially small compared to the remaining ones when $|\arg t - \pi| \geq \delta \, (\delta > 0)$ and thus can be dropped.

The asymptotic formulas (A.1.27–30) for $u(t)$ do not enjoy the same self-consistency as do (A.1.23–26). This is because the function $u(t)$ is not described by a contour integral over one of the paths $\mathcal{L}_j$.

The asymptotic formulas (A.1.23–30) allow us to characterize the behavior of the Airy functions $v(t)$, $w_1(t)$, $w_2(t)$ and $u(t)$ in the complex t-plane, and at the same time to decide when to use each kind of Airy function.

We divide the t-plane into three sectors:

Sector I : $-\pi/3 < \arg t < \pi/3$

Sector II : $\pi/3 < \arg t < \pi$

Sector III : $-\pi < \arg t < -\pi/3$.   (A.1.32)

As is evident from (A.1.25), the function $v(t)$ decays exponentially as $|t| \to \infty$ in sector I. In sectors II and III, $v(t)$ grows exponentially as $|t| \to \infty$ [Eq. (A.1.25), as we have noted, can be used for $|\arg t| \le \pi - \delta$, $\delta > 0$]. The fastest decay $(|v(t)| \sim \frac{1}{2}|t|^{-1/4} \exp(-\frac{2}{3}|t|^{3/2}))$ occurs when $\arg t = 0$, i.e., along the bisector of sector I. The fastest growth of $v(t)$ $(|v(t)| \sim \frac{1}{2}|t|^{-1/4} \exp(\frac{2}{3}|t|^{3/2}))$ occurs on the bisectors of sectors II and III, i.e., on $\arg t = \pm 2\pi/3$. In Fig. A.1.2, the sector of decay, sector I, is shaded, while the directions of fastest growth are indicated by thick lines. On the boundaries of sector I, i.e., $\arg t = \pm \pi/3$, the function $v(t)$ decays only algebraically, its real and imaginary parts oscillating with a phase shift of $\pi/2$:

$$v(t) \sim \tfrac{1}{2} e^{\mp \pi i/12} |t|^{-1/4} \exp\left(\mp i\tfrac{2}{3}|t|^{3/2}\right) \quad .$$

These directions are indicated on Fig. A.1.2 by the symbol $\sim$. On the negative real axis ($\arg t = \pi$) the function $v(t)$ is real, decays as $|t|^{-1/4}$, and its asymptotics (as well as the function itself) change sign an infinite number of times. Hence it follows that $v(t)$ has an infinite number of zeroes on the negative real axis. It will be shown below that all the zeroes of $v(t)$ are real and negative.

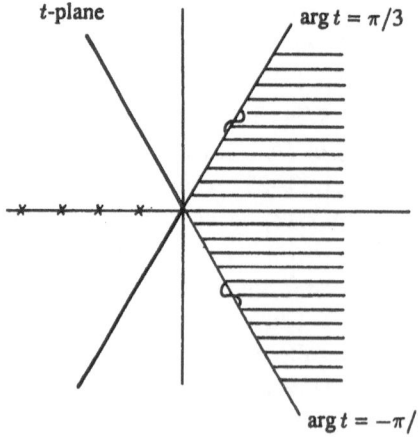

t-plane

$\arg t = \pi/3$

$\arg t = -\pi/3$

Fig. A.1.2. Behavior of $v(t)$ in the complex t-plane

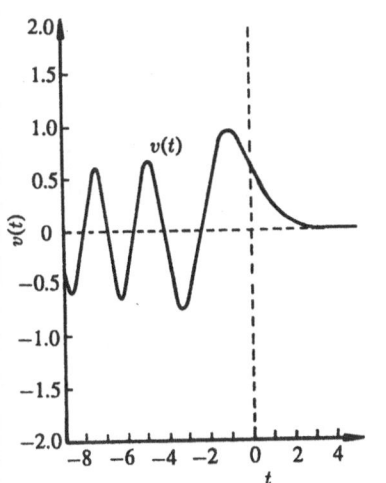

Fig. A.1.3. Behavior of $v(x)$ for real x

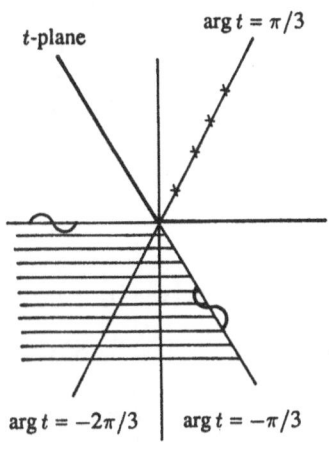

Fig. A.1.4. Behavior of $w_1(t)$ in the complex t-plane

Thus, $v(t)$ oscillates on the real axis as t goes from $-\infty$ to 0, and its amplitude increases, taking on a maximum positive value in the neighborhood of $t = 0$, and then monotonically decreasing as t increases for $t > 0$. This behavior of $v(t)$ for real t is shown in Fig. A.1.3.

The Airy function $v(t)$ is used anytime we wish to describe a smooth transition between a region of oscillation and a region of decay. Clearly, $v(t)$ is the simplest entire function which behaves this way on the real axis.

The behavior of the functions $w_1(t)$ and $w_2(t)$ in the complex t-plane (cf. Figs. A.1.4 and A.1.5) is completely analogous to that of $v(t)$, and differs from it only by the rotation of the t-plane by an angle of $-2\pi/3$ for $w_1(t)$, or by an angle of $2\pi/3$ for $w_2(t)$. In Figs. A.1.4 and A.1.5, the same notations are used as in Fig. A.1.2 to denote the sector of decay, the directions of fastest growth, and the directions of oscillating real and imaginary parts. The crosses in all these figures mark the rays on which the zeroes of the corresponding Airy functions are

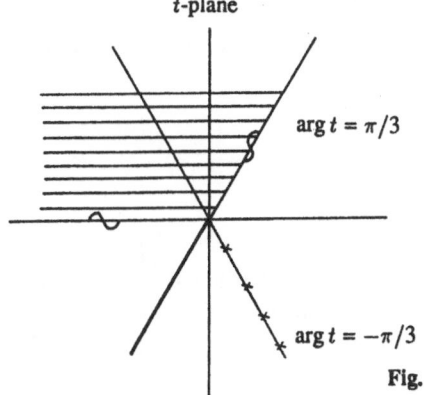

Fig. A.1.5. Behavior of $w_2(t)$ in the complex t-plane

409

located. We direct the reader's attention to the asymptotics of $w_1(t)$ and $w_2(t)$ for real positive t. Formulas (A.1.23) and (A.1.24) for $\arg t = 0$ lead to purely real values for these functions. But in fact the functions $w_1(t)$ and $w_2(t)$ for $\arg t = 0$ are complex-valued, although their imaginary parts are exponentially small, and thus do not show up in the asymptotic description. Because $\mathrm{Im}\{w_1(t)\} = v(t)$ and $\mathrm{Im}\{w_2(t)\} = -v(t)$ for real t, the asymptotics of $\mathrm{Im}\{w_1(t)\}$ and $\mathrm{Im}\{w_2(t)\}$ for $\arg t = 0$ have the form

$$\mathrm{Im}\{w_1(t)\} = -\mathrm{Im}\{w_2(t)\} = \frac{\exp\left(-\frac{2}{3}t^{3/2}\right)}{2t^{1/4}}(1 + O(t^{-3/2})) \quad .$$

The asymptotic behaviors of $w_1(t)$ and $w_2(t)$ as $t \to -\infty$ are

$$w_1(t) \sim |t|^{-1/4} \exp\left[i\left(\frac{2}{3}|t|^{3/2} + \frac{\pi}{4}\right)\right]$$

$$w_2(t) \sim |t|^{-1/4} \exp\left[-i\left(\frac{2}{3}|t|^{3/2} + \frac{\pi}{4}\right)\right]$$

[cf. (A.1.23) and (A.1.24)], and these dictate the use of these functions in situations when traveling waves at infinity must be described. When describing outgoing waves at infinity with a time convention $e^{-i\omega T}$ (where here we take T to be time), we use w_1; for a time factor of $e^{i\omega T}$, we use w_2.

For the function $u(t)$, all three sectors of (A.1.32) are sectors of growth. The zeroes of $u(t)$ are located on the ray $\arg t = \pi$, that is, on the negative real axis; as $t \to \infty$ the function $u(t)$ grows exponentially. The use of $u(t)$ or $Bi(t)$ is convenient only in situations when it is necessary to have a second linearly-independent solution of Airy's equation in addition to $v(t)$ which is also real for real t.

Let us now prove that all zeroes of $v(t)$ and of its derivative are located on the ray $\arg t = \pi$. Let $-t_s$ be a root of $v(t)$. Then the function $y_s(t) = v(-t_s + t)$ obeys the conditions

$$-y_s'' + t y_s = t_s y_s \quad , \quad y_s\big|_{t=0} = 0 \quad , \quad y_s\big|_{t=+\infty} = 0 \quad , \tag{A.1.33}$$

i.e. the t_s can be considered as eigenvalues of the boundary value problem (A.1.33). Multiplying the equation for $y_s(t)$ by the complex conjugate functions $\bar{y}_s(t)$ and integrating by parts, we get

$$t_s \int_0^\infty |y_s|^2 \, dt = \int_0^\infty (t|y_s|^2 + |y_s'|^2) \, dt \quad .$$

(The boundary terms vanish by virtue of the boundary condition $y_s\big|_{t=0} = 0$ and the asymptotic behavior of y_s as $t \to \infty$.) The latter equality implies that t_s is real and positive. Thus, the roots of $v(t)$ take the form $-t_s$, where $t_s > 0$.

The negativity of roots $-t_s'$ of the derivative of $v(t)$ is proved analogously. The asymptotic expansion of the function $v(t)$ as $t \to -\infty$ implies that $v(t)$ has

410

infinitely many roots [this fact can also be easily derived from the discreteness of the spectrum of the problem (A.1.33)].

We give here numerical values of the first several roots of the functions $v(-t)$ and $v'(-t)$:

s	t_s	t'_s
1	2.33811	1.01879
2	4.08795	3.24820
3	5.52056	4.82010
4	6.78671	6.16331
5	7.94417	7.37218

Based on the asymptotic expansion of the function v, we readily get the asymptotic expansions of t_s and t'_s as $s \to \infty$:

$$t_s = \left[\frac{3}{2}\pi\left(s - \frac{1}{4}\right) + O\left(\frac{1}{s}\right)\right]^{2/3}$$

$$= \left[\frac{3}{2}\pi\left(s - \frac{1}{4}\right)\right]^{2/3} + O\left(\frac{1}{s^{4/3}}\right) \ , \tag{A.1.34}$$

$$t'_s = \left[\frac{3}{2}\pi\left(s - \frac{3}{4}\right) + O\left(\frac{1}{s}\right)\right]^{2/3}$$

$$= \left[\frac{3}{2}\pi\left(s - \frac{3}{4}\right)\right]^{2/3} + O\left(\frac{1}{s^{4/3}}\right) \ , \tag{A.1.35}$$

Asymptotic formulas (A.1.34) and (A.1.35) give a satisfactory approximation for t_s and t'_s as early as $s = 5$. In fact

$$\left[\frac{3}{2}\pi\left(5 - \frac{1}{4}\right)\right]^{2/3} \approx 7.9425 \ , \quad \left[\frac{3}{2}\pi\left(5 - \frac{3}{4}\right)\right]^{2/3} \approx 7.3748 \ .$$

With increasing s, the accuracy of the asymptotic formulas (A.1.34) and (A.1.35) improves rapidly.

Clearly, the roots of the functions $w_1(t)$ and $w'_1(t)$ have the form $t_s e^{i\pi/3}$ and $t'_s e^{i\pi/3}$ respectively, while the roots of $w_2(t)$ and $w'_2(t)$ have the form $t_s e^{-i\pi/3}$ and $t'_s e^{-i\pi/3}$.

Let us finally show how the asymptotic expansions given above for the functions v, w_1 and w_2 can be derived. As the starting point in this task we use the integral representations (A.1.6) for the Airy functions $W_1(t)$, $W_2(t)$ and $W_3(t)$. These integral representations have the form

$$\text{const} \int_{\mathcal{L}} e^{tz - z^3/3} \, dz \ ,$$

where $\mathcal{L}$ is the integration contour.

Setting $t = \varrho e^{i\theta}$, $0 \le \theta < 2\pi$, and making the change of variable $z = \sqrt{\varrho}\,\zeta$, we reduce the latter integral to the classical form of the integral of the method of steepest descents (cf. [A.1.7] or [A.1.8]):

$$\text{const} \int_L e^{tz - z^3/3} \, dz = \sqrt{\varrho} \int_L \exp\left[\varrho^{3/2}\left(e^{i\theta}\zeta - \tfrac{1}{3}\zeta^3\right)\right] d\zeta \quad . \qquad (A.1.36)$$

To find the asymptotic expansions of the Airy functions for complex t as $|t| \to \infty$, it is sufficient to find the asymptotic expansion of this integral as $\varrho \to +\infty$ ($0 \le \theta < 2\pi$). The saddle points, as implied by the equation

$$\frac{d}{d\zeta}\left(e^{i\theta}\zeta - \tfrac{1}{3}\zeta^3\right) = e^{i\theta} - \zeta^2 = 0 \quad , \quad \text{are equal to } \zeta_{1,2} = \pm e^{i\theta/2} \quad .$$

The contours of steepest descent (ascent) for the function $\text{Re}\{e^{i\theta}\zeta - \tfrac{1}{3}\zeta^3\}$ which pass through the saddle points, cannot terminate at a finite point of the ζ-plane in view of the regularity of the phase function $e^{i\theta}\zeta - \tfrac{1}{3}\zeta^3$. The asymptotes for $|\zeta| \to \infty$ of any curve

$$\text{Im}\left\{e^{i\theta}\zeta - \tfrac{1}{3}\zeta^3\right\} - C = 0$$

are determined by the leading term on the left side, and take the form $\text{Im}\,\zeta^3 = 0$, or

$$\arg\zeta = 0, \frac{\pi}{3}, \frac{2\pi}{3}, \pi, \frac{4\pi}{3}, \frac{5\pi}{3} \quad .$$

Thus the contours of steepest descent (ascent) for $\text{Re}\{e^{i\theta}\zeta - \tfrac{1}{3}\zeta^3\}$ which pass through the saddle points, go out to infinity, approaching the bisectors of the sectors where $\text{Re}\{-\zeta^3\}$ is negative (or positive), i.e. the half-lines $\arg\zeta = 0$, $2\pi/3$, $4\pi/3$ (or $\arg\zeta = \pi/3$, π, $5\pi/3$).

Consider the paths of steepest descent that pass through the saddle points $\pm e^{i\theta/2}$. Depending on what asymptotes the path has, the integral over that path from $(1/\sqrt{\pi})\sqrt{\varrho}\exp[\varrho^{3/2}(e^{i\theta}\zeta - \tfrac{1}{3}\zeta^3)]$ will be equal to w_1, w_2 or $2iv$. The asymptotic expansion of the integral over the path of steepest descent as $\varrho \to +\infty$ is easily found by standard methods. Thus to find the asymptotic expansions of the Airy functions we need to know the behavior of the steepest-descent contours as $\zeta \to \infty$, to which we now turn.

The paths of steepest descent will be continuously deformed as θ increases from 0 to 2π. Their equations take the form

$$\text{Im}\left\{e^{i\theta}\zeta - \tfrac{1}{3}\zeta^3\right\} - \text{Im}\left\{e^{i\theta}\zeta - \tfrac{1}{3}\zeta^3\right\}\Big|_{\zeta = \pm\exp(i\theta/2)}$$

$$\equiv \tfrac{1}{3}\eta^3 - \xi^2\eta + \left(\xi_j^2 - \eta_j^2\right)\eta + 2\xi_j\eta_j\xi - \tfrac{1}{3}\eta_j^3$$

$$+ \xi_j^2\eta_j - \left(\xi_j^2 - \eta_j^2\right)\eta_j - 2\xi_j^2\eta_j = 0$$

$$\left(\zeta = \xi + i\eta; \quad \xi_1 + i\eta_1 = \exp\left(i\frac{\theta}{2}\right); \quad \xi_2 + i\eta_2 = -\exp\left(i\frac{\theta}{2}\right); \quad j = 1, 2\right) \quad .$$

$$(A.1.37)$$

412

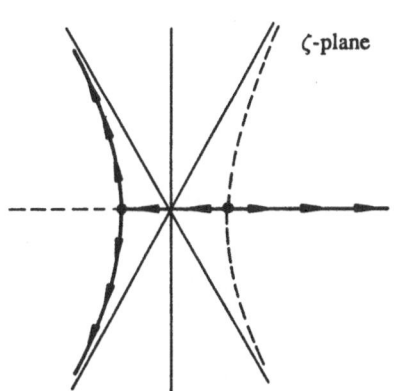

Fig. A.1.6. Steepest descent path for $\theta = 0$

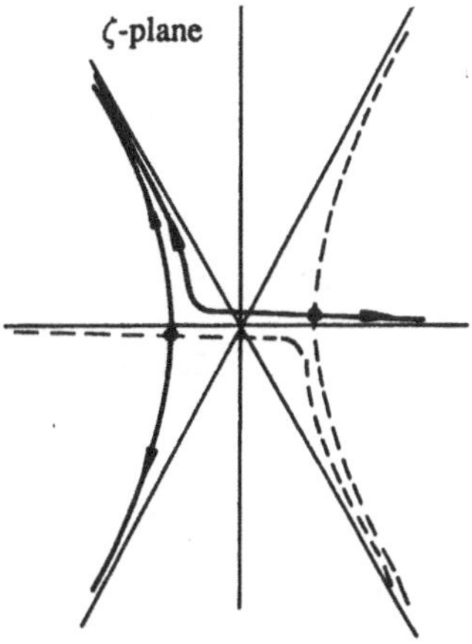

Fig. A.1.7. Steepest descent paths for $0 < \theta < 2\pi/3$; θ small

Setting $\theta = 0$ in (A.1.37), i.e., $\xi_j + i\eta_j = \pm 1$, we get the degenerate curve

$$\eta\left(\tfrac{1}{3}\eta^2 - \xi^2 + 1\right) = 0 \quad .$$

This curve is shown in Fig. A.1.6.

No matter what θ is ($0 \leq \theta < 2\pi$), there will be four lines going out from each saddle point (see Figs. A.1.6–14) – the four branches of the curve (A.1.37). Along two of the lines, $\mathrm{Re}(e^{i\theta}\zeta - \tfrac{1}{3}\zeta^3)$ decreases with increasing distance from the saddle point, while along the other two it increases. Paths of decrease are shown everywhere by heavy lines, paths of increase – by broken lines, while asymptotes and coordinate axes are shown by light unbroken lines. The arrows point in the direction of decreasing $\mathrm{Re}(e^{i\theta}\zeta - \tfrac{1}{3}\zeta^3)$. At small $\theta > 0$ the contour no longer degenerates[3] (Fig. A.1.7). It is evident that the path of steepest descent passing through $\zeta_1 = e^{i\theta/2}$, will correspond to the Airy functon $w_2(t) = w_2(\varrho\, e^{i\theta})$, while that passing through $\zeta_2 = -e^{i\theta/2}$, will correspond to the function $v(t) = v(\varrho\, e^{i\theta})$.

With a further increase in θ the paths of steepest descent are deformed (Fig. A.1.8), and finally when $\theta = 2\pi/3$ the two sections of the stationary contour merge between the saddle points so that both saddle points are connected by a path of steepest descent. In this case, equation (A.1.37) takes the form

$$(\eta - \sqrt{3}\xi)\left(\frac{1}{3}\eta^2 + \frac{1}{\sqrt{3}}\eta\xi - \frac{1}{2}\right) = 0$$

and the steepest-descent path degenerates into a hyperbola and a straight line (Fig. A.1.9). As θ increases further, the contour that passes through the point $-e^{i\theta/2}$ has endpoints at $+\infty\, e^{4\pi i/3}$ and $+\infty$ (Figs. A.1.10 and A.1.11). At $\theta = 4\pi/3$ equation (A.1.37) again degenerates:

---

[3] Here as elsewhere we will leave out the cumbersome though elementary analysis of the form of the contours.

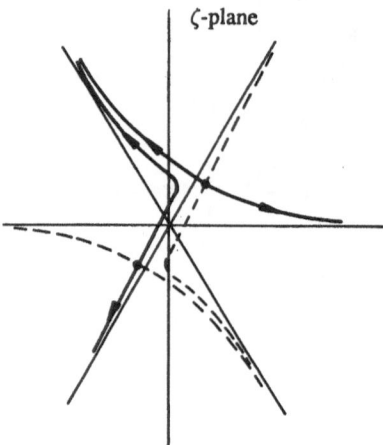

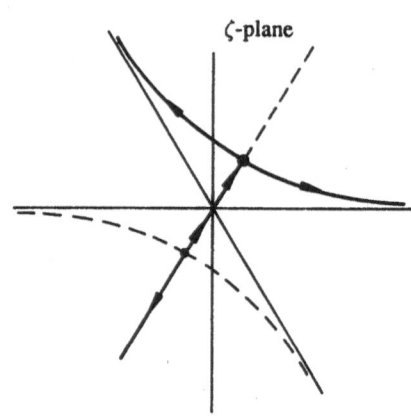

Fig. A.1.8. Steepest descent paths for $0 < \theta < 2\pi/3$; $2\pi/3 - \theta$ small

Fig. A.1.9. Steepest descent path for $\theta = 2\pi/3$

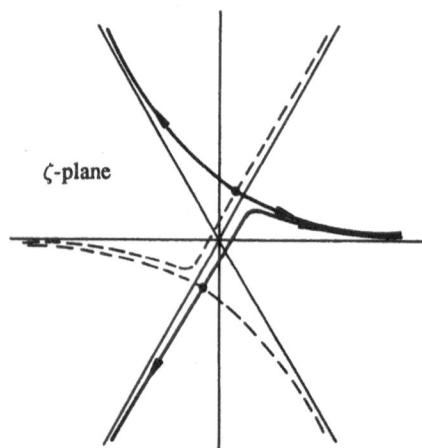

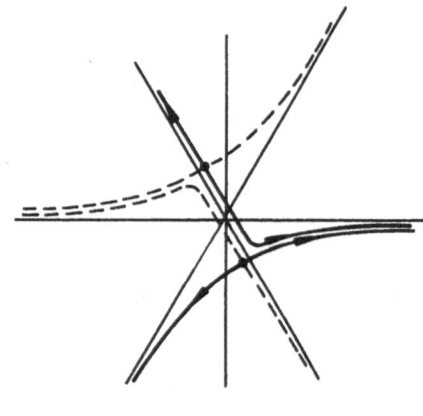

Fig. A.1.10. Steepest descent paths for $2\pi/3 < \theta < 4\pi/3$; $\theta - 2\pi/3$ small

Fig. A.1.11. Steepest descent paths for $2\pi/3 < \theta < 4\pi/3$; $4\pi/3 - \theta$ small

$$(\eta + \sqrt{3}\xi)\left(\frac{1}{3}\eta^2 - \frac{1}{\sqrt{3}}\eta\xi - \frac{1}{2}\right) = 0$$

(Fig. A.1.12), and with a further increase in θ the contour passing through $e^{i\theta/2}$ now joins $+\infty\, e^{4\pi i/3}$ and $+\infty\, e^{2\pi i/3}$ while the contour passing through $-e^{i\theta/2}$ goes from $+\infty\, e^{4\pi i/3}$ to $+\infty$ (Figs. A.1.13 and A.1.14). Finally, at $\theta = 2\pi$, we return once more to Fig. A.1.6.

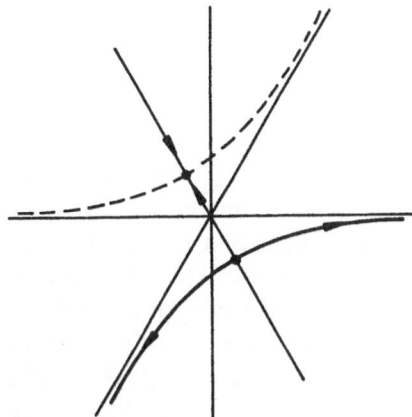

Fig. A.1.12. Steepest descent path from $\theta = 4\pi/3$

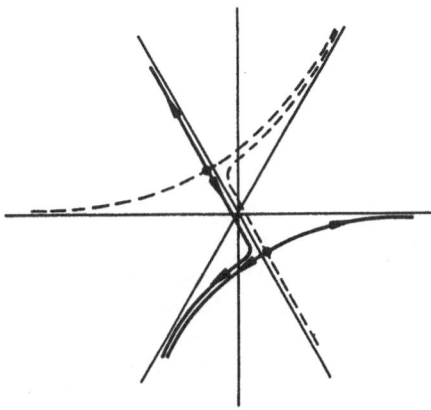

Fig. A.1.13. Steepest descent paths for $4\pi/3 < \theta < 2\pi$; $\theta - 4\pi/3$ small

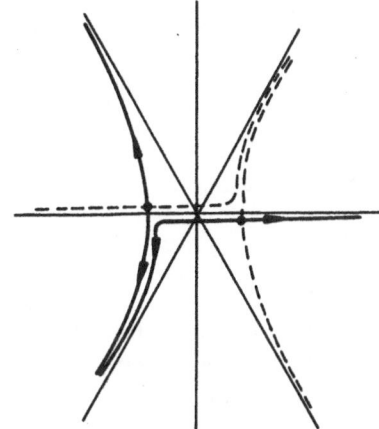

Fig. A.1.14. Steepest descent paths for $4\pi3 < \theta < 2\pi$; $2\pi - \theta$ small

Let us look briefly at how to find the asymptotic expansion of the function

$$w_1(t) = \frac{1}{\sqrt{\pi}} \int\limits_{+\infty\exp(4\pi i/3)}^{+\infty} e^{tz - z^3/3}\, dz$$

$$= \frac{\sqrt{\varrho}}{\sqrt{\pi}} \int\limits_{+\infty\exp(4\pi i/3)}^{+\infty} \exp\left[\varrho^{3/2}\left(e^{i\theta}\zeta - \tfrac{1}{3}\zeta^3\right)\right] d\zeta$$

$$(\varrho = |t| \quad,\quad e^{i\theta} = t/\varrho) \quad. \tag{A.1.38}$$

When $\theta = 0$, we deform the path of integration so that it coincides with the steepest-descent path.

Using the method of steepest descent, we get

$$w_1(t) \sim e^{(2/3)t^{3/2}}\, t^{-1/4} \left(1 + \sum_{j=1}^{\infty} \frac{c_j}{(t^{3/2})^j}\right) \, . \tag{A.1.39}$$

(Here t, $t^{3/2}$, $t^{-1/4}$, $(t^{3/2})^j$ are positive, and the series $\sum_{j=1}^{\infty}$ is asymptotic.)

When $0 > \arg t \geq -(2\pi - 2\pi/3) = -4\pi/3$ or (what amounts to the same thing) when $2\pi/3 \leq \arg t < 2\pi$ we can take as the integration path in (A.1.38) the steepest-desent path that passes through the point $-e^{i\theta/2}$, $2\pi/3 \leq \theta < 2\pi$ (Figs. A.1.9–14). Use of the method of steepest descent again gives (A.1.39), where $t^{3/2}$, $t^{-1/4}$, $(t^{3/2})^j$ should be understood as the analytic continuations of these functions which are positive when $t > 0$ into the region $0 > \arg t \geq -4\pi/3$. When $-4\pi/3 > \arg t > -2\pi$ (or $0 < \theta < 2\pi/3$; $t = \varrho e^{i\theta}$, $\varrho > 0$), we take as the path of integration in (A.1.38) the branch of the path of steepest descent that goes from $+\infty \exp(4\pi i/3)$ to $+\infty \exp(2\pi i/3)$, and then the branch that goes from $+\infty \exp(2\pi i/3)$ to $+\infty$ (Figs. A.1.7 and A.1.8).

The first branch contains the saddle point $-e^{i\theta/2}$, $0 < \arg\theta < 2\pi/3$, and the second contains $e^{i\theta/2}$.

Applying the method of steepest descent gives

$$w_1(t) = t^{-1/4}\, e^{(2/3)t^{3/2}} \left(1 + \sum_{j=1}^{\infty} \frac{c_j}{t^{3/2}}\right) \Bigg|_{t=\varrho\, e^{i\theta},\, 0<\theta<2\pi/3}$$

$$+\, t^{-1/4}\, e^{(2/3)t^{3/2}} \left(1 + \sum_{j=1}^{\infty} \frac{c_j}{t^{3/2}}\right) \Bigg|_{t=\varrho\, e^{i\theta'},\, \theta'=\theta-2\pi} \, .$$

An expression for the coefficient c_j can be found from the following considerations: since the asymptotic series obtained by the method of steepest descents can be differentiated, the right side of the asymptotic relation

$$w_1(t) \;=\; \frac{1}{\sqrt{\pi}} \int\limits_{+\infty \exp(4\pi i/3)}^{+\infty} e^{tz - z^3/3}\, dz$$

$$\sim t^{-1/4}\, e^{(2/3)t^{3/2}} \left(1 + \sum_{j=1}^{\infty} \frac{c_j}{t^{3j/2}}\right) \tag{A.1.40}$$

must be a formal solution of the Airy equation. Substituting the expression

$$t^{-1/4}\, e^{(2/3)t^{3/2}} \left(1 + \sum_{j=1}^{\infty} \frac{c_j}{t^{3j/2}}\right)$$

in the Airy equation $w'' - tw = 0$, and requiring that the coefficients of like powers of t^{-1} vanish, we get recurrence relations for c_j from which they are uniquely determined, and we arrive at formula (A.1.23). The asymptotics of $v(t)$ and $w_2(t)$ are found analogously.

A.2. Nonorthogonal Curvilinear Coordinate Systems[1]

We introduce curvilinear coordinates $q^1, q^2, \ldots, q^n$, assuming that

$$x^i = x^i(q^1, q^2, \ldots, q^n) \quad , \quad i = 1, 2, \ldots, n \quad , \tag{A.2.1}$$

and that equations (A.2.1) are uniquely solvable with respect to the q^j in some region of variation of the cartesian coordinates $x_1, x_2, \ldots, x_n$ under investigation. All functions are assumed to be sufficiently smooth. Let the set of numbers $A_i, i = 1, \ldots, n$ be given at some point M (the A_i depend on the choice of the curvilinear coordinate system). We say that these numbers form a *covariant vector* if upon conversion from the coordinate system $(q^1, q^2, \ldots, q^n)$ to a new system $(q^{1'}, q^{2'}, \ldots, q^{n'})$ the numbers A_i are transformed in accordance with the formula

$$A_{i'} = \frac{\partial q^i}{\partial q^{i'}} A_i \quad . \tag{A.2.2}$$

Here $A_{i'}$ is the same set of numbers in the coordinate system $q^{i'}$, while the derivatives $\partial q^i/\partial q^{i'}$ are calculated at M (it is assumed that $q^i = q^i(q^{1'}, q^{2'}, \ldots, q^{n'})$ and that these equations are uniquely solvable with respect to $q^{i'}$). The summation is carried out in (A.2.2) over the repeated indices.

In much the same way we define a *contravariant vector* $A^1, A^2, \ldots, A^n$. The definition of a contravariant vector differs from that of a covariant vector in the way the numbers A^i are transformed. Instead of (A.2.2), we have the formula

$$A^{i'} = \frac{\partial q^{i'}}{\partial q^i} A^i \tag{A.2.3}$$

(as always, summation is understood over repeated indices). A more general concept than the vector is the *tensor*. Let the set of numbers

$$T^{i_1 i_2 \ldots i_s}_{j_1 j_2 \ldots j_r}, \quad i_k, j_l = 1, 2, \ldots, n$$

be given at some point M in each curvilinear coordinate system. This set of numbers is said to be a tensor that is *covariant of order r* and *contravariant of order s* if, when it is converted to a new coordinate system $q^{i'}$, it is transformed in accordance with the law

$$T^{i_1' i_2' \ldots i_s'}_{j_1' j_2' \ldots j_r'} = \frac{\partial q^{i_1'}}{\partial q^{i_1}} \frac{\partial q^{i_2'}}{\partial q^{i_2}} \cdots \frac{\partial q^{i_s'}}{\partial q^{i_s}} \frac{\partial q^{j_1}}{\partial q^{j_1'}} \cdots \frac{\partial q^{j_r}}{\partial q^{j_r'}} T^{i_1 i_2 \ldots i_s}_{j_1 j_2 \ldots j_r} \quad . \tag{A.2.4}$$

When a vector or tensor is assigned at each point of a region under consideration, we use the term *vector field* or *tensor field*.

If in the region under consideration a certain number is put into correspondence with M that is defined only by the point M and is independent of the

---

[1] A more detailed exposition of the theory of nonorthogonal coordinate systems can be found in the book by *Raschewskii* [A.2.1].

coordinate system being considered, then we say that a *scalar field* is given, or that a *scalar* is assigned at each point M. It is sometimes convenient to consider a scalar as an 0-covariant and 0-contravariant tensor. If several tensor or vector fields are defined, then they can be used to construct new scalar, vector or tensor fields. For instance it can be easily proved that if we have two tensors

$$T^{i_1\ldots i_s}_{j_1\ldots j_r} \quad \text{and} \quad S^{k_1,\ldots,k_p}_{l_1,\ldots,l_t}$$

then the set of numbers

$$R^{i_1\ldots i_s k_1\ldots k_p}_{j_1\ldots j_r l_1\ldots l_t} = T^{i_1,\ldots,i_s}_{j_1,\ldots,j_r} \cdot S^{k_1,\ldots,k_p}_{l_1,\ldots,l_t}$$

is an $(r+t)$-covariant and $(s+p)$-contravariant tensor. It is just as easily proved that the set of numbers (tensors)

$$T^{\alpha i_2\ldots i_s}_{\alpha j_2\ldots j_r} = U^{i_2\ldots i_s}_{j_2\ldots j_r} \tag{A.2.5}$$

will be covariant of order $(r-1)$ and contravariant of order $(s-1)$. The operation of forming a new tensor by summing with respect to a single contravariant index (superscript) and a single covariant index (subscript) is called *contraction*. Naturally contraction can be carried out immediately with respect to β-contravariant and β-covariant indices ($\beta \le r$; $\beta \le s$). In particular, if A_i is a covariant vector and B^i is a contravariant vector, then $A_i B^i$ will be a scalar (called the *inner product of the vectors A_i and B^i*).

If T_{ij} is a second-order covariant-tensor and A^i is a contravariant vector, then

$$T_{ij} A^i A^j \tag{A.2.6}$$

will be a scalar. We have the

Theorem. *If in each coordinate system we assign a set of numbers T_{ij}, expression (A.2.6) being a scalar for any contravariant vector, then T_{ij} is a second-order covariant tensor.*

The proof of this theorem is simple, and we omit it.

Let us next give some examples of scalars, vectors and tensors. Let $\varphi = \varphi(q^1, q^2, \ldots, q^n)$ be come scalar field. It is easily shown that the set of numbers

$$A_i = \frac{\partial \varphi}{\partial q^i}$$

forms a covariant vector called the *gradient* of φ, denoted by grad φ.

An example of a tensor that is covariant of order 1 and contravariant of order 1 is the Kronecker delta δ^j_i:

$$\delta^j_i = \begin{cases} 1, & i = j \quad, \\ 0, & i \ne j \quad. \end{cases}$$

A contravariant vector is formed by the coordinate differentials

$$A^i = dq^i \quad.$$

Let us introduce the so-called *metric tensor*. An element of length in Euclidean space is expressed by the formula $ds^2 = (dx^1)^2 + (dx^2)^2 + \ldots + (dx^n)^2$. Substituting (A.2.1) for the x^i, we get

$$ds^2 = g_{ij} dq^i dq^j \quad , \qquad \text{where}$$

$$g_{ij} = \sum_{s=1}^{n} \frac{\partial x^s}{\partial q^i} \cdot \frac{\partial x^s}{\partial q^j} \quad .$$

By virtue of the fact that ds^2 is a scalar, g_{ij} is a second-order tensor called the *metric tensor*.

We now form the matrix g^{ij} inverse to g_{ij} (this can be done because g_{ij} is obviously positive definite). The matrix g^{ij} is a contravariant second-order tensor.

By using the tensors g^{ij} and g_{ij} we can carry out operations called raising and lowering of indices. For instance if A_i is a covariant vector, then the numbers

$$A^i = g^{ij} A_j$$

form a contravariant vector. The numbers A^i are called the *contravariant components of the vector* $\vec{A} = (A_1, \ldots, A_n)$. Similarly, if B^i is a contravariant vector, then the numbers

$$B_i = g_{ij} B^j$$

form a covariant vector and are called the *covariant components of the vector* $\vec{B} = (B^1, \ldots, B^n)$.

An important formula involving components of the metric tensor is the formula for change of variables in a multiple integral

$$\int_{\Omega} f(x^1, \ldots, x^n) dx^1 \ldots dx^n = \int_{\Sigma} f(q^1, \ldots, q^n) \sqrt{g} dq^1 \ldots dq^n \quad . \qquad (A.2.7)$$

Here $g = \det \| g_{ij} \|$, Σ is the region in which the $(q^1, \ldots, q^n)$ are to be varied so that the point $(x^1, \ldots, x^n)$ runs over the region Ω. An element of region Ω in $(q^1, \ldots, q^n)$ coordinates has the form $d\Omega = \sqrt{g}\, dq^1 \ldots dq^n$. Formula (A.2.7) plays a role of some importance in what follows.

Let us give a nontrivial example of a scalar. Let $\varphi(q^1, \ldots, q^n)$ be an arbitrary smooth function that is equal to zero in the vicinity of the boundary of region Σ, and let $A^i (q^1, \ldots, q^n)$ be some vector field. Using integration by parts, we can easily get the equality

$$-\int_{\Sigma} A^i \frac{\partial \varphi}{\partial q^i} \sqrt{g}\, dq^1 \ldots dq^n = \int_{\Sigma} \frac{1}{\sqrt{g}} \frac{\partial}{\partial q^i} (\sqrt{g} A^i) \varphi \sqrt{g}\, dq^1 \ldots dq^n \quad . \qquad (A.2.8)$$

Let us introduce the notation

$$\frac{1}{\sqrt{g}} \frac{\partial}{\partial q^i} (\sqrt{g} A^i) = \operatorname{div} \vec{A}$$

and show that div $\vec{A}$ is a scalar, i.e. upon passing to the new coordinates $(q^{1'}, q^{2'}, \ldots, q^{n'})$:

$$q^1 = q^1(q^{1'}, \ldots, q^{n'}), \ldots, q^n = q^n(q^{1'}, q^{2'}, \ldots, q^{n'}) \quad,$$

we have

$$\frac{1}{\sqrt{g}} \frac{\partial}{\partial q^i}(\sqrt{g} A^i)\Big|_{q^i = q^i(q^{1'}, \ldots, q^{n'})} = \frac{1}{\sqrt{g'}} \frac{\partial}{\partial q^{i'}}(\sqrt{g'} A^{i'})\Big|_{q^{1'}, \ldots, q^{n'}} \quad.$$

(A.2.9)

Here the $A^{i'}$ are the components of the vector under consideration, in the coordinate system $(q^{1'}, \ldots, q^{n'})$; $\sqrt{g'} = \sqrt{\det\|g_{i'j'}\|}$; $g_{i'j'}$ are the components of the metric tensor in $(q^{1'}, \ldots, q^{n'})$ coordinates. To prove (A.2.9), let us return to (A.2.8). We change variables of integration to $q^{1'}, \ldots, q^{n'}$ from $q^1, \ldots, q^n$. Using the fact that $A^i \partial\varphi/\partial q^i$ is a scalar, we get

$$\int_{\Sigma'} \frac{1}{\sqrt{g}} \frac{\partial}{\partial q^i}(\sqrt{g} A^i)\Big|_{q^i = q^i(q^{1'}, \ldots, q^{n'})} \varphi(q^{1'}, \ldots, q^{n'})\sqrt{g'} dq^{1'} \ldots dq^{n'}$$

$$= -\int_{\Sigma'} A^{i'} \frac{\partial\varphi}{\partial q^{i'}} \sqrt{g'} dq^{1'} \ldots dq^{n'} \quad.$$

Integrating by parts on the right side of the equation, we get

$$0 = \int_{\Sigma'} \frac{1}{\sqrt{g}} \frac{\partial}{\partial q^i}(\sqrt{g} A^i)\Big|_{q^i = q^i(q^{1'}, \ldots, q^{n'})} \varphi\sqrt{g'} dq^{1'} \ldots dq^{n'}$$

$$- \int_{\Sigma'} \frac{1}{\sqrt{g'}} \frac{\partial}{\partial q^{i'}}(\sqrt{g'} A^{i'})\varphi\sqrt{g'} dq^{1'} \ldots dq^{n'}$$

$$= \int_{\Sigma'} \left[\frac{1}{\sqrt{g}} \frac{\partial}{\partial q^i}(\sqrt{g} A^i)\Big|_{q^i = q^i(q^{1'}, \ldots, q^{n'})}\right.$$

$$\left. - \frac{1}{\sqrt{g'}} \frac{\partial}{\partial q^{i'}}(\sqrt{g'} A^{i'})\right]\varphi\sqrt{g'} dq^{1'} \ldots dq^{n'} \quad.$$

From the arbitrariness of φ and the fundamental lemma of the variational calculus it follows that the expression in brackets in the integrand is identically zero, which is equivalent to (A.2.9). The scalar

$$\frac{1}{\sqrt{g}} \frac{\partial}{\partial q^i}(A^i \sqrt{g}) = \mathrm{div}\, \vec{A}$$

(A.2.10)

is called the *divergence* of the vector $\vec{A} = \{A^1, \ldots, A^n\}$. If $\vec{A}$ is given by its covariant components: $\vec{A} = \{A_1, \ldots, A_n\}$, then its divergence is defined by (A.2.10), where now the A^i must be given by

$$A^i = g^{ij} A_j \quad.$$

(A.2.11)

In particular, if $u(q^1 \ldots q^n)$ is a scalar, then the divergence of its gradient will be a scalar as well. The divergence of the gradient of a scalar is called the *Laplacian*, and is ordinarily denoted by the symbol Δ:

$$\Delta u = \operatorname{div} \operatorname{grad} u = \frac{1}{\sqrt{g}} \frac{\partial}{\partial q^i} \left(g^{ij} \sqrt{g} \frac{\partial u}{\partial q^j} \right) \quad . \tag{A.2.12}$$

In cartesian coordinates, by virtue of the fact that $g_{ij} = 1$ when $i = j$, and $g_{ij} = 0$ when $i \neq j$, expressions (A.2.10) and (A.2.11) for the divergence, and (A.2.12) for the Laplacian take the form[2]

$$\operatorname{div} \vec{A} = \sum_{i=1}^{n} \frac{\partial A^i}{\partial x_i} \quad , \quad \Delta u = \sum_{i=1}^{n} \frac{\partial^2 u}{\partial x_i^2} \quad .$$

Thus formulas (A.2.10) and (A.2.12) give expressions for the divergence and the Laplacian in an arbitrary curvilinear coordinate system. Note that in proving that $\operatorname{div} \vec{A}$ is a scalar, we used only the metric tensor and the formula $\sqrt{g} dq^1 \ldots dq^n$ for a volume element. Therefore the expressions

$$\frac{1}{\sqrt{g}} \frac{\partial}{\partial q^i} (A^i \sqrt{g}) \quad \text{and} \quad \frac{1}{\sqrt{g}} \frac{\partial}{\partial q^i} \left(g^{ij} \sqrt{g} \frac{\partial u}{\partial q^j} \right)$$

will be scalars for an arbitrary Riemannian manifold as well. In this case they are also called the divergence and the Laplacian. In particular, any smooth surface can be considered as a Riemannian manifold. The coefficients of the first Gaussian quadratic form in this case will constitute the metric tensor. If the coordinates on the surface are denoted by q^1 and q^2, then an element of arc length will be

$$
\begin{aligned}
ds &= \sqrt{g_{11}(dq^1)^2 + 2g_{12} dq^1 dq^2 + g_{22}(dq^2)^2} \\
&= \sqrt{E(dq^1)^2 + 2F dq^1 dq^2 + G(dq^2)^2} \\
&\quad (E = g_{11} \quad , \quad F = g_{12} \quad , \quad G = g_{22}) \quad ,
\end{aligned}
$$

and the Laplacian is written as

$$
\begin{aligned}
\Delta u = \frac{1}{\sqrt{EG - F^2}} \Bigg[&\frac{\partial}{\partial q^1} \frac{G(\partial u/\partial q^1)}{\sqrt{EG - F^2}} - \frac{\partial}{\partial q^1} \frac{F(\partial u/\partial q^2)}{\sqrt{EG - F^2}} \\
&- \frac{\partial}{\partial q^2} \frac{F(\partial u/\partial q^1)}{\sqrt{EG - F^2}} + \frac{\partial}{\partial q^2} \frac{E(\partial u/\partial q^2)}{\sqrt{EG - F^2}} \Bigg] \quad .
\end{aligned}
$$

In particular, if the coordinate system is orthogonal, then

$$ds^2 = E(dq^1)^2 + G(dq^2)^2 \quad ,$$

$$\Delta u = \frac{1}{\sqrt{EG}} \left[\frac{\partial}{\partial q^1} \left(\sqrt{\frac{G}{E}} \frac{\partial u}{\partial q^1} \right) + \frac{\partial}{\partial q^2} \left(\sqrt{\frac{E}{G}} \frac{\partial u}{\partial q^2} \right) \right] \quad . \tag{A.2.13}$$

---

[2] In cartesian coordinates covariant components are the same as the contravariant ones; in particular $x^i = x_i$.

Formula (A.2.13) is applicable, of course, on a plane as well. If q^1 and q^2 are orthogonal coordinates on the plane, then ordinarily we assume $\sqrt{E} = h_1$; $\sqrt{G} = h_2$, and formulas (A.2.13) take the form

$$\Delta u = \frac{1}{h_1 h_2}\left[\frac{\partial}{\partial q^1}\left(\frac{h_2}{h_1}\frac{\partial u}{\partial q^1}\right) + \frac{\partial}{\partial q^2}\left(\frac{h_1}{h_2}\frac{\partial u}{\partial q^2}\right)\right] \ .$$

The functions $h_i = h_i(q^1, q^2)$ $(i = 1, 2)$ are called *Lamé coefficients*.

A.3. Solution of the Equation $y''(s) + K(s)y(s) = y^{-3}(s)$ [1]

The solution of

$$y''(s) + K(s)y(s) = y^{-3}(s) \tag{A.3.1}$$

can be reduced to the solution of the linear equation

$$y''(s) + K(s)y(s) = 0 \ . \tag{A.3.2}$$

More precisely, we have the following:

Theorem. *Let y_1 and y_2 be a fundamental system of solutions of* (A.3.2), *and let $\|a_{rt}\|$, $r, t = 1, 2$, be a symmetric constant matrix such that*

$$\det\|a_{rt}\| W^2(y_1, y_2) = 1 \tag{A.3.3}$$

(here $W(y_1, y_2) = y_1 y_2' - y_1' y_2$ is the Wronskian of y_1, y_2). Then the function

$$Y = \left(\sum_{r,t=1}^{2} a_{rt} y_r y_t\right)^{1/2} \tag{A.3.4}$$

will be a solution of (A.3.1), *and conversely any solution Y of* (A.3.1) *can be represented in the form* (A.3.4), *where $\|a_{rt}\|$ is a symmetric matrix that depends on the choice of y_1, y_2, Y, and satisfies* (A.3.3). *If, in addition, Y, y_1, y_2 are real, then a_{rt} is a real positive definite matrix satisfying* (A.3.3).

In fact,

$$Y' = Y^{-1} \sum_{r,t=1}^{2} a_{rt} y_r y_t'$$

$$Y'' = Y^{-1} \sum_{r,t=1}^{2} \left(a_{rt} y_r' y_t' + a_{rt} y_r y_t''\right) - Y^{-3}\left(\sum_{r,t=1}^{2} a_{rt} y_r y_t'\right)^2 \ .$$

---

[1] This equation was studied in connection with diffraction problems by *Babič* and *Lazutkin* [A.3.1]. Formula (A.3.4) for the solution was proposed by V.F. Lazutkin. This same reduction of the solution of $y'' + Ky = y^{-3}$ to the solution of $y'' + Ky = 0$ was first carried out by V.P. Ermakov in the 1880's, and much later (and independently) by the American physicist E. Pinney. A review of the work associated with this equation and a number of new results can be found in the paper by *Berkovič* and *Rozov* [A.3.2].

Substituting $-K(s)y_t$ for y_t'', and putting the remaining terms over a common denominator, we get

$$Y'' + K(s)Y = \frac{Q(s)}{Y^3} \quad ,$$

where

$$Q = \sum_{r,t=1}^{2} a_{rt}y_r y_t \sum_{m,n=1}^{2} a_{mn}y_m' y_n' - \sum_{r,t=1}^{2} a_{rt}y_r y_t' \sum_{m,n=1}^{2} a_{mn}y_m y_n'$$

$$\tag{A.3.5}$$

$$= \sum_{r,t,m,n=1}^{2} a_{rt}a_{mn}y_r y_n'(y_t y_m' - y_t' y_m) \quad . \tag{A.3.6}$$

Since $y_t y_m' - y_t' y_m = 0$ when $t = m$ and $y_1 y_2' - y_1' y_2 = W$,

$$Q = W \sum_{r,n=1}^{2} (a_{r1}a_{2n} - a_{r2}a_{1n})y_r y_n'$$

$$= W\left[(a_{11}a_{22} - a_{12}a_{12})y_1 y_2' + (a_{21}a_{21} - a_{22}a_{11})y_2 y_1'\right]$$

$$= W^2 \det\|a_{rt}\| \quad .$$

From this and from (A.3.3) and (A.3.5) it follows that Y is a solution of (A.3.1).

To prove that any solution of equation (A.3.1) can be represented in the form (A.3.4), it is sufficient to show that any solution of the Cauchy problem

$$Y(s_0) = c_0 \quad (c_0 \neq 0) \quad ; \quad Y'(s_0) = c_1 \tag{A.3.7}$$

can be represented in form (A.3.4) [(where Y is a solution of equation (A.3.1)]. Let us show that the symmetric matrix $\|a_{rt}\|(r, t = 1, 2)$ is determined by the initial data (A.3.7). To do this, we substitute (A.3.4) for Y in the initial conditions (A.3.7). We obtain after some calculation

$$\sum_{r,t=1}^{2} a_{rt}y_r y_t = c_0^2 \quad , \quad \sum_{r,t=1}^{2} a_{rt}y_r y_t' = c_0 c_1$$

$$(y_r = y_r(s_0) \quad , \quad y_t' = y_t'(s_0)) \quad . \tag{A.3.8}$$

We multiply the first of (A.3.8) by $\sum_{m,n=1}^{2} a_{mn}y_m' y_n'$, square the second, and subtract the results. After simple calculations, and considering (A.3.3), we get one further equation:

$$\sum_{r,t=1}^{2} a_{rt}y_r' y_t' = c_1^2 + \frac{1}{c_0^2} \quad . \tag{A.3.9}$$

The determinant of the system of three equations (A.3.8) and (A.3.9) in the three unknowns a_{11}, $a_{12} = a_{21}$ and a_{22} is equal to $[W(y_1, y_2)]^2 \neq 0$, as can be shown by simple calculations. It remains for us to verify that the a_{11}, a_{12}, a_{22}, found from (A.3.8) and (A.3.9) satisfy (A.3.3). Actually, (A.3.6) implies that

$$W^2 \det\|a_{rt}\| = Q = c_0^2\left(c_1^2 + \frac{1}{c_0^2}\right) - (c_0 c_1)^2 = 1 \quad.$$

These results apply both to the real and to the complex case. Let us consider the real case. If c_0, c_1, $y_1(s)$ and $y_2(s)$ are real, then the matrix $\|a_{rt}\|$, determined from equations (A.3.8) and (A.3.9) will be real and positive definite. In fact, the fact that $\|a_{rt}\|$ is real is obvious. Furthermore, the determinant of this matrix satisfies condition (A.3.3), and therefore is positive. The matrix $\|a_{rt}\|$, like any second-order real matrix with positive determinant, will be sign-definite. By virtue of the first equation in (A.3.8) it can only be positive ($c_0^2 > 0$).

A.4. Computation of and Tables for the Function $G_M(\gamma)$

The evaluation of the function

$$G_M(\gamma) = \int\limits_{\infty\,e^{-2\pi i/3}}^{\infty\,e^{\pi i/3}} \frac{1}{v(t)w_2(t)} \left[\frac{w_1(t)}{w_2(t)}\right]^M e^{i\gamma t}\, dt \tag{A.4.1}$$

was carried out on a BESM-2 computer. The integral (A.4.1) was broken up into two integrals:

$$\int\limits_{\infty\,e^{-2\pi i/3}}^{\infty\,e^{\pi i/3}} = \int\limits_{\infty\,e^{-2\pi i/3}}^{0} + \int\limits_{0}^{\infty\,e^{\pi i/3}} \quad,$$

which were reduced by changes of variables to integrals over the interval $(0, \infty)$. The formulas (see Appendix A.1)

$$w_2(t) = -e^{2\pi i/3} w_1(e^{2\pi i/3} t) \quad,$$
$$w_1(e^{\pi i/3} t) = 2 e^{\pi i/6} v(-t) \quad,$$
$$w_2(e^{-2\pi i/3} t) = 2 e^{-\pi i/6} v(t)$$

enable us to separate the real and imaginary parts of the integrands, after which the calculation of (A.4.1) reduces to the calculation of two real integrals for $\mathrm{Re}\, G_M(\gamma)$ and $\mathrm{Im}\, G_M(\gamma)$.

The argument γ of the function $G_M(\gamma)$ and the integer M are related by the inequality (12.5.26):

$$\frac{\gamma}{2\sqrt{\Delta}}\left(\frac{\omega P}{2c_0}\right)^{-1/21} - 1 < M \leq \frac{\gamma}{2\sqrt{\Delta}}\left(\frac{\omega P}{2c_0}\right)^{-1/21} \quad,$$

in which the coefficient $\Delta = O(1)$ can be varied within certain limits. When calculating the field in specific problems, the coefficient Δ must be chosen to minimize the discontinuity in the field (which occurs because of the errors in the asymptotic formulas) as the geometrical optics waves become distinguished. As shown by numerically comparing the asymptotic formulas with the exact solution for the case of a parabolic boundary [A.4.1] for values of the large parameter $\omega P/2c_0$ between $3 \sim 5$, the first geometrical-optic wave becomes distinguished at $\gamma \sim 1.8 - 2.0$, the second at $\gamma \sim 2.8 - 3.0$, which corresponds to $\Delta \sim 1$ and $\Delta \sim 0.5$ respectively.

Based upon this rough estimate of the values of γ for which the geometrical-optic waves become distinguished, the functions $G_M(\gamma)$ were tabulated over the following ranges:[1]

for $M = 0$: $0.2 \le \gamma \le 3.5$

for $M = 1$: $1.5 \le \gamma \le 6.5$

for $M = 2$: $2.0 \le \gamma \le 8.5$

for $M = 3$: $3.5 \le \gamma \le 10.5$.

The tables consist of the values of $\mathrm{Re}\, G_M(\gamma)$, $\mathrm{Im}\, G_M(\gamma)$, $|G_M(\gamma)|$ and $\arg G_M(\gamma)$ for γ in steps of 0.1.

Table A.1

γ	$\mathrm{Re}\{G_M(\gamma)\}$	$\mathrm{Im}\{G_M(\gamma)\}$	$|G_M(\gamma)|$	$\arg G_M(\gamma)$
		$M = 0$		
0.2	− 7.283	15.841	17.435	2.002
0.3	− 5.984	10.074	11.717	2.107
0.4	− 4.511	7.116	8.425	2.136
0.5	− 3.313	5.478	6.402	2.115
0.6	− 2.422	4.562	5.165	2.059
0.7	− 1.758	4.019	4.387	1.983
0.8	− 1.244	3.676	3.881	1.897
0.9	− 0.818	3.446	3.542	1.804
1.0	− 0.443	3.283	3.313	1.705
1.1	− 0.092	3.159	3.161	1.600
1.2	0.250	3.056	3.066	1.489
1.3	0.596	2.958	3.018	1.372
1.4	0.953	2.853	3.008	1.248
1.5	1.325	2.728	3.032	1.119
1.6	1.713	2.569	3.088	0.982
1.7	2.118	2.363	3.173	0.840
1.8	2.533	2.094	3.287	0.691
1.9	2.951	1.749	3.430	0.535
2.0	3.356	1.311	3.603	0.372
2.1	3.730	0.767	3.808	0.203
2.2	4.046	0.106	4.047	0.026
2.3	4.268	− 0.680	4.322	−0.158
2.4	4.355	− 1.587	4.635	−0.349
2.5	4.258	− 2.603	4.990	−0.549
2.6	3.923	− 3.697	5.390	−0.756
2.7	3.295	− 4.820	5.838	−0.971
2.8	2.327	− 5.897	6.339	−1.195

[1] The limitation to $\gamma \ge 0.2$ is determined by the convergence of the integral (A.4.1) on the upper part of the integration contour [cf. (12.4.26)].

γ	$\mathrm{Re}\{G_M(\gamma)\}$	$\mathrm{Im}\{G_M(\gamma)\}$	$|G_M(\gamma)|$	$\arg G_M(\gamma)$
		$M = 0$		
2.9	0.986	− 6.824	6.895	−1.427
3.0	− 0.734	− 7.472	7.508	−1.669
3.1	− 2.795	− 7.689	8.181	−1.919
3.2	− 5.099	− 7.312	8.915	−2.180
3.3	− 7.478	− 6.192	9.708	−2.450
3.4	− 9.681	− 4.218	10.560	−2.731
3.5	−11.382	− 1.366	11.463	−3.022

γ	$\mathrm{Re}\{G_M(\gamma)\}$	$\mathrm{Im}\{G_M(\gamma)\}$	$|G_M(\gamma)|$	$\arg G_M(\gamma)$
		$M = 1$		
1.5	− 2.388	1.651	2.903	2.537
1.6	− 2.320	1.579	2.807	2.544
1.7	− 2.256	1.524	2.669	2.552
1.8	− 2.219	1.484	2.648	2.547
1.9	− 2.130	1.459	2.582	2.541
2.0	− 2.066	1.447	2.522	2.531
2.1	− 2.000	1.448	2.470	2.515
2.2	− 1.931	1.461	2.422	2.494
2.3	− 1.859	1.485	2.380	2.467
2.4	− 1.782	1.521	2.343	2.435
2.5	− 1.689	1.566	2.310	2.397
2.6	− 1.608	1.619	2.281	2.353
2.7	− 1.508	1.680	2.258	2.302
2.8	− 1.399	1.747	2.238	2.246
2.9	− 1.278	1.819	2.223	2.183
3.0	− 1.145	1.894	2.212	2.114
3.1	− 0.997	1.969	2.207	2.039
3.2	− 0.833	2.043	2.206	1.958
3.3	− 0.653	2.112	2.211	1.871
3.4	− 0.455	2.174	2.221	1.777
3.5	− 0.238	2.224	2.236	1.677
3.6	− 0.002	2.258	2.258	1.561
3.7	0.253	2.272	2.286	1.460
3.8	0.525	2.260	2.320	1.342
3.9	0.813	2.217	2.361	1.219
4.0	1.113	2.137	2.409	1.091
4.1	1.421	2.014	2.465	0.956
4.2	1.732	1.844	2.529	0.817
4.3	2.036	1.620	2.602	0.672
4.4	2.326	1.338	2.683	0.522
4.5	2.590	0.994	2.775	0.366
4.6	2.815	0.589	2.876	0.206
4.7	2.985	0.123	2.988	0.041
4.8	3.085	− 0.399	3.111	−0.129
4.9	3.098	− 0.970	3.246	−0.304
5.0	3.006	− 1.576	3.394	−0.483
5.1	2.793	− 2.200	3.556	−0.667
5.2	2.445	− 2.818	3.731	−0.856
5.3	1.952	− 3.401	3.922	−1.050
5.4	1.310	− 3.914	4.128	−1.248
5.5	0.522	− 4.319	4.350	−1.450

γ	$\text{Re}\{G_M(\gamma)\}$	$\text{Im}\{G_M(\gamma)\}$	$|G_M(\gamma)|$	$\arg G_M(\gamma)$
		$M = 1$		
5.6	$-$ 0.399	$-$ 4.573	4.590	-1.658
5.7	$-$ 1.428	$-$ 4.632	4.848	-1.870
5.8	$-$ 2.527	$-$ 4.457	5.123	-2.086
5.9	$-$ 3.642	$-$ 4.012	5.418	-2.308
6.0	$-$ 4.706	$-$ 3.273	5.732	-2.534
6.1	$-$ 5.639	$-$ 2.232	6.065	-2.765
6.2	$-$ 6.353	$-$ 0.904	6.417	-3.000
6.3	$-$ 6.755	0.671	6.788	3.043
6.4	$-$ 6.756	2.424	7.178	2.797
6.5	$-$ 6.280	4.250	7.583	2.547

γ	$\text{Re}\{G_M(\gamma)\}$	$\text{Im}\{G_M(\gamma)\}$	$|G_M(\gamma)|$	$\arg G_M(\gamma)$
		$M = 2$		
2.0	$-$ 2.585	$-$ 0.325	2.605	-3.017
2.1	$-$ 2.505	$-$ 0.417	2.539	-2.976
2.2	$-$ 2.428	$-$ 0.497	2.479	-2.940
2.3	$-$ 2.356	$-$ 0.565	2.422	-2.906
2.4	$-$ 2.287	$-$ 0.621	2.370	-2.876
2.5	$-$ 2.222	$-$ 0.667	2.320	-2.850
2.6	$-$ 2.163	$-$ 0.703	2.274	-2.827
2.7	$-$ 2.108	$-$ 0.729	2.231	-2.809
2.8	$-$ 2.058	$-$ 0.746	2.190	-2.794
2.9	$-$ 2.014	$-$ 0.756	2.151	-2.782
3.0	$-$ 1.974	$-$ 0.757	2.115	-2.775
3.1	$-$ 1.940	$-$ 0.751	2.080	-2.772
3.2	$-$ 1.911	$-$ 0.738	2.048	-2.773
3.3	$-$ 1.886	$-$ 0.718	2.018	-2.778
3.4	$-$ 1.866	$-$ 0.692	1.990	-2.786
3.5	$-$ 1.850	$-$ 0.639	1.964	-2.799
3.6	$-$ 1.837	$-$ 0.620	1.939	-2.816
3.7	$-$ 1.828	$-$ 0.574	1.916	-2.837
3.8	$-$ 1.822	$-$ 0.522	1.895	-2.862
3.9	$-$ 1.818	$-$ 0.464	1.876	-2.892
4.0	$-$ 1.815	$-$ 0.400	1.858	-2.925
4.1	$-$ 1.813	$-$ 0.329	1.843	-2.962
4.2	$-$ 1.811	$-$ 0.251	1.828	-3.004
4.3	$-$ 1.808	$-$ 0.167	1.816	-3.049
4.4	$-$ 1.804	$-$ 0.076	1.805	-3.099
4.5	$-$ 1.796	0.021	1.796	3.130
4.6	$-$ 1.784	0.125	1.789	3.072
4.7	$-$ 1.768	0.235	1.784	3.009
4.8	$-$ 1.745	0.351	1.780	2.943
4.9	$-$ 1.714	0.473	1.778	2.872
5.0	$-$ 1.674	0.600	1.778	2.797
5.1	$-$ 1.623	0.731	1.780	2.719
5.2	$-$ 1.560	0.865	1.784	2.635
5.3	$-$ 1.484	1.000	1.790	2.548
5.4	$-$ 1.393	1.137	1.799	2.457
5.5	$-$ 1.286	1.272	1.809	2.362
5.6	$-$ 1.162	1.404	1.822	2.262
5.7	$-$ 1.019	1.529	1.837	2.158

γ	$\text{Re}\{G_M(\gamma)\}$	$\text{Im}\{G_M(\gamma)\}$	$\|G_M(\gamma)\|$	$\arg G_M(\gamma)$
		$M = 2$		
5.8	− 0.857	1.646	1.856	2.051
5.9	− 0.675	1.750	1.876	1.939
6.0	− 0.474	1.840	1.900	1.823
6.1	− 0.255	1.909	1.926	1.703
6.2	− 0.017	1.956	1.956	1.579
6.3	0.236	1.974	1.988	1.452
6.4	0.501	1.961	2.024	1.320
6.5	0.776	1.913	2.064	1.185
6.6	1.056	1.824	2.108	1.046
6.7	1.335	1.692	2.155	0.903
6.8	1.606	1.515	2.207	0.756
6.9	1.861	1.289	2.264	0.606
7.0	2.091	1.015	2.325	0.452
7.1	2.391	0.693	2.390	0.294
7.2	2.440	0.326	2.462	0.133
7.3	2.537	− 0.081	2.538	−0.032
7.4	2.568	− 0.521	2.621	−0.200
7.5	2.524	− 0.984	2.709	−0.372
7.6	2.395	− 1.459	2.804	−0.547
7.7	2.174	− 1.928	2.906	−0.726
7.8	1.856	− 2.376	3.015	−0.907
7.9	1.440	− 2.780	3.131	−1.093
8.0	0.929	− 3.119	3.255	−1.281
8.1	0.330	− 3.371	3.387	−1.473
8.2	− 0.343	− 3.510	3.527	−1.668
8.3	− 1.072	− 3.517	3.676	−1.866
8.4	− 1.829	− 3.370	3.835	−2.068
8.5	− 2.585	− 3.056	4.003	−2.273

γ	$\text{Re}\{G_M(\gamma)\}$	$\text{Im}\{G_M(\gamma)\}$	$\|G_M(\gamma)\|$	$\arg G_M(\gamma)$
		$M = 3$		
3.5	− 0.597	− 1.891	1.982	−1.876
3.6	− 0.540	− 1.876	1.953	−1.851
3.7	− 0.490	− 1.860	1.924	−1.828
3.8	− 0.447	− 1.843	1.897	−1.809
3.9	− 0.411	− 1.825	1.871	−1.792
4.0	− 0.381	− 1.806	1.846	−1.779
4.1	− 0.357	− 1.787	1.822	−1.768
4.2	− 0.340	− 1.767	1.799	−1.761
4.3	− 0.327	− 1.747	1.778	−1.756
4.4	− 0.321	− 1.728	1.757	−1.754
4.5	− 0.320	− 1.708	1.738	−1.756
4.6	− 0.324	− 1.688	1.719	−1.760
4.7	− 0.333	− 1.668	1.701	−1.768
4.8	− 0.346	− 1.649	1.685	−1.778
4.9	− 0.365	− 1.628	1.669	−1.791
5.0	− 0.388	− 1.608	1.654	−1.808
5.1	− 0.415	− 1.586	1.640	−1.827
5.2	− 0.447	− 1.564	1.626	−1.849
5.3	− 0.483	− 1.540	1.614	−1.875
5.4	− 0.523	− 1.514	1.602	−1.903

γ	Re$\{G_M(\gamma)\}$	Im$\{G_M(\gamma)\}$	$\|G_M(\gamma)\|$	arg $G_M(\gamma)$
		$M = 3$		
5.5	$-$ 0.566	$-$ 1.487	1.591	-1.934
5.6	$-$ 0.613	$-$ 1.457	1.581	-1.969
5.7	$-$ 0.663	$-$ 1.425	1.572	-2.006
5.8	$-$ 0.717	$-$ 1.390	1.563	-2.047
5.9	$-$ 0.773	$-$ 1.350	1.556	-2.091
6.0	$-$ 0.831	$-$ 1.307	1.549	-2.137
6.1	$-$ 0.892	$-$ 1.259	1.543	-2.187
6.2	$-$ 0.954	$-$ 1.206	1.538	-2.240
6.3	$-$ 1.017	$-$ 1.148	1.533	-2.296
6.4	$-$ 1.080	$-$ 1.084	1.530	-2.354
6.5	$-$ 1.143	$-$ 1.013	1.527	-2.417
6.6	$-$ 1.205	$-$ 0.935	1.526	-2.482
6.7	$-$ 1.266	$-$ 0.850	1.525	-2.550
6.8	$-$ 1.323	$-$ 0.758	1.525	-2.621
6.9	$-$ 1.377	$-$ 0.658	1.526	-2.696
7.0	$-$ 1.425	$-$ 0.550	1.528	-2.773
7.1	$-$ 1.468	$-$ 0.434	1.531	-2.854
7.2	$-$ 1.503	$-$ 0.311	1.535	-2.938
7.3	$-$ 1.529	$-$ 0.180	1.540	-3.024
7.4	$-$ 1.546	$-$ 0.042	1.546	-3.114
7.5	$-$ 1.550	0.102	1.554	3.075
7.6	$-$ 1.542	0.252	1.562	2.979
7.7	$-$ 1.518	0.406	1.572	2.880
7.8	$-$ 1.479	0.564	1.583	2.777
7.9	$-$ 1.423	0.722	1.595	2.672
8.0	$-$ 1.348	0.879	1.609	2.563
8.1	$-$ 1.253	1.034	1.624	2.451
8.2	$-$ 1.138	1.183	1.641	2.337
8.3	$-$ 1.002	1.323	1.660	2.219
8.4	$-$ 0.845	1.452	1.680	2.098
8.5	$-$ 0.668	1.565	1.701	1.974
8.6	$-$ 0.470	1.660	1.725	1.847
8.7	$-$ 0.255	1.732	1.750	1.717
8.8	$-$ 0.023	1.778	1.778	1.584
8.9	0.222	1.794	1.808	1.448
9.0	0.476	1.777	1.840	1.309
9.1	0.736	1.723	1.874	1.167
9.2	0.996	1.630	1.910	1.022
9.3	1.251	1.496	1.950	0.874
9.4	1.492	1.319	1.991	0.724
9.5	1.714	1.099	2.036	0.570
9.6	1.908	0.838	2.084	0.414
9.7	2.066	0.538	2.134	0.255
9.8	2.179	0.203	2.188	0.099
9.9	2.240	$-$ 0.162	2.246	-0.072
10.0	2.241	$-$ 0.547	2.307	-0.239
10.1	2.175	$-$ 0.944	2.371	-0.410
10.2	2.037	$-$ 1.342	2.440	-0.582
10.3	1.824	$-$ 1.728	2.513	-0.758
10.4	1.535	$-$ 2.086	2.590	-0.936
10.5	1.170	$-$ 2.401	2.671	-1.117

References

Items originally published in Russian are denoted by an asterisk (*). Where translations into English or other Western languages are available, these have been indicated in square brackets [ ], as are alternative transliterations of the authors' names.

Chapter 1

1.1 *V.I. Smirnov: *A Course of Higher Mathematics*, Vol. 4 (Fizmatgiz, Moscow 1958) [Addison-Wesley, Reading, MA 1964]

1.2 *D.M. Eidus: The principle of limiting absorption. Mat. Sb. **57 (99)**, 13–44 (1962) [Am. Math. Soc. Transl., Ser. 2, **47**, 157–191 (1965)]

1.3 *Yu.A. Kravtsov: A modification of the method of geometrical optics. Izv. Vyssh. Uchebn. Zaved., Radiofiz. **7**, 664–673 (1964)

1.4 *Yu.L. Gazaryan: The geometrical-acoustic approximation of the field in the neighborhood of a nonsingular section of caustic. Vopr. Din. Teor. Rasprostr. Seism. Voln, **5**, 77–89 (1961)

1.5 J.B. Keller, S.I. Rubinow: Asymptotic solution of eigenvalue problems. Annals Phys. **9**, 24–75 (1960); errata, ibid. **10**, 303–305 (1960)

1.6 *L.A. Vainshtein: Open resonators with spherical mirrors. Zh. Exp. Teor. Fiz. **45**, 684–697 (1963) [Sov. Phys.–JETP **18**, 471–479 (1964)]

1.7 *L.A. Vainshtein: Ray flux in a triaxial ellipsoid. Elektron. Bolsh. Moshchn. **4**, 93–105 (1965)

1.8 *L.A. Vainshtein [Weinstein]: *Open Resonators and Open Waveguides* (Sovetskoe Radio, Moscow 1966) [Golem, Boulder, CO 1969]

1.9 *V.P. Bykov: Geometrical optics of three-dimensional oscillations in open resonators. Elektron. Bolsh. Moshchn. **4**, 66–92 (1965)

1.10 *V.P. Bykov: Ray theory of open resonators and open waveguides whose oscillations are bounded by caustic surfaces. Radiotekh. Elektron. **11**, 477–487 (1966) [Radio Eng. Electron. Phys. **11**, 401–410 (1966)]

1.11 *M.A. Leontovich [Leontovič], V.A. Fock [Fok]: Solution of the problem of propagation of electromagnetic waves along the earth's surface by the method of parabolic equations. Zh. Exp. Teor. Fiz. **16**, (7), 557–573 (1946) [J. Phys. USSR **10**, 13–24 (1946)];
V.A. Fock [Fok]: *Electromagnetic Diffraction and Propagation Problems*, 2nd ed. (Sovetskoe Radio, Moscow 1970) [1st ed., Pergamon, Oxford 1965] Chap. 11

1.12 *V.M. Babich [Babič], N.Ya. Kirpichnikova [Kirpičnikova]: *The Boundary-Layer Method in Diffraction Problems* (Izdat. LGU, Leningrad 1974) [Springer Ser. Electrophys, Vol. 3 (Springer, Berlin, Heidelberg 1979)]

1.13 *P.Ya. Ufimtsev: Transverse diffusion in wedge diffraction. Radiotekh. Elektron. **10**, 1013–1023 (1965) [Radio Eng. Electron. Phys. **10**, 866–875 (1965)]

1.14 *T.F. Pankratova: Eigenfunctions of the Laplace operator on the surface of a triaxial ellipsoid and in the region exterior to it. Zap. Nauchn. Semin. LOMI AN SSSR **9**, 192–212 (1968) [Semin. Math. **9**, 87–95 (1970)]

Chapter 2

2.1 *I.M. Gel'fand, S.V. Fomin: *Calculus of Variations* (Fizmatgiz, Moscow 1961) [Prentice-Hall, Englewood Cliffs, NJ 1963]

2.2 D. Laugwitz: *Differential and Riemannian Geometry* (Academic, New York 1965)

2.3 *P.K. Rashevskii [Raschewskii]: *Riemannsche Geometrie und Tensoranalysis*, 3rd ed. (Nauka, Moscow 1967) [1st ed., VEB Deutscher Verlag der Wissenschaften, Berlin 1959]

2.4 H.W. Guggenheimer: *Differential Geometry* (Dover, New York 1977)

2.5 J.B. Keller: "A Geometrical Theory of Diffraction", in *Calculus of Variations and Its Applications*, ed. by L.M. Graves (McGraw-Hill, New York 1958) pp. 27–52

2.6 *V.I. Smirnov: *A Course of Higher Mathematics*, Vol. 3, Pt. 2 (Nauka, Moscow 1969) [Addison-Wesley, Reading, MA 1964]

2.7 *V.I. Smirnov: *A Course of Higher Mathematics*, Vol. 4 (Nauka, Moscow 1974) [Addison-Wesley, Reading, MA 1964]

2.8 J. Hadamard: *Lectures on Cauchy's Problem in Linear Partial Differential Equations* (Dover, New York 1952)

2.9 J. Hadamard: *Le Problème de Cauchy et les Equations aux Dérivées Partielles Linéaires Hyperboliques* (Hermann, Paris 1932)

2.10 L.B. Felsen, N. Marcuvitz: *Radiation and Scattering of Waves* (Prentice-Hall, Englewood Cliffs, NJ 1973)

2.11 *V.P. Maslov: *Théorie des Perturbations et Méthodes Asymptotiques*, 2nd ed. (Nauka, Moscow 1988) [1st ed., Dunod, Paris 1972]

2.12 *V.P. Maslov: *The Complex WKB Method in Nonlinear Equations* (Nauka, Moscow 1977)

2.13 V. Guillemin, S. Sternberg: *Geometric Asymptotics* (Am. Math. Soc., Providence, RI 1977)

2.14 *V.A. Borovikov, B.E. Kinber: *The Geometrical Theory of Diffraction* (Svyaz', Moscow 1978)

2.15 *Yu.A. Kravtsov, Yu.I. Orlov: *Geometrical Optics of Inhomogeneous Media* (Nauka, Moscow 1980) [Springer Ser. Wave Phenom., Vol. 6 (Springer, Berlin, Heidelberg 1990)]

2.16 J.B. Keller, R.M. Lewis, B.D. Seckler: Asymptotic solution of some diffraction problems. Commun. Pure Appl. Math. 9, 207–265 (1956)

2.17 Yu.A. Kravtsov: "Rays and Caustics as Physical Objects", in *Progress in Optics*, Vol. 26, ed. by E. Wolf (Elsevier, Amsterdam 1988) pp. 227–348

2.18 R.G. Kouyoumjian: The Geometrical Theory of Diffraction and Its Application, in *Numerical and Asymptotic Techniques in Electromagnetics*, ed. by R. Mittra (Springer, Berlin, Heidelberg 1975) pp. 165–215

2.19 *S.G. Mikhlin (ed.): *Linear Equations of Mathematical Physics* (Nauka, Moscow 1964) [Holt, Rinehart and Winston, New York 1967]

2.20 *S.M. Rytov: La diffraction de la lumière par les ultra-sons. Izv. Akad. Nauk SSSR, Ser. Fiz. 2, 223–259 (1937) [Actualités Scient. Industr., No. 613 (Hermann, Paris 1938)]

2.21 K. Bochenek, J. Plebański: On a certain optical interpretation of the Helmholtz equation. Bull. Acad. Pol. Sci., Ser. Sci. Tech. 4, 179–186 (1956)

2.22 K. Bochenek, J. Plebański: O metodach optyki mikrofalowej. Arch. Elektrotech. (Warsaw) 5, 293–323 (1956)

2.23 S. Pogorzelski: Zagadnienia optyki parageometrycznej. Arch. Elektrotech. (Warsaw) 11, 49–75 (1962)

2.24 S. Pogorzelski: "An Asymptotic Expansion of Electric Vector Fields with Complex Phase Function", in *Electromagnetic Theory and Antennas*, ed. by E.C. Jordan (Pergamon, Oxford 1963) pp. 119–121

2.25 R. Car, G.M. Cicuta, D. Zanon, F. Riva: High-energy, Rytov, eikonal expansions. Nuovo Cimento 39A, 253–271 (1977)

2.26 *S. M. Rytov: Sur la transition de l'optique ondulatoire à l'optique géometrique. Dokl. Akad. Nauk SSSR 18, 263–266 (1938) [C. R. Acad. Sci. URSS 18, 263–266 (1938)]

2.27 M. Kline, I.W. Kay: *Electromagnetic Theory and Geometrical Optics* (Wiley-Interscience, New York 1965)

2.28 M. Born, E. Wolf: *Principles of Optics* (Pergamon, Oxford 1975)

2.29 M. Kline: An asymptotic solution of Maxwell's equations. Commun. Pure Appl. Math. 4, 225–262 (1951)

2.30 R.K. Luneberg: *Mathematical Theory of Optics* (University of California Press, Berkeley, CA 1965)

2.31 W.J. Trjitzinsky: Analytic theory of parametric linear partial differential equations. Mat. Sb. 15 (57), 179–242 (1944)

2.32 J.B. Keller: Diffraction by a convex cylinder. IRE Trans. Ant. Prop. 4, 312–321 (1956)

2.33 *V.S. Buslaev: Shortwave asymptotic formulas in the problem of diffraction by convex bodies. Vestn. Leningr. Univ., Ser. Mat., Mekh., Astron. 3 (13), 5–21 (1962)

2.34 F. Ursell: On the short-wave asymptotic theory of the wave equation $(\nabla^2 + k^2)\phi = 0$. Proc. Cambridge Philos. Soc. 53, 115–133 (1957)

2.35 *V.M. Babich [Babič]: The shortwave asymptotic behavior of the Green's function for the region exterior to a finite convex region. Dokl. Akad. Nauk SSSR 146, 571–573 (1962) [Sov. Phys.–Dokl. 7, 792–794 (1963)]

2.36 *V.M. Babich [Babič]: The shortwave asymptotic behavior of the Green's function for Helmholtz's equation. Mat. Sb. 65, 576–630 (1964)

2.37 R. Grimshaw: High-frequency scattering by finite convex regions. Commun. Pure Appl. Math. 19, 167–198 (1966)

2.38 *V.M. Babich [Babič]: Rigorous justification of the shortwave approximation in the three-dimensional case. Zap. Nauchn. Semin LOMI AN SSSR 34, 23–51 (1973) [J. Sov. Math. 6, 488–509 (1976)]

2.39 *V.S. Buslaev: Potential theory and geometrical optics. Zap. Nauchn. Semin LOMI AN SSSR 22, 175–180 (1971) [J. Sov. Math. 2, 204–209 (1974)]

2.40 *V.S. Buslaev: On the asymptotic behavior of the spectral characteristics of exterior problems for Schrödinger equations. Izv. Akad. Nauk SSSR, Ser. Mat. 39, 149–235 (1975) [Math. USSR Izv. 9, 139–223 (1975)]

Chapter 3

3.1 G. N. Watson: *Theory of Bessel Functions* (Cambridge University Press, Cambridge 1966)

3.2 F.W.J. Olver: The asymptotic solutions of linear differential equations of the second order for large values of a parameter. Philos. Trans. R. Soc. London A 247, 307–327 (1954)

3.3 F.W.J. Olver: The asymptotic expansion of Bessel functions of large order. Philos. Trans. R. Soc. London A 247, 328–368 (1954)

3.4 F.W.J. Olver: *Asymptotics and Special Functions* (Academic, New York 1974)

3.5 *V.I. Smirnov: *A Course of Higher Mathematics*, Vol. 4 (Nauka, Moscow 1974) [Addison-Wesley, Reading, MA 1964]

3.6 D. Laugwitz: *Differential and Riemannian Geometry* (Academic, New York 1965)

3.7 *P.K. Rashevskii: *A Course in Differential Geometry* (Gostekhizdat, Moscow 1959)

3.8 *G.M. Fikhtengol'ts [Fichtenholz]: *Differential- und Integralrechnung*, Vol. 1 (Nauka, Moscow 1966) [VEB Deutscher Verlag der Wissenschaften, Berlin 1972]

3.9 H.W. Guggenheimer: *Differential Geometry* (Dover, New York 1977)

3.10 *G.I. Petrashen' [Petrašen'], N.S. Smirnova, G.I. Makarov: Asymptotic representations of the cylinder functions. Uch. Zap. Leningrad. Univ., No. 170 (27), 7–95 (1953); addenda, ibid., No. 246 (32), 347–352, 352–364 (1958)

3.11 *L.D. Landau, E.M. Lifshitz [Lifšits]: *The Classical Theory of Fields* (Nauka, Moscow 1973) [Pergamon, Oxford 1975]

3.12 *Yu.L. Gazaryan: The geometrical-acoustic approximation of the field in the neighborhood of a nonsingular section of caustic. Vopr. Din. Teor. Rasprostr. Seism. Voln 5, 77–89 (1961)

3.13 *Yu.A. Kravtsov: A modification of the method of geometrical optics. Izv. Vyssh. Uchebn. Zaved., Radiofiz. **7**, 664–673 (1964)

3.14 D. Ludwig: Uniform asymptotic expansions at a caustic. Commun. Pure Appl. Math. **19**, 215–250 (1966)

3.15 *V.M. Babich [Babič], N.Ya. Kirpichnikova [Kirpičnikova]: *The Boundary-Layer Method in Diffraction Problems* (Izdat. LGU, Leningrad 1974) [Springer Ser. Electrophys., Vol. 3 (Springer, Berlin, Heidelberg 1979)]

3.16 *V.P. Maslov: *Théorie des Perturbations et Méthodes Asymptotiques*, 2nd ed. (Nauka, Moscow 1988) [1st ed., Dunod, Paris 1972]

3.17 *V. S. Buslaev: The generating integral and the canonical Maslov operator in the WKB method. Funkts. Anal. Prilozh. **3** (3), 17–31 (1969) [Funct. Anal. Appl. **3**, 181–193 (1969)]

Chapter 4

4.1 H. Seifert, W. Threlfall: *A Textbook of Topology* (Academic, New York 1980)

4.2 Yu. Borisovich, N. Bliznyakov, Ya. Izrailevich, T. Fomenko: *Introduction to Topology* (Mir, Moscow 1985)

4.3 *V.I. Arnol'd: On a theorem of Liouville concerning integrable problems of dynamics. Sib. Mat. Zh. **4**, 471–474 (1963) [Am. Math. Soc. Transl., Ser. 2, **61**, 292–296 (1967)]

4.4 *V.A. Fock [Fok]: *Tables of Airy Functions* (GITTL, Moscow 1946)

4.5 G. Salmon: *A Treatise on Conic Sections* (Chelsea, New York 1954)

4.6 *L.A. Vainshtein [Weinstein]: *Open Resonators and Open Waveguides* (Sovetskoe Radio, Moscow 1966) [Golem, Boulder, CO 1969]

4.7 *V.I. Smirnov: *A Course of Higher Mathematics*, Vol. 4 (Nauka, Moscow 1974) [Addison-Wesley, Reading, MA 1964]

4.8 J.B. Keller, S. I. Rubinow: Asymptotic solution of eigenvalue problems. Annals Phys. **9**, 24–75 (1960); errata, ibid. **10**, 303–305 (1960)

4.9 A. Einstein: Zum Quantensatz von Sommerfeld und Epstein. Verh. Dtsch. Phys. Ges. **19**, 82–92 (1917)

4.10 *D.K. Ozerov: On a ray method for obtaining dispersion equations for interference SH-waves propagating in media with spherical boundaries. Vopr. Din. Teor. Rasprostr. Sejsm. Voln **8**, 60–69 (1966)

4.11 *D. K. Ozerov, A. S. Alekseev: Constructing hodographs of SH-waves from the dispersion curves of Love waves. Vopr. Din. Teor. Rasprostr. Sejsm. Voln **8**, 69–75 (1966)

4.12 *L.A. Vainshtein: Ray flux in a triaxial ellipsoid. Elektron. Bolsh. Moshchn. **4**, 93–105 (1965)

4.13 *V.P. Bykov: Geometrical optics of three-dimensional oscillations in open resonators. Elektron. Bolsh. Moshchn. **4**, 66–92 (1965)

4.14 *L.A. Vainshtein [Weinstein]: *Open Resonators and Open Waveguides* (Sovetskoe Radio, Moscow 1966) [Golem, Boulder, CO 1969]

4.15 *V.P. Maslov: *Théorie des Perturbations et Méthodes Asymptotiques*, 2nd ed. (Nauka, Moscow 1988) [1st ed., Dunod, Paris 1972]

4.16 *V.I. Arnol'd: A characteristic class entering into quantization conditions. Funkts. Anal. Prilozh. **1** (1), 1–14 (1967) [Funct. Anal. Appl. **1** (1), 1–13 (1967)]

4.17 *V.F. Lazutkin: *Convex Billiards and Eigenfunctions of the Laplacian* (Izdat. LGU, Leningrad 1981)

Chapter 5

5.1 *G.M. Fikhtengol'ts [Fichtenholz]: *Differential- und Integralrechnung*, Vol. 1 (Nauka, Moscow 1966) [VEB Deutscher Verlag der Wissenschaften, Berlin 1972]

5.2 J.B. Keller, S.I. Rubinow: Asymptotic solution of eigenvalue problems. Annals Phys. **9**, 24–75 (1960); errata, ibid. **10**, 303–305 (1960)

5.3   *V.I. Arnol'd: Small denominators and problems of stability of motion in classical and celestial mechanics. Usp. Mat. Nauk **18** (**6**), 91–192 (1963) [Russ. Math. Surv. **18** (**6**), 85–191 (1963)]

5.4   J. Moser: On invariant curves of area-preserving mapping of an annulus. Nachr. Akad. Wiss. Göttingen, Math.-Phys. Kl: II, 1–20 (1962)

5.5   *D.Sh. Mogilevskii: The instability of closed ray congruences. Vestn. Leningr. Univ., Ser. Mat., Mekh., Astron. **4** (**19**), 89–91 (1969)

5.6   *V.S. Buldyrev: Shortwave asymptotic behavior of the eigenfunctions of the Helmholtz equation. Dokl. Akad. Nauk SSSR **163**, 853–856 (1965) [Sov. Phys.–Dokl. **10**, 718–720 (1966)]

5.7   *V.S. Buldyrev, M.M. Popov: Application of the ray method to the calculation of the natural frequencies of multiple-mirror resonators. Opt. Spektrosk. **20**, 905–908 (1966) [Opt. Spectrosc. **20**, 500–501 (1966)]

5.8   *M.M. Popov: Geometrical optics and the eigenfrequencies of circular resonators. Vestn. Leningr. Univ., Ser. Fiz. Khim., No. 4, 42–51 (1967)

5.9   *V.P. Bykov: Geometrical optics of three-dimensional oscillations in open resonators. Elektron. Bolsh. Moshchn. **4**, 66–92 (1965)

5.10 *L.A. Vainshtein [Weinstein]: *Open Resonators and Open Waveguides* (Sovetskoe Radio, Moscow 1966) [Golem, Boulder, CO 1969]

Chapter 6

6.1   *V.I. Ivanov: Diffraction of short waves by a smooth convex cylinder. Nauchn. Dokl. Vyssh. Shk., Fiz.-Mat. Nauk **6**, 192–196 (1958)

6.2   *V.A. Fock [Fok]: *Electromagnetic Diffraction and Propagation Problems*, 2nd ed. (Sovetskoe Radio, Moscow 1970) [1st ed., Pergamon, Oxford 1965]

6.3   *V.I. Smirnov: *A Course of Higher Mathematics*, Vol. 3, Pt. 2 (Nauka, Moscow 1969) [Addison-Wesley, Reading, MA 1964]

6.4   *N.N. Lebedev: *Special Functions and Their Applications* (Fizmatgiz, Moscow 1963) [Dover, New York 1972]

6.5   *M.A. Leontovich [Leontovič]: On a method for solving the problem of electromagnetic wave propagation along the earth's surface. Izv. Akad. Nauk SSSR, Ser. Fiz. **8**, 16–22 (1944)

6.6   *V.A. Fock [Fok]: The field of a plane wave near the surface of a conducting body. Izv. Akad. Nauk SSSR, Ser. Fiz. **10** (**2**), 171–186 (1946) [J. Phys. USSR **10**, 399–409 (1946)]; also in [6.8], Chap. 5

6.7   *M.A. Leontovich [Leontovič], V.A. Fock [Fok]: Solution of the problem of propagation of electromagnetic waves along the earth's surface by the method of parabolic equations. Zh. Exp. Teor. Fiz. **16**, 557–573 (1946) [J. Phys. USSR **10**, 13–24 (1946)]; also in [6.8], Chap. 11

6.8   *V.S. Buslaev: Shortwave asymptotic formulas in the problem of diffraction by convex bodies. Vestn. Leningr. Univ., Ser. Mat., Mekh., Astron. **3** (**13**), 5–21 (1962)

6.9   *V.S. Buldyrev: Asymptotics of eigenfunctions of the Helmholtz equation for plane convex regions. Vestn. Leningr. Univ. **20** (**22**), 38–51 (1965)

6.10 *I.A. Molotkov: Wave propagation in an inhomogeneous halfspace whose refractive index depends on two coordinates. Probl. Difraktsii Rasprostr. Voln **6**, 89–104 (1966)

6.11 *L.A. Vainshtein: Open resonators with spherical mirrors. Zh. Exp. Teor. Fiz. **45**, 684–697 (1963) [Sov. Phys.–JETP **18**, 471–479 (1964)]

6.12 *V.F. Lazutkin: Parabolic equations and asymptotic eigenfunctions of the Helmholtz equation for three-dimensional regions. Vestn. Leningr. Univ. **20** (**22**), 52–57 (1965)

6.13 *V.M. Babich [Babič], V. F. Lazutkin: Eigenfunctions concentrated near a closed geodesic. Probl. Mat. Fiz. **2**, 15–25 (1967) [Top. Math. Phys. **2**, 9–18 (1968)]

6.14 L. Levey, L.B. Felsen: On incomplete Airy functions and their application to diffraction problems. Radio Sci. **4**, 959–969 (1969)

6.15 *P.Ya. Ufimtsev: Transverse diffusion in wedge diffraction. Radiotekh. Elektron. **10**, 1013–1023 (1965) [Radio Eng. Electron. Phys. **10**, 866–875 (1965)]

6.16   F.D. Tappert: "The Parabolic Approximation Method", in *Wave Propagation and Underwater Acoustics*, ed. by J.B. Keller, J.S. Papadakis (Springer, Berlin, Heidelberg 1977) pp. 224–287

6.17   *V.M. Babich [Babič], N.Ya. Kirpichnikova [Kirpičnikova]: *The Boundary-Layer Method in Diffraction Problems* (Izdat. LGU, Leningrad 1974) [Springer Ser. Electrophys., Vol. 3 (Springer, Berlin, Heidelberg 1979)]

Chapter 7

7.1   T.M. Cherry: Uniform asymptotic formulae for functions with transition points. Trans. Am. Math. Soc. **68**, 224–257 (1950)

7.2   *V.I. Smirnov: *A Course of Higher Mathematics*, Vol. 4 (Nauka, Moscow 1974) [Addison-Wesley, Reading, MA 1964]

7.3   H. Reichardt: Ausstrahlungsbedingungen für die Wellengleichung. Abh. Math. Sem. Univ. Hamburg **24**, 41–53 (1960)

7.4   *O.A. Ladyzhenskaya: *The Mixed Problem for a Hyperbolic Equation* (Gostekhizdat, Moscow 1953)

7.5   Lord Rayleigh (J.W. Strutt): The problem of the whispering gallery. Philos. Mag., Ser. 6, **20**, 1001–1004 (1910)

7.6   *P.E. Krasnushkin: The method of normal waves applied to waveguides. Vestn. Mosk. Univ., No. 1, 37–55 (1946)

7.7   *P.E. Krasnushkin, E.R. Mustel': On clinging of electromagnetic waves to a concave metal surface. C. R. (Dokl.) Acad. Sci. URSS **54**, 211–214 (1946)

7.8   *M.I. Vishik, L.A. Lyusternik: Regular degeneration and boundary layers for linear differential equations with small parameter. Usp. Mat. Nauk **12**, No. 5 (77), 3–122 (1957) [Am. Math. Soc. Transl., Ser. 2, **20**, 239–364 (1962)]

7.9   *V.S. Buldyrev: Wave propagation near the curved surface of an inhomogeneous body. Probl. Mat. Fiz. **2**, 61–84 (1967) [Top. Math. Phys. **2**, 47–65 (1968)]

7.10   F.W.J. Olver: The asymptotic solutions of linear differential equations of the second order for large values of a parameter. Philos. Trans. R. Soc. London A **247**, 307–327 (1954)

7.11   R.M. Lewis, N. Bleistein, D. Ludwig: Uniform asymptotic theory of creeping waves. Commun. Pure Appl. Math. **20**, 295–328 (1967)

7.12   *I.V. Mukhina, I.A. Molotkov: Propagation of Rayleigh waves in an elastic half-space which is inhomogeneous in two coordinates. Izv. Akad. Nauk SSSR, Fiz. Zemli, No. 4, 3–8 (1967) [Izv., Acad. Sci., USSR, Phys. Solid Earth 209–211 (1967)]

7.13   *N.Ya. Kirpichnikova [Kirpičnikova]: Rayleigh waves concentrated near a ray on the surface of an inhomogeneous elastic body. Zap. Nauchn. Semin. LOMI AN SSSR **15**, 91–115 (1969) [Semin. Math. **15**, 49–62 (1971)]

7.14   *V.F. Lazutkin: Asymptotics of the eigenfunctions of the Laplacian concentrated near the boundary of a region. Zh. Vychisl. Mat. Mat. Fiz. **7**, 1237–1249 (1967) [USSR Comp. Math. Math. Phys. **7** (6), 37–52 (1967)]

7.15   *V.I. Ivanov: Diffraction of short waves by a smooth convex cylinder. Nauchn. Dokl. Vyssh. Shk., Fiz.-Mat. Nauk **6**, 192–196 (1958)

7.16   *V.S. Buslaev: Shortwave asymptotic formulas in the problem of diffraction by convex bodies. Vestn. Leningr. Univ., Ser. Mat., Mekh., Astron. 3 (13), 5–21 (1962)

7.17   *V.S. Buldyrev: Asymptotics of eigenfunctions of the Helmholtz equation for plane convex regions. Vestn. Leningr. Univ. **20** (22), 38–51 (1965)

7.18   *I.A. Molotkov: Diffraction on a convex contour with smoothly varying radius of curvature and surface impedance. Probl. Mat. Fiz. **2**, 124–132 (1967) [Top. Math. Phys. **2**, 103–110 (1968)]

7.19   *V.M. Babich [Babič]: Analytic continuation to the second sheet of the resolvent of the exterior problem for the Laplacian. Teor. Funkts. Funkts. Anal. Prilozh. **3**, 151–157 (1966)

7.20 P.D. Lax, C.S. Morawetz, R.S. Phillips: Exponential decay of solutions of the wave equation in the exterior of a star-shaped obstacle. Commun. Pure Appl. Math. **16**, 477–486 (1963)

7.21 *V.M. Babich [Babič], N. S. Grigor'eva: Asymptotic properties of solutions to three-dimensional wave problems. Zap. Nauchn. Semin. LOMI AN SSSR **51**, 20–78 (1975) [J. Sov. Math. **11**, 372–413 (1979)]

7.22 *V.P. Maslov: *Théorie des Perturbations et Méthodes Asymptotiques*, 2nd ed. (Nauka, Moscow 1988) [1st ed., Dunod, Paris 1972]

Chapter 8

8.1 *V.I. Smirnov: *A Course of Higher Mathematics*, Vol. 3, Pt. 2 (Nauka, Moscow 1969) [Addison-Wesley, Reading, MA 1964]

8.2 *V.S. Buldyrev: The asymptotic behavior of the solutions of the wave equation concentrated near the axis of a two-dimensional waveguide in an inhomogeneous medium. Probl. Mat. Fiz. **3**, 5–30 (1968) [Top. Math. Phys. **3**, 1–23 (1969)]

8.3 *V.F. Lazutkin: "Asymptotic Expansion for the Eigenfunctions of the Laplacian Concentrated Near Some One-Dimensional Cycle"; Candidate's dissertation, Leningrad Section of the Mathematical Institute (1967)

8.4 *V.F. Lazutkin: Construction of an asymptotic series for eigenfunctions of the "bouncing ball" type. Tr. Mat. Inst. Steklov **95**, 106–118 (1968) [Proc. Steklov Inst. Math. **95**, 125–140 (1971)]

8.5 *V.S. Buldyrev: "The Etalon Problem Method in the Theory of Wave Diffraction and Propagation"; Doctoral dissertation, Leningrad University (1969)

8.6 *V.F. Lazutkin: An equation for the natural frequencies of a nonconfocal resonator with cylindrical mirrors which takes mirror aberration into account. Opt. Spektrosk. **24**, 453–454 (1968) [Opt. Spectrosc. **24**, 236 (1968)]

8.7 R. Smith: Bouncing ball waves. SIAM J. Appl. Math. **26**, 5–14 (1974)

Chapter 9

9.1 *P.K. Rashevskii [Raschewskii]: *Riemannsche Geometrie und Tensoranalysis*, 3rd ed. (Nauka, Moscow 1967) [1st ed., VEB Deutscher Verlag der Wissenschaften, Berlin 1959]

9.2 O. Veblen: *Invariants of Quadratic Differential Forms* (Cambridge University Press, Cambridge 1927)

9.3 R.L. Bishop, R.J. Crittenden: *Geometry of Manifolds* (Academic, New York 1967)

9.4 H. Seifert, W. Threlfall: *A Textbook of Topology* (Academic, New York 1980)

9.5 J. Milnor: *Morse Theory* (Princeton University Press, Princeton 1963)

9.6 *L.S. Pontryagin: *Ordinary Differential Equations* (Nauka, Moscow 1965) [Addison-Wesley, Reading, MA 1962]

9.7 N. Bourbaki: *Eléments de Mathématique*, Livre 4, Fonction d'une variable réelle (Hermann, Paris 1961) Chap. 4

9.8 *V.M. Babich [Babič]: Eigenfunctions concentrated in the neighborhood of a closed geodesic. Zap. Nauchn. Semin. LOMI AN SSSR **9**, 15–64 (1968) [Semin. Math. **9**, 7–26 (1970)]

9.9 *V.M. Babich [Babič], N.Ya. Kirpichnikova [Kirpičnikova]: *The Boundary-Layer Method in Diffraction Problems* (Izdat. LGU, Leningrad 1974) [Springer Ser. Electrophys., Vol. 3 (Springer, Berlin, Heidelberg 1979)]

9.10 *P.K. Rashevskii: *A Course in Differential Geometry* (Gostekhizdat, Moscow 1959)

9.11 *V.I. Smirnov: *A Course of Higher Mathematics*, Vol. 4 (Nauka, Moscow 1974) [Addison-Wesley, Reading, MA 1964]

9.12 *L.A. Vainshtein: Ray flux in a triaxial ellipsoid. Elektron. Bolsh. Moshchn. **4**, 93–105 (1965)

9.13 *L.A. Vainshtein [Weinstein]: *Open Resonators and Open Waveguides* (Sovetskoe Radio, Moscow 1966) [Golem, Boulder, CO 1969]

9.14 *V.P. Bykov: Geometrical optics of three-dimensional oscillations in open resonators. Elektron. Bolsh. Moshchn. **4**, 66–92 (1965)

9.15 *V.M. Babich [Babič]: Analytic continuation to the second sheet of the resolvent of the exterior problem for the Laplacian. Teor. Funkts. Funkts. Anal. Prilozh. **3**, 151–157 (1966)

9.16 *V.M. Babich [Babič]: Localization Concepts in Shortwave Diffraction Problems, in *Third All-Union Symposium on Wave Diffraction*, Tblisi, USSR 1964 (Nauka, Moscow 1964) pp. 78–79

9.17 *V.M. Babich [Babič], V.F. Lazutkin: Eigenfunctions concentrated near a closed geodesic. Probl. Mat. Fiz. **2**, 15–25 (1967) [Top. Math. Phys. **2**, 9–18 (1968)]

9.18 *M.G. Krein: "Fundamental Aspects of the Theory of λ-Zones of Stability for a Canonical System of Linear Differential Equations with Periodic Coefficients", in *Pamyati A.A. Andronova* (Izdat. Akad. Nauk SSSR, Moscow 1955) pp. 413–498

9.19 I.E. Segal: Foundations of the theory of dynamical systems of infinitely many degrees of freedom, II. Can. J. Math. **13**, 1–18 (1961)

9.20 *N.A. Chernikov: The system whose Hamiltonian is a time-dependent quadratic form in $\hat{x}$ and $\hat{p}$. Zh. Exp. Teor. Fiz. **53**, 1006–1017 (1967) [Sov. Phys.–JETP **26**, 603–608 (1968)]

9.21 *I.M. Gel'fand, S.V. Fomin: *Calculus of Variations* (Fizmatgiz, Moscow 1961) [Prentice-Hall, Englewood Cliffs, NJ 1963]

9.22 *V.F. Lazutkin: Spectral degeneracy and "small denominators" in the asymptotics of eigenfunctions of the "bouncing ball" type. Vestn. Leningr. Univ., Ser. Mat. Mekh., No. 7, 23–34 (1969) [Vestn. Leningr. Univ. Math. **2**, 103–116 (1975)]

9.23 *M.F. Pyshkina: Asymptotic behavior of eigenfunctions of the Helmholtz equation concentrated near a closed geodesic. Zap. Nauchn. Semin. LOMI AN SSSR **15**, 142–154 (1969) [Semin. Math. **15**, 88–92 (1971)]

9.24 *V.F. Lazutkin: "Asymptotic Expansion for the Eigenfunctions of the Laplacian Concentrated Near Some One-Dimensional Cycle"; Candidate's dissertation, Leningrad Section of the Mathematical Institute (1967)

9.25 *N.Ya. Kirpichnikova [Kirpičnikova]: Rayleigh waves concentrated near a ray on the surface of an inhomogeneous elastic body. Zap. Nauchn. Semin. LOMI AN SSSR **15**, 91–115 (1969) [Semin. Math. **15**, 49–62 (1971)]

9.26 *V.S. Buldyrev: Concentrated waves and the parabolic equation method in magnetohydrodynamics. Probl. Mat. Fiz. **6**, 68–76 (1973)

9.27 *V.P. Maslov: *The Complex WKB Method in Nonlinear Equations* (Nauka, Moscow 1977)

9.28 *V.M. Babich [Babič], V.V. Ulin: Complex ray solutions and eigenfunctions concentrated in a neighborhood of a closed geodesic. Zap. Nauchn. Semin. LOMI AN SSSR **104**, 6–13 (1981) [J. Sov. Math. **20**, 1749–1753 (1982)]

9.29 *V.E. Nomofilov: Asymptotic solutions of a system of equations of second order which are concentrated in the neighborhood of a ray. Zap. Nauchn. Semin. LOMI AN SSSR **104**, 170–179 (1981) [J. Sov. Math. **20**, 1854–1860 (1982)]

Chapter 10

10.1 *P.K. Rashevskii: *A Course in Differential Geometry* (Gostekhizdat, Moscow 1959)

10.2 *I.M. Gel'fand, S.V. Fomin: *Calculus of Variations* (Fizmatgiz, Moscow 1961) [Prentice-Hall, Englewood Cliffs, NJ 1963]

10.3 *L.S. Pontryagin: *Ordinary Differential Equations* (Nauka, Moscow 1965) [Addison-Wesley, Reading, MA 1962]

10.4 E.T. Whittaker: *A Treatise on the Analytical Dynamics of Particles and Rigid Bodies* (Cambridge University Press, Cambridge 1937)

10.5 *M.M. Popov: Resonators for lasers with rotated directions of the principal curvatures. Opt. Spektrosk. **25**, 314–316 (1968) [Opt. Spectrosc. **25**, 170–171 (1968)]

10.6 *M.M. Popov: The asymptotic behavior of certain subsequences of eigenvalues of boundary value problems for the Helmholtz equation in higher dimensions. Dokl. Akad. Nauk SSSR **184**, 1076-1079 (1969) [Sov. Phys.–Dokl. **14**, 108–110 (1969)]

10.7   *M.M. Popov: Natural oscillations of multimirror resonators. Vestn. Leningr. Univ., Ser. Fiz. Khim. **24 (22)**, 42–54 (1969)

10.8   *V.F. Boitsov: Eigenfrequencies and field distributions on mirrors of an astigmatic resonator in the geometrical optics approximation. Opt. Spektrosk. **25**, 311–314 (1968) [Opt. Spectrosc. **25**, 168–169 (1968)]

10.9   *B.N. Semenov: Asymptotic behavior of the eigenfunctions and eigenfrequencies of a multimirror resonator. Zap. Nauchn. Semin. LOMI AN SSSR **15**, 176–187 (1969) [Semin. Math. **15**, 102–109 (1971)]

Chapter 11

11.1   *V.I. Smirnov: *A Course of Higher Mathematics*, Vol. 4 (Nauka, Moscow 1974) [Addison-Wesley, Reading, MA 1964]

11.2   *V.B. Filippov: Rigorous justification of the shortwave asymptotic theory of diffraction in a shadow zone. Zap. Nauchn. Semin. LOMI AN SSSR **34**, 142–205 (1973) [J. Sov. Math. **6**, 577–626 (1976)]

11.3   *A.B. Zayaev, V.B. Filippov: Rigorous justification of the Friedlander-Keller formulas. Zap. Nauchn. Semin. LOMI AN SSSR **140**, 49–60 (1984) [J. Sov. Math. **32**, 134–143 (1986)]

11.4   *A.B. Zayaev: A rigorous proof of the asymptotic expansion of the Green's function in the shadow zone for diffraction at a convex body. Zap. Nauchn. Semin. LOMI AN SSSR **148**, 79–88 (1985) [J. Sov. Math. **38**, 1612–1619 (1987)]

11.5   F.G. Friedlander: *Sound Pulses* (Cambridge University Press, Cambridge 1958)

11.6   F. Ursell: On the short-wave asymptotic theory of the wave equation $(\nabla^2 + k^2)\phi = 0$. Proc. Cambridge Philos. Soc. **53**, 115–133 (1957)

11.7   *V.A. Fock [Fok]: *Tables of Airy Functions* (GITTL, Moscow 1946)

11.8   F.G. Friedlander, J.B. Keller: Asymptotic expansions of solutions of $(\nabla^2 + k^2)U = 0$. Commun. Pure Appl. Math. **8**, 387–394 (1955)

11.9   *G.D. Malyuzhinets, L.A. Vainshtein: Transverse diffusion in diffraction by an impedance cylinder of large radius I & II. Radiotekh. Elektron. **6**, 1247–1258, 1489–1495 (1961) [Radio Eng. Electron. Phys. **6**, 1106–1116, 1324–1330 (1961)]

11.10 *V.A. Fock [Fok]: *Electromagnetic Diffraction and Propagation Problems*, 2nd ed. (Sovetskoe Radio, Moscow 1970) [1st ed., Pergamon, Oxford 1965]

11.11 *M.A. Leontovich [Leontovič], V.A. Fock [Fok]: Solution of the problem of propagation of electromagnetic waves along the earth's surface by the method of parabolic equations. Zh. Exp. Teor. Fiz. **16** (7), 557–573 (1946) [J. Phys. USSR **10**, 13–24 (1946)]; also in [11.13], Chap. 11

11.12 W.P. Brown: On the asymptotic behavior of electromagnetic fields scattered from convex cylinders near grazing incidence. J. Math. Anal. Appl. **15**, 355–385 (1966)

11.13 *V.M. Babich [Babič]: A high-frequency point source of oscillations near a concave mirror. Zap. Nauchn. Semin. LOMI AN SSSR **51**, 5–20 (1975) [J. Sov. Math. **11**, 361–371 (1979)]

11.14 *V.M. Babich [Babič]: A point source of oscillations on the boundary of a region. Zap. Nauchn. Semin. LOMI AN SSSR **62**, 3–21 (1976) [J. Sov. Math. **11**, 665–676 (1979)]

11.15 *V.S. Buslaev: Potential theory and geometrical optics. Zap. Nauchn. Semin. LOMI AN SSSR **22**, 175–180 (1971) [J. Sov. Math. **2**, 204–209 (1974)]

11.16 *V.S. Buslaev: On the asymptotic behavior of the spectral characteristics of exterior problems for Schrödinger equations. Izv. Akad. Nauk SSSR, Ser. Mat. **39**, 149–235 (1975) [Math. USSR Izv. **9**, 139–223 (1975)]

11.17 *V.A. Fock [Fok], L.A. Vainshtein: Transverse diffusion in shortwave diffraction by a convex cylinder of smoothly varying curvature, I, II. Radiotekh. Elektron. **8**, 363–376, 377–388 (1963) [Radio Eng. Electron. Phys. **8**, 317–330, 330–341 (1963)]; also in [11.13], Chap. 9

11.18 *I.A. Molotkov: Diffraction on a convex contour with smoothly varying radius of curvature and surface impedance. Probl. Mat. Fiz. **2**, 124–132 (1967) [Top. Math. Phys. **2**, 103–110 (1968)]

11.19 *V.M. Babich [Babič]: Localization concepts in shortwave diffraction problems, in *Third All-Union Symposium on Wave Diffraction*, Tblisi, USSR 1964 (Nauka, Moscow 1964) pp. 78–79

11.20 *V.S. Buldyrev: Shortwave interference in diffraction by a nonuniform cylinder of arbitrary cross section. Izv. Vyssh. Uchebn. Zaved., Radiofiz. **10**, 699–711 (1967) [Radiophys. Quantum Electron. **10**, 383–389 (1967)]

11.21 *I.A. Molotkov: Green's function for the diffraction problem on a convex cylinder with variable impedance. Tr. Mat. Inst. Steklov **95**, 119–131 (1968) [Proc. Steklov Inst. Math. **95**, 141–157 (1971)]

11.22 *I.A. Molotkov: Wave propagation in an inhomogeneous halfspace whose refractive index depends on two coordinates. Probl. Difraktsii Rasprostr.Voln **6**, 89–104 (1966)

Chapter 12

12.1 L.B. Felsen, N. Marcuvitz: *Radiation and Scattering of Waves* (Prentice-Hall, Englewood Cliffs, NJ 1973)

12.2 *I.M. Gel'fand, G.E. Shilov: *Generalized Functions*, Vol. 1 (Gos. Izdat. Fiz. Mat. Lit., Moscow 1958) [Academic, New York 1964]

12.3 *A.I. Lanin: The calculation of interference waves for diffraction by a cylinder and a sphere. Zap. Nauchn. Semin. LOMI AN SSSR **9**, 64–104 (1968) [Semin. Math. **9**, 27–44 (1970)]

12.4 *V.S. Buldyrev: Wave propagation near the curved surface of an inhomogeneous body. Probl. Mat. Fiz. **2**, 61–84 (1967) [Top. Math. Phys. **2**, 47–65 (1968)]

12.5 *V.S. Buldyrev, A.I. Lanin: "The Investigation of the Interference Wave on the Surface of an Elastic Sphere", in *Chislennye Metody Resheniya Zadach Matem. Fiziki* (Nauka, Moscow 1966) pp. 131–143

12.6 *L.M. Brekhovskikh, I.D. Ivanov: Extending the range of applicability of ray theory in the study of wave propagation in layered media. Dokl. Akad. Nauk SSSR **83**, 545–548 (1952)

12.7 *L.M. Brekhovskikh: *Waves in Layered Media* (Nauka, Moscow 1973) [Academic, New York 1980]

12.8 T. Ishihara, L.B. Felsen, A. Green: High-frequency fields excited by a line source located on a perfectly conducting concave cylindrical surface. IEEE Trans. Ant. Prop. **26**, 757–767 (1978)

12.9 T. Ishihara, L.B. Felsen: High-frequency fields excited by a line source on a concave cylindrical impedance surface. IEEE Trans. Ant. Prop. **27**, 172–179 (1979)

12.10 E. Topuz, E. Niver, L.B. Felsen: Electromagnetic fields near a concave perfectly conducting cylindrical surface. IEEE Trans. Ant. Prop. **30**, 280–292 (1982)

12.11 *V.S. Buldyrev: Investigation of the Green's function in the diffraction problem for a transparent circular cylinder (I), in *Chislennye Metody Resheniya Differents. i Integral'n. Uravnenii i Kvadraturnye Formuly* (Nauka, Moscow 1964) pp. 275–286

12.12 *V.S. Buldyrev, A.I. Lanin: Investigation of the Green's function in the diffraction problem for a transparent circular cylinder (II). Zh. Vychisl. Mat. Mat. Fiz. **6**, 90–105 (1966) [USSR Comp. Math. Math. Phys. **6** (1), 128–149 (1966)]

12.13 *V.S. Buldyrev, V.E. Grikurov: Interference waves in two contacting elastic media. Vestn. Leningr. Univ., Ser. Mat. Mekh. No. 1, 46–54 (1975) [Vestn. Leningrad Univ. Math. **8**, 31–41 (1980)]

12.14 *V.S. Buldyrev, V.E. Grikurov: Interference waves in the theory of elasticity. Probl. Mat. Fiz. **8**, 16–29 (1976)

12.15 *V.E. Grikurov: Theoretical seismograms of interference wave fronts. Izv. Akad. Nauk SSSR, Fiz. Zemli, No. 6, 38–42 (1980) [Izv., Acad. Sci., USSR, Phys. Solid Earth **16**, 413–416 (1981)]

12.16 *V.S. Buldyrev, A.I. Lanin: Asymptotic formulas for a wave propagating along a concave surface: limits of their applicability. Radiotekh. Elektron. **20**, 49–58 (1975) [Radio Eng. Electron. Phys. **20** (1), 19–27 (1975)]

12.17 *V.S. Buldyrev, A.I. Lanin: Computing the function $G_M(\gamma)$. Zap. Nauchn. Semin. LOMI AN SSSR **51**, 85–92 (1975) [J. Sov. Math. **11**, 418–424 (1979)]

12.18 *V.M. Babich [Babič]: A high-frequency point source of oscillations near a concave mirror. Zap. Nauchn. Semin. LOMI AN SSSR **51**, 5–20 (1975) [J. Sov. Math. **11**, 361–371 (1979)]

12.19 *V.M. Babich [Babič]: A point source of oscillations on the boundary of a region. Zap. Nauchn. Semin. LOMI AN SSSR **62**, 3–21 (1976) [J. Sov. Math. **11**, 665–676 (1979)]

12.20 *V.M. Babich [Babič]: A point source of oscillations near a concave mirror, II. Zap. Nauchn. Semin. LOMI AN SSSR **89**, 3–13 (1979) [J. Sov. Math. **19**, 1279–1288 (1982)]

Chapter 13

13.1 *V.I. Smirnov: *A Course of Higher Mathematics*, Vol. 3, Pt. 2 (Nauka, Moscow 1969) [Addison-Wesley, Reading, MA 1964]

13.2 T.M. Cherry: Uniform asymptotic formulae for functions with transition points. Trans. Am. Math. Soc. **68**, 224–257 (1950)

13.3 F.W.J. Olver: The asymptotic solutions of linear differential equations of the second order for large values of a parameter. Philos. Trans. R. Soc. London A **247**, 307–327 (1954)

13.4 F.W.J. Olver: The asymptotic expansion of Bessel functions of large order. Philos. Trans. R. Soc. London A **247**, 328–368 (1954)

13.5 D. Ludwig: Uniform asymptotic expansion of the field scattered by a convex object at high frequencies. Commun. Pure Appl. Math. **20**, 103–138 (1967)

13.6 *M.V. Fedoryuk: *The Saddle-Point Method* (Nauka, Moscow 1977)

13.7 N. Bleistein, R.A. Handelsman: *Asymptotic Expansions of Integrals* (Holt, Rinehart and Winston, New York 1975)

13.8 *V.A. Fock [Fok]: The field of a plane wave near the surface of a conducting body. Izv. Akad. Nauk SSSR, Ser. Fiz. **10** (2), 171–186 (1946) [J. Phys. USSR **10**, 399–409 (1946)]; also in [13.9], Chap. 5

13.9 *V.A. Fock [Fok]: *Electromagnetic Diffraction and Propagation Problems*, 2nd ed. (Sovetskoe Radio, Moscow 1970) [1st ed., Pergamon, Oxford 1965]

13.10 C.S. Morawetz, D. Ludwig: An inequality for the reduced wave operator and the justification of geometrical optics. Commun. Pure Appl. Math. **21**, 187–203 (1968)

13.11 J.B. Keller: Diffraction by a convex cylinder. IRE Trans. Ant. Prop. **4**, 312–321 (1956)

13.12 B.R. Levy, J.B. Keller: Diffraction by a smooth object. Commun. Pure Appl. Math. **12**, 159–209 (1959)

13.13 *V.S. Buslaev: Shortwave asymptotic formulas in the problem of diffraction by convex bodies. Vestn. Leningr. Univ., Ser. Mat., Mekh. Astron. **3** (13), 5–21 (1962)

13.14 *V.M. Babich [Babič]: D. Ludwig's method and the boundary-layer method in the problem of diffraction by a smooth body. Zap. Nauchn. Semin. LOMI AN SSSR **27**, 17–33 (1972) [J. Sov. Math. **3**, 395–407 (1975)]

13.15 *V.M. Babich [Babič], N.Ya. Kirpichnikova [Kirpičnikova]: *The Boundary-Layer Method in Diffraction Problems* (Izdat. LGU, Leningrad 1974) [Springer Ser. Electrophys., Vol. 3 (Springer, Berlin, Heidelberg 1979)]

Appendix

A1.1 *V.I. Smirnov: *A Course of Higher Mathematics*, Vol. 3, Pt. 2 (Nauka, Moscow 1969) [Addison-Wesley, Reading, MA 1964]

A1.2 *N.N. Lebedev: *Special Functions and Their Applications* (Fizmatgiz, Moscow 1963) [Dover, New York 1972]

A1.3 *V.A. Fock [Fok]: *Tables of Airy Functions* (GITTL, Moscow 1946)

A1.4 *G.D. Yakovleva: *Tables of Airy Functions and Their Derivatives* (Nauka, Moscow 1969)

A1.5 J.C.P. Miller: *The Airy Integral* (Cambridge University Press, Cambridge 1946)

A1.6 *V.A. Fock [Fok]: *Electromagnetic Diffraction and Propagation Problems*, 2nd ed. (Sovetskoe Radio, Moscow 1970) [1st ed., Pergamon, Oxford 1965]

A1.7 *M.V. Fedoryuk: *The Saddle-Point Method* (Nauka, Moscow 1977)

A1.8 N. Bleistein, R.A. Handelsman: *Asymptotic Expansions of Integrals* (Holt, Rinehart and Winston, New York 1975)

A2.1 *P.K. Rashevskii [Raschewskii]: *Riemannsche Geometrie und Tensoranalysis*, 3rd ed. (Nauka, Moscow 1967) [1st ed., VEB Deutscher Verlag der Wissenschaften, Berlin 1959]

A3.1 *V.M. Babich [Babič], V.F. Lazutkin: Eigenfunctions concentrated near a closed geodesic. Probl. Mat. Fiz. **2**, 15–25 (1967) [Top. Math. Phys. **2**, 9–18 (1968)]

A3.2 *L.M. Berkovič, N.Kh. Rozov: Some remarks on differential equations of the form $y'' + a_0(x)y = \varphi(x)y^\alpha$. Differents. Uravnen. **8**, 2076–2079 (1972) [Differ. Equ. **8**, 1609–1612 (1972)]

A4.1 *V.S. Buldyrev, A.I. Lanin: Computing the function $G_M(\gamma)$. Zap. Nauchn. Semin. LOMI AN SSSR **51**, 85–92 (1975) [J. Sov. Math. **11**, 418–424 (1979)]

Subject Index